CHEMISTRY AND TECHNOLOGY OF ALTERNATE FUELS

CHEMISTRY AND TECHNOLOGY OF ALTERNATE FUELS

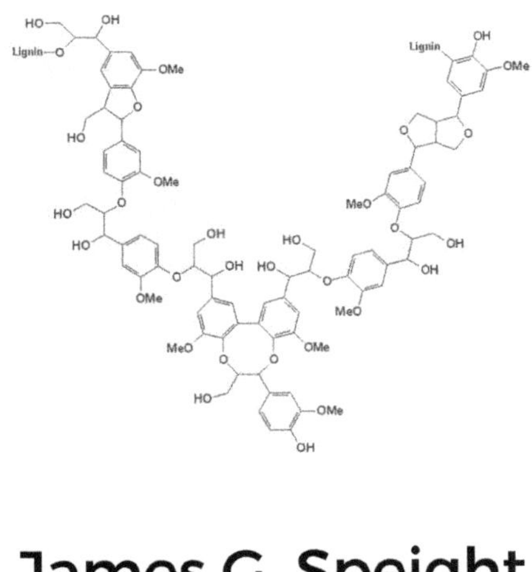

James G. Speight

CD & W Inc., USA

World Scientific

NEW JERSEY • LONDON • SINGAPORE • BEIJING • SHANGHAI • HONG KONG • TAIPEI • CHENNAI • TOKYO

Published by

World Scientific Publishing Co. Pte. Ltd.
5 Toh Tuck Link, Singapore 596224
USA office: 27 Warren Street, Suite 401-402, Hackensack, NJ 07601
UK office: 57 Shelton Street, Covent Garden, London WC2H 9HE

British Library Cataloguing-in-Publication Data
A catalogue record for this book is available from the British Library.

CHEMISTRY AND TECHNOLOGY OF ALTERNATE FUELS

ISBN 978-981-120-364-0 (hardcover)
ISBN 978-981-120-365-7 (ebook for institutions)
ISBN 978-981-120-366-4 (ebook for individuals)

For any available supplementary material, please visit
https://www.worldscientific.com/worldscibooks/10.1142/11366#t=suppl

Desk Editor: Herbert Moses

Typeset by Stallion Press
Email: enquiries@stallionpress.com

Preface

Crude oil-based fuels are well-established products that have served industrial and domestic consumers for more than 100 years. In the foreseeable future, most of these fuels will still be largely based on liquid hydrocarbons. The specifications of such fuels will however continue to be adjusted as they have been adjusted till now to meet the changing demands from consumers. Traditional crude oil refining has undergone increasing levels of sophistication to produce fuels of appropriate specifications. Increasing operating costs continuously exerted pressure on refining margins, but it remains problematic to convert all refinery streams into products with acceptable specifications at a reasonable return.

On the other hand, alternate fuels, such as gasoline and diesel, from other sources, are making progress toward the fuel balance. For example, naphtha, the typical starting liquid for automotive fuels, and biodiesel from plant sources are similar to naphtha and kerosene, but may have differences that include a different distribution of the constituents. At the moment, the potential for producing liquid fuels from various types of biomass is also receiving considerable attention.

However, the time is running out and crude oil, once considered inexhaustible, is now being depleted at a rapid rate. As the amount of available crude oil decreases, the need for alternate technologies to produce liquid fuels that could potentially help prolong the liquid fuel culture and mitigate the forthcoming effects of the shortage of transportation fuels has been suggested to occur under the Hubbert peak oil theory. To mitigate the influence of the oil peak and the subsequent depletion of supplies, unconventional (or non-crude oil-derived) fuels are becoming a major issue in the consciousness of oil-importing countries.

In the near term, the ability of conventional fuel sources and technologies to support the global demand for energy will depend on how efficiently the energy sector can match available energy resources with the end user and how efficiently and cost effectively the energy can be delivered. These factors are directly related to the continuing evolution of a truly global energy market. In the long term, a sustainable energy future cannot be created by treating energy as an independent topic (Zatzman, 2012). Rather, the role of energy and the inter-relationship of the energy market with other markets and various aspects of market infrastructure deserve further attention and consideration. Achievement of greater energy efficiency will depend on the developing world market's ability to integrate energy resources within a common structure.

However, the production of liquid fuels from sources other than crude oil has a checkered history. The on-again/off-again efforts that are the results of political maneuvering have seen to it that the race to secure self-sufficiency by the production of non-conventional fuels never gets much farther than the starting point! This is due in no small part to the price fluctuations in the price of crude oil and the lack of foresight on the part of various levels of government. It must be realized that for decades, the price of crude oil (and the resulting fuel products) has always been maintained at a level that was sufficiently low to discourage the establishment of an alternate fuels industry. However, we are close to the time when the lack of being prepared for the production of non-conventional fuels may set any national government on its heels.

Following the nomenclature and definitions presented in the United States Energy Policy Act of 1992 (Section 301), in the context of the present book, alternate fuels (alternative fuels) are defined as "methanol, denatured ethanol, and other alcohols; mixtures containing 85 percent or more (or such other percentage, but not less than 70 percent, as determined by the Secretary, by rule, to provide for requirements relating to cold start, safety, or vehicle functions) by volume of methanol, denatured ethanol, and other alcohols with gasoline or other fuels; natural gas; liquefied petroleum gas; hydrogen; coal-derived liquid fuels; fuels (other than alcohol) derived from biological materials; electricity (including electricity from solar energy); and any other fuel the Secretary determines, by rule, is substantially not crude oil and would yield substantial energy security benefits and substantial environmental benefits." It is this definition that is used to guide the contents of this book and show that sources that are "*substantially not petroleum*" are available as sources of fuels.

Thus, for the purposes of this text, the term *alternate fuels* is used to represent those fuels that are produced from plant sources and from other sources such as waste. Thus, biofuels are *bio-materials* that are made from renewable biological sources, which are now contemplated as grown specifically for the purpose of providing useful heat upon combustion. Biofuels are produced from sources, such as corn, soybeans, flaxseed, rapeseed, sugarcane, palm oil, sugar beet, raw sewage, food scraps, animal parts, and rice.

The dynamics are now coming into place for the establishment of an alternate fuels industry and it is up to various levels of government not just to promote the establishment of such an industry but to lead the way in recognizing that it is not only supply and demand but also the available and variable technology. For example, the technology of the tar sand industry and the oil shale industry is not the same as it was in the 1970s. The processes for recovery of the raw materials and the processing options have changed in an attempt to increase the efficiency of oil production. Various national events (for the United States) and international events (for other countries) have made it essential that we move ahead to develop fuels from non-conventional sources.

Voices are being raised for the establishment of an industry that produces and develops liquid fuels from non-conventional sources, but there is still a long way to go. Incentives are still needed to develop such resources. However, there still remains a cone of silence in many government capitals that covers the cries to develop non-conventional fuel sources. Hopefully, the silence will end within the near future before it is too late.

This book is written to assist the reader in understanding the options that are available for the production of fuels — often referred to as alternate fuels — from non-conventional sources and to compare these alternate fuels with the fuels from conventional (fossil fuel) sources.

For the purposes of this book, non-conventional sources are any sources of gaseous, liquid, and solid fuels other than crude oil and heavy oil. In addition, the book includes tables that present the properties of these alternate fuels. It is essential that the properties of such materials be presented in order to assist the researcher understand the nature of the feedstocks as well as the nature of the products. If a product cannot be employed for its hoped-for-use, it is not a desirable product and must be changed accordingly. Such plans can only be made when the properties of the original product are understood.

About the Author

Dr. James G. Speight has a BSc and PhD in Chemistry; he also holds a DSc in Geological Sciences and a PhD in Petroleum Engineering. He has more than 50 years of experience in the following areas: (i) the properties, recovery, and refining of conventional petroleum, heavy oil, and tar sand bitumen, (ii) the properties and refining of natural gas, and (iii) the properties and refining of biomass, biofuels, biogas, and the generation of bioenergy. His work has also focused on environmental effects, environmental remediation, and safety issues associated with the production and use of fuels and biofuels. He is the author (and co-author) of more than 85 books in petroleum science, petroleum engineering, biomass and biofuels, and environmental sciences.

Although he has always worked in private industry which focused on contract-based work, Dr. Speight has served as a Visiting Professor in the College of Science, University of Mosul (Iraq), Visiting Professor in Chemical Engineering at the Technical University of Denmark, and the University of Trinidad and Tobago as well as held adjunct appointments at various universities. He has also served as a thesis examiner for more than 25 theses.

As a result of his work, Dr. Speight has been honored as the recipient of the following awards:

- Diploma of Honor, United States National Petroleum Engineering Society. *For Outstanding Contributions to the Petroleum Industry,* 1995.
- Gold Medal of the Russian Academy of Sciences. *For Outstanding Work in the Area of Petroleum Science*, 1996.
- Einstein Medal of the Russian Academy of Sciences. *In recognition of Outstanding Contributions and Service in the Field of Geologic Sciences*, 2001.
- Gold Medal — Scientists without Frontiers, Russian Academy of Sciences. *In recognition of His Continuous Encouragement of Scientists to Work Together across International Borders*, 2005.
- Gold Medal — Giants of Science and Engineering, Russian Academy of Sciences. *In recognition of Continued Excellence in Science and Engineering*, 2006.
- Methanex Distinguished Professor, University of Trinidad and Tobago. *In Recognition of Excellence in Research*, 2007. In 2018, he received the American Excellence Award for Excellence in Client Solutions from the United States Institute of Trade and Commerce, Washington, DC.

Contents

Preface v

About the Author ix

Part 1 The Need for Alternate Fuels **1**

Chapter 1 Current Fuel Sources and Alternate Fuel Sources 3

1.1 Introduction 3

1.2 Fossil Fuel Sources 10

1.2.1 Natural gas 11

1.2.2 Gas hydrates 12

1.2.3 The crude oil family 13

1.2.3.1 Opportunity crude oil 14

1.2.3.2 High acid crude oil 15

1.2.3.3 Foamy oil 15

1.2.3.4 Heavy crude oil 16

1.2.4 Unconventional fossil fuel resources 17

1.2.4.1 Extra heavy crude oil and tar sand bitumen 17

1.2.4.2 Coal 21

1.2.4.3 Oil shale 22

1.3 Non-Fossil Fuel Energy Sources 24

1.3.1 Biomass 27

1.3.2 Geothermal energy 29

1.3.3 Hydrogen energy 30

1.3.4 Nuclear energy 31

1.3.5 Solar energy 32
1.3.6 Tidal energy 32
1.3.7 Wind energy 33
1.4 Energy Supply 34
1.4.1 The peak oil theory 34
1.4.1.1 Reserve depletion 35
1.4.1.2 Effects and consequences 38
1.4.2 Economic factors 44
1.4.3 Geopolitical factors 44
1.4.4 Physical factors 45
1.4.5 Technological factors 47
1.5 Energy Independence and Energy Security 48
1.5.1 Energy independence 49
1.5.2 Energy security 53
References 56

Chapter 2 Conventional Fuels 63
2.1 Introduction 63
2.2 Gaseous Fuels 72
2.2.1 Composition 73
2.2.2 Manufacture 74
2.2.3 Properties and uses 74
2.3 Liquid Fuels 77
2.3.1 Naphtha 77
2.3.1.1 Manufacture 78
2.3.1.2 Composition 82
2.3.1.3 Properties and uses 83
2.3.2 Gasoline 85
2.3.2.1 Manufacture 86
2.3.2.2 Composition 88
2.3.2.3 Properties and uses 90
2.3.3 Kerosene 96
2.3.3.1 Manufacture 97
2.3.3.2 Composition 98
2.3.3.3 Properties and uses 99
2.3.4 Fuel oil 99
2.3.4.1 Manufacture 100
2.3.4.2 Composition 102
2.3.4.3 Properties and uses 103

2.4 Solid Fuels 105
2.4.1 Wax 105
2.4.1.1 Manufacture 106
2.4.1.2 Composition 108
2.4.1.3 Properties and uses 108
2.4.2 Resid 109
2.4.2.1 Manufacture 109
2.4.2.2 Composition 111
2.4.2.3 Properties and uses 112
2.4.3 Coke 112
2.4.3.1 Composition 112
2.4.3.2 Properties and uses 113
2.5 Conclusions 115
References 115

Chapter 3 Alternate Fuels 119
3.1 Introduction 119
3.2 Composition and Properties 123
3.3 Classification of Biofuels 126
3.3.1 First-generation biofuels 127
3.3.1.1 Sugars and starch 127
3.3.1.2 Cellulose 130
3.3.1.3 Lignin 131
3.3.2 Second-generation biofuels 131
3.3.3 Third-generation biofuels 132
3.4 Types of Biofuels 133
3.4.1 Gaseous fuels 133
3.4.1.1 Biogas 134
3.4.1.2 Low-Btu gas 136
3.4.1.3 Medium-Btu gas 137
3.4.1.4 High heat-content gas 138
3.4.1.5 Landfill gas 139
3.4.1.6 Synthesis gas 140
3.4.1.7 Hydrogen 141
3.4.2 Liquid fuels 141
3.4.2.1 Alcohols 143
3.4.2.2 Biodiesel 149
3.4.2.3 Hydrocarbons 157
3.4.2.4 Bio-oil 158

3.4.2.5 Bioethers 161
3.4.2.6 Liquid fuels from synthesis gas 162
3.4.3 Solid fuels 165
3.5 Property Variation with Source 166
3.5.1 General observations 166
3.5.2 Fuel specifications and performance 170
References 175

Part 2 Feedstocks **181**

Chapter 4 Feedstocks for Fossil Fuels 183
4.1 Introduction 183
4.2 Natural Gas 184
4.2.1 History 185
4.2.2 Formation and occurrence 186
4.2.2.1 Associated gas 190
4.2.2.2 Non-associated gas 191
4.2.3 Unconventional gas 192
4.2.3.1 Coalbed methane 192
4.2.3.2 Gas hydrates 194
4.2.3.3 Shale gas 195
4.2.4 Composition and properties 196
4.2.4.1 Composition 196
4.2.4.2 Properties 198
4.3 The Crude Oil Family 204
4.3.1 Conventional crude oil 205
4.3.2 Opportunity crude oil 208
4.3.3 High acid crude oil 209
4.3.4 Foamy oil 209
4.3.5 Tight oil 210
4.3.6 Heavy crude oil 211
4.4 Extra Heavy Crude Oil and Tar Sand Bitumen 213
4.4.1 Extra heavy crude oil 214
4.4.2 Tar sand Bitumen 214
4.5 Coal 216
4.6 Oil Shale 220
References 225

Chapter 5 Feedstocks for Alternate Fuels 229
5.1 Introduction 229
5.2 Crops 236
5.2.1 Energy crops 240
5.2.1.1 Cordgrass and switchgrass 243
5.2.1.2 Jerusalem artichoke 244
5.2.1.3 Miscanthus 245
5.2.1.4 Reed plants 246
5.2.1.5 Residual herbaceous biomass 246
5.2.1.6 Short-rotation coppice 248
5.2.1.7 Sorghum 248
5.3 Wood 249
5.3.1 History 249
5.3.2 Composition and properties 253
5.3.2.1 Chemical composition 256
5.3.3 Types of wood 267
5.3.3.1 Hardwood 269
5.3.3.2 Softwood 270
5.3.4 Uses 271
5.4 Waste 274
5.4.1 Domestic and industrial waste 278
5.4.2 Effects of waste 280
References 283

Chapter 6 Production of Alternate Fuels 287
6.1 Introduction 287
6.2 Production of Alternate Fuels 291
6.2.1 Anaerobic digestion 293
6.2.2 Biophotolysis 298
6.2.3 Combustion 299
6.2.3.1 Types of combustors 302
6.2.3.2 Combustion plants 306
6.2.4 Fermentation 310
6.2.5 Gasification 311
6.2.6 Pyrolysis 318
6.2.7 Other processes 324
6.2.7.1 Acid hydrolysis 324

6.2.7.2 Briquetting 327
6.2.7.3 Enzymatic hydrolysis 328
6.2.7.4 Transesterification 330
References 332

Chapter 7 Landfill Gas 335
7.1 Introduction 335
7.2 Landfill Classification 336
7.2.1 Bioreactor landfill sites 336
7.2.2 Construction and demolition debris landfill sites 338
7.2.3 Conventional municipal solid waste landfill sites 339
7.2.4 Hazardous waste landfill sites 339
7.2.5 Manual landfill sites 340
7.2.6 Sanitary landfill sites 340
7.2.7 Surface impoundment sites 341
7.3 Landfill Gas 341
7.4 Formation of Landfill Gas 353
7.4.1 Organic reactions 355
7.4.1.1 Hydrolysis and aerobic degradation 355
7.4.1.2 Hydrolysis and fermentation 355
7.4.1.3 Acetogenesis 356
7.4.1.4 Methanogenesis 356
7.4.1.5 Oxidation 356
7.4.2 Inorganic reactions 357
7.4.3 Landfill leachate 357
7.4.4 Landfill gas monitoring 359
7.4.5 Landfill mining 360
7.5 Gas Migration 361
7.5.1 Diffusion 362
7.5.2 Pressure 363
7.5.3 Permeability 363
7.5.4 Factors affecting gas migration 363
7.5.4.1 Landfill cover 364
7.5.4.2 Natural and anthropogenic pathways 364
7.5.4.3 Wind speed and direction 364
7.5.4.4 Moisture 364
7.5.4.5 Groundwater levels 364
7.5.4.6 Temperature 365
7.5.4.7 Barometric and soil gas pressure 365

7.6 Uses 365
7.6.1 Gas collection 365
7.6.2 Gas pretreatment 367
7.6.3 Environmental impact 368
7.6.3.1 Ash, slag, heavy metals, and dust 368
7.6.3.2 Flue gas 369
7.6.4 Gasification of landfill waste 369
References 370

Chapter 8 Synthesis Gas 373
8.1 Introduction 373
8.2 Coal Gasification 376
8.2.1 Chemistry 377
8.2.2 Processes 379
8.2.3 Gasifiers 380
8.3 Gasification of Crude Oil Fractions 383
8.3.1 Feedstocks 383
8.3.2 Chemistry 386
8.3.3 Commercial processes 388
8.3.3.1 Heavy residue gasification and combined cycle power generation 388
8.3.3.2 Hybrid gasification process 389
8.3.3.3 Hydrocarbon gasification 390
8.3.3.4 Hypro process 390
8.3.3.5 Pyrolysis processes 391
8.3.3.6 Shell gasification (partial oxidation) process 392
8.3.3.7 *Steam–methane reforming* *392*
8.3.3.8 Steam–naphtha reforming 396
8.3.3.9 Synthesis gas generation 396
8.3.3.10 Texaco gasification process 397
8.4 Gasification of Other Feedstocks 398
8.5 Fuels and Petrochemicals 400
8.5.1 Gaseous fuels and chemicals 406
8.5.1.1 Ammonia 406
8.5.1.2 Hydrogen 408
8.5.1.3 Synthetic natural gas 408
8.5.2 Liquid fuels and chemicals 409
8.5.2.1 Fischer–Tropsch liquids 409
8.5.2.2 Methanol 410

8.5.2.3 Dimethyl ether 410
8.5.2.4 Methanol-to-gasoline and olefins 411
8.5.2.5 Other processes 411
8.6 The Future 412
References 420

Chapter 9 Gas Cleaning 425
9.1 Introduction 425
9.2 Gas Streams 429
9.3 Gas Cleaning 435
9.3.1 Effect of gas composition 437
9.3.2 Process types 438
9.4 Water Removal 441
9.4.1 Absorption 442
9.4.2 Adsorption 444
9.4.3 Cryogenics 446
9.4.4 Methanol-based processes 447
9.5 Acid Gas Removal 448
9.5.1 Adsorption 449
9.5.2 Absorption 449
9.5.3 Chemisorption 451
9.5.4 Other processes 457
9.6 Removal of Condensable Hydrocarbons 461
9.6.1 Extraction 464
9.6.2 Absorption 465
9.6.3 Fractionation 466
9.6.4 Enrichment 467
9.7 Tar Removal 469
9.7.1 Physical methods 469
9.7.2 Thermal methods 470
9.8 Other Methods of Gas Cleaning 471
9.8.1 Nitrogen removal 471
9.8.2 Ammonia removal 473
9.8.3 Particulate matter removal 473
9.8.4 Siloxane removal 474
9.8.5 Alkali metal salt removal 475
9.8.6 Biological methods 475
9.8.6.1 Biofiltration 476
9.8.6.2 Bio-scrubbing 479
9.8.6.3 Biooxidation 480

9.9 Tail Gas Cleaning 480
9.9.1 Claus process 481
9.9.2 SCOT process 482
References 483

Chapter 10 The Fischer–Tropsch Process 489
10.1 Introduction 489
10.2 Production of Synthesis Gas 496
10.2.1 Feedstocks 498
10.3 The Fischer–Tropsch Process 503
10.3.1 Chemistry 503
10.3.2 Process principles 508
10.3.3 Product distribution 511
10.4 Reactors and Catalysts 513
10.4.1 Reactors 513
10.4.2 Catalysts 516
10.5 Products and Product Quality 520
10.5.1 Products 521
10.5.2 Product quality 521
10.5.3 Refining Fischer–Tropsch products 525
References 528

Part 3 Properties of Alternate Fuels and the Future **531**

Chapter 11 Properties of Alternate Fuels 533
11.1 Introduction 533
11.2 Gaseous Fuels 535
11.2.1 Natural gas 535
11.2.1.1 Composition 536
11.2.1.2 Properties 539
11.2.2 Synthetic gases 546
11.2.2.1 Producer gas 546
11.2.2.2 Blue water gas 548
11.2.2.3 Carbureted water gas 548
11.2.2.4 Coal and coke oven gas 549
11.2.2.5 Biogas 549
11.2.2.6 Wood gas 551
11.2.2.7 Refuse gas 552
11.3 Liquid Fuels 553
11.3.1 Methanol 553

11.3.2 Ethanol 555
11.3.3 Propanol and butanol 556
11.3.4 Bio-diesel 557
11.3.5 Hydrocarbon fuels 563
11.3.5.1 Naphtha and gasoline 564
11.3.5.2 Kerosene and diesel 567
11.3.5.3 Other fuels 570
11.4 Solid Fuels 573
11.4.1 Fuelwood 574
11.4.2 Logs and wood chips 575
11.4.3 Pellets and briquettes 576
11.4.3.1 Briquette manufacture 578
11.4.3.2 Bagasse briquettes 579
11.4.3.3 Sawdust briquettes 580
11.4.3.4 Urban waste briquettes 580
11.4.4 Charcoal 581
11.4.5 Coke 586
11.5 Fuel Quality 588
11.5.1 Bulk properties 589
11.5.2 Specifications 594
References 595

Chapter 12 Production of Fuels in a Biorefinery 597
12.1 Introduction 597
12.2 A Biorefinery 599
12.3 Feedstocks 602
12.3.1 Carbohydrates 605
12.3.2 Vegetable oils 607
12.3.3 Plant fibers 608
12.3.4 Waste 609
12.4 Process Options 610
12.4.1 Direct combustion 612
12.4.2 Fermentation and hydrolysis 614
12.4.3 Gasification 617
12.4.3.1 Gasification chemistry 617
12.4.3.2 Gasifiers 619
12.4.3.3 Synthesis gas 623
12.4.4 Pyrolysis 624
12.4.5 Anaerobic digestion 625

12.4.6 Transesterification 630
12.4.6.1 Feedstocks 630
12.4.6.2 Transesterification 631
12.4.6.3 Catalytic transesterification 632
12.4.6.4 Transesterification using supercritical methanol 634
12.4.6.5 Effect of process parameters 635
12.5 Alternate Fuels 636
12.5.1 Gaseous fuels 637
12.5.2 Liquid fuels 638
12.5.3 Solid fuels 640
12.6 Environmental Benefits and the Future 641
References 644

Conversion Factors 651

Glossary 661

Index 711

Part 1

The Need for Alternate Fuels

Chapter 1

Current Fuel Sources and Alternate Fuel Sources

1.1 Introduction

The Earth contains a finite supply of fossil fuels (the major fossil fuels are natural gas, crude oil, and coal) although there are doubts about the real amount of fossil fuels remaining on the Earth. By inference, the fossil fuels of the Earth can only provide a limited supply of gaseous, liquid, and solid fuels. Fossil fuels are those fuels (namely coal, crude oil (including heavy crude oil), tar sand bitumen, natural gas, and oil shale) produced by the decay of plant remains over geological time; they represent an unrealized potential, with liquid fuels derived from crude oil being only a fraction of those that could be ultimately produced from heavy crude oil, extra heavy crude oil, and tar sand bitumen (Speight, 1990, 1997, 2011a, 2013d, 2013e, 2014a).

The best current estimates of the longevity of each fossil fuel are obtained from the reserves/production ratio (BP, 2019), which gives an indication (in years) of how long each fossil fuel will last at the current rates of production. Thus, estimates vary from at least 50 years of crude oil at current rates of consumption to 300 years of coal at current rates of consumption, with natural gas varying between the two extremes. In addition, the amounts of natural gas and crude oil located in tight sandstone formations and shale formations have added a recent but exciting twist to the amount of fossil fuels remaining on the Earth. Peak energy theory proponents are inclined to discount the tight formations and shale

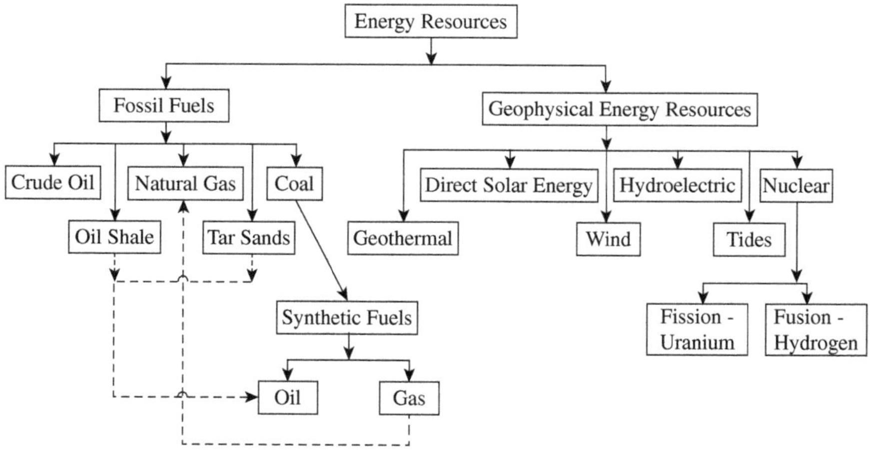

Figure 1.1: The energy resources of the world.

formation as a mere aberration (or a hiccup) in the depletion of these resources, while opponents of the peak energy theory take the opposite view and consider tight formations and shale formations as prolonging the longevity of natural gas and crude oil by a substantial time period. In addition, some areas of the Earth are still relatively unexplored or they have been poorly analyzed, and (using crude oil as an example) the knowledge of in-ground resources will increase dramatically as an oil reservoir is exploited.

Energy sources — or at least some of the energy sources (Figure 1.1) — have been used since the beginning of recorded history, and fossil fuel resources will continue to be recognized as major sources of energy in at least the foreseeable future (Crane *et al.*, 2010; World Energy Council, 2008; Gudmestad *et al.*, 2010; Speight, 2011a, 2011b, 2014a; Khoshnaw, 2013; BP, 2019). In fact, at present, the majority of the energy consumed worldwide comes from fossil fuels (crude oil: ca. 38–40%, coal: ca. 31–35%, natural gas: ca. 20–25%), with the remainder of the energy requirements coming from nuclear and hydroelectric sources. As a result, fossil fuels (in varying amounts depending on the source of information) are projected to be the major sources of energy for the next 50 years (Crane *et al.*, 2010; World Energy Council, 2008; Gudmestad *et al.*, 2010; Speight, 2011a, 2011b, 2014a; Khoshnaw, 2013; BP, 2019).

Alternate fuels (also known as alternative fuels, non-conventional fuels, or advanced fuels) are any materials that can be used as fuels, other than conventional fuels that are defined in this text as fuels derived from natural gas, crude oil, coal, or oil shale. Nuclear fuels such as uranium and thorium and the artificial radioisotope fuels that are made in nuclear reactors are not included in this definition. For example, common alternate fuels include bio-hydrogen, biogas, bio-alcohol, biodiesel, refuse-derived fuel, non-fossil fuel methane, and related hydrocarbon derivatives (Lago *et al.*, 2019). The prefix *bio* as used in this text indicates that the fuel was obtained from a biological source and did not necessarily exist, or occur naturally, in the fuel form.

Alternate fuels produced from sources other than known fossil fuel sources (such as natural gas, crude oil, heavy crude oil, extra heavy oil, tar sand bitumen, coal, and oil shale) are making some progress in facing the fuel demand. For example, diesel fuel from plant sources (commonly referred to as biodiesel) is similar in performance to diesel fuel from crude oil and has the added advantage of a higher cetane rating than that of crude oil-derived diesel. However, the production of liquid fuels from sources other than crude oil has a checkered history. The on-again/off-again efforts, which are a consequence of the inability of political decision-makers to formulate meaningful policies, have caused the production of non-conventional fuels to move slowly, if at all (Yergin, 1993; Bower, 2009; Wihbey, 2009; Speight, 2011a, 2011b, 2014a; Yergin, 2011).

This is due, in no small part, to the price fluctuations of natural gas and crude oil, which lead to price fluctuations (which tend to be very upward) of the commonly used fuel products (i.e. gasoline and diesel fuel), and the lack of planning and associated foresight on the part of various levels of government. It must be realized that for decades, the price of crude oil produced in the crude oil-exporting nations has always been maintained at a level that is sufficiently low to discourage the establishment of a domestic synthetic fuel industry in many of the crude oil-consuming countries. However, in spite of additional supplies of crude oil and natural gas coming from tight formations and shale formations, the time will come when the lack of being prepared for the production of non-conventional fuels may set many a national government on its heels. It is not a matter of "Will the lack of preparedness come to fruition?" but "When will the lack of preparedness come to fruition?" This leads to the theory of *peak energy production*, which is the subject of considerable discussion and debate (Speight and Islam, 2016).

In the short term, the ability of conventional fuel sources and technologies to support the global demand for energy will depend on how efficiently the energy sector can match available energy resources with the end user and how efficiently and cost effectively the energy can be delivered. These factors are directly related to the continuing evolution of a truly global energy market. In the long term, a sustainable energy future cannot be created by treating energy as an independent topic (Zatzman, 2012). Rather, the role of energy and the inter-relationship of the energy market with other markets and various aspects of market infrastructure demand further attention and consideration. Achievement of greater energy efficiency will depend on the ability of the developing world's markets to integrate energy resources within a common structure (Gudmestad *et al.*, 2010; Speight, 2011b; Khoshnaw, 2013).

World petro-politics are now in place (Bentley, 2002; Speight, 2011a; Speight and Islam, 2016) insofar as the countries that use fuels derived from crude oil are not only subject to any changes in the policies of the natural gas- and crude oil-producing countries but also subject to any random or planned acts of terrorism. In addition to the depletion of the reserves of natural gas and crude oil, this emphasizes the need for energy independence in the form of the establishment of a synthetic fuel (including biofuels) industry and, without being unduly dismissive of such efforts, in fact, it is up to the various levels of government to not just promote the establishment of such an industry but also lead the way in recognizing that it is not only supply and demand but the available and variable technology. Unfortunately, although there may be sufficient crude oil remaining on the Earth to maintain the Petroleum Age (i.e. the age in which the developed countries of the world would not operate) for another fifty years (Speight, 2011a, 2011b), the time to prepare is *now*. The world is not yet on the precipice of energy deficiency (as many alarmists claim), but it is necessary that the politicians in the various levels of (national) government of oil-consuming nations look beyond the next election with an eye to the future. It should also be the focus of the proponents of biofuel production and use to ensure that sufficient feedstocks are available to successfully operate a biofuel refinery, thereby contributing alternate fuels to the gradual (but not drastic) reduction of crude oil-based fuels (Speight, 2008; Giampietro and Mayumi, 2009; Speight, 2011a, 2011b). However, it is time for procrastination to cease and procrastination will not help in getting beyond the depletion of crude oil and natural gas resources, and the various levels of government must start being serious about looking to the future for other sources of energy to supplement and even replace the current source of hydrocarbon fuels.

In addition and in keeping with the preferential use of light (low-density) crude oil as well as the maturation effect in the reservoir, crude oils available currently to the refinery is somewhat different in composition and properties from those available approximately 50 years ago (Parkash, 2003; Gary *et al.*, 2007; Speight, 2008, 2011a, 2014a, 2015b, 2017; Siefried and Witzel, 2010; Hsu and Robinson, 2017). The current crude oils are somewhat heavier insofar as they have higher proportions of non-volatile (asphaltic) constituents. In fact, by the standards of yesteryear, many of the crude oils currently in use would have been classified as heavy (high-density or viscous) feedstocks, bearing in mind that they may not approach the definitions that should be used based on the method of recovery. Achievement of changes in feedstock character, such as this tendency toward more viscous (heavier) crude oils, require adjustments in refinery operations to handle these heavier crude oils in order to reduce the amount of coke formed during processing and to balance the overall product slate (Speight, 2011a, 2014a).

As the 21st century matures, there will continue to be an increased demand for energy to support the needs of the commerce industry and residential uses — in fact, as the 2040–2049 decade approaches, commercial and residential energy demand is expected to rise considerably — by approximately 30% — over the current energy demand. This increase is due, in part, to the developing countries, where national economies are expanding and the move away from rural living to city living is increasing. In addition, the fuel of the rural population (biomass) is giving way to the fuel of the cities (transportation fuels and electric power) as the lifestyles of the population of the developing countries have changed from agrarian to metropolitan. Furthermore, the increased population of the cities requires more effective public transportation systems as the rising middle class seeks private means of transportation (automobiles). As a result, fossil fuels will continue to be the predominant source of energy for at least the next 50 years.

However, there are several variables that can impact energy demand from fossil fuels. For example, coal (as a source of electrical energy) faces significant challenges from governmental policies to reduce greenhouse gas emissions and fuels from crude oil can also face similar legislation (Speight, 2013a, 2013b, 2014a, 2017). In addition to the types of application and use, location and regional resources, cost of energy, cleanness and environmental factors, safety of generation and utilization, socioeconomic factors, as well as global and regional politics play a major role (Speight, 2011a; Speight and Islam, 2016). More particularly, the recovery of natural gas and crude oil from tight sandstone and shale formations faces challenges in hydraulic fracturing.

Briefly, hydraulic fracturing is an extractive method used by crude oil and natural gas companies to open pathways in tight (low-permeability) geologic formations so that the oil or gas trapped within can be recovered at a higher flow rate. When used in combination with horizontal drilling, hydraulic fracturing has allowed industry to access natural gas reserves previously considered uneconomical, particularly in shale formations. Although hydraulic fracturing creates access to more natural gas supplies, the process requires the use of large quantities of water and fracturing fluids, which are injected underground at high volumes and pressure. Oil and gas service companies design fracturing fluids to create fractures and transport sand or other granular substances to prop open the fractures. The composition of these fluids varies by formation, ranging from a simple mixture of water and sand to more complex mixtures with a multitude of chemical additives. Hydraulic fracturing has opened access to vast domestic reserves of natural gas that could provide an important stepping stone to a clean energy future. Yet questions related to the safety of hydraulic fracturing persist and this technology has been the subject of both enthusiasm and increasing environmental and health concerns in recent years, especially in relation to the possibility (some would say *reality*) of contaminated drinking water because of the chemicals used in the process and the disturbance of the geological formations (Speight, 2015a).

The danger of the peak energy theory is that the world is approaching an energy precipice in which the crude oil 1 year will not be available the next year. On the other hand, the peak energy opponents take a more realistic view in that the depletion of fossil fuels will occur gradually and, and with the current trends in considering other sources of energy, the concept of the energy precipice is not logical. In fact, the most unrealistic variable in the peak energy scenario arises from the misuse of data that supposedly indicate that the world is approaching the energy precipice in which fossil fuels will no longer be available for use as energy sources — the date of the energy precipice is only wildly speculative and, in many cases, totally unrealistic. Fossil fuel energy sources will undoubtedly reach a depletion point in the future when these energy sources are no longer available — but not at the moment or even in the present century. At the same time, new gas-fired generating units use highly efficient technologies and are supported by abundant gas supplies. As a result, gas is increasingly viewed as the most economical fossil fuel choice for electricity generation for the United States. Finally, a word on reserve estimation. There are a number of different methods by which crude oil and natural gas reserves can be calculated. These methods can be grouped into three general categories: (i) volumetric methods, (ii) the materials balance method, and (iii) the decline curve method or the production performance method.

The methods termed *volumetric methods* represent attempts to determine the amount of oil in-place by using the size of the reservoir as well as the physical properties of the reservoir rock(s) and the reservoir fluids. In the calculation process, a recovery factor is assumed, using data (and assumptions) from other crude oil and natural gas fields with similar characteristics to the field under evaluation. Based on these assumptions, the estimated amount of crude oil or natural gas in-place is multiplied by the recovery factor (derived from the other (similar) fields to arrive at an estimate of the reserves in-place. Current recovery factors for oil fields around the world typically range between 10% and 60% v/v of the crude oil and natural gas in-place, while some recovery factors are in excess of 80% v/v of the crude oil and natural gas in-place. The wide variance is due largely to the diversity of fluid and reservoir characteristics for different deposits. The method is most useful early in the life of the reservoir, before significant production has occurred. However, site specificity, which arises because of the differences in reservoir character (for example, reservoir mineralogy, porosity, and permeability) and the character of the reservoir fluids must also be given serious consideration; otherwise the estimation of the reserves in-place may be in error (by an order of magnitude above or below the real amount in-place). Such is the difficulty in estimating the reserves.

In addition, the materials *balance method* for a crude oil field or a natural gas field uses an equation (or derivation thereof) that relates (in the case of a crude oil reservoir with associated natural gas) the volume of crude oil, water and gas that has been produced from a reservoir, and the change in reservoir pressure, to calculate the remaining crude oil or natural gas. The calculation uses the assumption that as fluids from the reservoir are produced, there will be a change in the reservoir pressure that depends on the remaining volume of oil and gas. The method requires extensive pressure–volume–temperature analysis as well as an accurate pressure history of the field. If the pressure history of the field is not available, the calculation requires some production to occur (or to have occurred) (typically 5–10% v/v of ultimate recovery), unless reliable pressure history can be used from a field with similar reservoir rock characteristics as well as the characteristics of the reservoir fluids.

The *decline curve method* (also known as the *production performance method*) uses known production data to fit a decline curve and estimate future oil production — the three most common forms of decline curves are *exponential*, *hyperbolic*, and *harmonic*. The decline curve analysis is a long-established tool for developing a future outlook for crude oil and/or natural gas production from an individual well or from an entire oil field. Depletion has a fundamental role in the extraction of finite resources and is one of the driving mechanisms for oil flows within a reservoir and the depletion rate can be connected to decline curves. Consequently, depletion analysis is a useful tool for analyzing and forecasting crude oil and natural gas production.

In the calculation, it is assumed that the production will decline on a reasonably smooth curve, and so allowances must be made for wells shut-in and production restrictions. The curve can be expressed mathematically or plotted on a graph to estimate future production. It has the advantage of (implicitly) including all reservoir characteristics. However, the method requires a sufficient well or reservoir history to establish a statistically significant trend, ideally when production is not curtailed by regulatory or other artificial conditions.

Generally, the initial estimates of the size of newly discovered oil fields are usually too low. As crude oil and/or natural gas production continues, successive estimates of the ultimate recovery of fields tend to increase. The term *reserve growth* refers to the typical increases in estimated ultimate recovery that occur as oil fields are developed and produced. Reserve growth has now become an important part of estimating the total potential reserves of an individual province or country. As the world's crude oil reserves continue to decline, there is a need for the reserve estimates to be more precise through the application of the reserve-growth concept. In fact, the concept could be applied even to the so-called undiscovered resources with some qualifications as to the inherent risk (Speight, 2011a, 2014a).

1.2 Fossil Fuel Sources

Global energy consumption is increasing and is expected to rise by 41% over the period up to 2035 — compared to a 52% rise over the last twenty years and a 30% rise over the last decade. About 95% of the growth in demand is expected to come from the emerging economies, while energy use in the advanced economies of North America, Europe, and Asia as a group is expected to grow only very slowly — and begin to decline in the later years of the forecast period (BP, 2019). The data for reserve estimates indicate that there are sufficient reserves to cover this trend at least to and even beyond the year 2035, perhaps even to the year 2050.

1.2.1 *Natural gas*

It is rare that crude oil and also heavy crude oil do not occur without an accompanying cover of gas (Speight, 2014a, 2019). It is therefore important, when describing reserves of crude oil, to also acknowledge the occurrence, properties, and character of the natural gas. In recent years, natural gas has gained popularity among a variety of industrial sectors. Natural gas burns cleaner than coal or crude oil, thus providing environmental benefits. While natural gas is distributed mainly via pipelines, some of the natural gas is transported in the liquid phase (LNG) across oceans by specially designed tankers.

Approximately 45% v/v of the recoverable natural gas reserves of the world are *unconventional*, composed mainly of unconventional gas, i.e. gas from shale formations (tight gas) and coal-bed methane. The International Energy Agency (IEA) estimates that the global gas demand will increase by more than 50% between 2010 and 2035, and unconventional gas will make up 32% of the total supply. While Russia and countries of the Middle East hold the largest reserves of conventional gas, the available sources of unconventional gas are spread across the world and can be found in countries that are currently net importers, such as the United States and China. But the ease of accessibility will vary from well to well — extracting gas from tight shale reservoirs and extracting this gas involves pressurized fracture of the reservoir rock with water and water-containing various chemicals (*hydraulic fracturing*, *fracking*) (Speight, 2015a).

Assuming that the current level of natural gas consumption of the world is maintained, the reserve would be enough to last for another 64 more years. However, in this estimation of natural gas longevity, factors such as the increase in annual consumption, the discovery of new reservoirs, and advances in discovery/recovery technology, and the utilization of natural gas hydrates are not included. As a result of discoveries of gas in tight shale formations — which has offset more than the annual consumption — the world reserves of natural gas have been in a generally upward trend, due to discoveries of major natural gas fields. However, it should also be remembered that the total gas resource base (like any fossil fuel or mineral resource base) is dictated by economics. Therefore, when resource data are quoted, some attention must be given to the cost of recovering those resources. Most important, the economics must also include a cost factor that reflects the willingness to secure total, or a specific degree of, energy independence.

Natural gas liquids (NGLs) — which are the higher boiling hydrocarbon constituents of natural gas separated from natural gas at a gas processing plant, and includes ethane, propane, butane, and pentanes — have taken on a new prominence as shale gas production has increased and prices have fallen (Ratner and Tiemann, 2014). As a result, most producers are accepting the challenges with the opportunism and have shifted production to tight formations, such as the Bakken formation in North Dakota and Montana, to capitalize on the occurrence of NGLs in shale gas development (Speight, 2013f, 2015a; Sandrea, 2014).

1.2.2 *Gas hydrates*

Methane hydrates (also called methane clathrates) are a resource in which a large amount of methane is trapped within a crystal structure of water, forming a solid similar to ice (Chapter 1) (Collett, 2009). Methane hydrates exist as methane (the chief constituent of natural gas) trapped in cage-like lattices of ice which, if either warmed or depressurized, revert back to water and natural gas. When brought to the surface of the Earth, one cubic meter of gas hydrate releases 164 cubic meters of natural gas (solid-to-volume ratio = 1–164). Hydrate deposits may be several hundred meters thick and generally occur in two types of settings: under Arctic permafrost and beneath the ocean floor.

Gas hydrates occur in two discrete geological situations: (i) marine shelf sediments and (ii) on-shore Polar Regions beneath permafrost (Kvenvolden, 1993; Kvenvolden and Lorenson, 2000). These two hydrates occur in these two types of settings because these are the settings where the pressure–temperature conditions are within the hydrate stability field (Lerche and Bagirov, 1998). Gas hydrates can be detected seismically as well as by well logs (Goldberg and Saito, 1998; Hornbach *et al.*, 2003).

Global estimates of the methane hydrate resources/reserves vary considerably but vary from plentiful to immense and the energy content of methane occurring in hydrate form potentially exceeds the combined energy content of all other known fossil fuels (BOEM, 2012). When drilling in crude oil-bearing and gas-bearing formations submerged in deep water, the reservoir gas may flow into the well bore and form gas hydrates owing to the low temperatures and high pressures found during deep water drilling. The gas hydrates may then flow upward with drilling mud or other discharged fluids. When the hydrates rise, the pressure in the annulus decreases and the hydrates dissociate into gas and water. The rapid gas expansion ejects fluid from the well, reducing the pressure further, which leads to more hydrate dissociation and further fluid ejection.

1.2.3 *The crude oil family*

Crude oil and the equivalent term petroleum cover a wide assortment of materials consisting of mixtures of hydrocarbon derivatives and other compounds containing variable amounts of sulfur, nitrogen, and oxygen, which may vary widely in volatility, specific gravity, and viscosity. Metal-containing constituents, notably those compounds that contain vanadium and nickel, usually occur in the more viscous crude oils in amounts up to several thousands of parts per million and can have serious consequences during processing of these feedstocks. Because crude oil is a mixture of widely varying constituents and proportions, its physical properties also vary widely and the color from colorless to black. The crude oil family consists of various types of crude oil: (i) conventional crude oil, (ii) crude oil from tight formations, (iii) opportunity crude oils, (iv) high-acid crude oil, (v) foamy oil, and (vi) heavy crude oil.

To date, crude oil production on a worldwide basis has come almost exclusively from what are considered to be conventional crude oil reservoirs from which crude oil can be produced using tried-and-true recovery technologies compared with non-conventional sources that require more complex or more expensive technologies to extract — examples of such resources are tar sand bitumen, liquids from coal, liquids from biomass, and liquids from oil shale (Lee, 1990, 1991; Scouten, 1990; Speight, 2008, 2011b, 2012, 2013a, 2014b, 2016).

Generally, *crude oil from tight formations* (sometimes referred to as unconventional tight oil resources) are found at considerable depths in sedimentary rock formations that are characterized by very low permeability. While some of the tight oil reserves produce oil directly from shales, tight oil resources are also produced from low-permeability siltstone formations, sandstone formations, and carbonate formations that occur in close association with a shale source rock. Furthermore, crude oil from tight shale formation is characterized by a low content of high-boiling (resid) constituents, low-sulfur content, and a significant molecular weight distribution of the paraffinic wax content (Speight, 2014a, 2015b). Finally, the properties of crude oils from tight formations are highly variable. Density and other properties can show wide variation even within the same field. The Bakken crude oil is a light sweet (low density, low sulfur) crude oil with an API of 42° and a sulfur content of 0.19% w/w. Similarly, the Eagle Ford crude oil is also a light sweet crude oil with a published API gravity as high as 62° API but more usually between 40° and 55° API and a sulfur content of the order of 0.1% w/w.

There is also the need for a refinery to be configured to accommodate *opportunity crude oils* and/or *high acid crude oils*, which, for many purposes, are often included with heavy (high-density or viscous) feedstocks (Speight, 2014a, 2014b; Yeung, 2014).

1.2.3.1 *Opportunity crude oil*

Opportunity crude oils are either new crude oils with unknown or poorly understood properties relating to processing issues or existing crude oils with well-known properties and processing concerns (Ohmes, 2014). Opportunity crude oils are often, but not always, heavy crude oils, but in either case they are more difficult to process owing to high levels of solids (and other contaminants) produced with the oil, high levels of acidity, and high viscosity. These crude oils may also be incompatible with other oils in the refinery feedstock blend and cause excessive equipment fouling when processed either in a blend or separately (Speight, 2015b).

Opportunity crude oils, while offering initial pricing advantages, may have composition problems, which can cause severe problems at the refinery, harming infrastructure, yield, and profitability. Before refining, there is the need for comprehensive evaluations of opportunity crudes, giving the potential buyer and seller the needed data to make informed decisions regarding fair pricing and the suitability of a particular opportunity crude oil for a refinery. This will assist the refiner to manage the ever-changing crude oil quality input to a refinery — including quality and quantity requirements and situations, crude oil variations, contractual specifications, and risks associated with such opportunity crudes.

1.2.3.2 *High acid crude oil*

High acid crude oils are crude oils that contain considerable proportions of naphthenic acids, which, as commonly used in the crude oil industry, refer collectively to all of the organic acids present in the crude oil (Shalaby, 2005; Speight, 2014b). In many instances, high-acid crude oils are actually the heavier crude oils (Speight, 2014a, 2014b). The total acid matrix is therefore complex and it is unlikely that a simple titration, such as the traditional methods for the measurement of the total acid number, can give meaningful results to use in predictions of problems. An alternative way of defining the relative organic acid fraction of crude oils is therefore a real need in the oil industry, both upstream and downstream.

High-acid crude oils cause corrosion in the refinery — corrosion is predominant at temperatures in excess of 180°C (355°F) (Kane and Cayard, 2002; Ghoshal and Sainik, 2013; Speight, 2014c) — and occurs particularly in the atmospheric distillation unit (the first point of entry of the high-acid crude oil) and also in the vacuum distillation units. In addition, overhead corrosion is caused by the mineral salts, magnesium, calcium and sodium chloride, which are hydrolyzed to produce volatile hydrochloric acid, causing a highly corrosive condition in the overhead exchangers. Therefore, these salts present a significant contamination with opportunity crude oils. Other contaminants in opportunity crude oils which are shown to accelerate the hydrolysis reactions are inorganic clay minerals and organic acids.

1.2.3.3 *Foamy oil*

Foamy oil is oil-continuous foam that contains dispersed gas bubbles produced at the well head from the heavy crude oil reservoirs under solution gas drive (Maini, 1999, 2001; Sheng *et al.*, 1999). The nature of the gas dispersions in oil distinguishes the foamy oil behavior from conventional heavy crude oil. The gas that comes out of the solution in the reservoir does not coalesce into large gas bubbles nor into a continuous flowing gas phase. Instead, it remains as small bubbles entrained in the crude oil, keeping the effective oil viscosity low while providing expansive energy that helps drive the oil toward the producing well. Foamy oil accounts for unusually high production in the heavy crude oil reservoirs under solution gas drive.

During primary production of heavy crude oil from solution-gas-drive reservoirs, the oil is pushed into the production wells by energy supplied by the dissolved gas. As fluid is withdrawn from the production wells, the pressure in the reservoir declines and the gas that was dissolved in the oil at high pressure starts to come out of the solution (*foamy oil*). As pressure declines further with continued removal of fluids from the production wells, more gas is released from the solution and the gas already released expands in volume. The expanding gas, which at this point is in the form of isolated bubbles, pushes the oil out of the pores and provides energy for the flow of oil into the production well. This process is very efficient until the isolated gas bubbles link up and the gas itself starts flowing into the production well. Once the gas flow starts, the oil has to compete with the gas for the available flow energy. Thus, in some heavy crude oil reservoirs, due to the properties of the oil and sand and also due to the production methods, the released gas forms foam with the oil and remains subdivided in the form of dispersed bubbles for much longer.

1.2.3.4 *Heavy crude oil*

Heavy crude oil is a type of crude oil that is different from conventional crude oil insofar as it is much more difficult to recover from the subsurface of the reservoir. Heavy crude oils, particularly heavy crude oil formed by the biodegradation of organic deposits, are found in shallow reservoirs formed by unconsolidated sands. This characteristic, which causes difficulties during well drilling and completion operations, may become a production advantage due to higher permeability. In simple terms, heavy crude oil is a type of crude oil which is very viscous and does not flow easily. The common characteristic properties (relative to conventional crude oil) are high specific gravity, low hydrogen to carbon ratios, high carbon residues, and high contents of asphaltenes, heavy metal, sulfur, and nitrogen. Specialized refining processes are required to produce more useful fractions, such as naphtha, kerosene, and gas oil.

1.2.4 *Unconventional fossil fuel resources*

While crude oil and natural gas are conventional (fossil fuel) fuel sources, non-conventional sources (also called *alternative fuel sources*) are any materials or substances (other than conventional fuels) which can be used as fuels. Thus, before commencing a presentation on the non-fossil (unconventional) fuel sources, it is necessary to mention those fossil fuel resources that are generally considered unconventional resources in terms of the production of gaseous fuels, liquid fuels, and solid fuels. These resources are (i) extra heavy crude oil and tar sand bitumen, (ii) coal, and (iii) oil shale. This will allow to place the production of fuels from biomass in a proper perspective. This section presents an overview of the production of fuels from these sources and allows the reader to place the production of fuels from biomass in a proper perspective.

1.2.4.1 *Extra heavy crude oil and tar sand bitumen*

Two types of non-conventional oil reserves can be discerned: (i) extra heavy crude oil and (ii) tar sand bitumen and oil. Both of the viscous materials isolated from these sources have an AI gravity of the order of (under ambient conditions) 5–10° with a viscosity in excess of several hundred thousand centipoises. Typically, these materials are non-mobile (at best, they are difficult to be mobile) under deposit conditions — as long as the deposit temperature is below the pour point of the material (Speight, 2014a, 2107).

In addition to conventional crude oil and heavy crude oil, there remains two other, even more viscous materials, which offers some relief to the potential shortfalls in supply (Meyer and De Witt, 1990; Meyer and Attanasi, 2003; Speight, 2014a; BP, 2019). They are the *bitumen* known as extra heavy crude oil and the bitumen found in *tar sand* (*oil sand*) deposits (Speight, 2014a). However, many of these reserves are only available with some difficulty and optional refinery scenarios will be necessary for the conversion of these materials to liquid products because of the differences in character between conventional crude oil and tar sand bitumen (Speight, 2000, 2014a).

Extra heavy crude oil is a non-descript term (related to viscosity) of little scientific meaning that is usually applied to tar sand bitumen, which is generally capable of free flow under reservoir conditions (Speight, 2014a, 2017). Thus, the general difference is that extra heavy crude oil (which may have properties similar to tar sand bitumen in the laboratory), unlike tar sand bitumen in the deposit, has some degree of mobility in the reservoir or deposit (Delbianco and Montanari, 2009; Speight, 2014a, 2017). Extra heavy crude oils can flow at reservoir temperature and can be produced economically, without additional viscosity-reduction techniques, through variants of conventional processes such as long horizontal wells, or multilaterals. This is the case, for instance, in the Orinoco basin (Venezuela) or in offshore reservoirs of the coast of Brazil but, once outside of the influence of the high reservoir temperature, these oils are too viscous at the surface to be transported through conventional pipelines and require heated pipelines for transportation. Alternatively, the oil must be partially upgraded or fully upgraded or diluted with a low-boiling hydrocarbon fraction (such as aromatic naphtha) to create a mix that is suitable for transportation (Parkash, 2003; Gary *et al.*, 2007; Speight, 2014a, 2017; Hsu and Robinson, 2017).

Tar sand (referred to as *oil sand* in Canada) deposits are found in various countries throughout the world but in vast quantities in Alberta and Venezuela. There have been many attempts to define tar sand deposits and the bitumen contained therein. In order to define conventional crude oil, heavy crude oil, and bitumen, the use of a single physical parameter such as viscosity is not sufficient. Other properties such as API gravity, elemental analysis, composition, and, most of all, the properties of the bulk deposit must also be included in any definition of these materials. Only then will it be possible to classify crude oil and its derivatives.

In fact, the most appropriate and workable definition of *tar sand* is found in the writings of the United States government (United States Congress, 1976), which is not subject to any variation in chemical or physical properties that can vary depending on the method of property determination and the accuracy of that method (Speight, 2014a, 2016), viz.:

> Tar sands are the several rock types that contain an extremely viscous hydrocarbon which is not recoverable in its natural state by conventional oil well production methods including currently used enhanced recovery techniques. The hydrocarbon-bearing rocks are variously known as bitumen-rocks oil, impregnated rocks, oil sands, and rock asphalt.

This definition speaks of the character of the bitumen through the method of recovery (Speight, 2014a, 2016). Thus, the bitumen found in tar sand deposits is an extremely viscous material that is *immobile under reservoir conditions* and cannot be recovered through a well by the application of secondary or enhanced recovery techniques.

By inference and by omission, conventional crude oil and heavy crude oil are also indicated in this definition — crude oil is the material that can be recovered by conventional oil well production methods, whereas heavy crude oil is the material that can be recovered by enhanced recovery methods. Extra heavy crude oil can also be accommodated under this definition because the properties of extra heavy crude oil approximates the properties and behavior of tar sand bitumen under ambient conditions but is mobile because the reservoir temperature is higher than the pour point of the oil (Speight, 2014a). In the early days of resource development, the sole method for the recovery of the bitumen from tar sand formations was by the use of a mining process followed by separation of the bitumen by the hot water process. Current methods also include the use of advanced *in situ* processes that are not covered by the above definition. After recovery, the bitumen is then used to produce hydrocarbon derivatives by a conversion process.

Extra heavy crude oil is found predominantly in Venezuela, while tar sand deposits are found in various countries throughout the world but in vast quantities in Alberta and Venezuela. In relation to the bitumen in the Alberta tar sand deposits, approximately 15–20% v/v of the bitumen is recoverable, which amounts to approximately 75% v/v of the crude oil reserves in North America. Furthermore, it is only recently that the deposits of tar sand have been recognized as viable energy sources and there have been changes in the way the world views oil reserves, and tar sand reserves have come to be included in the tally of global crude oil and heavy crude oil reserves (Levant, 2010).

In spite of the high estimations of the reserves of bitumen, two conditions of vital concern for the economic development of tar-sand deposits are the concentration of the resource, or the percentage bitumen saturation, and its accessibility, usually measured by the overburden thickness. Recovery methods are based either on mining combined with some further processing or on operation on the tar sands *in situ*. The mining methods are applicable to shallow deposits, characterized by an overburden ratio (i.e. overburden depth to thickness of tar-sand deposit) that is deposit specific. For example, indications are that for the Athabasca deposit, no more than 15–20% v/v of the in-place deposit is available within current concepts of the economics and technology of open-pit mining and advanced non-mining recovery techniques (Speight, 2014a, 2019).

The largest accumulation of extra heavy crude oil is the Venezuelan Orinoco heavy-oil belt, which contains approximately 90% v/v of the world's heavy crude oil reserves when measured on an oil-in-place basis. In addition to extra-heavy Orinoco oil, South America also has technically recoverable heavy crude oil, so that, in total, approximately 61% v/v of the known technically recoverable heavy crude oil is in South America. However, the actual production data for heavy crude oil recovery are difficult to assess because of the loose and often arbitrary definitions that are employed.

For the United States, the prediction of peak oil turned out to be correct and, after the US production peaked in 1971 and thus lost its excess production capacity, OPEC was able to manipulate oil prices. Since then, oil production in several other countries has also peaked. However, for a variety of reasons, it is difficult to predict the oil peak in any given region. Based on available production data, proponents have previously (and incorrectly) predicted the peak in world oil production to be in the years 1989 and 1995, or at some time in the period 1995–2000, and in 2007 and beyond, at which time supplies of crude oil would diminish drastically.

However, the distinction between *conventional recovery methods* and *non-conventional recovery methods* is further complicated, because what is considered to be a mainstream technology can change over time. The wide application of hydraulic fracturing techniques to the recovery of otherwise inaccessible natural gas and crude oil has changed the dynamics of reserve estimates (Speight, 2015a). In addition, offshore oil deposits and reservoirs were considered to be a non-conventional source of crudes oil 50 years ago but with the advancement of technology many of these resources are considered to be conventional (Speight, 2015b). This is consistent with the inclusion of tar sand deposits and oil shale in the category *non-conventional sources*.

At present, crude oil provides approximately 40% (natural gas 25%) of the non-solar energy of the world, and most future assessments indicate that the demand for oil will increase substantially. Predictions of impending oil shortages are as old as the industry itself, and the literature is full of arguments between *optimists* and *pessimists* related to the amount of crude oil that remains and what other resources might be available. There are four factors which affect the economics of oil production and use that need to be understood to assess the availability of oil for the future: (i) the quality of the reserves, (ii) the quantity of the reserves, (iii) the recoverability of the reserves, and (iv) the patterns of exploitation of the resource over time, i.e. reservoir management.

The majority of crude oil reserves identified to date are located in a relatively small number of very large fields (giant fields) and, in fact, approximately 300 of the largest oil fields contain almost 75% v/v of the available crude oil. Although most of the nations of the world produce at least minor amounts of crude oil, the primary concentrations are in Saudi Arabia, Russia, the United States (chiefly Texas, California, Louisiana, Alaska, Oklahoma, and Kansas), Iran, China, Norway, Mexico, Venezuela, Iraq, Great Britain, the United Arab Emirates, Nigeria, and Kuwait. The largest known reserves are in the Middle East. Furthermore, this has put issues such as energy security, energy trade, and alternative energies at the forefront of the political agenda worldwide (Luciani, 2013). Until recently, crude oil prices were on an upward path and the more recent (at the time of writing) downward price spiral highlights the interconnected nature of the energy markets. In fact, it is only a matter of time before energy-consuming countries decide that the way around this is to produce energy from other sources. And this has already commenced. Once this happens in a substantial way, OPEC (Organization of the Petroleum Exporting Countries; the oil-reducing countries) will be at the mercy of the oil-consuming countries.

1.2.4.2 *Coal*

Coal is an organic sedimentary rock that is formed from the accumulation and preservation of plant materials, usually in a swamp environment (Speight, 2013a, 2013b). Coal is a combustible rock and along with crude oil and natural gas, it is one of the three most important fossil fuels, such as for the generation of electricity, and provides approximately 40% of the electricity production on a worldwide basis. In many countries, these data are much higher: Poland relies on coal for approximately 94% of its electricity; South Africa relies on coal for approximately 92% of its electricity; China relies on coal for approximately 77% of its electricity; and Australia relies on coal for approximately 76% of its electricity. Total recoverable reserves of coal around the world are estimated at 891,531 million tons — the United States has sufficient coal reserves to last (at current rates of consumption) in excess of 250 years (BP, 2019).

Coal has been studied extensively for conversion into gaseous and liquid fuels as well as hydrocarbon feedstocks. Largely thanks to its relative abundance and stable fuel price in the market, coal has been a focal target for synthetic conversion into other forms of fuels, i.e. synfuels.

Research and development work on coal conversion has seen peaks (highs) and valleys (lows) due to external factors including the comparative fossil fuel market as well as the international energy outlook of the era. Coal can be gasified, liquefied, pyrolyzed, and co-processed with other fuels including oil, biomass, scrap tires, and municipal solid wastes (MSWs) (Speight, 2008, 2011b, 2011c, 2013a, 2014b). Secondary conversion of coal-derived gas and liquids can generate a wide array of petrochemical products as well as alternative fuels.

During the past two centuries, coal has played this important role — providing coal gas for lighting and heating and for electricity generation along with the accompanying importance of coal as an essential fuel for steel and cement production, as well as a variety of other industrial activities. On a worldwide basis, in excess of 4 billion tons (4.0×10^9 tons) of coal is consumed by a variety of sectors — including power generation (steam coal and/or lignite), iron and steel production (coking coal), cement manufacturing, and as a solid fuel or a source of liquid fuels (Speight, 2013a, 2103b). In fact, coal remains an important source of energy in many countries and is used to provide approximately 40% of the electricity worldwide, but this does not give a true picture of the use of coal in electricity production. During the mentioned time, the coal industry has been pressurized to take into consideration the environmental aspects of coal use and has responded with a variety of on-stream coal-cleaning and gas-cleaning technologies (Speight, 2013a).

As a result of the adaptation by the industry of various environmental regulations that affect coal use, coal will continue to play a key role in the energy mix of the world, with the demand in certain regions set to grow rapidly. Growth in both the steam and coking coal markets will be strongest in the developing Asian countries, where the demand for electricity and the need for steel in construction and car production, and demands for household appliances will increase as incomes rise.

1.2.4.3 *Oil shale*

Just like the term *oil sand* (tar sand in the United States), the term *oil shale* is a misnomer, since the mineral does not contain oil nor is it always *shale*. The organic material is chiefly *kerogen* and the *shale* is usually a relatively hard rock, called marl. *Oil shale* is a complex and intimate mixture of organic and inorganic materials that vary widely in composition and properties. In general terms, oil shale is a fine-grained sedimentary rock that is rich in organic matter and yields oil when heated. Some oil shale is genuine shale, but others have been misclassified and are actually siltstones, impure limestone, or even impure coal. Oil shale does not contain oil and only produces oil when it is heated to approximately 500°C (approximately 930°F), when some of the organic material is transformed into a distillate similar to crude oil (Lee, 1990; Scouten, 1990; Speight, 2008, 2012).

Thus, when properly processed, kerogen can be converted into a substance somewhat similar to crude oil, which is often better than the lowest grade of oil produced from conventional oil reservoirs but of lower quality than conventional light crude oil. *Shale oil* (*retort oil*) is the liquid oil condensed from the effluent in oil shale retorting and typically contains appreciable amounts of water and solids, in addition to having an irrepressible tendency to form sediments. Oil shale is an inorganic, non-porous sedimentary marlstone rock containing various amounts of solid organic material (known as *kerogen*) that yields hydrocarbon derivatives, along with non-hydrocarbon derivatives, and a variety of solid products, when subjected to pyrolysis (a treatment that consists of heating the rock at high temperature) (Lee, 1990; Scouten, 1990; Speight, 2008, 2012).

Oil production potential from oil shale is measured by a laboratory pyrolysis method called the Fischer Assay (Speight, 1994, 2008, 2012) and is reported in barrels per ton (42 US gallons per barrel, approximately 35 imperial gallons per barrel). Rich oil shale zones can yield more than 40 US gallons per ton, while most shale zones produce 10–25 US gallons per ton. Yields of shale oil in excess of 25 US gallons per ton are generally regarded as the most economically attractive and hence the most favorable for initial development. Thus, oil shale has, however, a definite potential for meeting the energy demand in an environmentally acceptable manner (Lee, 1990; Scouten, 1990; Bartis *et al.*, 2005; Andrews, 2006; Speight, 2008, 2012).

The oil shale deposits in the western United States contain approximately 15% w/w organic material (kerogen) (Lee, 1990; Scouten, 1990; Speight, 2012). By heating oil shale to high temperatures (>500°C, >930°F), the kerogen is decomposed and converted to a volatile liquid product. However, shale oil is sufficiently different from crude oil, and refining of shale oil presents some unusual problems but, nevertheless, shale oil can be refined into a variety of liquid fuels, gases, and high-value products for the petrochemical industry.

The United States has vast known oil shale resources that are concentrated mainly in the Green River Formation in the states of Colorado, Wyoming and Utah, which account for nearly three-quarters of this potential (Lee, 1990; Scouten, 1990; Bartis *et al.*, 2005; Andrews, 2006; Speight, 2008, 2012). Because of the abundance and geographic concentration of the known resource, oil shale has been recognized as a potentially valuable United States energy resource since as early as 1859, the same year Colonel Drake completed his first oil well in Titusville, Pennsylvania (Chapter 1). Early products derived from shale oil included kerosene and lamp oil, paraffin, fuel oil, lubricating oil and grease, naphtha, illuminating gas, and ammonium sulfate fertilizer.

Since the beginning of the 20th century, when the United States Navy converted its ships from coal to fuel oil and the economy of the United States was transformed by gasoline-fueled automobiles and diesel-fueled trucks and trains, concerns have been raised related to assuring adequate supplies of liquid fuels at affordable prices to meet the growing needs of the nation and its consumers. Thus, it is not surprising that the abundant resources of oil shale in the United States were given consideration as a major source of these fuels. In fact, the Mineral Leasing Act of 1920 made crude oil and oil shale resources on Federal lands available for development under the terms of federal mineral leases. This enthusiasm for oil shale resources was mitigated to a large extent by the discoveries of more economically producible and easy-to-refine liquid crude oil in commercial quantities.

Despite the huge resources, oil shale is an under-utilized energy resource. In fact, one of the issues that arise when dealing with fuels from oil shale is the start–stop–start episodic nature of the various projects. The projects have varied in time and economic investment and viability. The reasons include competition from cheaper energy sources, heavy front-end investments, and, of late, an unfavorable environmental record. Oil shale has, however, a definite potential for meeting the energy demand in an environmentally acceptable manner (Bartis *et al.*, 2005; Andrews, 2006).

1.3 Non-Fossil Fuel Energy Sources

All fossil fuels are non-renewable and as such they will, eventually, be depleted to the point where recovery from the formations/reservoirs/

deposits will be prohibitively expensive or even impossible. Furthermore, energy generation from fossil fuels requires combustion, thus increasing the potential for damage to the environment through pollutants and greenhouse gas emission (Speight and Lee, 2000). In order to sustain the future of the world with a clean environment and non-depletive energy, the use of renewable energy sources is a move in the right direction.

Fuel sources (gas, liquid, and solid) are those sources that can be used to produce fuels (gas, liquid, and solid), which are combustible or energy-generating molecular species that can be harnessed to create mechanical energy. Crude oil and natural gas are the well-known conventional fuel sources that produce well-established products that have served industry and consumers for more than 100 years. Over the past four decades, in spite of the energy shocks of the 1970s, there has been rapid escalation in fuel demand to the point that many countries are net importers of crude oil and crude oil products and this is projected to continue. However, the time is running out and these fuel sources, once considered inexhaustible, are now being depleted at a rapid rate. In fact, there is little doubt that the supplies of crude oil are being depleted with each year that passes, but it is unclear just how long it will take to reach the bottom of the well!

In the current context, alternative fuels include in this analysis non-conventional crude oil and synfuels for the transportation sector. While the main focus of this text is on transportation fuels, it also includes upstream crude oil products used in transportation fuel production, lubricant products, and resids for asphalt. Briefly, a refinery produces a mix of low-boiling (low-density) products (mainly transportation fuels) and high-boiling (high density) products. To date, large-scale alternatives for oil products (gasoline, diesel, and LPG) do not exist and the shift in the refinery's product mix to transportation fuels of increasing quality results in an increase of the refinery's energy use. Therefore, transportation fuel demands will increasingly determine the demand for crude oil products.

The impact of the depletion of natural gas and crude oil resources can be overcome by serious planning for the world *Beyond Petroleum* (the slogan used by BP, formerly British Petroleum), but it is a trade-off between having a plentiful supply of liquid fuels and the higher cost (initially with a fall in production costs as technology advances) of the crude oil replacements. The flaw in this plan, of course, is its acceptance by the various levels of government in natural gas- and crude oil-consuming nations as the politicians think more of their reelection than the future of their country. And so, the matter falls into the hands of consumers and requires recognition that the price of fuels will, more than likely, continue to rise.

Non-fossil fuels are alternative sources of energy that do not rely on continued consumption of the limited supplies of crude oil, coal, and natural gas. Examples of non-fossil fuel energy sources include (i) biomass, (ii) geothermal, (iii) hydrogen, (iv) nuclear, (v) solar, (vi) tidal, and (vii) wind sources (Nersesian, 2007; Speight, 2008, 2011c) — in the current context, the focus is on bio-sources. Such resources are considered to be extremely important to the future of energy generation, because they are renewable energy sources that could be exploited continuously and do not suffer depletion. In addition, energy production using non-fossil-based sources is claimed to generate much less pollution than fossil fuel energy sources. This is considered crucial by many governments who are looking for ways to reduce the amount of pollution produced by their countries. The advantages of fossil fuel resources are often considered to include the know-how and the ease of production; many opponents of the use of fossil fuels cite the adverse effects on the environment (Speight, 2008, 2013a, 2013b, 2014a) and consider non-fossil fuels as a much better way of generating energy.

Furthermore, an alternative fuel or a synthetic fuel is defined according to the context of its usage. In the context of substitutes for crude oil-based fuel, the term "alternative fuel" or "synthetic fuel" implies any available fuel or energy source and may also refer to a fuel derived from a renewable energy sources. However, in the context of environmental sustainability, "alternative fuel" often implies an ecologically benign renewable fuel. Thus, as the amount of available crude oil decreases, the need for alternate technologies to produce liquid fuels grows; such alternate liquid fuels could arise from the use of biomass as the source material. These fuels could potentially help prolong the liquid fuel culture and mitigate the forthcoming effects of the shortage of transportation fuels that is predicted to occur.

Although there are now available methods for burning gas and similar products in a very efficient manner, for example, clean fossil fuels, a certain amount of pollution is still generated. Accordingly, various initiatives now exist, especially in Western countries, to encourage corporations and energy companies to invest in methods for producing energy from renewable (non-fossil fuel) resources. These resources are listed below — alphabetically rather than in any form of preference.

This section presents an overview of the production of fuels from unconventional sources in order to place these unconventional sources in the correct context of a growing biomass industry. Renewable energy sources include (alphabetically rather than by common use or preference) (i) biomass, (ii) geothermal energy, (iii) hydrogen energy, (iv) nuclear energy, (v) solar energy, (vi) tidal energy, and (vii) wind energy. However, although it is necessary to mention the potential sources of alternate fuel (or alternate energy), the focus of this book remains on biomass.

1.3.1 *Biomass*

By definition, biomass is material that comes from plants. Plants use the light energy from the sun to convert water and carbon dioxide into sugars that can be stored through a process called photosynthesis. Some plants, such as sugarcane and sugar beets, store the energy as simple sugars, which are mostly used for food. Other plants store the energy as more complex sugars (starches) and include grains such as corn and are also used for food. Another type of plant matter — cellulosic biomass — is made up of very complex sugar polymers (complex polysaccharides), and is not generally used as a food source. This type of biomass is increasing in use as a feedstock for ethanol (sometimes, in this case, referred to as bioethanol) production. Other feedstocks of interest include agricultural and forestry residues, organic urban wastes, food processing and other industrial wastes, and energy crops.

Thus, biomass is a renewable resource that has received considerable attention due to environmental considerations and the increasing demands of energy worldwide (Brown, 2003; NREL, 2003; Wright *et al.*, 2006; Tsai *et al.*, 2007; Speight, 2008, 2011c; Langeveld *et al.*, 2010; Lee and Shah, 2013; Hornung, 2014). Biomass is produced by a photosynthetic process (photosynthesis) that involves chemical reactions occurring on the Earth between sunlight and green plants within the plants in the form of chemical energy. In the process, solar energy is absorbed by green plants and some microorganisms to synthesize organic compounds from low energy carbon dioxide (CO_2) and water (H_2O). For example,

$$6CO_2 + 6H_2 \rightarrow \underset{\text{Biomass}}{6CH_2O} + 3O_2$$

For convenience here, the general formula of the organic material produced during the photosynthesis process is $(CH_2O)_n$, which is mainly carbohydrate material. Some of the simple carbohydrates involved in this process are the simple carbohydrates glucose ($C_6H_{12}O_6$) and sucrose ($C_{12}H_{22}O_{11}$) and constitute biomass, which is a renewable energy source because of its natural and repeated occurrence in the environment in the presence of sunlight. The amount of biomass that can be grown certainly depends on the availability of sunlight to drive the conversion of carbon dioxide and water into carbohydrates. In addition to the limitation of sunlight, there is a limit placed by the availability of appropriate land, temperature, climate, and nutrients (namely nitrogen, phosphorus, and trace minerals) in the soil.

Biomass is clean for it has negligible content of sulfur, nitrogen, and ash, which cause lower emissions of sulfur dioxide, nitrogen oxides and soot than conventional fossil fuels. Biomass resources are many and varied and include (i) forest and mill residues, (ii) agricultural crops and wastes, (iii) wood and wood wastes, (iv) animal wastes, (v) livestock operation residues, (vi) aquatic plants, (vii) fast-growing trees and plants, and (viii) municipal and industrial wastes. The role of wood and forestry residues in terms of energy production is as old as fire itself, and in many societies, wood is still the major source of energy. In general, biomass can include anything that is not a fossil fuel that is based on bio-organic materials other than natural gas, crude oil, heavy crude oil, extra heavy crude oil, and tar sand bitumen (Lucia *et al.*, 2006; Speight, 2008, 2011c; Lee and Shah, 2013).

There are many types of biomass resources that can be used and replaced without irreversibly depleting reserves and the use of biomass will continue to grow in importance as the replacements for fossil fuel sources and as feedstocks for a range of products (Narayan, 2007; Speight, 2008, 2011c; Lee and Shah, 2013). Some biomass materials also have particular unique and beneficial properties which can be exploited in a range of products including pharmaceuticals and certain lubricants. In this context, the increased use of biofuels should be viewed as *one of a range of possible measures* for achieving self-sufficiency in energy, rather than a panacea to completely replace the fossil fuels (Crocker and Crofcheck, 2006; Worldwatch Institute, 2006; Freeman, 2007; Nersesian, 2007).

The world's biofuel production is still small, being approximately 70 billion (70×10^9 tons) oil equivalent. With the volatility of the prices of crude oil, the relative competitiveness of renewable and alternative fuels is drastically improving. Furthermore, technological advances in the alternative renewable energy areas as well as public awareness backed by strong governmental support and incentives make the outlook of the alternative and renewable energy very promising (Energy Security Leadership Council, 2013). There are seven countries that can be considered the leaders in biofuel (particularly bioethanol) production (as a percentage of the total production): United States (43.5%), Brazil (24%), France (3%), Germany (4%), Argentina (3%), China (3%), and Indonesia (2.5%). But this still represents a minor amount of the total energy consumption in these countries.

Ethanol (bio-ethanol) from corn has been used at an accelerated pace as gasoline blending fuel as well as a new brand of fuel *E85*, which contains 85% ethanol and 15% gasoline. The majority of the gas stations in the United States supply E85 fuels regularly and many automakers are offering multiple lines of automobiles that can be operated on either conventional gasoline or E85. This has not only contributed to cleaner burning fuels but also supplemented the amount of gasoline sold.

Energy generation utilizing biomass and (MSWs is also promising in the regions where landfill space is very limited. Technological advances in the fields have made this option efficient and environmentally safe, possibly even supplementing refinery feedstocks as sources of energy through the installation of gasification units (Speight, 2008, 2011a, 2011b, 2011c, 2013b).

1.3.2 *Geothermal energy*

At its most basic, geothermal power is about extracting energy from the ground around us. Thus, geothermal energy is the heat from the Earth and the resources for geothermal energy range from the shallow ground to the hot water and hot rock found a few miles beneath the surface of the Earth, and down even deeper to the extremely high temperatures of molten rock called magma (Turcotte and Schubert, 2002; Lay *et al.*, 2008). This energy source has immense potential but does little to disrupt the land. However, the heavy upfront costs of creating geothermal power plants has led to slower adoption than may have been expected for a fuel source with so much promise.

In the United States, most geothermal reservoirs of hot water are located in the western states, Alaska, and Hawaii. Wells can be drilled into underground reservoirs for the generation of electricity. Some geothermal power plants use the steam from a reservoir to power a turbine/generator, while others use the hot water to boil a working fluid that vaporizes and then turns a turbine. Hot water near the surface of the Earth can be used directly for heat. Direct-use applications include heating buildings, growing plants in greenhouses, drying crops, heating water at fish farms, and several different industrial processes.

1.3.3 *Hydrogen energy*

Unlike natural gas, hydrogen is a completely clean burning fuel — the product is water (H_2O) and there are no oxides of sulfur or carbon that can cause environmental pollution. The major issue with this form of alternative energy is that it is mostly derived from the use of natural gas and fossil fuels (Speight, 2014, 2017; Speight and Islam, 2016). As such, it could be argued that the emissions created to extract it counteract the benefits of its use. The process of electrolysis, which is essential for the splitting of water into hydrogen and oxygen, makes this less of an issue. However, electrolysis still ranks below the previously mentioned methods for obtaining hydrogen, although research continues to make it more efficient and cost-effective.

Hydrogen is the simplest element — an atom of hydrogen consists of only one proton and one electron and is also the most plentiful element in the universe. Despite its simplicity and abundance, hydrogen does not occur naturally as a gas on the Earth — it is always combined with other elements in the form of naturally-occurring materials. Hydrogen is also found in many organic compounds, notably natural gas, crude oil, heavy crude oil, extra heavy crude oil, and tar sand bitumen, which are sources of the currently-used liquid fuels. Hydrogen can be separated from hydrocarbon derivatives through the application of heat — a process known as *reforming*. Currently, most hydrogen is made this way from natural gas (steam-methane reforming). An electrical current can also be used to separate water into its components of oxygen and hydrogen (electrolysis), and various species of algae and bacteria, using sunlight as their energy source, even give off hydrogen under certain conditions.

The hydrogen-energy-based economy is a proposed system for delivering energy using hydrogen (Tester *et al.*, 2005). The advocates of it promote hydrogen as a potential fuel for motive power (including cars and boats), the energy needs of buildings, and portable electronics. Free hydrogen does not occur naturally in quantity, but can be generated by steam reforming of hydrocarbon derivatives, electrolysis of water, and a variety of other methods.

In the future, hydrogen could also join electricity as an important energy carrier. An energy carrier moves and delivers energy in a usable form to consumers. Renewable energy sources, such as the sun and wind, cannot produce energy all the time. But they could, for example, produce electric energy and hydrogen, which can be stored until it is needed. Hydrogen can also be transported (like electricity) to locations where it is needed.

Public awareness of clean burning and energy-efficient hydrogen has propelled unprecedented interest in hydrogen technology and fuel cell research and development. Many experts in the world predict the future as *hydrogen economy*. For the hydrogen economy to be fully arriving, a long list of technological advances must be accomplished, which include technologies for inexpensive generation, safe distribution and storage, safe and efficient materials for hydrogen handling, hydrogen internal combustion engine, hydrogen fuel cells, and loss prevention.

1.3.4 *Nuclear energy*

Nuclear energy (nuclear power) is the use of exothermic nuclear processes to generate useful heat and electricity and includes nuclear fission, nuclear decay, and nuclear fusion (Lutz *et al.*, 2009). Nuclear energy provided approximately 6% of the energy requirements worldwide and 13% of the electricity production worldwide in 2012. In 2013, the International Atomic Energy Agency (IAEA) reported that there are 437 operational nuclear power reactors in 31 countries, although not every reactor is producing electricity. In addition, there are approximately 140 naval vessels using nuclear propulsion in operation, powered by some 180 reactors. As of 2013, attaining a net energy gain from sustained nuclear fusion reactions, excluding natural fusion power sources such as the Sun, remains an ongoing area of international physics and engineering research.

Nuclear power is amongst the most abundant forms of alternative energy. It creates a number of direct benefits in terms of emissions and efficiency, while also boosting the economy by creating jobs in plant creation and operation. Many countries have relied on nuclear power to produce at least a quarter of their electricity over the past half-decade, and there are approximately 500 plants in operation throughout the world. The drawback with the generation of nuclear energy is that when there is a malfunction in a nuclear power plant the potential for catastrophe exists — the situations in Chernobyl and Fukushima are examples.

1.3.5 *Solar energy*

Solar energy is radiant light and heat from the sun that is harnessed using a range of technologies such as solar heating, solar photovoltaics, solar thermal electricity, solar architecture, and artificial photosynthesis. Solar technologies are broadly characterized as either (i) passive or (ii) active, depending on the method used to capture, convert, and distribute solar energy. Active solar techniques include the use of photovoltaic panels and solar thermal collectors to harness the energy. Passive solar techniques include (i) orienting the solar panels placed on a building towards the Sun, (ii) selecting materials with favorable thermal mass or light dispersing properties, and (iii) designing spaces that naturally circulate air (Mills, 2004).

This technology has evolved considerably over the past three decades and is now used for large-scale energy production and power generation for a single home. This energy source is completely renewable and the costs of installation are outweighed by the money saved in energy bills from traditional suppliers. Nevertheless, solar cells are prone to deterioration over large periods of time and are not as effective in unideal weather conditions. However, the development of affordable, inexhaustible and clean solar energy technologies will have huge longer-term benefits. The use of solar energy technologies will increase energy security for energy-importing countries through (i) reliance on an indigenous, inexhaustible and mostly import-independent resource, (ii) enhancement of energy sustainability, and (iii) reduction in pollution.

1.3.6 *Tidal energy*

Historically, tidal energy (in the form of tide mills) has been used both in Europe and on the Atlantic coast of North America. The incoming water was contained in large storage ponds, and as the tide went out, it turned waterwheels that used the mechanical power it produced to mill grain. The earliest occurrences date from the Middle Ages, or even from Roman times leading, in the 19th century, to the process of using falling water and spinning turbines to create energy.

While tidal energy uses the power of water to generate energy, much like hydroelectric methods, its application actually has more in common with wind turbines. The most common form of tidal energy generation is the use of tidal stream generators that use the kinetic energy of the ocean to power turbines, without producing the waste generated by the use of fossil fuels or being as susceptible to the elements as other forms of alternative energy.

Although not yet widely used, tidal energy (tidal power) has potential for future electricity generation (Turcotte and Schubert, 2002; Evans, 2007; Douglas *et al.*, 2008). Tides are more predictable than wind energy and solar power. Among sources of renewable energy, tidal power has traditionally suffered from the relatively high cost and limited availability of sites with sufficiently high tidal ranges or flow velocities, thus constricting its total availability. However, many recent technological developments and improvements, in both design (such as dynamic tidal power and tidal lagoons) and turbine technology (such as axial turbines and cross-flow turbines), indicate that the total availability of tidal power may be much higher than previously assumed, and that economic and environmental costs may be brought down to competitive levels.

Another form of energy that is often included in the category of tidal energy is wave energy. Wave energy converters hold an advantage over tidal energy sources, because they (the wave energy converters) can be placed in the ocean in various situations and locations. Much like tidal energy, the benefits of using wave energy lie in the lack of waste produced.

1.3.7 *Wind energy*

Wind energy is a free, renewable resource, so no matter how much is used, there will still be the same supply in the future (Fthenakis and Kim, 2009). This form of energy generation has become increasingly popular in recent years. It offers much the same benefits as many other alternative fuel sources do in that it makes use of a renewable source and generates no waste.

Unlike conventional power plants, wind plants emit no air pollutants or greenhouse gases. Approximately 80% of the cost of energy generation from wind is the machinery, with the balance being site preparation and installation. If wind generating systems are compared with fossil-fueled systems on a *life-cycle* cost basis (counting fuel and operating expenses for the life of the generator), current wind energy installations power more than twenty million homes in the United States. Unfortunately, this form of energy generation also presents challenges insofar as wind turbines restrict views and have the potential to be dangerous to some forms of wildlife.

1.4 Energy Supply

Energy supply (energy flow) is the delivery of fuels or transformed fuels to the point of consumption and the term potentially encompasses the extraction, transmission, generation distribution, and storage of fuels. This supply of energy can be disrupted by several factors, including imposition of higher energy prices due to action by OPEC (in the case of crude oil) or other cartel, war, or political disputes, economic disputes, or physical damage to the energy infrastructure because of terrorism. The security of energy supply is a major concern of national security and energy independence and can fall under the influence of any one or more of the following factors: (i) depletion of crude oil reserves, (ii) economic factors, often referred to as the peak oil theory, (iii) geopolitical factors, (iv) physical factors, and (v) technological factors.

1.4.1 *The peak oil theory*

The peak oil theory is based on the fact that the total amount of crude oil is indeed is finite and, therefore, production will one day reach a peak and then begin to decline (Speight and Islam, 2016). The theory relates to the point in time when the maximum rate of global crude oil recovery is realized, after which the rate of production enters a terminal decline (Deffeyes, 2002, 2005). However, peak oil is *not* related to running out of oil but to the peaking and subsequent decline of the production rate of oil. Many observers have interpreted peak oil to mean that the world is running out of oil at the time of the peak and the downside (the decreased production side of the curve) is often thought to be a precipice down which the oil-consuming nations (the oil-importing nations) speed to oblivion.

This is common sense, as explained in the resource depletion theory, which, in this case, assumes that reserves of crude oil will not be replenished (i.e. that abiogenic replenishment is negligible) and future world oil production must inevitably reach a peak and then decline as these reserves are exhausted (Hubbert, 1956, 1962). There is no doubt that crude oil and natural gas are being consumed at a steady rate, but whether or not the Hubbert peak oil theory will affect the consumption of crude oil is another issue. It is a theory that is based on reserve estimates and reserve consumption. No one will disagree that hydrocarbon resources (in the form of crude oil and natural gas) are finite resources and will run out at some future point in time, but the proponents of an energy precipice must recognize that this will not be the case, at least not for now (Speight and Islam, 2016). The issue is the timing of this event — it remains to be seen whether it is tomorrow, next week, next month, next year, or in 50 or more years. Current evidence (Speight, 2011a, 2011c; BP, 2019) favors a lifetime of at least 50 years for the current reserves of crude oil and natural gas, or perhaps longer if hydraulic fracturing continues to play a dominant role in crude oil and natural gas production (Speight, 2015a).

1.4.1.1 *Reserve depletion*

The peak oil theory (also known as the *Hubbert peak theory*) that originated in 1956 predicted that oil production in the United States (and the world) would reach a maximum (peak) between 1965 and 1970 (Hubbert, 1956). The concept is based on the observed production rates of individual oil wells and the combined production rate of a field of related oil wells. The aggregate production rate from an oil field over time appears to grow exponentially until the rate reaches a peak and then declines, sometimes rapidly, until the field is depleted.

According to the Hubbert postulate, the production rate of a limited resource, such as crude oil, will follow a bell-shaped curve representing the increase in production followed by a diminished rate of production, and then a decline in production. Furthermore, the Hubert model has been shown to be descriptive of the peaking and decline of production from oil fields in many countries (Brandt, 2007). In fact, Hubbert predicted in 1974 that peak oil would occur (if the trends prior to 1974 continued) in 1995. However, in the late 1970s and early 1980s, global oil consumption actually dropped (due to the shift to energy-efficient cars and the shift to electricity and natural gas for heating), and then rebounded to a lower level of growth in the mid-1980s. Thus oil production did not reach a peak in 1995, and has climbed to more than double the rate that was initially projected. This underscores the fact that the only reliable way of identifying the timing of peak oil may be in retrospect — hindsight is usually 20–20! Thus, predictions (of necessity) have been tweaked through the years as up-to-date information becomes more readily available, such as new reserve growth data. Predictions of the timing of peak oil by the peak oil theory proponents include the alternate possibilities that (i) peak oil production has occurred recently, (ii) peak oil production will occur in the very near future, or (iii) a plateau of oil production will sustain supply for several decades. *None* of these predictions dispute the peaking of oil production, but where they actually disagree is the time when peak oil production will occur.

However, the argument for peak oil production has not been presented in the context of a systematic evaluation of the available data. The concept of the Hubbert-derived bell curve was a heads-up warning and started in the realization that crude oil was a finite resource, which has been accepted as a fact by many of those peak oil proponents who failed to recognize the potential for a skewed bell curve (Figure 1.2(b)) through the potential of technological advances and the impact of heavy crude oil and tar sand bitumen — which was (at the time the original concept) developed in the mid-to-late 1950s. Indeed, the post-peak reservoir decline curve assumptions are rebutted by the observation that the geometry of a so-called *typical* bell-curve oil field production profile is often distinctly asymmetrical and does not generally show a precipitous mirror-image decline in production after an apparent peak.

In fact, the *peak oil theory*, because it has been approached as an emotional issue, is causing confusion, leading to inappropriate actions that turn attention away from the real issues. Crude oil is critical to the global economy, and emotional outbursts should not replace careful analysis related to the very real challenges in delivering liquid fuels to meet the needs of growing economies. However, the global decline will begin by 2020 or later, and one assumes that major investments in energy alternatives will occur before a crisis, without requiring major changes in the lifestyle of heavily oil-consuming nations (Jackson, 2006). Indeed, historically (or hysterically) this is the fifth time that the world is said to be running out of oil. Each time (whether it was the shortage of gasoline at the end of World War I or the gasoline shortage of the 1970s), advances in technology have managed to offset the decline of available oil. In fact, there is sufficient crude oil remaining in the ground that will become available as more technological advances are explored (Chapter 1). And this does not include liquid fuels produced from tar sand, coal, oil shale, and gas-to-liquid processes (Speight, 2008, 2013a, 2013b, 2013c, 2013d, 2014a, 2014b). The global resource base of conventional and unconventional oils is likely to grow (Jackson, 2006) — the operative word being *likely* — but subject to advances in technology for making unavailable residual oil available, *likely* has the distinct chance of becoming *real*; such an oil does not include the vague reserve classification of the so-called *yet-to-be-discovered* crude oil (Speight, 2008, 2011a, 2014a).

Also, there is some question related to the validity of the reserves estimates given by oil-producing (OPEC) countries (Campbell and Laherrère, 1998; Cohen, 2007). Many of the so-called reserves may actually be inflated and are, in fact, resources — they are not delineated and are inaccessible, so they are unavailable for production. Nevertheless, it is thought that global crude oil production will follow an *undulating plateau* for one or more decades before declining slowly and the slope of decline will become more gradual and much less steep as the rate of increase of oil discoveries increases (Jackson, 2006; Cohen, 2007). It is anticipated that increasing oil prices will support the current production plateau for (at least) another three to four decades, but beyond that, it is difficult to predict the actual events (Campbell, 2004a, 2004b, 2005). Furthermore, the touted production pattern (for the original Hubbert prediction) does not include the impact of heavy crude oil and tar sand bitumen on the availability of oil by the conversion of the latter (tar sand bitumen) into synthetic crude oil. In addition, technologies for the production of gas-related liquids (condensate and NGL), gas-to-liquids (GTL), coal-to-liquids (CTL), and biomass-to-liquids (BTL), as well as waste-to-liquids (WTL) (Speight, 2008, 2011b, 2013) will also be needed to bridge the potential gap in the supply of liquid fuels. In short, before the panic (that the world is running out of oil) continues, all of the options must be explored on a logical and systematic basis rather than based on the emotion of the moment (Bentley, 2002).

On the other hand, the rigid belief in undiscovered oil resources should not detract us from the reality of the depletion of proven oil reserves. The importance of tar sand resources in Canada and similar resources of the Orinoco tar belt and oil shale resources provides a global inventory that, when developed, will allow the productive capacity of hydrocarbon-based fuels to continue to expand and continue well into the foreseeable future. In fact, unconventional sources of liquid fuels (especially tar sand bitumen and extra heavy crude oil) have only recently been included as part of oil reserves. However, oil companies can book them as proven reserves after opening a mine (for tar sand and, thence, bitumen recovery) or a thermal facility for recovery of modified bitumen. However, these unconventional sources are not as efficient to produce and require extra energy to refine, resulting in higher production costs and more greenhouse gas emissions per barrel equivalent. While the energy used, the resources needed, and environmental effects of extracting unconventional sources have traditionally been prohibitively high, the three major unconventional oil sources being considered for large-scale production are the extra heavy crude oil and bitumen in the Orinoco Belt (Venezuela), the tar sand (oil sand) at Athabasca (Alberta), and the oil shale of the Green River formation (Colorado, Utah, and Wyoming, USA) in the United States. Taken together, these resource occurrences are approximately equal to the *identified reserves* of conventional crude oil accredited to the Middle East (Speight, 2008, 2011a).

Nevertheless, despite the large quantities of oil available in non-conventional sources, limitations on production prevent them from becoming an immediate replacement for conventional crude oil, and peak oil is a recognizable reality. Even under highly optimistic assumptions, production of synthetic crude oil from the Alberta tar sand could reach 5 million barrels per day by 2030 provided that there is an immediate program for development (Söderbergh *et al.*, 2007) making it possible that within the foreseeable future, the supply of crude oil will be supplemented by oil (synthetic crude oil) from sources other than crude oil (unconventional sources) (Speight, 2008, 2011c).

1.4.1.2 *Effects and consequences*

The widespread use of fossil fuels has been one of the most important stimuli of economic growth and has allowed the consumption of energy at a greater rate than the rate at which it is being replaced and presents an unprecedented risk management problem (Yergin, 1993; Hirsch, 2005; Hirsch *et al.*, 2005; Yergin, 2011). A peak in the production of crude oil will occur, but it is uncertain whether it will occur slowly or abruptly — given appropriate warnings, the latter is likely to be the case. The adoption of alternate technologies to supplant the deficit in oil production will require a substantial time period of the order of 10–20 years. If actions are taken before the onset of *peak oil*, economic upheaval is not inevitable, and the problems can be solved with existing technologies but government intervention is necessary.

The production of fuels from fossil fuel sources (especially the crude oil-based fuels) gives well-established products that have served industry and domestic consumers for more than 100 years, and in the foreseeable future, various fuels will still be largely based on hydrocarbon fuels derived from crude oil. Although the theory of peak oil is questionable, there is no doubt that crude oil, once considered inexhaustible, is being depleted at a measurable rate. The supposition by peak oil proponents is that supplies of crude oil are approaching a precipice in which fuels that are currently available may, within a foreseeable short time frame, become no longer available. While such a scenario is considered to be unlikely, there is a need to consider alternate technologies to produce liquid fuels that could mitigate the forthcoming effects of the shortage of transportation fuels.

Whatever the type of crude oil, it is the product of the burial and transformation of biomass over the last 200 million years (Speight, 2014a) and, historically, has led the way in usability as an energy source. However, the total amount of crude oil is indeed is finite, and, therefore, production will one day reach a peak and then begin to decline.

Thus, controversy surrounds the theory — not so much from the theory itself, which is quite realistic but from the way in which the theory is used by varying groups of alarmists — since predictions of the time of the global peak are dependent on the past production and the discovery data used in the calculation. Furthermore, a common method of representing resource availability and depletion is often depicted by a bell curve in which the ascending side and descending side are equal (Figure 1.2(a)). However, in terms of resource depletion, the crucial question is related to whether or not the ascending side is steeper than the descending side (Figure 1.2(b)) — the precipice depletion theory, which means that supply cannot keep up with growing demand. Alternatively, the supply–demand differential leading to the rate of resource depletion can be slowed down (Figure 1.2(c)) by voluntary reductions in crude oil consumption before physical limits to continued supply growth emerge. Not surprisingly, concerns have arisen in recent years related to the relationship between (i) the growing consumption of crude oil and the availability of crude oil reserves and (ii) the impact of potentially dwindling supplies of crude oil and rising prices on the various national economies. Following a peak in world oil

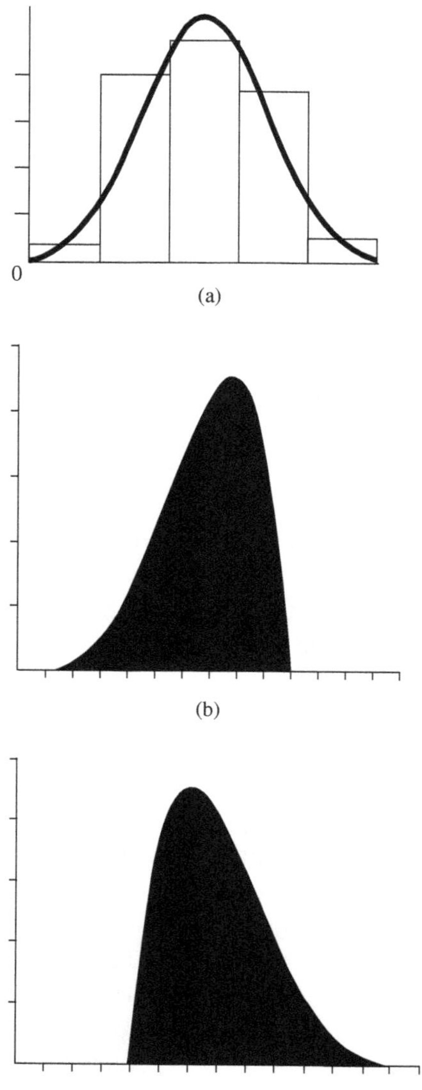

Figure 1.2: (a) The typical resource bell curve. (b) Bell curve skewed to the lower end. (c) Bell curve skewed to the higher end.

production, the rate of crude oil production would eventually decrease and, necessarily, so would the rate of consumption of oil.

Estimations of the years for which crude oil can be supplied and consumed at the current consumption rate have often been made by professionals and policy makers, but the numbers have been inconsistent and fluctuating from year to year — some of the numbers have seemed realistic and others have been totally illogical. This uncertainty comes from the difficulty of estimating the future recoverable amount of crude oil from all the proved and unproved reserves. The Society of Petroleum Engineers (SPE, United States) and the World Petroleum Council (WPC, London, United Kingdom) have developed and approved several definitions of the terminology used to describe crude oil reserves in an effort to facilitate consistency among professionals using these terms (SPE, 2007). They are: (i) proved reserves, also referred to as proven reserves, (ii) unproved reserves, also referred to as unproven reserves, which may be further sub-divided into (iii) probable reserves and (iv) possible reserves (Table 1.1).

In terms of world resources, the Western Hemisphere has approximately 69% v/v of the technically recoverable (using current technology) heavy crude oil and approximately 82% v/v of the technically recoverable (using current technology) tar sand bitumen (Meyer and Attanasi, 2003). In contrast, the Eastern Hemisphere has approximately 85% v/v of the conventional crude oil reserves of the world. On an international note, the bitumen in tar sand deposits represents a potentially large supply of energy. However, many of the reserves are available only with some difficulty and optional refinery scenarios will be necessary for the conversion of these materials into liquid products because of the substantial differences in character between conventional crude oil, extra heavy crude oil, and tar sand bitumen (Table 1.2). Because of the diversity of available information and the continuing attempts to delineate the various world tar sand deposits, it is virtually impossible to present accurate numbers that reflect the extent of the reserves in terms of the barrel unit. Indeed, investigations into the extent of many of the various deposits spread throughout the world are continuing at such a rate that the numbers vary from one year to the next.

The issue of peak oil can be overcome by either (i) waiting until the world oil production reaches a peak before taking crash program action — this would leave the world with a significant liquid fuel deficit for more than two decades, or (ii) initiating an urgent program for

Table 1.1: General definitions of reserves.

Reserve type	Definition
Proved reserves	The quantities of crude oil which, by analysis of geological and engineering data, can be estimated with reasonable certainty to be commercially recoverable, from a given date forward, from known reservoirs and under current economic conditions, operating methods, and government regulations. Proved reserves can be categorized as developed or undeveloped.
Unproved reserves	Estimates based on geologic and/or engineering data similar to that used in estimates of proved reserves; but technical, contractual, economic, or regulatory uncertainties preclude such reserves being classified as proved. Unproved reserves may be further classified as probable reserves and possible reserves.
Probable reserves	Unproved reserves which analysis of geological and engineering data suggests are more likely than not to be recoverable. If a probabilistic interpretation has to be given, there should be at least a 50% probability that the quantities actually recovered will equal or exceed the sum of estimated proved plus probable reserves.
Possible reserves	Unproved reserves which analysis of geological and engineering data suggests are less likely to be recoverable than probable reserves. If a probabilistic interpretation has to be given, there should be at least a 10% probability that the quantities actually recovered will equal or exceed the sum of estimated proved plus probable plus possible reserves.

alternate fuel source development ten years before the available crude oil in the world reaches a peak — this would leave a world liquid fuel shortfall for several years (perhaps a decade) after peak oil, or (iii) initiate an urgent program for alternate fuel source development 20 years before peak oil — this would offer the possibility of avoiding a world liquid fuel shortfall for the forecast period.

Thus it is the development of non-fossil fuel resources that is the focus of this book. In fact, the development of programs related to the production of fuel products (alternate fuels) will reduce the consumption of traditional crude oil sources as well as affect the timing of peak oil and the shape of the Hubbert curve.

Table 1.2: Observable differences between conventional crude oil, tight oil, heavy crude oil, extra heavy crude oil, and tar sand bitumen.[a]

Conventional Crude Oil

- Mobile in the reservoir; API gravity: >25°
- High-permeability reservoir
- Primary recovery
- Secondary recovery

Tight Crude Oil (i.e. Crude Oil from Tight Formations)

- Similar properties to the properties of conventional crude oil; API gravity: >25°
- Immobile in the reservoir
- Low-permeability reservoir
- Horizontal drilling into reservoir
- Fracturing (typically multi-fracturing) to release fluids/gases

Heavy Crude Oil

- More viscous than conventional crude oil; API gravity: 10–20°
- Mobile in the reservoir
- High-permeability reservoir
- Secondary recovery
- Tertiary recovery (enhanced oil recovery — EOR; e.g. steam stimulation)

Extra Heavy Crude Oil

- Similar properties to the properties of tar sand bitumen; API gravity: <10°
- Mobile in the reservoir
- High-permeability reservoir
- Secondary recovery
- Tertiary recovery (enhanced oil recovery — EOR; e.g. steam stimulation)

Tar Sand Bitumen

- Immobile in the deposit; API gravity: <10°
- High-permeability reservoir
- Mining (often preceded by explosive fracturing)
- Steam assisted gravity draining (SAGD)
- Solvent methods (VAPEX)
- Extreme heating methods
- Innovative methods[b]

Notes:

[a]This list is not intended for use as a means of classification.

[b]Innovative methods excludes tertiary recovery methods and methods such as steam assisted gravity drainage (SAGD) and vapor assisted extraction (VAPEX) methods but does include variants or hybrids thereof.

1.4.2 *Economic factors*

Crude oil economics is, as might be expected if the peak oil theory is true, dominated by the so-called occurrences of the remaining reserves that are going to be depleted in the very near future. While there may be varied volatility in the price of crude oil, if there is a belief that the price in the future will be higher than the price for immediate delivery, there will be a flurry of economic activity to purchase short-term contracts. This, in turn, leads to a so-called equilibrium price level that is invariably higher than the previous price level, which can influence producers to defer production activities in anticipation of a rise in price (Fleming, 2000). In actual fact, the price circle is indeed a vicious circle insofar as one arc of the price circle affects the other arcs (of the price circle), leading to what might be called economic anarchy. This anarchy might easily be laid at the feet of the peak oil theorists who base their theory on faulty or uncertain data and speculation. Moreover, the economic principles, which explain how a market economy works, tend to break down when applied to natural resources such as oil. In fact, there are two ways in which the principles of market economics do not apply to crude oil: (i) the current price of oil has virtually no influence on the rate at which it is discovered, (ii) the rules of supply and demand do not always hold, and (iii) a rise in the price of crude oil does not always lead to an increase in production.

While the current high oil prices may encourage the development and adoption of alternatives to oil, if high oil prices are not sustained, efforts to develop and adopt alternatives may fall by the wayside. The high oil prices and fears of running out of oil in the 1970s and early 1980s encouraged investments in alternative energy sources (including synthetic fuels made from coal and oil shale), but when oil prices decrease, investments in these alternatives became uneconomic. In fact, the development of renewable energy systems needs to be supported by decisive, well-coordinated action by governments, in sustained multi-decade programs. Many oil-consuming nations are moving to alternate fuel development rather than be faced with a *destabilizing energy gap*.

1.4.3 *Geopolitical factors*

The true picture of oil supply may never be known. Difficult as it is because of a variety of factors, reporting data on oil production or oil reserves is a political act (Laherrère, 2001). The US SEC (United States

Securities and Exchange Commission) obliges the oil companies listed on the US stock market to report only *proved reserves* and to omit *probable reserves* that are reported in the rest of the world. This practice of reporting only proved reserves can lead to a strong reserve growth since 90% of the annual reserves additions come from revisions of old fields, showing that the assessment of the fields has been poorly reported. In fact, reporting of production is not much better and may give a false impression of oil abundance (Simmons, 2000) — technical data do exist and must be included in any peak theory model.

1.4.4 *Physical factors*

Crude oil reserves (Speight, 2011a, 2014a) are the estimated quantities of oil that are claimed to be recoverable under existing operating and economic conditions. However, because of reservoir characteristics and the limitations of current recovery technologies, only a fraction of this oil can be brought to the surface; it is this producible fraction that is considered to be *reserves*. Crude oil recovery varies greatly from oil field to oil field based on the characteristics of the field and the operating history as well as in response to changes in technology and economics.

Now that the predictions of crude oil production peaks in the years 1989 and 1995, or at some time in the period 1995–2000, and in 2007 have passed and crude oil still continues to flow, recent studies estimate that crude oil production will reach a peak at some time between now and 2040 and then decline rapidly (Figure 1.2(a)) as depleted reservoirs refuse to produce the remaining oil. Many of these projections cover a wide range of time, including two studies for which the range extends into the next century (Gordon, 2007). However, uncertainties in trying to determine the timing of peak oil are related to (i) the amount of recoverable crude oil throughout the world, (ii) advances in recovery technology as well as recovery economics, (iii) environmental challenges to produce crude oil, (iv) political and investment risk factors that may affect oil exploration and production, and last but certainly not the least, (v) the future world demand for oil. The uncertainties related to oil exploration and production also make it difficult to estimate the rate of decline after the peak.

However, predictions of the occurrence of an oil production peak and a resulting decline in production often use different estimates of the amount of oil that remains in the ground, and these differences explain some of the wide ranges of these predictions as well as the errors in prediction. Estimates of the amount of crude oil remaining in *depleted* reservoirs are highly uncertain because most of these data are self-reported and unverified by independent scientists/engineers/auditors. In addition, many parts of the world are yet to be fully explored for oil-containing reservoirs; and there is no comprehensive assessment of reserves from non-conventional sources. This uncertainty surrounding estimates of crude oil that is still in the ground contributes to the uncertainty that surrounds the estimation (and it is only *estimation*) of proven reserves as well as the uncertainty surrounding expected increases in these reserves and estimated future oil discoveries. The uncertainty of *proven reserves* showed itself when the system of country production quotas was introduced in the 1980s, partly based on reserves levels, and there have been dramatic increases in reported reserves among OPEC producers (Chapter 1).

According to current estimates, more than three-quarters of the oil reserves of the world are located in OPEC countries. The bulk of OPEC oil reserves is located in the Middle East, with Saudi Arabia, Iran, and Iraq contributing 41.8% to the OPEC total. OPEC member countries have made significant contributions to their reserves in recent years by adopting best practices in the industry (BP, 2019).

There has been astonishment at the OPEC estimates of proven reserves (Campbell and Laherrère, 1998) since OPEC estimates increased sharply in the 1980s, corresponding to a change in quota rules instituted by OPEC that linked a member production quota of a member country in part to its remaining proven reserves. Indeed, companies that are not subject to the federal securities laws in the United States and their related liability standards include companies wholly owned by various OPEC member countries where the majority of reserves are located. In addition, many OPEC countries' reported reserves remained relatively unchanged during the 1990s even as they continued high levels of oil production. The potential disbelief in the data reported by OPEC is problematic with respect to predicting the timing of a peak in oil production, because OPEC holds most of the current estimated proven oil reserves of the world.

The United States Geological Survey provides oil resources estimates, which are different from proved reserves estimates. Oil resources estimates are significantly higher, because they estimate the total oil resource base of the world, rather than just what is now proven to be economically producible. Estimates of the resource base by the United States Geological Survey include past production and current reserves as well as the potential for future increases in current conventional oil reserves (often referred to as *reserves growth*) and the amount of estimated conventional oil that has the potential to be added to these reserves. Estimates of reserves growth and those resources that have the potential to be added to oil reserves are important in determining when oil production may reach a peak.

Further contributing to the uncertainty of the timing of a peak is the lack of a comprehensive assessment of oil derived from nonconventional sources. For example, estimates of crude oil longevity have only recently started to include oil derived from non-conventional sources (BP, 2019) and oil derived from these sources was not included in early peak oil theories. Yet oil derived from non-conventional sources exists in substantial amounts, which could greatly delay the onset of a peak in production. However, challenges faced in this production (Speight, 2008, 2011a, 2014a, 2019) indicate that the amount of nonconventional oil that will eventually be produced is, like the peak oil theory, highly speculative. However, despite this apparent uncertainty, development and production of oil (synthetic crude oil) from the Alberta tar sands and production of Venezuelan extra-heavy crude oil are currently under way and refining technologies are being adapted to produce liquid fuels from these sources (Speight, 2008, 2013b, 2013c, 2013d, 2013c, 2014a).

1.4.5 *Technological factors*

In region after region, there are reports of (i) aging and depleted fields, (ii) poor quality crude oil, usually termed as heavy crude oil, (iii) the need for enhanced recovery methods, and (4) new areas turning out to be dry wells, leading to the claim that peak oil has arrived. For example, for whatever reason, the fields in Alaska, the former Soviet Union, Mexico, Venezuela, and Norway (North Sea) are all claimed to be past their peak. It is grudgingly admitted by the peak oil theorists that there is (or there may be) a (remote or even unlikely) possibility of new finds of oil fields off the coast of West Africa, but their development is still years away, and these new finds will not be on a scale that is capable of making a difference. It is also further claimed that the only producers with an oil resource that may be capable of keeping oil flowing into the world market at a roughly constant level are five Middle East OPEC countries — Saudi Arabia, Iran, Iraq, Kuwait and the United Arab Emirates (Fleming, 2000). However, because of much speculation on the part of the peak oil theorists, there is some difficulty when it comes to projecting the timing of a peak in oil production, because (i) technological advances, (ii) increased efficiency in recovery methods and, hence, (iii) reduced or stable recovery economics, and (iv) environmental challenges make it unclear how much additional oil can ultimately be recovered from proven reserves or from hard-to-reach locations and from non-conventional sources.

Furthermore, the potential increase in oil recovery (using advanced enhanced recovery methods) has been estimated (by the United States Department of Energy) to be of the order of 30–50% v/v of the oil remaining in many reservoirs. These technologies include injecting steam or heated water; gases, such as carbon dioxide; or chemicals into the reservoir to stimulate oil flow and allow for increased recovery with the ultimate potential recovery of at least an additional 2 million barrels per day in the decade 2020s. Such estimates are given little consideration by the peak oil theorists and, hence, the timing of peak oil is also difficult to estimate, leading to errors in the calculations and the extent of the tailing of the oil production data (Figure 1.2(b)). Globally, industry analysts report that deep water (depths of 1,000–5,000 feet) and ultra-deep water (5,000–10,000 feet) drilling efforts are concentrated offshore in Africa, Latin America, and North America, and capital expenditures for these efforts are expected to grow throughout at least the decade 2020s (Speight, 2015b).

Ultimately, however, the consequences of the occurrence of a peak and a permanent decline in oil production could be even more prolonged and severe than those of past oil supply shocks. Even then the decline rate is the subject of speculation, but like death and taxes, the decline rate is happening! The most important variable is the amount of oil left in the reservoirs but, even then, this is subject to debate and error, leaving the decline rate for fields in production difficult to assess (Eagles, 2006; Gerdes, 2007; Jackson, 2007). At best, generalities can be calculated. For example, for the current fields in production, a low decline rate (2%/year) would result in peak oil around 2018, while a more moderate decline rate (4.5%/year) would result in peak oil around 2015 and a high decline rate (8%/year) should have peaks four years ago.

1.5 Energy Independence and Energy Security

A measure of dependence on crude oil energy security is a concern (or should be a concern) of all countries that import natural gas and crude oil (or, for that matter, any form of energy). In the United States, increasing crude oil imports are considered a threat to national security but there is also the line of thinking that the level of imports has no significant impact on energy security, or even national security. However, the issue becomes a problem when import vulnerability increases as crude oil imports rise, which occurs when oil-consuming countries increase the share of crude oil imports from politically unstable areas of the world. More generally, there are four measures of crude oil dependence: (i) crude oil imports as a percentage of total crude oil consumption, (ii) the number of days total crude oil stocks cover crude oil imports, (iii) the number of days total stocks cover consumption, and (iv) the percentage of crude oil in total energy consumption (Speight, 2011; Speight and Islam, 2016).

1.5.1 *Energy independence*

Energy independence has been a political non-issue in the United States since the first Arab oil embargo in 1973. Since then, the speeches of various presidents to the Congress of the United States have continued to call for an end to the US dependence on foreign oil. Nevertheless, the United States has grown more dependent on foreign oil with no end in sight. For example, in 1970 the United States imported approximately one third of the daily oil requirement. Currently, the amount of imported oil is two thirds of the daily requirement! The congressional rhetoric of energy independence continues, but meaningful suggestions of how to address this issue remain few and far between. The economy of the United States feeds on oil and the country consumes far more oil than it can produce.

Generally, the concept of energy independence for the United States runs contrary to the trend of the internationalization of trade. The government of the United States continues to reduce trade barriers through policies such as the North American Free Trade Agreement (NAFTA), the elimination of tariffs, as well as other free trade agreements. As a result, the percentage of the US economy that comes from international trade is steadily rising.

Increased world trade is beneficial both economically and politically, insofar as it is supposed to help establish amicable relations between countries. Through mutually beneficial exchange, a great deal of this increased interrelationship will, in theory, establish opportunities for personal ties that make military war (or other forms of military action) less likely. However, there are also contrary cases where countries acquire the means to be more destructive, if they so choose, through expanded economic opportunities. This type of argument has been used with regard to Iran.

Economic interdependence also makes the domestic economy more susceptible to disruptions in distant and unstable regions of the globe, such as the Middle East, South America, and Africa. In fact, in many countries with proven reserves, oil production could be shut down by wars, strikes, and other political events, thus reducing the flow of oil to the world market. If these events occurred repeatedly or in many different locations, they could constrain oil exploration and production, resulting in a peak despite the existence of proven oil reserves.

Furthermore, the oil industry itself has been a story of vast swings between periods of overproduction, when low prices and profits caused oil producers to devise ways to restrict output and raise prices, and periods when oil supplies seemed to be on the brink of exhaustion, stimulating a global search for new supplies. This cycle may now be approaching the end. It appears that world oil supplies may truly be reaching their natural limits. With proven world oil reserves anticipated to last for less than 40 years, the age of oil that began near Titusville may be coming to an end. In the years to come, the search for new sources of oil will be transformed into a quest for entirely new sources of energy.

Political and investment risk factors continue to affect future oil exploration and production and, ultimately, the timing of peak oil production. These factors include changing political conditions and investment climates in many countries that have large proven oil reserves. These factors are important in affecting future oil exploration and production. Even in the United States, political considerations may affect the rate of oil exploration and production. For example, environmental restrictions may effect a peak in oil production by barring oil exploration and production in environmentally sensitive areas. The government must adopt policies that ensure our energy independence. The United States Congress is no longer believable when the members of the Congress lay the blame on foreign governments or events for an impending crisis. The Congress needs to look north and note the positive role played by the Government of Canada in the early 1960s when the decision was made to encourage development of the Alberta tar sands. Synthetic crude oil production is now in excess of 1,000,000 million barrels per day — less than 6% of the daily liquid fuel requirement in the United States but a much higher percentage of the daily liquid fuel requirement in Canada. In the United States, the issue to be faced is not so much oil reserves but oil policies.

The United States is the only oil-importing country with significant production, which has been a net importer of crude oil and crude oil products but, due to the development of crude oil resources in tight (low-to-no permeability) formations, it is now a net exporter of crude oil and crude oil products. It is also unique in that among the net importers, it has the lowest dependence in terms of net imports as a percentage of consumption, but it also has the highest absolute level of imports. The geo-political and economic interests and commitments raise the level of concern by US policy-makers about the dependence on imported oil.

In addition, commercial oil stocks in the United States have been at their lowest level in three decades. Total crude oil inventories, which include commercial stocks and stocks in the Strategic Crude Oil Reserve, are relatively low, in terms of daily coverage. Current commercial inventories are near the level at which spot shortages can occur. The past decade has seen scenarios in which the decline in commercial stocks is greater than the increase in stocks in the Strategic Petroleum Reserve and the capacity of the Strategic Petroleum Reserve and commercial stocks to deal with a crisis is less than what it was before the refilling program began (Speight, 2011; Speight and Islam, 2016). Moreover, the premature release of crude oil from the Strategic Petroleum Reserve can jeopardize national security in case of continued political problems in the oil-producing countries and weaken the ability of the United States to respond to real shortages.

Although some of the oil-importing countries have made progress in reducing their dependence on oil, the dependence of the United States on crude oil has increased in recent years from 38% of the total energy consumption in 1995 to approximately 40% at the present time. This indicates two possible areas of concern regarding the extent to which crude oil influences energy security: (i) the increase in the crude oil share of energy use, and (ii) the inability or unwillingness of the United States to reduce dependence on oil.

It might be argued that the degree of dependence has no impact on energy security as long as foreign oil is imported from secure sources. However, if the degree of dependence on non-secure sources increases, energy security would be in jeopardy. In this case, vulnerability would increase and economic and national security of individual oil-importing countries would be compromised.

The percentage of imports from the top five suppliers can be used as a measure of the supply's vulnerability to an interruption from one or more key suppliers. This is an important measure of the vulnerability of the United States to supply-disruption, because it reveals the high level of import concentration by importing from few suppliers. The top five suppliers to oil-importing countries are Saudi Arabia, the former USSR, Norway, Venezuela, and Mexico. The United States has a unique arrangement as well as location with Canada and Mexico, but the political stability of Saudi Arabia is always open to discussion and question. In addition, continuing problems of crude oil from various countries (using Iran, Iraq, and Venezuela as current (at the time of writing) examples) underline the United States' vulnerability to an interruption from a major supplier. Venezuela has had severe problems and the potential for an interruption in oil supply always exists. Iraq crude has been off the market for several years (oil supply from it is only now starting again) and the conflict in Nigeria has significantly influenced oil output.

Another important measure of vulnerability is the share of world crude coming from the Gulf region. The Gulf region has been viewed historically as a politically unstable area. Incidents in the region led to the three energy crises in 1973, 1979, and the two recent Gulf Wars. While a smaller share of imported oil from the Gulf producers means lesser vulnerability, it may increase vulnerability by having to rely on other sources, such as Venezuela.

Whichever measure is used to assess energy dependence, the United States remains susceptible to an energy crisis because of the high dependence on imported oil. The United States has not made a significant reduction in dependence on imports since the mid-1980s and continues to import almost 70% of the total crude oil consumption.

Dependence and vulnerability to oil imports in the United States and, for that matter, in other oil-importing countries can be reduced by diversification of suppliers and by energy diversification. In addition, diversification of suppliers has the potential to lower the relative impact of supply disruption on most countries. The political instability that swings back and forth in countries such as Venezuela, Nigeria, and Iraq emphasizes the need to diversify the suppliers and thereby remove the reliance on a small number of oil-producing countries.

As the issues that result from the use of traditional fossil fuels become more prominent, alternative fuel sources like the ones mentioned here are likely to gain further importance. The benefits alleviate many of the problems caused by fossil fuel use, particularly when it comes to emissions. However, the advancement of some of these technologies has slowed down due to the amount of investment needed to make these technologies viable.

1.5.2 *Energy security*

Energy security is the continuous and uninterrupted availability of energy to a specific country or region. The security of energy supply plays a crucial role in decisions that are related to the formulation of energy policy strategies. The economies of many countries are dependent on the energy imports in the sense that their balance of payments is affected by the magnitude of the vulnerability that the countries have in crude oil.

The subject of energy security has been for many years an important concern among energy policy-makers. The devastating short- and long-term effects of the oil crisis of 1973 on the global economy indicated, since then, that the need to guarantee the availability of energy resource supply in a sustainable and timely manner with the energy price being at a level that will not adversely affect the economic performance of the European continent is of utmost importance (Asia Pacific Research Centre 2007). The continuous destabilization of the Middle East, growing fears related to further military intervention in this fragile geopolitical area, environmental catastrophes, the advent of organized terrorist operations across the globe, political risks, and legal reforms have profoundly increased the possibility of potential energy disruptions that will have detrimental effects, considering the dependence of Europe on external energy suppliers.

The fact that oil-producing countries in the Middle East provide more than 50% of the global consumption of crude oil (El-Genk, 2008) is indicative of the low diversification of energy sources and the accompanying risks to smooth energy supply. The diversification that is offered by the alternative supplies from Russia and Africa cannot provide a sound solution for a supply disruption that may occur in the Middle East region. An overview of the oil market and the related risks and incidents clearly indicates that the risks associated with energy supply are many. War and civil conflicts might have been replaced, to some extent, by weather conditions and monopolistic practices, but they are still playing a crucial role in energy supply (Doukas *et al.*, 2008). Therefore, the high dependence that most countries have in energy imports has made it essential for the policy-makers to focus on the concept of security of energy supply. In this context, there is a need to assess the current energy system and the risk of energy disruptions in order to better design and adopt the required policies.

Uncertainty related to the future demand for oil — which will influence how quickly the remaining oil is used — will contribute to the uncertainty related to the timing of peak oil production. It is very likely that crude oil will continue to be a major source of energy well into the future and world consumption of crude oil products may even grow during the nest four decades. Future world oil demand will depend on uncertain factors such as world economic growth, future government policy, and consumer choices, which can affect oil demand and thereby influence the timing of a peak.

If the peak occurs in the more distant future or the decline following a peak is less severe, alternative technologies have a greater potential to mitigate the consequences. The United States Department of Energy projects that the alternative technologies have the potential to displace up to an equivalent of 34% of annual US consumption of crude oil products in the 2025–2030 time frame. However, the United States Department of Energy also considers these projections optimistic — since the assumption is that sufficient time and effort are dedicated to the development of these technologies to overcome the challenges they face.

The prospect of a peak in oil production presents problems of global proportion whose consequences will depend critically on our preparedness. The consequences would be most dire if a peak occurs soon, without warning, and is followed by a sharp decline in oil production because alternative energy sources, particularly for transportation, are not yet available in large quantities. Such a peak would require sharp reductions in oil consumption, and the competition for increasingly scarce energy would drive up prices, possibly to unprecedented levels, causing severe economic damage. While these consequences would be felt globally, the United States, being the largest consumer of oil and a nation that is heavily dependent on oil for transportation, is especially vulnerable.

The popularity of the *energy risk-premium* concept has led to the formulation of a vast pool of knowledge encompassing an abundance of derivative models. Traders have the ability to hedge against various risks and create risk-neutral portfolios using a diversified mix of energy derivative securities. However, up to now, the risk-premium concept has not been used in the energy domain to quantify an energy security indicator. The main reason for this is that current techniques used in the energy domain do not incorporate the necessary probabilistic models that reflect on risk parameters associated with rare *catastrophic* events that cause adverse movements of the spot price of the underlying instrument.

A catastrophic event in general (or catastrophe) is an event that causes *severe losses, injury, or property damage that affects a large population of exposures and is caused by natural or handmade events* (Fisher and Kist, 2001). Examples of catastrophic events include natural disasters (hurricanes, earthquakes, and floods) and terrorist attacks. Over the last 20 years, natural catastrophes have been happening with increasing intensity.

Catastrophic events in the energy context have a slightly different meaning. They can be events with low frequency of occurrence that cause the spot price of the energy commodity to soar (*price volatility*). Usually, a price increase due to the catastrophic event does not have a lasting effect and the spot price tends to return to (or close to) its initial value. To combat such events, it will be necessary for the non-oil producing nations to commence development of sources of energy other than oil. Once all risk indicators associated with catastrophic events have been identified and properly estimated in terms of frequency of occurrence and impact on the underlying spot price, the respective premium can be calculated under the common assumptions of derivatives pricing.

In the longer term, there are many possible alternatives to using oil, including using biofuels and by improving the automotive fuel efficiency, but these alternatives will require large investments and in some cases major changes in infrastructure, or breakthrough technological advances. In the past, the private sector has responded to higher oil prices by investing in alternatives, and it is doing so now.

In summary, the natural gas and crude oil industries (gas and oil from tight formations notwithstanding) are at the verge of a major decision period with the onset of the processing of high volumes of heavy crude oil and residua. Several technology breakthroughs have made this possible but many technical challenges remain. Some would argue that the periods when there was a decline in the prices of natural gas and crude oil were the impetus to start the development of technologies related to the production of alternate energy. This has not been apparent.

References

Alberta Energy. 2014. Government of the Province of Alberta, Edmonton, Alberta, Canada. http://www.energy.alberta.ca/; http://www.energy.alberta.ca/ourbusiness/oilsands.asp.

Bentley, R.W. 2002. Global oil and gas depletion: An overview. *Energy Policy* 30: 189–205.

BOEM. 2012. Assessment of In-Place Gas Hydrate Resources of the Lower 48 United States Outer Continental Shelf. Fact Sheet RED-2012-01. Bureau of Ocean Energy Management, United States Department of the Interior, Washington, DC.

Bower, T. 2009. *Oil: Money, Politics, and Power in the 21st Century*. Grand Central Publishing, Hachette Book Group, Inc., New York.

BP. 2019. BP Statistical Review of World Energy. BP PLC, London, United Kingdom. https://www.bp.com/content/dam/bp/business-sites/en/global/corporate/pdfs/energy-economics/statistical-review/bp-stats-review-2019-full-report.pdf.

Brandt, A.R. 2007. Testing Hubbert. *Energy Policy* 35(5): 3074–3088.

Campbell, C.J. 1997. A European view of oil reserves. *Oil & Gas Journal* 95(52): 33–37.

Campbell, C.J. and Laherrère, J.H. 1998. The end of cheap oil. *Scientific American* 278: 78–83.

Campbell, C.J. 2004a. *The Essence of Oil & Gas Depletion*. Multi-Science Publishing, Brentwood, Essex, United Kingdom.

Campbell, C.J. 2004b. *The Coming Oil Crisis*. Multi-Science Publishing, Brentwood, Essex, United Kingdom.

Campbell, C.J. 2005. *Oil Crisis*. Multi-Science Publishing, Brentwood, Essex, England.

Cohen, D. 2007. The Perfect Storm. ASPO-USA/Energy Bulletin. October 31. http://www.energybulletin.net/node/36510.

Collett, T.S. 2009. Natural Gas Hydrates: A Review. In: *Natural Gas Hydrates — Energy Resource Potential and Associated Geologic Hazards*, T.S. Collett, A.H. Johnson, C.C. Knapp and R. Ray Boswell (Editors). AAPG Memoir 89. American Association of Petroleum Geologists, Tulsa, Oklahoma, pp. 146–219.

Crane, H.D., Kinderman, E.M., and Malhotra, R. 2010. *A Cubic Mile of Oil: Realties and Options for Averting the Looming Global Energy Crises.* Oxford University Press, Oxford, United Kingdom.

Deffeyes, K.S. 2002. *Hubbert's Peak: The Impending World Oil Shortage.* Princeton University Press, Princeton, New Jersey.

Deffeyes, K.S. 2005. *Beyond Oil: The View from Hubbert's Peak.* Hill and Wang, New York.

Douglas, C.A., Harrison, G.P., and Chick, J.P. 2008. Life cycle assessment of the Seagen marine current turbine. *Proceedings. Institution of Mechanical Engineers, Part M: Journal of Engineering for the Maritime Environment* 222(1): 1–12.

Eagles, L. 2006. Medium Term Oil Market Report. OECD/International Transport Forum Roundtable. International Energy Agency, Paris, France.

Energy Security Leadership Council. 2013. A National Strategy for Energy Security: Harnessing American Resources and Innovation. Energy Security Leadership Council, Washington, DC.

Evans, R. 2007. *Fueling Our Future: An Introduction to Sustainable Energy.* Cambridge University Press, Cambridge, United Kingdom.

Fleming, D. 2000. After oil. *Prospect Magazine* November(57): 12–13. http://www.prospect-magazine.co.uk.

Fthenakis, V. and Kim, H.C. 2009. Land use and electricity generation: A life-cycle analysis. *Renewable and Sustainable Energy Reviews* 13(6–7): 1465.

Gary, J.G., Handwerk, G.E., and Kaiser, M.J. 2007. *Petroleum Refining: Technology and Economics*, 5th Edition. CRC Press, Taylor & Francis Group, Boca Raton, Florida.

Gerdes, J. 2007. *Modest Non-OPEC Supply Growth Underpins $60+ Oil Price.* SunTrust Robinson Humphrey, Atlanta, Georgia.

Giampietro, M. and Mayumi, K. 2009. *The Biofuel Delusion: The Fallacy of Large-Scale Agro-biofuel Production.* TJ International, Padstow, Cornwall, United Kingdom; also published by Earthscan, London, United Kingdom.

Goldberg, D. and Saito, S. 1998. Detection of Gas Hydrates Using Downhole Logs. In: *Hydrates — Relevance to World Margin Stability and Climate Change Gas*, J.-P. Henriet and J. Mienert (Editors). The Geological Society, London, United Kingdom, pp. 129–132.

Gordon, 2007. Crude Oil: Uncertainty about Future Oil Supply Makes It Important to Develop a Strategy for Addressing a Peak and Decline in Oil Production. Report No. GAO-07-283. Report to Congressional Requesters. General Accountability Office, Washington, DC. February.

Gudmestad, O.T., Zolotukhin, A.B., and Jarlsby, E.T. 2010. *Petroleum Resources with Emphasis on Offshore Fields*. WIT Press, Billerica, Massachusetts.

Hirsch, R.L. 2005. The inevitable peaking of world oil production. *The Atlantic Council of the United States Bulletin* XVI(3): 1–9.

Hirsch, R.L., Bezdek, R., and Wendling, R. 2005. Peaking Of World Oil Production: Impacts, Mitigation, and Risk Management. February. http://www.netl.doe.gov/publications/ others/pdf/Oil_Peaking_NETL.pdf.

Hornbach, M.J., Holbrook, W.S., Gorman, A.R., Hackwith, K.L., Lizarralde, D., and Pecher, I. 2003. Direct seismic detection of methane hydrate on the blake ridge. *Geophysics* 68: 92–100.

Hornung, A. 2014. *Transformation of Biomass: Theory to Practice*. John Wiley & Sons Inc., Chichester, West Sussex, United Kingdom.

Hsu, C.S. and Robinson, P.R. (Editors) 2017. *Handbook of Petroleum Technology*. Springer International Publishing AG, Chaim, Switzerland.

Hubbert, M.K. 1956. *Nuclear Energy and the Fossil Fuels — Drilling and Production Practice*. Institute, Washington, DC.

Hubbert, M.K. 1962. *Energy Resources*. Report to the Committee on Natural Resources, National Academy of Sciences, Washington, DC.

Jackson, P.M. 2006. *Why the Peak Oil Theory Falls Down: Myths, Legends, and the Future of Oil Resources*. Cambridge Energy Research Associates, Cambridge, Massachusetts. November 14.

Jackson, P.M. 2007. *Finding the Critical Numbers: What Are the Real Decline Rates for Global Oil Production*? Cambridge Energy Research Associates, Cambridge, Massachusetts.

Khoshnaw, F.M. (Editor). 2013. *Crude oil and Mineral Resources*, WIT Press, Billerica, Massachusetts.

Kvenvolden, K.A. 1993. Gas Hydrates as a Potential Energy Resource — A Review of Their Methane Content. In: *The Future of Energy Gases*, D.G. Howell (Editor). Professional Paper No. 1570. United States Geological Survey, Washington, DC, pp. 555–561.

Kvenvolden, K.A. and Lorenson, T.D. 2000. The Global Occurrence of Natural Gas Hydrate. In: *Natural Gas Hydrates: Occurrence, Distribution, and Dynamics*, C.K. Paull and W.P. Dillon (Editors). AGU Monograph, American Geophysical Union, Washington, DC, p. 55.

Lago, C., Caldés, N., and Lechón, Y. 2019. *The Role of Bioenergy in the Emerging Bioeconomy: Resources, Technologies, Sustainability and Policy*. Elsevier BV, Amsterdam, Netherlands.

Laherrère, J.H. 2001. Paper presented at the EMF/IEA/IEW Meeting IIASA, Laxenburg, Austria. Plenary Session I: Resources, June 19.

Langeveld, H., Sanders, J., and Meeusen, M. 2010 (Editors). *The Biobased Economy: Biofuels, Materials, and Chemicals in the Post-oil Era.* TJ International, Padstow, Cornwall, United Kingdom; also published by Earthscan, London, United Kingdom.

Lay, T., Hernlund, J., and Buffett, B.A. 2008. Core-mantle boundary heat flow. *Nat. Geosci.* 1: 25.

Lee, S. 1990. *Oil Shale Technology*. CRC Press, Taylor & Francis Group, Boca Raton, Florida.

Lee, S. and Shah, Y.T. 2013. *Biofuels and Bioenergy: Processes and Technologies*. CRC Press, Taylor & Francis Group, Boca Raton Florida.

Lerche, I. and Bagirov, E. 1998. Guide to gas hydrate stability in various geological settings. *Mar. Petrol. Geol.* 15: 427–438.

Levant, E. 2010. *Ethical Oil: The Case for Canada's Oil Sands*. McClelland & Stewart Ltd., Toronto, Ontario, Canada.

Luciani, G. 2013. *Security of Oil Supplies: Issues and Remarks*. Claeys & Casteels, Deventer, Netherlands.

Lutz, M., Schneider, M., and Thomas, S. (Editors). 2009. *International Perspectives of Energy Policy and the Role of Nuclear Power*. Multi-Science Publishing Co. Ltd, Hocklery, Essex, United Kingdom.

Meyer, R.F. and Attanasi, E.D. 2003. *Heavy Oil and Natural Bitumen — Strategic Petroleum Resources*. Fact Sheet 70-03. United States Geological Survey, Washington, DC. http://pubs.usgs.gov/fs/fs070-03/fs070-03.html.

Mills, D. 2004. Advances in solar thermal electricity technology. *J. Sol. Energy* 76(1–3): 19–31.

Nersesian, R.L. 2007. *Energy for the 21st Century: A Comprehensive Guide to Conventional and Alternative Fuel Sources*. M.E. Sharpe Inc., Armonk, New York.

Parkash, S. 2003. *Refining Processes Handbook*. Gulf Professional Publishing, Elsevier, Amsterdam, Netherlands.

Poveda, C.A. and Lipsett, M.G. 2014. *Surface Mining Operations in Oil Sands: Establishing Sustainable Development Indicators*. WIT Press, Billerica, Massachusetts.

Ratner, M. and Tiemann, M. 2014. An Overview of Unconventional Oil and Natural Gas: Resources and Federal Actions. CRS Report. Congressional Research Service, Washington, DC.

Sandrea, I. 2014. *US Shale Gas and Tight Oil: Industry Performance: Challenges and Opportunities*. Oxford Institute for Energy Studies, University of Oxford, Oxford, United Kingdom. http://www.oxfordenergy.org/wpcms/wp-content/uploads/2014/03/US-shale-gas-and-tight-oil-industry-performance-challenges-and-opportunities.pdf; accessed October 14, 2014.

Scouten, C.S. 1990. Oil Shale. In: *Fuel Science and Technology Handbook*, J.G. Speight (Editor), Chapters 25–31. Marcel Dekker, New York. pp. 795–1053.

Siefried, D. and Witzel, A.W. 2010. *Renewable Energy — The Facts. London, United Kingdom.* Formerly published by Energieagentur Regio Freiburg, Freiburg, Germany.

Simmons, M.R. 2000. Fighting Rising Demand and Rising Decline Curves: Can the challenge be met? In: *SPE Asia Pacific Oil and Gas Conference*, Yokohama, April 25. http://www.simmonscointl.com/web/downloads/spe.pdf

Söderbergh, B., Robelius, F., and Aleklett, K. 2007. A crash program scenario for the Canadian oil sands industry. *Energy Policy* 35(3): 1931–1947.

SPE. 2007. Guidelines for Application of the Petroleum Resources Management System (PRMS). Society of Petroleum Engineers, Richardson, Texas. http://www.spe.org/industry/reserves.php; http://www.world-petroleum.org/index.php?/Reserves-Resources/petroleum-resources-classification-system-and-definitions.html.

Speight, J.G. (Editor). 1990. *Fuel Science and Technology Handbook.* Marcel Dekker, New York.

Speight, J.G. 1997. Tar Sands. *Kirk-Othmer Encyclopedia of Chemical Technology*, 4th Edition, vol. 23, p. 717.

Speight, J.G. and Lee, S. 2000. *Environmental Technology Handbook.* Taylor & Francis, Philadelphia, Pennsylvania.

Speight, J.G. 2008. *Synthetic Fuels Handbook: Properties, Processes, and Performance.* McGraw-Hill, New York.

Speight, J.G. 2009. *Enhanced Recovery Methods for Heavy Oil and Tar Sands.* Gulf Publishing Company, Houston, Texas.

Speight, J.G. 2011a. *An Introduction to Petroleum Technology, Economics, and Politics.* Scrivener Publishing, Salem, Massachusetts.

Speight, J.G. (Editor). 2011b. *Biofuels Handbook.* Royal Society of Chemistry, London, United Kingdom.

Speight, J.G. 2011c. *The Refinery of the Future.* Gulf Professional Publishing, Elsevier, Oxford, United Kingdom.

Speight, J.G. 2012. *Shale Oil Production Processes.* Gulf Professional Publishing, Elsevier, Oxford, United Kingdom.

Speight, J.G. 2013a. *The Chemistry and Technology of Coal*, 3rd Edition. CRC Press, Taylor & Francis Group, Boca Raton, Florida.

Speight, J.G. 2013b. *Coal-Fired Power Generation Handbook.* Scrivener Publishing, Salem Massachusetts.

Speight, J.G. 2013c. *Heavy and Extra Heavy Oil Upgrading Technologies.* Gulf Professional Publishing, Elsevier, Oxford, United Kingdom.

Speight, J.G. 2013d. *Oil Sand Production Processes*. Gulf Professional Publishing, Elsevier, Oxford, United Kingdom.

Speight, J.G. 2013e. *Heavy Oil Production Processes*. Gulf Professional Publishing, Elsevier, Oxford, United Kingdom.

Speight, J.G. 2013f. *Shale Gas Production Processes*. Gulf Professional Publishing, Elsevier, Oxford, United Kingdom.

Speight, J.G. 2014a. *The Chemistry and Technology of Petroleum*, 5th Edition. CRC Press, Taylor & Francis Group, Boca Raton, Florida.

Speight, J.G. 2014b. *The Gasification of Unconventional Feedstocks*. Gulf Professional Publishing, Elsevier, Oxford, United Kingdom.

Speight, J.G. 2015a. *Handbook of Hydraulic Fracturing*. John Wiley & Sons Inc., Hoboken, New Jersey.

Speight, J.G. 2015b. *Offshore Oil and Gas*. Gulf Professional Publishing, Elsevier, Oxford, United Kingdom.

Speight, J.G. and Islam, M.R. 2016. *Peak Energy — Myth or Reality*. Scrivener Publishing, Beverly, Massachusetts.

Speight, J.G. 2017. *Handbook of Petroleum Refining*. CRC Press, Taylor & Francis Group, Boca Raton, Florida.

Speight, J.G. 2019. *Natural Gas: A Basic Handbook*, 2nd Edition. Gulf Publishing Company, Elsevier, Cambridge, Massachusetts.

Tester, J.W., Drake, E.M., Driscoll, M.J., Golay, M.W., and Peters, W.A. 2005. *Sustainable Energy: Choosing Among Options*. MIT Press, Cambridge, Massachusetts.

Turcotte, D.L. and Schubert, G. 2002. *Geodynamics*, 2nd Edition. Cambridge University Press, Cambridge, United Kingdom, pp. 136–137.

US DOE. 2014. United States Department of Energy, Washington, DC. http://energy.gov/fe/services/petroleum-reserves; accessed September 15, 2014. http://www.fossil.energy.gov/programs/reserves/npr/Heavy_Oil_Fact_Sheet.pdf; accessed September 15, 2014.

Wihbey, P.M. 2009. *The Rise of the New Oil Order: The Facts behind the New Massive Supplies, the Exciting New Technologies, and the Emerging Powers of the XXIst Century*. Academy & Finance SA, Geneva, Switzerland.

World Energy Council. 2008. *World Energy Outlook: Global Trends to 2030*. International Energy Agency, Paris, France.

Yergin, Y. 1991. *The Prize: The Epic Quest for Oil, Money and Power*. Free Press, New York.

Yergin, D. 2011. *The Quest: Energy Security and the Making of the Modern World*. Penguin Press HC, New York.

Zatzman, G.M. 2012. *Sustainable Resource Development*. Scrivener Publishing LLC, Beverly, Massachusetts.

Chapter 2

Conventional Fuels

2.1 Introduction

Conventional fuels are gaseous liquids, and solid fuels produced from natural gas and crude oil are commonly categorized as *crude oil products* or *petroleum products* — natural gas, typically but not always, arises from the same reservoirs as crude oil. By definition, crude oil products, in contrast to *petrochemicals*, are those bulk fractions that are derived from crude oil and have commercial value as a bulk product. In the strictest sense, petrochemicals are also crude oil products, but they are individual chemicals that are used as the basic building blocks of the chemical industry (Parkash, 2003; Gary *et al.*, 2007; Speight, 2011a, 2014, 2017, 2019; Hsu and Robinson, 2017).

The use of crude oil and crude oil products was established in pre-Christian times and is known largely through documentation by many of the older civilizations (Table 2.1). However, there have been many changes in product demand since crude oil first came into use about five to six millennia ago (Parkash, 2003; Gary *et al.*, 2007; Speight, 2011a, 2014, 2017, 2019; Hsu and Robinson, 2017). These changes in product demand are largely responsible for the evolution of the industry from bitumen used in ancient times to gasoline and other liquid fuels of the modern era of crude oil utilization (Table 2.2).

Crude oil is an extremely complex mixture of hydrocarbon compounds, usually with minor amounts of nitrogen-, oxygen-, and sulfur-containing compounds as well as trace amounts of metal-containing compounds (Parkash, 2003; Gary *et al.*, 2007; Speight, 2011a, 2014,

Table 2.1: Crude oil and its various derivatives, such as bitumen, have been known and used for almost 6000 years.

3800 BC	First documented use of asphalt for caulking reed boats.
3500 BC	Asphalt used as cement for jewelry and for ornamental applications.
3000 BC	Use of asphalt as a construction cement by Sumerians; also believed to be used as a road material; asphalt was used to seal a bathing pool or a water tank at Mohenjo Daro.
2500 BC	Use of asphalt and other petroleum liquids (oils) in the embalming process; asphalt believed to be widely used for caulking boats.
1500 BC	Use of asphalt for medicinal purposes and (when mixed with beer) as a sedative for the stomach; continued reference to use of asphalt liquids (oil) as an illuminant in lamps.
1000 BC	Use of asphalt as a waterproofing agent by lake dwellers in Switzerland.
500 BC	Use of asphalt mixed with sulfur as an incendiary device in Greek wars; also use of asphalt liquid (oil) in warfare.
350 BC	Occurrence of flammable oils in wells in Persia.
300 BC	Use of asphalt and liquid asphalt as an incendiary device (Greek fire) in warfare.
250 BC	Occurrences of asphalt and oil seepage in several areas of the Fertile Crescent (Mesopotamia); repeated use of liquid asphalt (oil) as an illuminant in lamps.
750 AD	Use in Italy of asphalt as a color in paintings.
950 AD	Report of destructive distillation of asphalt to produce distillate; reference to distillate as nafta (naft, naphtha).
1500 AD	Discovery of asphalt deposits in the Americas; first attempted documentation of the relationship of asphalt to naphtha (petroleum).
1600 AD	Asphalt used for a variety of tasks; relationship of asphalt to coal and wood tar studied; asphalt studied; used for paving; continued documentation of the use of naphtha as an illuminant and the production of naphtha from asphalt; importance of naphtha as fuel realized.
1859 AD	Discovery of petroleum in North America; birth of modern-day petroleum science and refining.

Table 2.2: Examples of general types of fuels.

	Primary (natural)	Secondary (manufactured)
Gaseous fuels	Natural gas	Hydrogen, methane, ethane, propane, butane
Solid fuels	Wood, dung	Coke, charcoal
Liquid fuels	Crude oil	Naphtha, gasoline, kerosene, fuel oil

2017, 2019; Hsu and Robinson, 2017). In addition, the properties of crude oil vary widely (Parkash, 2003; Gary *et al.*, 2007; Speight, 2011a, 2014, 2017, 2019; Hsu and Robinson, 2017). Thus, crude oil is not used in its raw state. A variety of processing steps are required for converting crude oil from its raw state to products that have well-defined properties and that meet the specifications for sale of the product(s) (ASTM, 2019).

The constant demand for products, such as liquid fuels, is the main driving force behind the crude oil industry. Other products, such as lubricating oils, waxes, and asphalt (produced from the crude oil residuum), have also added to the popularity of crude oil as a national resource. Indeed, such fuel products derived from crude oil supply more than half of the global total supply of energy. Gasoline, kerosene, and diesel oil provide fuel for automobiles, tractors, trucks, aircraft, and ships. Fuel oil and natural gas are used to heat homes and commercial buildings as well as to generate electricity. Crude oil products are the basic materials for the manufacture of synthetic fibers for clothing and in plastics, paints, fertilizers, insecticides, soaps, and synthetic rubber. The use of crude oil as a source of raw material in manufacturing is central to the functioning of modern industry.

Product complexity has made this industry unique among industries. Product complexity becomes even more meaningful when various fractions from different types of crude oil as well as fractions from synthetic crude oil are blended with the corresponding crude oil stock. The implications for refining the fractions to salable products increase. However, most of the time, the crude oil industry was inspired by the development of automobiles and the continued demand for gasoline and other fuels. This demand was accompanied by the demand for other products: diesel fuel for engines, lubricants for engine parts and machinery parts, fuel oil to provide power for industrial complexes, and asphalt for roadways.

Unlike processes, products are more difficult to place on an individual evolutionary scale. Processes have changed and evolved to accommodate the demand for, say, higher-octane fuels, longer-lasting asphalt, or lower-sulfur coke. In this section, a general overview of some crude oil products is presented in order to show the *raison d'être* of the industry. Another consideration that must be acknowledged is the change in character and composition of the original crude oil feedstock (Parkash, 2003; Gary *et al.*, 2007; Speight, 2011a, 2014, 2017, 2019; Hsu and Robinson, 2017). In the early days of the crude oil industry, several products were obtained by distillation and could be used without further treatment. In the modern refinery, the different character and composition of the crude oil dictate that any liquids obtained by distillation must go through one or more of the several available product-improvement processes (Parkash, 2003; Gary *et al.*, 2007; Speight, 2011a, 2014, 2017, 2019; Hsu and Robinson, 2017). Such changes in feedstock character and composition have caused the refining industry to evolve in such a way that changes in crude oil can be accommodated. It must be recognized that adequate storage facilities for the gases, liquids, and solids produced during the refining operations are also an essential part of a refinery. Without such facilities, refineries would be incapable of operating efficiently.

The customary processing of crude oil does not usually involve the separation and handling of pure hydrocarbon derivatives. Indeed, crude oil-derived products are always mixtures: occasionally simple, but more often very complex. Thus, for the purposes of this chapter, materials such as the gross fractions of crude oil (e.g. gasoline, naphtha, kerosene, and the like), which are usually obtained by distillation and/or refining, are classified as *crude oil products*; asphalt and other solid products (such as wax, resid, and coke) are included in this division.

The demand for crude oil products (Table 2.3) — particularly transportation fuels (gasoline and diesel) and petrochemical feedstocks

Table 2.3: Petroleum products.

Product	Lower carbon limit	Upper carbon limit	Lower boiling point (°C)	Upper boiling point (°C)	Lower boiling point (°F)	Upper boiling point (°F)
Refinery gas	C_1	C_4	−161	−1	−259	31
Liquefied petroleum gas	C_3	C_4	−42	−1	−44	31
Naphtha	C_5	C_{17}	36	302	97	575
Gasoline	C_4	C_{12}	−1	216	31	421
Kerosene/diesel fuel	C_8	C_{18}	126	258	302	575
Aviation turbine fuel	C_8	C_{16}	126	287	302	548
Fuel oil	C_{12}	$>C_{20}$	216	421	>343	>649
Lubricating oil	$>C_{20}$	—	>343	—	>649	—
Wax	C_{17}	$>C_{20}$	302	>343	575	>649
Asphalt	$>C_{20}$	—	>343	—	>649	—
Coke	$>C_{50}$[a]	—	>1000[a]	—	>1832[a]	—

Note: [a]Carbon number and boiling point are difficult to assess; inserted for illustrative purposes only.

(such as aromatic derivatives and olefin derivatives) — is increasing throughout the world. Traditional markets such as North America and Europe are experiencing a moderate increase in demand, whereas emerging Asian markets such as India and China are witnessing a rapid surge in demand. This has stimulated a fresh technological approach for optimizing refinery efficiency and throughput. Major oil companies and technology suppliers/licensors are investing heavily in revamping their refining technologies in an effort to cater to the growing needs of customers.

A steady evolution in product specifications caused by endless waves of fresh environmental regulations plays a major role in the development of crude oil refining technologies. In the United States and Europe, gasoline and diesel specifications have changed radically in the past decades and will continue to do so in the future. Currently, reducing the sulfur levels of finished products is the dominant objective. Sulfur is ubiquitous in crude oil, and refiners are seeking technologies to achieve the mandated levels of sulfur in crude oil products.

As crude oil products are being shipped worldwide, they need to comply with stringent environment-related regulations prevalent in specific countries. Japan and Singapore have already implemented strict legislation/rules and many countries are likely to follow suit as they confront environmental issues such as smog. These changing rules also cause a negative impact on the market for high-boiling products such as fuel oil.

Refineries are eager to adapt themselves to changing circumstances and are amenable to trying new technologies that are radically different in character. This is evident from the increasing use of ultrasonic technology and novel separation methods. Currently, refineries are also looking forward to exploiting heavy (more viscous) crude oils, provided they have the refinery technology capable of handling them. Heavy crude oil, relative to low-density crude oil or conventional crude oil, is rich in higher-boiling constituents (Speight, 2014, 2017), and transforming the heavier components into low-boiling fractions is thus a necessity. However, this requires technological changes, including more effective use of hydrogen within the refinery, which in turn increases operating costs.

Heavy crude oil could also be contaminated with sulfur and metal particles, which must be removed to meet quality standards. A deeper understanding of how catalysts work, both chemically and physically, provides greater scope for technological improvements. Nanotechnology and combinatorial chemistry are among the techniques that are likely to help push forward the frontiers of efficiency and selectivity. Even though the nature of crude oil is changing, refineries are here to stay for the foreseeable future (Speight, 2011a), since crude oil products satisfy wide-ranging energy requirements/demands that are not fully catered to by natural gas, liquefied petroleum oil gas (LPG), or coal. At present, alternative energy schemes lack technological sophistication or economic sufficiency to be considered a substitute for crude oil products.

A significant group of products from crude oil, petrochemicals, form the basis of a major industry. They are, in the strictest sense, different from crude oil products insofar as petrochemicals are the basic building blocks of the chemical industry. They will not be considered here but will be dealt with separately in this work.

The modern crude oil industry began in 1859 with the discovery and subsequent commercialization of crude oil for the production of products in Pennsylvania (Bell, 1945). After completion of the first well (by Edwin Drake), the surrounding areas were immediately leased and extensive drilling took place. It is from this time that modern crude oil products have evolved.

The nomenclature of crude oil products is as diverse as the nomenclature applied to crude oil itself. For example, several names can be applied to one product and caution is advised when attempting to define a crude oil product (Parkash, 2003; Gary *et al.*, 2007; Speight, 2011a, 2014, 2017, 2019; Hsu and Robinson, 2017).

There is a myriad of other products that have evolved throughout the short life of the crude oil industry. The complexities of product composition have matched the evolution of the products (Hoffman, 1992). In fact, it is the complexity of product composition that has served the industry well and, at the same time, had an adverse effect on product use. Product complexity, and the means by which the product is evaluated, has made this industry unique among industries. But product complexity has also brought to the fore issues such as instability and incompatibility. In order to understand the evolution of the products, it is essential to have an understanding of the composition of the various products (Hoffman, 1992).

This type of classification separates this group of products from those obtained as crude oil chemicals (petrochemicals), for which the emphasis is on separation and purification of single chemical compounds, which are in fact starting materials for a host of other chemical products.

The modern crude oil refinery consists of a very complex mix of high-technology processes that efficiently convert a wide array of crude oils into products that are in common use. Each refinery has a unique processing configuration as a result of the logistics and associated economics related to its specific refinery feedstocks and the product slate. The refiner must continuously optimize the mix of product volumes and this is accomplished by executing decisions regarding parameters as varied as crude oil feedstock selection, adjustments in product cut-points, and reactor severities in individual processes. Additional options include changing the dispositions of intermediate product streams to alternative processing units or alternative finished product blends.

In fact, many refinery products are typically the result of the blending of several component streams or blending stocks, as they are commonly referred to. In most cases, product blending is accomplished by controlling the volumes of blend stocks from individual component storage tanks that are mixed in the finished product storage tank. Samples of the finished blend are then analyzed by laboratory testing for all product specifications prior to shipping. Alternatively, *in-line blending* refers to pipeline shipments in which the finished product is actually blended directly into the product pipeline (as opposed to a standing product storage tank). The most commonly recognized blending operations occur in the gasoline production and diesel production section of the refinery (Parkash, 2003; Gary *et al.*, 2007; Speight, 2011a, 2014, 2017, 2019; Hsu and Robinson, 2017). The various gasoline streams exist so that specifications (dependent on the geographic location, environmental regulations, and weather patterns) can be met.

Gasoline blending involves combining the components that make up motor gasoline. The components include the various hydrocarbon streams produced by distillation, cracking, reforming, polymerization, tetraethyl lead, and identification of color dye, as well as other special-purpose components, such as solvent oil and anti-icing compounds. The physical process of blending the components is simple, but how much of each component must be included in a blend is much more difficult to determine. The physical operation is carried out by simultaneously pumping all the components of a gasoline blend into a pipeline that leads to the gasoline storage tank, but the pumps must be set to deliver automatically a proper proportion of each component. Baffles in the pipeline are often used to mix the components as they travel to the storage tank.

Selection on the components and their proportions in a blend is the most complex problem in a refinery, and there must not be any incompatibility between the various blend stocks or other physical (or chemical) interactions in the finished blend. For example, many different hydrocarbon streams may need to be blended to produce quality gasoline. Each property of each stream is a variable, and the effect on the product is considerable. For example, the low-octane number of straight-run naphtha limits its use as a gasoline component, although its other properties may make it desirable. The problem is further complicated by changes in the properties of the component streams due to processing changes. For example, an increase in cracking temperature produces a smaller volume of higher-octane cracked naphtha, but before this cracked naphtha can be included in a blend, adjustments must be made to the proportions of the other hydrocarbon components. Similarly, the introduction of new processes and changes in the specifications of the finished gasoline dictate a reevaluation of the components that make up the gasoline (Gibbs, 1989).

Gasoline blending is not the sole blending operation and there are other product blending operations used in a refinery. The applicable specifications vary by product but typically include properties pertinent to the behavior of the product in use. Many product specifications do not blend linearly by component volumes. Under these circumstances, the finished blend properties are predicted using experience-based algorithms for the applicable blending components. Thus, from gasoline to kerosene to fuel oil, it is required that the various streams used for producing a blended liquid fuel must produce a stable blended product that meets the specifications required for sale of the blended product. This concept can also apply to gaseous fuels, especially those that are mixtures of various constituents and that may undergo stratification (during which the constituents form layers according to density) under physical conditions, when the departure of the gas has reduced temperature during storage.

Also, on the subject of blending different streams, it is very rare that the modern refinery processes only a single crude oil. In many cases, the refinery may process a blend of several crude oils from different reservoirs. Thus, when various crude oils are blended at the refinery gate, the raw crude oils must also form a stable blend prior to introduction into any of the refinery processes; otherwise a refinery shut-down may be necessary for removing any material that phase-separates from an unstable blend material (Speight, 2014, 2015a, 2017).

Historically, the properties of the various fuels (gaseous, liquid, and solid) have been continuously changing for various reasons, including crude oil prices, crude oil quality, refinery technologies, relative demand for diesel and gasoline fuel, and changing engine technologies. In the recent years, environmental considerations and emission legislation have been increasingly more important in the formulation and properties of fuels. The interaction mechanisms between fuel quality, engine technologies, and emissions need to be understood to find the most effective approach towards low-emission diesel engines. A number of research studies have been carried out to investigate the influence of fuel properties on emissions.

Also, it is necessary to mention that the fuels used for military purposes are also complex mixtures of hydrocarbon derivatives that are produced by distillation of crude oil. The actual composition of any given fuel varies depending on the source of the crude oil, refinery processes, and product specifications. The hydrocarbon derivatives in jet and diesel fuels are less volatile than those in gasoline. For example, JP-5 is a high-flash-point jet fuel developed by the United States Navy and is a specifically refined type of kerosene consisting of C9–C16 paraffin derivatives (53% v/v), cycloparaffin derivatives (31% v/v), aromatic derivatives (16% v/v), and olefin derivatives (0.5% v/v). The aromatic content of JP-5 fuel might vary from less than 2.5% v/v to greater than 22% v/v, but the benzene content of JP-5 is typically less than 0.02% v/v and a small amount of polycyclic aromatic hydrocarbons might be present in the fuel. Because water contamination in aviation fuels is a serious problem, a fuel-system icing inhibitor is added to the fuel to eliminate the formation of ice in aircraft systems. JP-8 is similar to commercial A-1 jet fuel and was developed for the United States Air Force to provide a safe kerosene-based jet fuel that would still have adequate reliability and an acceptable freezing point. The marine diesel fuel (diesel fuel, marine, DFM) is a blend of diesel fuel that is basically the same as kerosene to which high-boiling-point fractions and high-boiling-point residual oils have been added. Diesel fuels consist primarily of C9–C20 hydrocarbon derivatives, which, for the marine diesel fuel, are of the order of paraffin derivatives (13% v/v), aromatic derivatives (44% v/v), and 44% naphthalene derivatives (43% v/v), but the fuel may contain 10% v/v polycyclic aromatic hydrocarbon derivatives.

Thus, within any groups of liquid fuels, there are many fuels with varying physical and chemical properties, such as ignition quality, distillation curve, and chemical composition. These include conventional gasoline, conventional diesel fuel, gasoline–diesel blends, and the fuel oils. Certain fuel properties could affect the performance of the engine because combustion is dependent on mixture stratification as well as chemical kinetics. For example, there are direct impacts of fuel distillation and aromatic content on the soot emissions of engines, while there are also reports of a minimal effect.

As the drive to find alternative fuels to replace such as hydrocarbon fuels gasoline and diesel fuel continues and has already resulted in a wide range of different fuels (Chapter 3), some of which are commercially available in significant quantities while others are still in the investigative and experimental stages, they must, by their very nature, be flammable and as a result the fire, explosion, and ignition properties will need to be known. In order to make a valid comparison of the properties of such fuels with those of the well-known conventional fuels, it is necessary to compare the fuels from fossil-fuel sources and those from renewable sources. In addition, because of the complexity of the various fuels, in any text dealing with alternate fuels and alternate fuel sources, it is advantageous to include a description of the conventional fuels that are currently derived from fossil fuel sources. This will present to the reader a basis for comparison of standards and specifications that the alternate fuels must adhere to in order to satisfy the market demand.

2.2 Gaseous Fuels

In the context of this book, gaseous fuels are the combustible or energy-generating gases that can be used to create energy. The gaseous fuels presented in this section are those fuels that are produced from fossil fuels, such as natural gas, crude oil, coal, and oil shale.

Natural gas, which is predominantly methane but does contain a host of other hydrocarbon constituents (Table 2.4), occurs in underground reservoirs either separately or in association with crude oil (Parkash, 2003; Gary *et al*., 2007; Speight, 2011a, 2014, 2017, 2019; Hsu and Robinson, 2017). The principal types of gaseous fuels are oil (distillation) gas, reformed natural gas, and reformed propane or LPG.

LPG is the term applied to certain specific hydrocarbon derivatives and their mixtures, which exist in the gaseous state under atmospheric ambient conditions but can be converted to the liquid state under conditions of moderate pressure at ambient temperature. These are the low-boiling hydrocarbon derivative fractions of the paraffin series, derived from refinery processes, crude oil stabilization plants and natural gas processing plants comprising propane ($CH_3CH_2CH_3$), butane ($CH_3CH_2CH_2CH_3$), *iso*-butane ($CH_3CH(CH_3)CH_3$), and to a lesser extent propylene ($CH_3CH{=}CH_2$)

Table 2.4: General properties of the potential organic constituents of natural gas up to and including *n*-octane (n-C_8H_{18}) as well as toluene, ethylbenzene, and the xylene isomers.

Constituent	Molecular weight	Specific gravity	Vapor density air = 1	Boiling point (°C)	Ignition temperature (°C)	Flash point (°C)
Methane	16	0.553	0.56	−160	537	−221
Ethane	30	0.572	1.04	−89	515	−135
Propane	44	0.504	1.50	−42	468	−104
Butane	58	0.601	2.11	−1	405	−60
Pentane	72	0.626	2.48	36	260	−40
Hexane	86	0.659	3.00	69	225	−23
Benzene	78	0.879	2.80	80	560	−11
Heptane	100	0.668	3.50	98	215	−4
Octane	114	0.707	3.90	126	220	13
Toluene	92	0.867	3.20	161	533	4
Ethyl benzene	106	0.867	3.70	136	432	15
o-Xylene	106	0.88	3.7	144	463	17
m-Xylene	106	0.86	3.7	139	527	25
p-Xylene	106	0.861	3.7	138	528	25

or butylene ($CH_3CH_2CH{=}CH_2$). The most common commercial products are propane, butane, or some mixture of the two (Table 2.5) and are generally extracted from natural gas or crude oil. Propylene ($CH_3CH{=}CH_2$) and butylene derivatives ($CH_3CH{=}CHCH_3$ or $CH_3CH_2CH{=}CH_2$) result from cracking other hydrocarbon derivatives in a crude oil refinery and are two important chemical feedstocks.

Mixed gas is a gas prepared by adding natural gas or liquefied petroleum gas to a manufactured gas, giving a product of better utility and higher-heat content or Btu value.

2.2.1 *Composition*

The principal constituent of natural gas is methane (CH_4). Other constituents are paraffinic hydrocarbon derivatives, such as ethane (CH_3CH_3), propane ($CH_3CH_2CH_3$), and the butanes ($CH_3CH_2CH_2CH_3$ and/or

Table 2.5: Properties of propane and butane.

	Propane	Butane
Formula	C_3H_8	C_4H_{10}
Boiling point (F)	–44°	32°
Specific gravity — gas (air = 1.00)	1.53	2.00
Specific gravity — liquid (water = 1.00)	0.51	0.58
lb/gallon — liquid at 60°F	4.24	4.81
BTU/gallon — gas at 60°F	91690	102032
BTU/lb. — gas	21591	21221
BTU/ft.^{3} — gas at 60°F	2516	3280
Flash point (F)	−156	–96
Ignition temperature in air (F)	920–1020	900–1000
Maximum flame temperature in air (F)	3595	3615
Octane number (*iso*-octane = 100)	100+	92

$(CH_3)_3CH$). Many natural gases contain nitrogen (N_2) as well as carbon dioxide (CO_2) and hydrogen sulfide (H_2S). Trace quantities of argon, hydrogen, and helium may also be present. Generally, the hydrocarbon derivatives having a higher molecular weight than methane, carbon dioxide, and hydrogen sulfide are removed from natural gas prior to its use as a fuel. Gases produced in a refinery contain methane, ethane, ethylene, propylene, hydrogen, carbon monoxide, carbon dioxide, and nitrogen, with low concentrations of water vapor, oxygen, and other gases.

2.2.2 *Manufacture*

Unless produced specifically as a product (e.g. liquefied petroleum gas), the gaseous products of refinery operations are mixtures of various gases. Each gas is a by-product of a refining process. Thus, the compositions of natural, manufactured, and mixed gases can vary so widely that no single set of specifications could cover all situations.

2.2.3 *Properties and uses*

As already noted, the compositions of natural, manufactured, and mixed gases can vary so widely that no single set of specifications could cover all situations. The requirements are usually based on performances in burners and equipment, on minimum heat content, and on maximum sulfur content. Gas utilities in most states come under the supervision of state commissions or regulatory bodies and the utilities must provide a gas that is acceptable to all types of consumers and that will give satisfactory performance in all kinds of gas-consuming equipment.

Since natural gas as delivered to pipelines has practically no odor, the addition of an odorant is required by most regulations in order that the presence of the gas can be detected readily in case of accidents and leaks. This odorization is provided by the addition of trace amounts of some organic sulfur compounds to the gas before it reaches the consumer. The standard requirement is that a user will be able to detect the presence of the gas by odor when the concentration reaches 1% of gas in air. Since the lower limit of flammability of natural gas is approximately 5%, this 1% requirement is essentially equivalent to one-fifth of the lower limit of flammability. The combustion of these trace amounts of odorant does not create any serious problems of sulfur content or toxicity.

The different methods for gas analysis include absorption, distillation, combustion, mass spectroscopy, infrared spectroscopy, and gas chromatography (Speight, 2015a). Absorption methods involve absorbing individual constituents one at a time in suitable solvents and recording contraction in the volume measured. Distillation methods depend on the separation of constituents by fractional distillation and the measurement of the volumes distilled. In combustion methods, certain combustible elements are caused to burn to carbon dioxide and water, and the volume changes are used to calculate the composition. Infrared spectroscopy is useful for particular applications. For the most accurate analysis, mass spectroscopy and gas chromatography are the preferred methods.

The specific gravity of product gases, including liquefied petroleum gas, may be determined conveniently using a number of methods and a variety of instruments (Speight, 2015a). The heat value of gases is generally determined at constant pressure in a flow calorimeter in which the heat released by the combustion of a definite quantity of gas is absorbed by a measured quantity of water or air.

The lower and upper limits of *flammability* of organic compounds (Table 2.6) indicate the percentage of combustible gas in air below which and above which flame will not propagate. When flame is initiated in mixtures having compositions within these limits, it will propagate and therefore the mixtures are flammable. The knowledge of flammable limits and their use in establishing safe practices for handling gaseous fuels is

Table 2.6: Flammability limits of selected hydrocarbon derivatives.[a]

Compound	Limits of flammability	
	Lower volume (%)	Upper volume (%)
n-Amylene	1.42	8.70
Benzene	1.40	7.10
n-Butane	1.86	8.41
iso-butane	1.80	8.44
Butene-1	1.65	9.95
Butene-2	1.75	9.70
Cyclohexane	1.26	7.75
Cyclopropane	2.40	10.40
n-Decane	0.77	5.35
2,3-Dimethylpentane	1.12	6.75
2,2-Dimethylpropane	1.38	7.50
Ethane	3.00	12.50
Ethylene	2.75	28.60
n-Heptane	1.10	6.70
n-Hexane	1.18	7.40
Methane	5.00	15.00
Methylcyclohexane	1.15	6.70
n-Nonane	0.83	2.90
n-Octane	0.95	6.50
n-Pentane	1.40	7.80
iso-pentane (2-methylbutane)	1.32	7.60
Propane	2.12	9.35
Propylene	2.00	11.20
Toluene	1.27	6.75
Turpentine	0.80	—
o-Xylene	1.00	6.00
m-Xylene	1.10	7.00
p-Xylene	1.10	7.00

Note: [a]Listed alphabetically.

important: (i) when purging equipment is used in gas service, (ii) in controlling the factory or mine atmosphere, or (iii) in handling liquefied gases.

Many factors enter the experimental determination of flammable limits of gas mixtures, including the diameter and length of the tube or vessel used in the test, the temperature and pressure of the gases, and the direction of flame propagation — upward or downward. For these and other reasons, great care must be used in the application of the data. In monitoring closed spaces where small amounts of gases enter the atmosphere, the maximum concentration of the combustible gas is often limited to one-fifth of the concentration of the gas at the lower limit of flammability of the gas–air mixture.

2.3 Liquid Fuels

In the context of this book, liquid fuels are the combustible or energy-generating molecules that can be harnessed to create mechanical energy, usually producing kinetic energy. It is the fumes of liquid fuels that are flammable instead of the fluid. The liquid fuels presented here are those fuels that are produced from fossil fuels, such as natural gas, crude oil, coal, and oil shale.

Some common properties of liquid fuels are that they are easy to transport and can be handled with relative ease. The physical properties of liquid fuels vary by temperature, though not as greatly as for gaseous fuels. Some of these properties are as follows: (i) the flash point, which is the lowest temperature at which a flammable concentration of vapor is produced, (ii) the fire point, which is the temperature at which sustained burning of vapor will occur, (iii) the cloud point for diesel fuels, which is the temperature at which dissolved waxy compounds begin to coalesce, and (iv) the pour point, which is the temperature below which the fuel is too thick to pour freely. These properties affect the safety and handling of the fuel.

2.3.1 *Naphtha*

Naphtha (often referred to as *naft* in the older literature) is actually a generic term applied to a refined, a partly refined, or an unrefined crude oil fraction. In the strictest sense of the term, not less than 10% v/v of the material should distill below 175°C (345°F) and not less than 95% v/v of the material should distill below 240°C (465°F) under standardized distillation conditions (Speight, 2015a).

Generally (but this can be refinery-dependent), naphtha is an unrefined crude oil that distills below 240°C (465°F) and is (after the gaseous constituents) the most volatile fraction of the crude oil. Naphtha resembles gasoline in terms of boiling range and carbon number, being a precursor to gasoline. In some specifications, not less than 10% of the material should distil below approximately 75°C (167°F) (Pandey *et al.*, 2004). It is typically used as a precursor to gasoline or to a variety of solvents.

The term *crude oil solvent* (*petroleum solvent*) is used to describe the liquid hydrocarbon fractions obtained from petroleum and used in industrial processes and formulations. These fractions are also referred to as *naphtha* or *industrial naphtha*. By definition, the solvents obtained from the petrochemical industry such as alcohols, ethers, and the like are not included in this chapter. A refinery is capable of producing hydrocarbon derivatives of a high degree of purity, and at the present time, crude oil-derived solvents are available covering a wide range of solvent properties including both volatile and high-boiling qualities.

Naphtha has been available since the early days of the crude oil industry and is produced either by distillation of crude oil isolated from a surface seepage or (more likely) by destructive distillation of the bituminous material obtained from tar sand deposits. Other crude oil products boiling within the naphtha boiling range include (i) *industrial spirit* and (ii) *white spirit. Industrial spirit* comprises liquids distilling between 30°C and 200°C (−1°F and 390°F), with a temperature difference between the 5% volume and 90% volume distillation points, including losses, of not more than 60°C (140°F). There exist several (up to eight) grades of industrial spirit, depend on the position of the cut in the distillation range defined above. On the other hand, *white spirit* is an industrial spirit with a flash point above 30°C (99°F) and a distillation range of 135–200°C (275–390°F).

2.3.1.1 *Manufacture*

In general, naphtha may be prepared by any one of the following methods: (i) fractionation of straight-run, cracked, and reforming distillates, or even fractionation of crude oil, (ii) solvent extraction, (iii) hydrogenation of cracked distillate, (iv) polymerization of unsaturated compounds, i.e. polymerization of olefin derivatives, and (v) alkylation processes. In fact, the naphtha may be a combination of product streams from more than one of these processes.

The more common method of naphtha preparation is distillation. Depending on the design of the distillation unit, either one or two naphtha steams may be produced: (i) a single naphtha with an end point of approximately 205°C (400°F) and similar to straight-run gasoline or (ii) this same fraction divided into a low-boiling naphtha and a high-boiling naphtha. The end point of the low-boiling naphtha is varied to suit the subsequent subdivision of the naphtha into narrower boiling fractions and may be of the order of 120°C (250°F).

Before the naphtha is redistilled into a number of fractions with boiling ranges suitable for aliphatic solvents, the naphtha are usually treated to remove sulfur compounds, as well as aromatic hydrocarbon derivatives, which are present in sufficient quantity to cause an odor. Aliphatic solvents that are specially treated to remove aromatic hydrocarbon derivatives are known as *deodorized solvents*. *Odorless solvent* is the name given to the high-boiling alkylate used as an aliphatic solvent, which is a by-product in the manufacture of aviation alkylate.

Sulfur compounds are most commonly removed or converted to a harmless form by chemical treatment with lye, doctor solution, copper chloride, or similar treating agents. Hydrorefining processes are also often used in place of chemical treatment. Solvent naphtha are solvents selected for low-sulfur content, and the usual treatment processes, if required, remove only sulfur compounds (Parkash, 2003; Gary *et al.*, 2007; Speight, 2011a, 2014, 2017, 2019; Hsu and Robinson, 2017). Naphtha with a small aromatic content has a slight odor, but the aromatic constituents increase the solvent power of the naphtha; there is no need to remove aromatic derivatives unless an odor-free solvent is specified.

Naphtha, either naturally sweet (odorless) or treated until sweet, is separated into several fractions efficiently in fractional distillation towers, often called pipe stills, columns, or column steam stills (Parkash, 2003; Gary *et al.*, 2007; Speight, 2011a, 2014, 2017, 2019; Hsu and Robinson, 2017). A typical arrangement consists of primary and secondary fractional distillation towers and a stripper. High-boiling naphtha, for example, is heated by a steam heater and passed into the primary tower, which is usually operated under vacuum. The vacuum permits vaporization of the naphtha at the temperatures obtainable from the steam heater. The primary tower separates the naphtha into three parts: (i) a high-boiling material, (ii) a side-stream product of narrow boiling range, and (iii) an overhead product.

The high-boiling material is often removed as a bottom product (depending on the distillation unit) and sent to a cracking unit for conversion to lower-boiling products. The side stream product of narrow boiling range, after passing through the stripper, may be suitable for use as an aliphatic solvent. The overhead product is sent to the secondary (vacuum) tower where the overhead product from the primary tower is divided into an overhead and a bottom product in the secondary tower, which operates under a partial vacuum with steam injected into the bottom of the tower to assist in the fractionation. The overhead and bottom products are finished aliphatic solvents, or if the feed to the primary tower is low-boiling naphtha instead of high-boiling naphtha, other aliphatic solvents of different boiling ranges are produced.

Several methods, involving *solvent extraction* or *destructive hydrogenation* (*hydrocracking*) (Parkash, 2003; Gary *et al.*, 2007; Speight, 2011a, 2014, 2017, 2019; Hsu and Robinson, 2017), can accomplish the removal of aromatic hydrocarbon derivatives from naphtha. Using the latter method, aromatic hydrocarbon constituents are converted into odorless, straight-chain paraffin hydrocarbon derivatives that are required in aliphatic solvents.

The *Edeleanu* process (Parkash, 2003; Gary *et al.*, 2007; Speight, 2011a, 2014, 2017, 2019; Hsu and Robinson, 2017) was originally developed to improve the burning characteristics of kerosene by extraction of the smoke-forming aromatic compounds. Thus, it is not surprising that its use has been extended to the improvement of other products as well as the segregation of aromatic hydrocarbon derivatives for use as solvents. Naphtha fractions rich in aromatic derivatives may be treated by the Edeleanu process for the purpose of recovering the aromatic derivatives, or the product stream from a catalytic reformer unit — particularly when the unit is operated to produce maximum aromatic derivatives — may be Edeleanu treated to recover the aromatic derivatives. The other most widely used processes for this purpose are the extractive distillation process and the *Udex* process. Processes such as the *Arosorb* process and the cyclic adsorption process are used to a lesser extent.

Extractive distillation (Parkash, 2003; Gary *et al.*, 2007; Speight, 2011a, 2014, 2017, 2019; Hsu and Robinson, 2017) is used to recover aromatic hydrocarbon derivatives from, say, reformate fractions in the following manner. By means of preliminary distillation in a 65-tray pre-fractionator, a fraction containing a single aromatic can be separated from reformate, and this aromatic concentrate is then pumped to an extraction distillation tower near the top and the aromatic concentrate enters near the bottom. A reboiler in the extractive distillation tower induces the aromatic concentrate to ascend the tower, where it comes into contact with the descending solvent.

The solvent removes the aromatic constituents and accumulates at the bosom of the tower; the non-aromatic portion of the concentrate leaves the top of the tower and may contain approximately 1% v/v of the aromatic derivatives. The solvent and dissolved aromatic derivatives are conveyed from the bottom of the extractive distillation tower to a solvent stripper, where fractional distillation separates the aromatic derivatives from the solvent as an overhead product. The solvent is recirculated to the extractive distillation tower, whereas the aromatic stream is treated with sulfuric acid and clay to yield a finished product of high purity.

The *Udex* process (Parkash, 2003; Gary *et al.*, 2007; Speight, 2011a, 2014, 2017, 2019; Hsu and Robinson, 2017) is also employed to recover aromatic streams from reformate fractions. This process uses a mixture of water and diethylene glycol to extract aromatic derivatives. Unlike extractive distillation, an aromatic concentrate is not required and the solvent removes all the aromatic derivatives, which are separated from one another by subsequent fractional distillation.

The reformate is pumped into the base of an extractor tower. The feed rises in the tower countercurrent to the descending diethylene glycol–water solution, which extracts the aromatic derivatives from the feed. The non-aromatic portion of the feed leaves the top of the tower, and the aromatic-rich solvent leaves the bottom of the tower. Distillation in a solvent stripper separates the solvent from the aromatic derivatives, which are sulfuric acid and clay treated and then separated into individual aromatic derivatives by fractional distillation.

Silica gel (SiO_2) is an adsorbent for aromatic derivatives and has found use in extracting aromatic derivatives from refinery streams (Arosorb and cyclic adsorption processes) (Parkash, 2003; Gary *et al.*, 2007; Speight, 2011a, 2014, 2017, 2019; Hsu and Robinson, 2017). Silica gel is manufactured amorphous silica that is extremely porous and has the property of selectively removing and holding certain chemical compounds from mixtures. For example, silica gel selectively removes aromatic derivatives from a crude oil fraction, and after the non-aromatic portion of the fraction is drained from the silica gel, the adsorbed aromatic derivatives are washed from the silica gel by a stripper (or desorbent). Depending on the kind of feedstock, xylene, kerosene, or pentane may be used as the desorbent.

However, silica gel can be poisoned by contaminants, and the feedstock must be treated to remove water as well as nitrogen-, oxygen-, and sulfur-containing compounds by passing the feedstock through beds of alumina and/or other materials that remove impurities. The treated feedstock then enters one of several silica gel cases (columns) where the aromatic derivatives are adsorbed. The time period required for adsorption depends on the nature of the feedstock; for example, reformate product streams have been known to require substantially less treatment time than kerosene fractions.

2.3.1.2 *Composition*

Naphtha contains varying amounts of paraffin derivatives, olefin derivatives, naphthene derivatives, aromatic derivatives and olefin derivatives in different proportions in addition to potential isomers of paraffin that exist in the naphtha boiling range. As a result, naphtha is divided predominantly into two main types: (i) aliphatic naphtha and (ii) aromatic naphtha. The two types differ in two ways: (1) in the kind of hydrocarbon derivatives making up the solvent and (2) in the methods used for their manufacture. Aliphatic solvents are composed of paraffinic hydrocarbon derivatives and cycloparaffin derivatives (naphthene derivatives) and may be obtained directly from crude oil by distillation. The second type of naphtha contains aromatic derivatives, usually alkyl-substituted benzene, and is very rarely, if at all, obtained from crude oil as straight-run materials.

Stoddard solvent is a crude oil distillate widely used as a dry-cleaning solvent and a general cleaner and degreaser. It may also be used as a paint thinner, as a solvent in some types of photocopier toners, in some types of printing inks, and in some adhesives. Stoddard solvent is considered to be a form of mineral spirits, white spirits, and naphtha, however, not all forms of mineral spirits, white spirits, and naphtha are considered to be Stoddard solvents. The Stoddard solvent consists of linear alkane derivatives (30–50% v/v), branched alkane derivatives (20–40% v/v), cycloalkane derivatives (30–40% v/v), and aromatic hydrocarbon derivatives (10–20% v/v). The typical hydrocarbon derivatives range from C_7 to (at least) C_{12}.

2.3.1.3 *Properties and uses*

The boiling ranges of naphtha fractions are defined as follows: (i) low-boiling, which is the fraction boiling from 38°C to 150°C (100–300°F) and (ii) high-boiling, which is the fraction boiling between 150°C and 230°C (300°F and 450°F).

Naphtha is required to have a low level of odor to meet the specifications for its use (Pandey *et al.*, 2004), which is related to the chemical composition — generally, paraffin hydrocarbon derivatives possess the mildest odor, and the aromatic hydrocarbon derivatives have a much stronger odor. Naphtha containing higher proportions of aromatic constituents may be pale yellow — usually, naphtha is colorless (water white) and can be tested for the level of contaminants (Speight, 2015a).

Naphtha is used as automotive fuel, engine fuel, and jet-B (naphtha type). Broadly, naphtha is classified as *light naphtha* (i.e. low-boiling naphtha) and *heavy naphtha* (i.e. high-boiling naphtha) Light naphtha is used as a rubber solvent and a lacquer diluent, while high-boiling naphtha finds its application as a varnish solvent, and naphtha for the dyestuff industry and for the cleaning industry. More specifically, naphtha is valuable as a solvent owing to its good dissolving power. The wide range of naphtha available, from the ordinary paraffin straight-run to the highly aromatic types, and the varying degrees of volatility possible offer products suitable for many uses (Boenheim and Pearson, 1973; Hadley and Turner, 1973).

The main uses of naphtha fall into the following general areas: (i) solvents (diluents) for paints (ii) dry-cleaning solvents, (iii) solvents for cutback asphalt, (iv) solvents in the rubber industry, and (v) solvents for industrial extraction processes (Parkash, 2003; Gary *et al.*, 2007; Speight, 2011a, 2014, 2017, 2019; Hsu and Robinson, 2017). *Turpentine*, the older more conventional solvent for paints, has now been almost completely replaced with the discovery that the cheaper and more abundant crude oil naphtha is equally satisfactory. The differences in application are slight: naphtha causes a slightly greater decrease in viscosity than turpentine when added to some paints, and depending on the boiling range, may also show difference in evaporation rate.

The solvent power required for conventional paint diluents is low and can be reached by distillates from paraffinic crude oils, which are usually recognized as the poorest solvents in the crude oil naphtha group. In addition to solvent power and correct evaporation rate, a paint thinner should also be resistant to oxidation, i.e. the thinner should not develop bad color and odor during use. The thinner should be free of corrosive impurities and reactive materials, such as certain types of sulfur compounds, when employed with paints containing lead and similar metals. The requirements are best met by straight-run distillates from paraffinic crude oils that boil from 120°C to 205°C (250–400°F). The components of enamels, vanishes, nitrocellulose lacquers, and synthetic resin finishes are not as soluble in paraffinic naphtha as the materials in conventional paints, and hence naphthenic and aromatic naphtha are favored for such uses.

Dry cleaning is a well-established industry, and the standardized requirements for the solvent are usually met by straight-run naphtha from a low-sulfur, suitably refined paraffinic crude oil. An aromatic hydrocarbon content is not desirable, since it may cause removal of dyes from fabrics or too efficient removal of natural oils from wool, for example. Such a product is usually high-boiling and, hence, safe from fire risks, as well as stable enough for extensive reuse and reclaiming. It is especially important that dry-cleaning solvents leave no odor on the cloth, and for this reason (coupled with reuse and reclaiming) the solvents cannot be treated with sulfuric acid. The acid treatment leaves the oil with very small quantities of sulfonated hydrocarbon derivatives, which leave a residual odor on the cloth and render the solvent unstable when exposed to distillation temperatures.

Cutback asphalt is asphalt cement diluted with a crude oil distillate to make it suitable for direct application to road surfaces with little or no heating. Asphalt cement, in turn, is a combination of hard asphalt and either a high-boiling distillate or a viscous residuum of an asphaltic crude oil. The products are classified as rapid, medium, and slow-curing, depending on the rate of evaporation of the solvent. A rapid-curing product may contain 40–50% v/v of material distilling up to 360°C (680°F); a slow-curing mixture may have only 25% v/v of such material. Gasoline naphtha, kerosene, and low-boiling fuel oils boiling from 38°C to 330°C (100–30°F) are used in different products and for different purposes; the use may also dictate the nature of the residuum that can be used for asphalt production.

Naphtha is used in the rubber industry for dampening the play and tread stocks of automobile tires during manufacture to obtain better adhesion between the units of the tire. They are also consumed extensively in making rubber cements (adhesives) or are employed in the fabrication of rubberized cloth, hot-water bottles, bathing caps, gloves, overshoes, and toys. These cements are solutions of rubber and were formerly made with benzene, but crude oil naphtha is now preferred because of its less toxic character.

Crude oil distillates are also added in amounts up to 25% v/v or higher at various stages in the polymerization of butadiene-styrene to synthetic rubber. Those employed in oil extended rubber are of the aromatic type. These distillates are generally high-boiling fractions and preferably contain no wax, boil from 425°C to 510°C (800–950°F), have characterization factors of 10.5–11.6, a viscosity index lower than 0, bromine numbers of 6–30, and API gravity of 3–24 (Parkash, 2003; Gary *et al.*, 2007; Speight, 2011a, 2014, 2017, 2019; Hsu and Robinson, 2017).

Naphtha is used for extraction on a fairly wide scale, particularly for the extraction of residual oil from castor beans, soybeans, cottonseed, and wheat germ and for the recovery of *grease* from mixed garbage and refuse. The solvent employed in these cases is a hexane cut, boiling from approximately 65°C to 120°C (150–250°F). When the oils recovered are of edible grade or intended for refined purposes, stable solvents completely free of residual odor and taste are necessary, and straight-run streams from low-sulfur, paraffinic crude oils are generally satisfactory.

The recovery of wood resin by naphtha extraction of the resinous portions of dead trees of the resin-bearing varieties or stumps, for example, is also used in the wood industry. The chipped wood is steamed to distill out the resinous products recoverable in this way and then extracted with a naphtha solvent, usually a well-refined, low-sulfur, paraffinic product boiling from, say, 95°C to 150°C (200–300°F).

Crude oil distillates of various compositions and volatility are also employed as solvents in the manufacture of printing inks, leather coatings, diluents for dyes, and degreasing of wool fibers, polishes, and waxes, as well as rust-proofing and water-proofing compositions, mildew-proofing compositions, insecticides, and wood preservatives.

2.3.2 *Gasoline*

Gasoline, also called *gas* (United States and Canada) or *petrol* (Great Britain) or *benzine* (Europe), is a mixture of volatile, flammable liquid hydrocarbon derivatives derived from crude oil and is used as fuel for internal-combustion engines. It is also used as a solvent for oils and fats. Originally a by-product of the crude oil industry (kerosene being the principal product), gasoline became the preferred automobile fuel because of its high energy of combustion and its capacity to mix readily with air in a carburetor.

Gasoline is a mixture of hydrocarbon derivatives that usually boil below 180°C (355°F) or, at most, below 200°C (390°F). The hydrocarbon constituents in this boiling range are those that have four to twelve carbon atoms in their molecular structure and fall into three general types: paraffin derivatives (including the cycloparaffin derivatives and branched materials), olefin derivatives, and aromatic derivatives.

2.3.2.1 *Manufacture*

Gasoline was at first produced by distillation, simply separating the volatile, more valuable fractions of crude oil. Later processes, designed to raise the yield of gasoline from crude oil, decomposed higher-molecular weight constituents into lower-molecular weight products by a process known as *cracking*. Similar to typical gasoline, several processes produce the blending stocks for reformulated gasoline (Parkash, 2003; Gary *et al.*, 2007; Speight, 2011a, 2014, 2017, 2019; Hsu and Robinson, 2017). Up to and during the first decade of the present century, the gasoline produced was that originally present in crude oil or that could be condensed from natural gas. However, it was soon discovered that if the heavier portions of crude oil (such as the fraction that boiled higher than kerosene, e.g. gas oil) were heated to more severe temperatures, thermal degradation (or cracking) occurred to produce smaller molecules within the range suitable for gasoline. Therefore, gasoline that was not originally in the crude oil could be manufactured.

Thermal cracking, employing heat and high pressures, was introduced in 1913 but was replaced after 1937 by catalytic cracking, the application of catalysts that facilitate chemical reactions producing more gasoline. Other methods used to improve the quality of gasoline and increase its supply include *polymerization*, *alkylation*, *isomerization*, and *reforming* (Parkash, 2003; Gary *et al.*, 2007; Speight, 2011a, 2014, 2017, 2019; Hsu and Robinson, 2017).

Polymerization is the conversion of gaseous olefin derivatives, such as propylene and butylene, into larger molecules in the gasoline range. *Alkylation* is a process combining an olefin and paraffin such as *iso*-butane. *Isomerization* is the conversion of straight-chain hydrocarbon derivatives to branched-chain hydrocarbon derivatives. *Reforming* is the use of either heat or a catalyst to rearrange the molecular structure.

At first, cracked gasoline was regarded as an inferior product because of its comparative instability on storage, and special techniques had to be developed to combat this marked instability. As more and better quality gasoline was required, the crude oil industry focused on processes by which this material could be produced. These processes include catalytic cracking, thermal and catalytic reforming, hydrocracking, alkylation, and polymerization. The end result was the production of gasoline constituents with enhanced stability and performance.

However, despite the variations in the composition of the gasoline produced by the various available processes, this material is rarely, if ever, suitable for use as such. It is at this stage of a refinery operation that blending becomes important.

Aviation gasoline is a form of motor gasoline that has been especially prepared for use in aviation piston engines. It has an octane number suited to the engine, a freezing point of –60°C (–76°F), and a distillation range usually within the limit of 30–180°C (86–356°F) compared to −1–200°C (30–390°F) for automobile gasoline. The narrower boiling range ensures better distribution of the vaporized fuel through the more complicated induction systems of aircraft engines. Aircraft operate at altitudes at which the prevailing pressure is less than the pressure at the surface of the earth (pressure at 17,500 feet is 7.5 psi compared to 14.7 psi at the surface of the earth). Thus, the vapor pressure of aviation gasoline must be limited to reduce boiling in the tanks, fuel lines, and carburetors. Thus, the aviation gasoline does not usually contain the gaseous hydrocarbon derivatives (butane isomers) that give automobile gasoline the higher vapor pressures.

Aviation gasoline is strictly limited regarding hydrocarbon composition. The important properties of the hydrocarbon derivatives are the highest octane numbers economically possible, boiling points in the limited temperature range of aviation gasoline, maximum heat content per pound (high proportion of combined hydrogen), and high chemical stability to withstand storage. Aviation gasoline is composed of paraffin derivatives and *iso*-paraffin derivatives (50–60% v/v), moderate amounts of naphthene derivatives (20–30% v/v), small amounts of aromatic derivatives (10% v/v), and usually no olefin derivatives, whereas motor gasoline may contain up to 30% v/v olefin derivatives and up to 40% v/v aromatic derivatives.

Under the conditions of use in aircraft, olefin derivatives have a tendency to form gum, cause pre-ignition, and have relatively poor anti-knock characteristics under lean mixture (cruising) conditions; for these reasons olefin derivatives are detrimental to aviation gasoline. Aromatic derivatives have excellent anti-knock characteristics under rich mixture (takeoff) conditions, but are much like the olefin derivatives under lean mixture conditions; hence, the proportion of aromatic derivatives in aviation gasoline is limited. Some naphthene derivatives with suitable boiling temperatures are excellent aviation gasoline components but are not segregated as such in refinery operations. They are usually natural components of the straight-run naphtha (aviation base stocks) used in blending aviation gasoline. The lower-boiling paraffin derivatives (pentane and hexane), and both the high-boiling and low-boiling *iso*-paraffin derivatives (*iso*-pentane to *iso*-octane) are excellent aviation gasoline components. These hydrocarbon derivatives have high heat content per pound and are chemically stable, and the *iso*-paraffin derivatives have high-octane numbers under both lean and rich mixture conditions.

The manufacture of aviation gasoline is thus dependent on the availability and selection of fractions containing suitable hydrocarbon derivatives. The lower-boiling hydrocarbon derivatives are usually found in straight-run naphtha from certain crude oil. These fractions have high contents of *iso*-pentane and *iso*-hexane derivatives and provide the needed volatility, as well as high-octane number components. Higher-boiling *iso*-paraffin derivatives are provided by aviation alkylate, which consists mostly of branched octanes. Aromatic derivatives, such as benzene, toluene, and xylene, are obtained from catalytic reforming or a similar source.

To increase the proportion of higher-boiling octane components, such as aviation alkylate and xylenes, the proportion of lower-boiling components must also be increased to maintain the proper volatility. *Iso*-pentane and, to some extent, *iso*-hexane are the lower-boiling components used. *Iso*-pentane and *iso*-hexane may be separated from selected naphtha by superfractionators or synthesized from the normal hydrocarbon derivatives by isomerization. In general, most aviation gasolines are made by blending a selected straight-run naphtha fraction (aviation base stock) with *iso*-pentane and aviation alkylate.

2.3.2.2 *Composition*

Automotive gasoline is a mixture of low-boiling hydrocarbon compounds suitable for use in spark-ignited internal combustion engines and having an octane rating of at least 60. Additives that have been used in gasoline include alkyl tertiary butyl ethers (e.g. MTBE), ethanol (ethyl alcohol), methanol (methyl alcohol), tetramethyl-lead [$(CH_3)_4Pb$], tetraethyl-lead [$(C_2H_5)_4Pb$], ethylene dichloride ($ClCH_2CH_2Cl$), and ethylene dibromide ($BrCH_2CH_2Br$). Other categories of compounds that may be added to gasoline include anti-knock agents, antioxidants, metal deactivators, lead scavengers, anti-rust agents, anti-icing agents, upper-cylinder lubricants, detergents, and dyes.

Automotive gasoline typically contains approximately 200 hydrocarbon compounds, although nearly 1,000 have been identified. The relative concentration of the compounds vary considerably, depending on the source of crude oil, the refinery process, and product specifications. Typical hydrocarbon chain lengths range from C_4 to C_{12} with a general hydrocarbon distribution consisting of 4–8% v/v alkane derivatives, 2–5% v/v alkene derivatives, 25–40% v/v *iso*-alkane derivatives, 3–7% v/v cycloalkane derivatives, 1–4% v/v cycloalkene derivatives, and 20–50% v/v aromatic derivatives. However, these proportions vary greatly. Unleaded gasoline may have higher proportions of aromatic hydrocarbon derivatives than leaded gasoline.

Nevertheless, gasoline is manufactured to meet specifications and regulations and not to achieve a specific distribution of hydrocarbon derivatives by class and size. However, chemical composition often defines properties. For example, volatility is defined by the individual hydrocarbon constituents and the lowest boiling constituent(s) defines the volatile as determined by certain test methods.

At first, cracked gasoline was regarded as an inferior product because of its comparative instability on storage, but as more gasoline was required, the crude oil industry focused on processes by which this material could be produced (e.g. catalytic cracking, thermal and catalytic reforming, hydrocracking, alkylation, and polymerization) and thus the problem of storage instability was addressed and resolved.

The majority of members of the paraffin, olefin, and aromatic series (of which there are approximately 500) boiling below 200°C (390°F) have been found in the gasoline fraction of crude oil. However, it appears that the distribution of the individual members of straight-run gasoline (i.e. distilled from crude oil without thermal alteration) is not even.

Highly branched paraffin derivatives, which are particularly valuable constituents of gasoline(s), are not usually the principal paraffinic constituents of straight-run gasoline. The more predominant paraffinic constituents are usually the normal (straight-chain) isomers, which may dominate the branched isomer(s) by a factor of 2 or more. This is presumed to indicate the tendency to produce long uninterrupted carbon chains during crude oil maturation rather than those in which branching occurs. However, this trend is somewhat different for the cyclic constituents of gasoline, i.e. cycloparaffin derivatives (naphthene derivatives) and aromatic derivatives. In these cases, the preference appears to be for several short side chains rather than one long substituent.

Gasoline can vary widely in composition: even those with the same octane number may be quite different, not only in the physical makeup but also in the molecular structure of the constituents and the octane number regularly needs enhancement (Watanabe *et al.*, 2010). The variation in the content of aromatic derivatives as well as the variation in the content of normal paraffin derivatives, branched paraffin derivatives, cyclopentanes, and cyclohexane derivatives involve characteristics of any one individual crude oil and may in some instances be used for crude oil identification. Furthermore, straight-run gasoline generally shows a decrease in paraffin content with an increase in molecular weight, but the cycloparaffin derivatives (naphthene derivatives) and aromatic derivatives increase with increasing molecular weight. Indeed, the hydrocarbon type variation may also vary markedly from process to process.

The reduction of the lead content of gasoline and the introduction of reformulated gasoline have been very successful in reducing automobile emissions (Absi-Halabi *et al.*, 1997; Lappas *et al.*, 1999). Further improvements in fuel quality have been proposed for the years 2000 and beyond. These projections are accompanied by a noticeable and measurable decrease in crude oil quality and the reformulated gasoline will help meet environmental regulations on emissions for liquid fuels (Parkash, 2003; Gary *et al.*, 2007; Speight, 2011a, 2014, 2017, 2019; Hsu and Robinson, 2017).

2.3.2.3 *Properties and uses*

Despite the diversity of processes within a modern crude oil refinery, no single stream meets all the requirements of gasoline. Thus, the final step in gasoline manufacture is *blending* the various streams into a finished product (Parkash, 2003; Gary *et al.*, 2007; Speight, 2011a, 2014, 2017, 2019; Hsu and Robinson, 2017). It is not uncommon for the finished gasoline to be made up of six or more streams and several factors make this flexibility critical: (i) the requirements of the gasoline specification as presented in the standard test method and the regulatory requirements and (ii) performance specifications that are subject to local climatic conditions and regulations (Speight, 2015a).

The early criterion for gasoline quality was Baumé gravity or API gravity (Parkash, 2003; Gary *et al.*, 2007; Speight, 2011a, 2014, 2017, 2019; Hsu and Robinson, 2017). For example, 70° API gravity gasoline contained fewer, if any, heavier gasoline constituents than 60° API gasoline. Therefore, 70° API gasoline was a higher-quality and hence economically more valuable gasoline. However, apart from its use as a rough estimation of quality (for not only crude oil products but also crude oil), specific gravity is no longer of any significance as a true indicator of gasoline quality.

The gasoline performance and hence quality of an automobile gasoline is determined by its resistance to knock, for example, *detonation* or *ping* during service. The anti-knock quality of the fuel limits the power and economy that an engine using that fuel can produce: the higher the anti-knock quality of the fuel, the more the power and efficiency of the engine.

In the early days of automobile engine development there was a demand for more powerful engines and knocking was not a problem. This demand for more powerful engines was met at first by adding more and larger pistons to the engines; some were even built with 16 cylinders. However, the practical limit to the size to which an engine could be built meant that other ways to obtain increased power had to be explored. The obvious method for achieving more power from an engine of given size was to increase its compression ratio (a measure of the extent to which the gasoline–air mixture is compressed in the cylinder of an engine). The more the mixture is compressed before ignition, the more power the engine can deliver, but the increased performance was accompanied by an increased tendency for detonation or *knocking*. Eventually the cause of engine knock was traced to the gasoline. Since some gasoline seemed to cause more knock than another, and since there was no suitable testing procedure, it was difficult to determine the relative anti-knock characteristics of gasoline. It appeared, however, that cracked gasoline caused less knock than straight-run gasoline and, hence, the use of cracked gasoline as a fuel increased.

In 1922, tetraethyl lead was discovered to be an excellent anti-knock material when added in small quantities to gasoline, and gasoline containing tetraethyllead became widely available. However, the problem of how to increase the anti-knock characteristics of cracked gasoline became acute in the 1930s. One feature of the problem concerned the need to measure the anti-knock characteristics of gasoline accurately. This was solved in 1933 by the general use of a single-cylinder test engine, which allowed comparisons of the anti-knock characteristics of gasoline to be made in terms of octane numbers. The octane numbers formed a scale ranging from 0 to 100: the higher the number, the greater the anti-knock characteristics. In 1939, a second and less severe test procedure using the same test engine was developed, and the results obtained by this test were also expressed in octane numbers.

Octane numbers are obtained by either of the two test procedures; those obtained by the first method are called *motor octane numbers* (indicative of high-speed performance). Those obtained by the second method are called *research octane numbers* (indicative of normal road performance). Octane numbers quoted are usually, unless stated otherwise, research octane numbers (Speight, 2015a). In the test methods used to determine the anti-knock properties of gasoline, comparisons are made with blends of two pure hydrocarbon derivatives, *n*-heptane [$CH_3(CH2)_5CH_3$] and 2,2,4-trimethylpentane (also called *iso*-octane, $CH_3C(CH_3)_2CH_2CH(CH_3)CH_3$). *Iso*-octane has an octane number of 100 and is high in its resistance to knocking; *n*-heptane is quite low (with an octane number of 0) in its resistance to knocking.

Extensive studies of the octane numbers of individual hydrocarbon derivatives have given rise to some general rules. For example, normal paraffin derivatives have the least desirable knocking characteristics, and these become progressively worse as the molecular weight increases. *Iso*-paraffin derivatives have higher-octane numbers than the corresponding normal isomers, and the octane number increases as the degree of branching of the chain is increased. Olefin derivatives have markedly higher octane numbers than the related paraffin derivatives; naphthene derivatives are usually better than the corresponding normal paraffin derivatives, but they rarely have very high-octane numbers; aromatic derivatives usually have quite high-octane numbers.

Blends of *n*-heptane and *iso*-octane thus serve as a reference system for gasoline and provide a wide range of quality used as an anti-knock scale. The exact blend, which matches identically the anti-knock resistance of the fuel under test, is found, and the percentage of *iso*-octane in that blend is termed the *octane number* of the gasoline. For example, gasoline with a knocking ability which matches that of a blend of 90% v/v *iso*-octane and 10% v/v *n*-heptane has an octane number of 90. However, many pure hydrocarbon derivatives and even commercial gasoline have anti-knock quality above an octane number of 100. In this range it is common practice to extend the reference values by the use of varying amounts of tetraethyl lead in pure *iso*-octane.

With an accurate and reliable means of measuring octane numbers, it was possible to determine the cracking conditions — temperature, cracking time, and pressure — that caused increases in the anti-knock characteristics of cracked gasoline. In general, it was found that higher cracking temperatures and lower pressures produced higher-octane gasoline, but unfortunately more gas, cracked residua, and coke were formed at the expense of the volume of cracked gasoline.

To produce higher-octane gasoline, cracking coil temperatures were pushed up to 510°C (950°F), and pressures dropped from 1000 to 350 psi. This was the limit of thermal cracking units, for at temperatures over 510°C (950°F), coke formed so rapidly in the cracking coil that the unit became inoperative after only a short time on-stream. Hence, it was at this stage that the nature of the gasoline-producing process was reexamined, leading to the development of other processes, such as reforming, polymerization, and alkylation (Parkash, 2003; Gary *et al.*, 2007; Speight, 2011a, 2014, 2017, 2019; Hsu and Robinson, 2017) for the production of gasoline components having suitably high-octane numbers.

It is worth noting here that the continued decline in crude oil reserves and the issue of environmental protection have become extremely important in the search for alternatives to crude oil. Thus, oxygenates, either neat or as additives to fuels, appear to be the principal alternative fuel candidates for the crude oil refinery.

In the late 20th century, the rising price of crude oil (and hence of gasoline) led to the increasing use of *gasohol*, which is a mixture of 90% unleaded gasoline and 10% ethanol (ethyl alcohol). Gasohol burns well in gasoline engines and is a desirable alternative fuel for certain applications because of the availability of ethanol, which can be produced from grains, potatoes, and certain other plant matter. Methanol and a number of other alcohols and ethers are considered to be high-octane enhancers of gasoline (Mills and Ecklund, 1987; Gary and Alson, 1989). They can be produced from various hydrocarbon sources other than crude oil and may also offer environmental advantages insofar as the use of oxygenates would presumably suppress the release of vehicle pollutants into the air.

During the manufacture and distribution of gasoline, gasoline comes into contact with water and particulate matter and can become *contaminated* with such materials. Water is allowed to settle from the fuel in storage tanks and then it is withdrawn regularly and disposed of properly. Particulate matter is removed by filters installed in the distribution system (Speight, 2015a).

Adulteration differs from contamination insofar as unacceptable materials are deliberately added to gasoline for a variety of reasons not discussed here. Such activities may not only lower the octane number but also adversely affect volatility, which in turn also affects performance. In some countries, dyes and markers are used to detect adulteration and/or octane number testing may be required to detect adulteration.

Additives are gasoline-soluble chemicals that are mixed with gasoline to enhance certain performance characteristics or provide characteristics not inherent in the gasoline. Additives are generally derived from crude oil-based materials and their function and chemistry are highly specialized. They produce the desired effect at the parts-per-million (ppm) concentration range.

Oxidation inhibitors (antioxidants) are aromatic amines and hindered phenols that prevent gasoline components (particularly olefin derivatives) from reacting with oxygen in the air to form *peroxides* or *gums*. Peroxides can degrade the anti-knock quality, cause fuel-pump wear, and attack plastic or elastomeric fuel system parts, soluble gums can lead to engine deposits, and insoluble gums can plug fuel filters. Inhibition of oxidation is particularly important for the fuels used in modern fuel-injected vehicles, as their fuel recirculation design may subject the fuel to more temperature- and oxygen-exposure stress.

Corrosion inhibitors are carboxylic acids and carboxylates that prevent free water in gasoline from rusting or corroding pipelines and storage tanks. Corrosion inhibitors become less important once gasoline is in the vehicle. The metal parts in the fuel systems of modern vehicles are made of corrosion-resistant alloys or steel coated with corrosion-resistant coatings. More plastic parts are replacing metals in fuel systems and, in addition, service station systems and operations are designed to prevent free water from being delivered to a fuel tank.

Demulsifiers are polyglycol derivatives that improve the water-separating characteristics of gasoline by preventing the formation of stable emulsions.

Anti-knock compounds are compounds (such as tetraethyl lead) that increase the anti-knock quality of gasoline. Gasoline containing tetraethyl lead was first marketed in the 1920s and the average concentration of lead in gasoline was gradually increased until it reached a maximum of approximately 2.5 g/gal. in the late 1960s. After that, a series of events resulted in the use of less lead, and EPA regulations required the phased reduction of the lead content of gasoline since 1979. The EPA completely banned the addition of lead additives to on-road gasoline in 1996 and the amount of incidental lead may not exceed 0.05 g/gal.

Anti-icing additives are surfactants, alcohols, and glycols that prevent ice formation in the carburetor and fuel system. The need for this additive is being reduced as older-model vehicles with carburetors are replaced by vehicles with fuel injection systems.

Dyes are oil-soluble solids and liquids used to visually distinguish the batches, grades, or applications of gasoline products. For example, gasoline for general aviation, which is manufactured to different and more exacting requirements, is dyed blue to distinguish it from motor gasoline for safety reasons.

Markers are a means of distinguishing specific batches of gasoline without providing an obvious visual clue. A refiner may add a marker to its gasoline so it can be identified as it moves through the distribution system.

Drag reducers are high-molecular-weight polymers that improve the fluid flow characteristics of low-viscosity crude oil products. Drag reducers lower pumping costs by reducing friction between the flowing gasoline and the walls of the pipe.

Oxygenates are carbon-, hydrogen-, and oxygen-containing combustible liquids that are added to gasoline to improve performance. The addition of oxygenates to gasoline is not new since ethanol (ethyl alcohol or grain alcohol) has been added to gasoline for decades. Thus, *oxygenated gasoline* is a mixture of conventional hydrocarbon-based gasoline and one or more oxygenates. The current oxygenates belong to one of two classes of organic molecules: alcohols and ethers. The most widely used oxygenates in the United States are ethanol, methyl tertiary-butyl ether (MTBE) and tertiary-amyl methyl ether (TAME). Ethyl tertiary-butyl ether (ETBE) is another ether that could be used. Oxygenates may be used in areas of the United States where they are not required as long as concentration limits (as refined by environmental regulations) are observed.

Of all the oxygenates, methyl-*t*-butyl ether (MTBE) is attractive for a variety of technical reasons. It has a low-vapor pressure, can be blended with other fuels without phase separation, and has the desirable octane characteristics. If oxygenates achieve recognition as vehicle fuels, the biggest contributor will probably be methanol, the production of which is mostly from synthesis gas derived from methane or other sources (Parkash, 2003; Gary *et al.*, 2007; Speight, 2011a, 2014, 2017, 2019; Hsu and Robinson, 2017).

The higher alcohols also offer some potential as motor fuels. These alcohols can be produced at temperatures below 300°C (570°F) using copper oxide–zinc oxide–alumina catalysts promoted with potassium. *Iso*-butyl alcohol is of particular interest because of its high-octane rating, which makes it desirable as a gasoline-blending agent. This alcohol can be reacted with methanol in the presence of a catalyst to produce methyl-*t*-butyl ether. Although it is currently cheaper to make *iso*-butyl alcohol from *iso*-butylene, it can be synthesized from syngas with alkali-promoted zinc oxide catalysts at temperatures above 400°C (750°F).

2.3.3 *Kerosene*

Kerosene (*kerosine*), also called paraffin or paraffin oil, is a flammable pale yellow or colorless oily liquid with a characteristic odor. It is obtained from crude oil and is used in lamps and domestic heaters or furnaces, as a fuel or a fuel component for jet engines, and as a solvent for greases and insecticides.

Kerosene is intermediate in volatility between gasoline and gas/diesel oil. It is medium oil distilling between 150°C and 300°C (300°F and 570°F). It has a flash point of approximately 25°C (77°F) and is suitable for use as an illuminant when burned in a wide lamp. The term *kerosene* is also very often incorrectly applied to various fuel oils, but a fuel oil is actually any liquid or liquid crude oil product that produces heat when burned in a suitable container or that produces power when burned in an engine.

Kerosene was the main refinery product before the onset of the automobile age, but today kerosene can be considered as one of several secondary crude oil products after the primary refinery product — gasoline. Kerosene originated as a straight-run crude oil fraction that boils between approximately 205°C and 260°C (400°F and 500°F) (Walmsley, 1973). Some crude oils, for example, those from the Pennsylvania oil fields, contain kerosene fractions of very high quality, but other crude oils need to be thoroughly refined to remove aromatic derivatives and sulfur compounds before a satisfactory kerosene fraction can be obtained.

Jet fuels consist of gasoline- and kerosene-type jet fuels meeting the specifications for use in aviation turbine power units and are often referred to as *gasoline-type jet fuel* and *kerosene-type jet fuel*. The major jet fuels used by the military are JP-4, JP-5, JP-6, JP-7, and JP-8. Briefly, JP-4 is a wide-cut fuel developed for wide availability. JP-6 is a higher cut than JP-4 and is characterized by fewer impurities. JP-5 is a specially blended kerosene, and JP-7 is a high-flash point special kerosene used in advanced supersonic aircraft. JP-8 is a kerosene modeled on Jet A-l fuel (used in civilian aircraft). From what data are available, the typical hydrocarbon chain lengths characterizing JP-4 range from C_4 to C_{16}. Aviation fuels consist primarily of straight and branched alkane derivatives and cycloalkane derivatives. Aromatic hydrocarbon derivatives are limited to 20–25% v/v of the total mixture because they produce smoke when burned. A maximum of 5% v/v alkene derivatives is specified for JP-4. The approximate distribution by chemical class is as follows: (i) straight chain alkane derivatives, 32% v/v; branched alkane derivatives, 31% v/v; cycloalkane derivatives, 16% v/v; aromatic hydrocarbon derivatives, 21% v/v.

Gasoline-type jet fuel includes all low-boiling hydrocarbon oils for use in aviation turbine power units that distill between 100°C and 250°C (212°F and 480°F). It is obtained by blending kerosene and gasoline or naphtha in such a way that the aromatic content does not exceed 25% in volume. Additives can be included to improve fuel stability and combustibility.

Kerosene-type jet fuel is a medium distillate product that is used for aviation turbine power units. It has the same distillation characteristics and flash point as kerosene (between 150°C and 300°C, 300°F and 570°F, but not generally above 250°C, 480°F). In addition, it has particular specifications (such as freezing point) which are established by the International Air Transport Association (IATA).

2.3.3.1 *Manufacture*

Kerosene was first manufactured in the 1850s from coal tar, hence the name coal oil is often applied to kerosene, but crude oil became the major source after 1859. From that time, the kerosene fraction was and has remained a distillation fraction of crude oil. However, the quantity and quality vary with the type of crude oil, and although some crude oils yield excellent kerosene quite simply, others produce kerosene that requires substantial refining.

Kerosene is now largely produced by cracking the less volatile portion of crude oil at atmospheric pressure and elevated temperatures (Parkash, 2003; Gary *et al.*, 2007; Speight, 2011a, 2014, 2017, 2019; Hsu and Robinson, 2017). In the early days, the poorer quality kerosene was treated with large quantities of sulfuric acid to convert them to marketable products. However, this treatment resulted in high acid and kerosene losses, but the later development of the *Edeleanu* process overcame these problems.

Kerosene is a very stable product, and additives are not required to improve the quality. Apart from the removal of excessive quantities of aromatic derivatives by the Edeleanu process, kerosene fractions may need only a lye wash or a doctor treatment if hydrogen sulfide is present to remove mercaptans (Parkash, 2003; Gary *et al.*, 2007; Speight, 2011a, 2014, 2017, 2019; Hsu and Robinson, 2017).

2.3.3.2 *Composition*

Chemically, kerosene is a mixture of hydrocarbon derivatives; the chemical composition depends on its source, but it usually consists of approximately 10 different hydrocarbon derivatives, each containing 10–16 carbon atoms per molecule; the constituents include *n*-dodecane (n-$C_{12}H_{26}$), alkyl benzenes, and naphthalene and its derivatives. Kerosene is less volatile than gasoline; it boils between approximately 140°C (285°F) and 320°C (610°F).

Kerosene, because of its use as a burning oil, must be free of aromatic and unsaturated hydrocarbon derivatives, as well as free of the more obnoxious sulfur compounds. The desirable constituents of kerosene are saturated hydrocarbon derivatives, and it is for this reason that kerosene is manufactured as a straight-run fraction, not by a cracking process.

Although the kerosene constituents are predominantly saturated materials, there is evidence for the presence of substituted tetrahydronaphthalene. Dicycloparaffin derivatives also occur in substantial amounts in kerosene. Other hydrocarbon derivatives with both aromatic and cycloparaffin rings in the same molecule, such as substituted indan, also occur in kerosene. The predominant structure of the dinuclear aromatic derivatives appears to be that in which the aromatic rings are condensed, such as naphthalene whereas the *isolated* two-ring compounds, such as biphenyl, are only present in traces, if at all.

2.3.3.3 *Properties and uses*

Kerosene is by nature a fraction distilled from crude oil that has been used as a fuel oil from the beginning of the crude oil-refining industry. As such, low proportions of aromatic and unsaturated hydrocarbon derivatives are desirable to maintain the lowest possible level of smoke during burning. Although some aromatic derivatives may occur within the boiling range assigned to kerosene, excessive amounts can be removed by extraction; that kerosene is not usually prepared from cracked products almost certainly excludes the presence of unsaturated hydrocarbon derivatives.

The essential properties of kerosene are flash point, fire point, distillation range, burning, sulfur content, color, and cloud point. In the case of the flash point, the minimum flash temperature is generally placed above the prevailing ambient temperature. The fire point is related to the fire hazard associated with its handling and use. The boiling range is of less importance for kerosene than for gasoline, but it can be taken as an indication of the viscosity of the product, for which there is no requirement for kerosene. The ability of kerosene to burn steadily and cleanly over an extended period is an important property and gives some indication of the purity or composition of the product.

The significance of the total sulfur content of a fuel oil varies greatly with the type of oil and the use to which it is put. Sulfur content is of great importance when the oil to be burned produces sulfur oxides that contaminate the surroundings. The color of kerosene is of little significance, but a product darker than usual may have resulted from contamination or aging, and in fact a color darker than specified may be considered by some users as unsatisfactory. Finally, the cloud point of kerosene gives an indication of the temperature at which the wick may become coated with wax particles, thus lowering the burning qualities of the oil (Speight, 2015a).

2.3.4 *Fuel oil*

Fuel oil is classified in several ways but was formally divided into two main types: *distillate fuel oil* and *residual fuel oil*, each of which is a blend of two or more refinery streams (Figure 2.1). Distillate fuel oil is

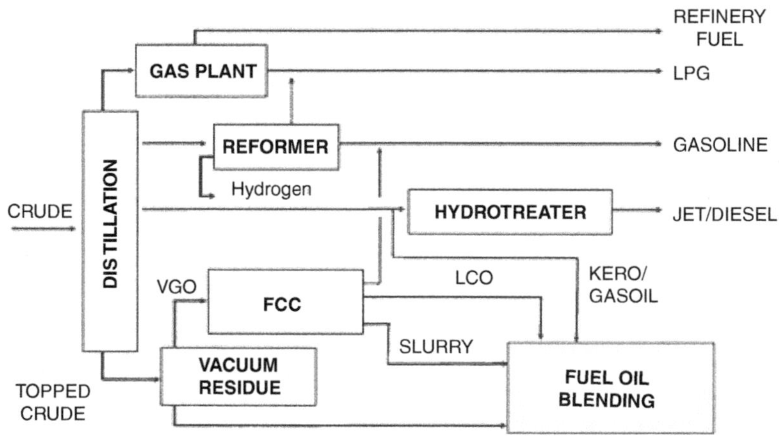

Figure 2.1: Fuel oil production by blending.

vaporized and condensed during a distillation process and thus has a definite boiling range, and does not contain high-boiling constituents. A fuel oil that contains an amount of the residue from crude distillation of thermal cracking is a residual fuel oil. The terms *distillate fuel oil* and *residual fuel oil* are losing their significance, since fuel oil is now made for specific uses and may be distillates or residuals or mixtures of the two. The terms *domestic fuel oil*, *diesel fuel oil*, and *heavy fuel oil* are more indicative of the uses of fuel oils.

2.3.4.1 *Manufacture*

The manufacture of fuel oils at one time largely involved using what was left after removing desired products from crude oil. Now, fuel oil manufacture is a complex matter of selecting and blending various crude oil fractions to meet definite specifications, and the production of a homogeneous, stable fuel oil requires experience backed by laboratory control.

Heavy fuel oil comprises all residual fuel oils and the constituents range from distillable constituents to residual (non-distillable) constituents that must be heated to 260°C (500°F) or more before they can be used. The kinematic viscosity is above 10 centistokes at 80°C (176°F). The flash point is always above 50°C (122°F) and the density is always higher than 0.900. In general, heavy fuel oil usually contains cracked residua, reduced crude, or cracking coil heavy product which is mixed (cut back) to a specified viscosity with cracked gas oils and fractionator bottoms. For some industrial purposes in which flames or flue gases come into contact with the product (ceramics, glass, heat treating, and open hearth furnaces), fuel oils must be blended to contain minimum sulfur content, and hence, low-sulfur residues are preferable for these fuels.

Stove oil, similar to kerosene, is always a straight-run fraction from suitable crude oils, whereas other fuel oils are usually blends of two or more fractions, one of which is usually cracked gas oil. The straight-run fractions available for blending into fuel oils are high-boiling naphtha, low-boiling and high-boiling gas oils, reduced crude, and pitch. Cracked fractions, such as low-boiling and high-boiling gas oils from catalytic cracking, cracking coil tar, and fractionator bottoms from catalytic cracking, may also be used as blends to meet the specifications of different fuel oils.

Since the boiling ranges, sulfur contents, and other properties of even the same fraction vary from crude oil to crude oil and with the way the crude oil is processed, it is difficult to specify which fractions are blended to produce specific fuel oils. In general, however, furnace fuel oil is a blend of straight-run gas oil and cracked gas oil to produce a product boiling in the 175–345°C (350–50°F) range.

Diesel fuel oil is essentially the same as furnace fuel oil, but the proportion of cracked gas oil is usually lower since the high aromatic content of cracked gas oil reduces the cetane value of diesel fuel.

Diesel fuels originally were straight-run products obtained from the distillation of crude oil. However, with the use of various cracking processes to produce diesel constituents, diesel fuels may also contain varying amounts of selected cracked distillates to increase the volume available for meeting the growing demand. Care is taken to select the cracked stocks in such a manner that specifications are met as simply as possible.

Under the broad definition of diesel fuel, many possible combinations of characteristics (such as volatility, ignition quality, viscosity, gravity, and stability) exist.

For clarification, biodiesel is not a true hydrocarbon product — it consists of the alkyl esters of long-chain fatty acids and is a transportation fuel that has gradually grown in popularity over the past decade (Speight, 2011b). Common sources of the feedstocks for biodiesel include soybean oil, algal oil (Lu *et al.*, 2009), higuereta oil (Ramon *et al.*, 2009), canola oil (Amanda *et al.*, 2007), vegetable oil (Rafael *et al.*, 2008), palm oil (Donato *et al.*, 2008), and corn oil (Hossein *et al.*, 2009).

2.3.4.2 *Composition*

All of the fuel oils described here are refined from crude oil and may be categorized either as distillate fuels or as residual fuels depending on the method of production. Fuel oil from biomass is not included here, having been presented in detail elsewhere (Speight, 2008, 2011b).

Fuel oil No. 1 and fuel oil No. 2 are distillate fuels that consist of distilled process streams. Residual fuel oil, such as fuel oil No. 4, is composed of the residuum remaining after distillation or cracking, or blends of such residues with distillates. Diesel fuel is almost similar to the fuel oils used for heating (fuel oil No. 1, fuel oil No. 2, and fuel oil No. 4).

All fuel oils consist of complex mixtures of aliphatic and aromatic hydrocarbon derivatives, their relative amounts depending on the source and grade of the fuel oil (Nielsen *et al.*, 2008). The aliphatic alkane derivatives (paraffin derivatives) and cycloalkane constituents (naphthene constituents) are hydrogen saturated and compose as much as 90% v/v of the fuel oil. Aromatic constituents (e.g. benzene) and olefin constituents compose up to 20% v/v and 1% v/v, respectively, of the fuel oils. Fuel oil No. 1 (straight-run kerosene) is a distillate that consists primarily of hydrocarbon derivatives in the C_9–C_{16} range, while fuel oil No. 2 is a higher-boiling, usually blended, distillate with hydrocarbon derivatives in the C_{11}–C_{20} range.

Diesel fuels predominantly contain a mixture of C_{10}–C_{19} hydrocarbon derivatives, which include aliphatic hydrocarbon derivatives (approximately 65% v/v), olefin hydrocarbon derivatives (up to 2% v/v), and aromatic hydrocarbon derivatives (up to 35% v/v). Jet fuels are based primarily on straight-run kerosene, as well as additives. All of the above fuel oils contain less than 5% v/v polycyclic aromatic hydrocarbon derivatives. Fuel No. 4 (also known as *marine diesel fuel*) is less volatile than diesel fuel No. 2 and may contain up to 15% v/v residual (high-boiling) streams, in addition to 5–10% v/v polycyclic aromatic hydrocarbon constituents.

In some countries, residual fuel oil is also known as *heavy fuel oil*, which also consists primarily of the residue from distillation or cracking units in the refinery. Historically, fuel oils were based on residua from the atmospheric distillation column and were known as straight-run fuels.

However, the increasing demand for transportation fuels, such as gasoline, kerosene, and diesel fuel, has led to an increased value for the atmospheric residue as a feedstock for vacuum distillation and for cracking processes. As a consequence, most heavy fuel oils are currently based on vacuum residua from thermal and catalytic cracking operations. These fuels differ in character from straight-run fuels in that the density and mean molecular weight are higher — the atomic hydrogen–carbon is lower, indicating a higher degree of aromaticity. The density of some heavy fuel oils can be above that of water (>1.000), which has environmental implications in the event of a spillage into water systems.

Residual fuel oil (and/or heavy fuel oil) is typically more complex in composition and impurities than distillate fuel oil. Therefore, a specific composition cannot always be determined — the sulfur content in residual fuel oil has been reported to vary up to 5% w/w.

Residual fuel oils are complex mixtures of high-molecular-weight compounds having a typical boiling range of 350–650°C (660–1200°F). They consist of aromatic, aliphatic, and naphthenic hydrocarbon derivatives, typically having carbon numbers from C_{20} to C_{50}, together with the asphaltene constituents and smaller amounts of heterocyclic compounds containing sulfur, nitrogen, and oxygen. They have chemical characteristics similar to liquid asphalt and, hence, are considered to be stabilized suspensions of asphaltene constituents in an oily medium (API, 2004).

Residual fuel oil also contains organo-metallic compounds from their presence in the original crude oil — the most important of which are nickel and vanadium. The metals (especially vanadium) are particularly of major significance for fuels burned in both diesel engines and boilers because when combined with sodium (perhaps from brine contamination from the reservoir or remaining after the refinery dewatering/desalting process) and other metallic compounds in critical proportions, they can lead to the formation of high-melting point ash, which is corrosive to engine parts. Other elements that occur in heavy fuel oils include iron, potassium, aluminum and silicon — the latter two metals are mainly derived from refinery catalyst fines.

2.3.4.3 *Properties and uses*

Domestic fuel oil is the fuel oil that is used primarily in home. This category of fuel oil includes kerosene, stove oil, and furnace fuel oil, which are distillate fuel oils.

Diesel fuel oil is also a distillate fuel oil that distils between 180°C and 380°C (356°F and 716°F). Different grades are available depending on the use: (i) diesel oil for diesel or compression-ignition engines and (ii) low-boiling heating oil for industrial and commercial use.

No. 1 fuel oil is a crude oil distillate that is one of the most widely used of all the fuel oil types. It is used in atomizing burners that spray fuel into a combustion chamber, where the tiny droplets burn while in suspension. It is also used as a carrier for pesticides, as a weed killer, and as a mold release agent in the ceramic and pottery industries, and in the cleaning industry. It is found in asphalt coatings, enamels, paints, thinners, and varnishes. No. 1 fuel oil is a low-boiling crude oil distillate (straight-run kerosene) consisting primarily of hydrocarbon derivatives in the range C_9–C_{16}. Fuel oil No. 1 is very similar in composition to diesel fuel; the primary difference lies in the additives.

No. 2 fuel oil is a crude oil distillate that may be referred to as domestic or industrial fuel oil. Domestic fuel oil is usually a lower boiling and straight-run product. It is used primarily in home heating. Industrial distillate is a cracked product or a blend of both. It is used in smelting furnaces, ceramic kilns, and packaged boilers. No. 2 fuel oil is characterized by hydrocarbon chain lengths in the C_{11}–C_{20} range. The composition consists of aliphatic hydrocarbon derivatives (straight chain alkane derivatives and cycloalkane derivatives) (64% v/v), 1–2% unsaturated hydrocarbon derivatives, i.e. alkene derivatives (1–2% v/v), and aromatic hydrocarbon derivatives (including alkyl benzenes and 2-ring, 3-ring aromatic derivatives) (35% v/v) but contains only small amounts of the polycyclic aromatic hydrocarbon derivatives (<5% v/v).

No. 6 fuel oil (also called *Bunker C oil* or *residual fuel oil*) is the residuum from crude oil after naphtha–gasoline, No. 1 fuel oil, and No. 2 fuel oil have been removed. No. 6 fuel oil can be blended directly to heavy fuel oil or used to produce asphalt. Residual fuel oil is more complex in composition and impurities than distillate fuels. Limited data are available on the composition of No. 6 fuel oil. Polycyclic aromatic hydrocarbon derivatives (including the alkylated derivatives) and metal-containing constituents are components of No. 6 fuel oil.

To characterize diesel fuel for use and thereby establish a framework of definition and reference, various classifications are used in different countries. For example, in the United States, grades No. 1D and 2D are distillate fuels, the types most commonly used in high-speed engines of the mobile type, medium-speed stationary engines, and railroad engines.

Grade 4D covers the class of more viscous distillates and, sometimes, blends of these distillates with residual fuel oils. No. 4D fuels are applicable in the low- and medium-speed engines employed in services involving sustained load and predominantly constant speed.

Cetane number is a measure of the tendency of a diesel fuel to knock inside a diesel engine. The scale is based on the ignition characteristics of two hydrocarbon derivatives: *n*-hexadecane (cetane) and 2,3,4,5,6,7,8-heptamethylnonane. Cetane has a short delay period during ignition and is assigned a cetane number of 100; heptamethylnonane has a long delay period and has been assigned a cetane number of 15. Just as the octane number is meaningful for automobile fuels, the cetane number is a means for determining the ignition quality of diesel fuels and is equivalent to the percentage by volume of cetane in the blend with heptamethylnonane, which matches the ignition quality of the test fuel (Speight, 2015a).

2.4 Solid Fuels

The term *solid fuel* as used in the context of this book refers to various forms of solid material that can be burned to release energy, generally providing heat and light through the combustion process. Solid fuels have been used throughout human history to create fire and are still in widespread use throughout the world.

2.4.1 *Wax*

Crude oil wax is of two general types: (i) *paraffin wax* in crude oil distillates and (ii) *microcrystalline wax* in crude oil residua. The melting point of wax is not directly related to its boiling point, because waxes contain hydrocarbon derivatives of different chemical nature. Nevertheless, waxes are graded according to their melting point and oil content. Microcrystalline waxes form approximately 1–2% w/w of crude oil and are a valuable product with numerous applications. These waxes are usually obtained from high-boiling lubricating oil distillates by solvent dewaxing and from tank bottom sludge by acid clay treatment (Agrawal *et al.*, 1997). However, these crude waxes usually contain an appreciable quantity (10–20% w/w) of residual oil and, as such, are not suitable for many applications, such as paper coating, electrical insulation, textile printing, and polishes.

2.4.1.1 *Manufacture*

Paraffin wax from solvent dewaxing operations (Parkash, 2003; Gary *et al.*, 2007; Speight, 2011a, 2014, 2017, 2019; Hsu and Robinson, 2017) is commonly known as *slack wax*, and the processes employed for the production of wax are aimed at deoiling the slack wax (crude oil wax concentrate).

Wax sweating was originally used in Scotland to separate wax fractions with various melting points from the wax obtained from shale oils. Wax sweating is still used to some extent but is being replaced by the more convenient wax recrystallization process. In wax sweating, a cake of slack wax is slowly warmed to a temperature at which the oil in the wax and the lower-melting waxes become fluid and drip (or sweat) from the bottom of the cake, leaving a residue of higher-melting wax. However, wax sweating can be carried out only when the residual wax consists of large crystals that have spaces between them, through which the oil and lower-melting waxes can percolate; it is therefore limited to wax obtained from low-boiling paraffin distillate.

The amount of oil separated by sweating is now much smaller than what it used to be owing to the development of highly efficient solvent dewaxing techniques (Parkash, 2003; Gary *et al.*, 2007; Speight, 2011a, 2014, 2017, 2019; Hsu and Robinson, 2017). In fact, wax sweating is now more concerned with the separation of slack wax into fractions with different melting points. A wax sweater consists of a series of approximately nine shallow pans arranged one above the other in a sweater house or oven, and each pan is divided horizontally by a wire screen. The pan is filled to the level of the screen with cold water. Molten wax is then introduced and allowed to solidify, and the water is then drained from the pan, leaving the wax cake supported on the screen.

A single sweater oven may contain more than 600 barrels of wax, and steam coils arranged on the walls of the oven slowly heat the wax cakes, allowing oil and the lower-melting waxes to sweat from the cakes and drip into the pans. The first liquid removed from the pans is called *foots oil*, which melts at 38°C (100°F) or lower, followed by *interfoots oil*, which melts in the range 38–44°C (100–112°F). Crude scale wax next drips from the wax cake and consists of wax fractions with melting points over 44°C (112°F).

When oil removal was an important function of sweating, the sweating operation was continued until the residual wax cake on the screen was free of oil. When the melting point of the wax on the screen has increased to the required level, allowing the oven to cool terminates sweating. The wax on the screen is a sweated wax with the melting point of a commercial grade of paraffin wax, which after a finished treatment becomes refined paraffinic wax. The crude scale wax obtained in the sweating operation may be recovered as such or treated to improve the color, in which case it is white crude scale wax. The crude scale wax and interfoots, however, are the sources of more waxes with lower melting points. The crude scale wax and interfoots are resweated several times to yield sweated waxes, which are treated to produce a series of refined paraffin waxes with melting points ranging from approximately 50°C to 65°C (125–150°F).

Sweated waxes generally contain small amounts of unsaturated aromatic and sulfur compounds, which are the source of unwanted color, odor, and taste that reduce the ability of the wax to resist oxidation; the commonly used method for removing these impurities is clay treatment of the molten wax.

Wax recrystallization, similar to wax sweating, separates slack wax into fractions, but instead of using the differences in melting points, it makes use of the different solubility of the wax fractions in a solvent, such as the ketone used in the dewaxing process (Parkash, 2003; Gary *et al.*, 2007; Speight, 2011a, 2014, 2017, 2019; Hsu and Robinson, 2017). When a mixture of ketone and slack wax is heated, the slack wax usually dissolves completely, and if the solution is cooled slowly, a temperature is reached at which a crop of wax crystals is formed. These crystals will all be of the same melting point, and if they are removed by filtration, a wax fraction with a specific melting point is obtained. If the clear filtrate is further cooled, a second crop of wax crystals with a lower melting point is obtained. Thus, by alternate cooling and filtration, the slack wax can be subdivided into a large number of wax fractions, each with different melting points.

This method of producing wax fractions is much faster and more convenient than sweating and results in a much more complete separation of the various fractions. Furthermore, recrystallization can also be applied to the microcrystalline waxes obtained from intermediate and high-boiling paraffin distillates, which cannot be sweated. Indeed, the microcrystalline waxes have higher melting points and differ in their properties from the paraffin waxes obtained from low-boiling paraffin distillates; thus wax recrystallization makes new kinds of waxes available.

2.4.1.2 *Composition*

Paraffin wax is a solid crystalline mixture of straight-chain (normal) hydrocarbon derivatives ranging from C_{20} to C_{30} and possibly higher, that is, CH_3 $(CH_2)_n$ CH_3, where $n \geq 18$. It is distinguished by its solid state at ordinary temperatures (25°C, 77°F) and low viscosity (35–45 SUS at 99°C, 210°F) when melted. However, in contrast to crude oil wax, petrolatum (*petroleum jelly*), although solid at ordinary temperatures, does in fact contain both solid and liquid hydrocarbon derivatives. It is essentially a low-melting, ductile, microcrystalline wax.

2.4.1.3 *Properties and uses*

The physical properties of microcrystalline waxes are greatly affected by the oil content (Kumar *et al.*, 2007) and hence by achieving the desired level of oil content, waxes of desired physical properties can be obtained. Deep deoiling of microcrystalline wax is comparatively difficult compared to paraffin wax (macrocrystalline wax) as the oil remains occluded in these and is difficult to separate by sweating. Also since wax and residual oil have similar boiling ranges, separation by distillation is difficult. However, these waxes can be deoiled by treatment with solvents at lower temperature that have high-oil miscibility and poor wax solubility and these have been used extensively to separate. These conventional solvent deoiling processes to upgrade the quality of high-oil content microcrystalline waxes involve agitation followed by extractive crystallization at low temperatures with solvents such as methyl *iso*-butyl ketone, methyl ethyl ketone–toluene mixtures, and dichloroethane — typically using drum filters.

The melting point of paraffin wax has both direct and indirect significance in most wax utilization. All wax grades are commercially indicated by a range of melting temperatures rather than a single value, and a range of 1°C (2°F) usually indicates a good degree of refinement. Other common physical properties that help to determine the degree of refinement of the wax are color, oil content, API gravity, flash point, and viscosity, although the last three properties are usually not given by the producer unless specifically requested (Speight, 2015a).

Crude oil waxes (and petrolatum) have found many uses in pharmaceuticals, cosmetics, paper manufacturing, candle making, electrical goods, rubber compounding, textiles, and many others, too numerous to mention here — for additional information, more specific texts on crude oil waxes should be consulted.

2.4.2 *Resid*

Bitumen is one of the oldest engineering materials known to humanity, having been used since the beginning of civilization. The bitumen used by early civilizations occurred naturally and was formed when crude oil seeped up through fractures and fissures to the surface of the earth and (after evaporation of the volatile constituents) appeared either as soft workable mortars or as hard brittle black veins of rock formations. In the period 5000–6000 BC, the Sumerians had a thriving ship-building industry that produced and used bitumen for caulking and waterproofing. In 2600 BC, the Egyptians used bitumen as a waterproofing material and also to impregnate the wrappings of mummies as a preservative. Other civilizations widely used bitumen as a mortar for building and paving blocks used in temples, irrigation systems, reservoirs, and highways.

2.4.2.1 *Manufacture*

The production of resid is the result of distilling all of the volatile material from the crude oil under atmospheric pressure or reduced pressure without raising the temperature so high as to cause cracking (thermal decomposition of the non-volatile material) On the other hand, the term "asphalt" refers to a product produced from the resid which is made up of non-volatile hydrocarbon derivatives in the feedstock, along with similar materials produced by thermal alteration during the distillation sequences, or they may be produced by air blowing residua.

Residuum manufacture is, in essence, a matter of distilling everything possible from crude oil until a residue with the desired properties is obtained. This is usually done by stages (Figure 2.2). Crude oil distillation at atmospheric pressure removes the lower-boiling fractions and yields reduced crude that may contain higher-boiling (lubricating) oils and even wax. Distillation of the reduced crude under vacuum removes the oils (and wax) as overhead products and the residuum remains as a bottom (or residual) product. The majority of the polar functionalities in the original crude oil turn out to be non-volatile and they concentrate in the vacuum residuum (Figure 2.3). It is this concentration effect that confers on the resid (and, by inference, any asphalt produced from the resid) some of its

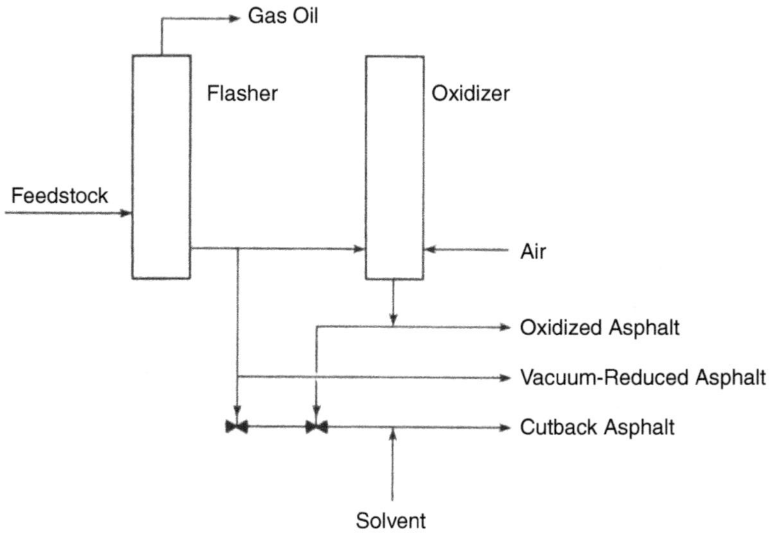

Figure 2.2: General representation of asphalt manufacture.

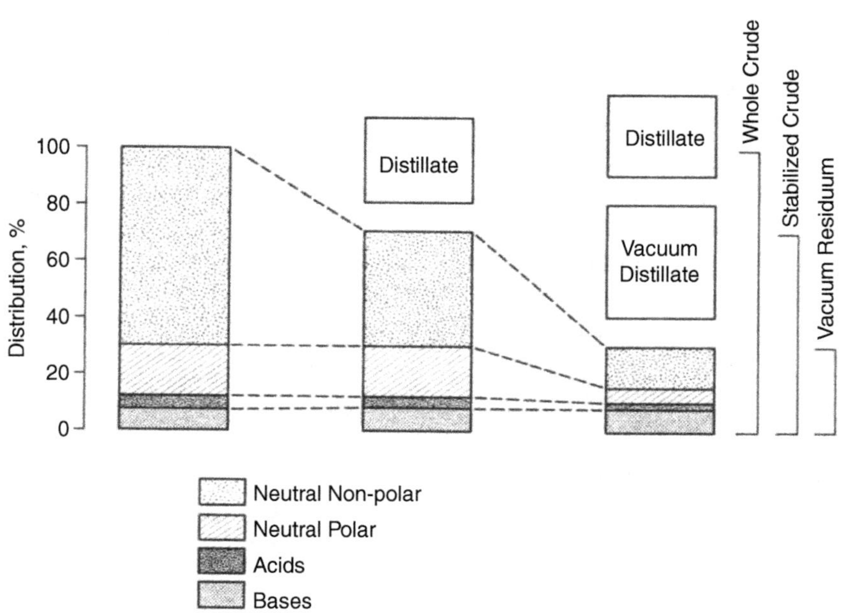

Figure 2.3: Accumulation of polar functions in residua during distillation.

unique properties. At this stage, the asphalt may be frequently, but incorrectly, referred to as pitch and has a softening point related to the amount of volatile material removed: the more volatile material distilled from the resid, the higher the softening point.

However, since there are wide variations in refinery operations and crude oil, asphalt with softening points ranging from 25°C to 55°C (80–130°F) may be produced. Blending with higher- and lower-softening point asphalt may produce asphalt of intermediate softening points. If lubricating oils are not required, the reduced crude may be distilled in a flash drum, which is similar to a bubble tower but has few, if any, trays. Asphalt descends to the base of the drum as the oil components pass out of the top of the drum. If the asphalt has a relatively low softening point, it can be hardened by further distillation with steam or by oxidation.

Resid — or a close proximity to the resid from the distillation unit — may also be produced by propane deasphalting (Parkash, 2003; Gary *et al.*, 2007; Speight, 2011a, 2014, 2017, 2019; Hsu and Robinson, 2017) and the asphalt so produced from the deasphalting unit may have a softening point of approximately 95°C (200°F). Softer grades are made by blending *hard asphalt* with the extract obtained by the solvent treatment of lubricating oils. If the asphalt requires a change in properties to a harder material, such as that required for roofing, pipe coating, or as an undersealant or water-proofing material, *soft asphalt* can be converted into harder asphalt by oxidation. This process produces a material that softens at a higher temperature than paving asphalts. It may be air blown at the refinery, at an asphalt processing plant, or at a roofing material plant. The asphalt is heated to approximately 260°C (500°F) and air is bubbled through the hot mix for 1–5 hours. The product, when cooled, remains in the liquid phase.

2.4.2.2 *Composition*

Resid is the residue of mixed-base and asphalt-base crude oils. It cannot be distilled even under the highest vacuum, because the temperatures required to do so promote formation of coke. Resid has complex chemical and physical compositions that usually vary with the source of the crude oil and is considered dispersion of particles (asphaltene constituents) in a high-boiling fluid composed of oil and resin constituents (Speight, 2014, 2015b).

2.4.2.3 *Properties and uses*

The properties of resid are determined by factors such as the properties of the feedstock (paraffinic or aromatic), as well as the nature and proportion of the asphaltene constituents and of the resin constituents (Pauli and Branthaver, 1998). The asphaltene constituents have been suggested to be lyophobic; the resin constituents are lyophilic, and the interaction of the resin constituents with the asphaltene constituents is responsible for asphaltene dispersion, which seems to exercise marked control on the nature of the resid (and the asphalt produced therefrom). The asphaltene constituents vary in character but are of sufficiently high molecular weight to require dispersion as micelles, which are peptized by the resin constituents. If the asphaltene constituents are relatively low in molecular weight, the resin constituents plentiful, and the medium aromatic in nature, the result may be viscous asphalt without anomalous properties. If, however, the medium is paraffinic and the resin constituents are scarce, and the asphaltene constituents are high in molecular (or micellar) weight (these conditions are encouraged by vacuum, steam reduction, or air blowing), the asphalt is of gel type and exhibits the properties that accompany such a structure. A high content of resin constituents imparts to a product desirable adhesive character and plasticity; high asphaltene content is usually responsible for the harder, more brittle, asphalt as evidenced from the structure and rheological properties of modified asphalt (Giavarini *et al.*, 2000).

2.4.3 *Coke*

Coke is the residue left by the destructive distillation of petroleum residua. That formed in catalytic cracking operations is usually non-recoverable, as it is often employed as fuel for the process.

2.4.3.1 *Composition*

The composition of crude oil coke varies with the source of the crude oil, but in general, large amounts of high-molecular-weight complex hydrocarbon derivatives (rich in carbon but correspondingly poor in hydrogen) make up a high proportion. The solubility of crude oil *coke* in carbon disulfide has been reported to be as high as 50–80% w/w, but this is in fact a misnomer, since coke is the insoluble, honeycomb material that is the end product of thermal processes.

2.4.3.2 *Properties and uses*

Three physical structures of coke can be produced by delayed coking: (i) shot coke, (ii) sponge coke, and (iii) needle coke.

Shot coke is an abnormal type of coke resembling small balls. Due to mechanisms not well understood, the coke from some coker feedstocks forms into small, tight, non-attached clusters that look like pellets, marbles, or ball bearings. It is usually a very hard coke, i.e. low-Hardgrove grindability index (Speight, 2013). Such a coke is less desirable to the end users because of the difficulties in handling and grinding. It is believed that feedstocks high in asphaltene constituents and low in API favor shot coke formation. Blending aromatic materials with the feedstock and/or increasing the recycle ratio reduces the yield of shot coke. Fluidization in the coke drums may cause formation of shot coke.

Occasionally, the smaller *shot coke* may agglomerate into ostrich egg-sized pieces. While shot coke may look like it is entirely made up of shot, most shot cokes are not 100% pure shot coke. Interestingly, even sponge coke may have some measure of embedded shot coke. The test for such a measure is not precise, but a low-shot-coke percentage is sometimes specified for anode grade of crude oil coke. In the case of fuel coke production, there is a general belief that shot structure of coke means that the coker is being run under the most economical conditions.

Sponge coke is the common type of coke produced by delayed coking units. It is in a form that resembles a sponge and has also been called honeycomb coke. Sponge coke, mostly used for anode grade, is dull and black, having a porous, amorphous structure.

Needle coke (*acicular coke*) is a special-quality coke produced from aromatic feed stocks, is silver gray, has a crystalline broken needle structure, and is believed to be chemically produced by cross-linking of condensed aromatic hydrocarbon derivatives during coking reactions. It has a crystalline structure with more unidirectional pores, is used in the production of electrodes for the steel and aluminum industries, and is particularly valuable because the electrodes must be replaced regularly.

Crude oil coke is employed for a number of purposes, but its chief use is in the manufacture of carbon electrodes for aluminum refining, which requires a high-purity carbon — low in ash and sulfur-free; the volatile matter must be removed by calcining. In addition to its use as a metallurgical reducing agent, crude oil coke is employed in the manufacture of carbon

Table 2.7: General characteristic of fuels and an ideal fuel.

Fuel type	General properties
Gaseous fuel	High heat content
	No ash or smoke
	Large storage tanks are required
Liquid fuel	High calorific value
	No dust ash and clinker
	Clean fuels
	Low furnace air
	Low furnace space
	Used in internal combustion engines
Solid fuel	Ash is high
	Low thermal efficiency
	Form clinker
	Low calorific value and require large excess air
	Cannot be used in internal combustion engines
Ideal fuel	High calorific value
	Moderate ignition temperature
	Low moisture content
	Low nitrogen-containing combustible matter
	Low sulfur-containing combustible matter
	Should burn in air with efficiency
	Combustion should be controllable
	No spontaneous combustion
	Products of combustion not harmful to the environment[a]
	Low cost
	Easy to transport
	Low storage cost

Note: [a]Includes humans and animals.

brushes, silicon carbide abrasives, and structural carbon (e.g. pipes and Rashig rings), as well as calcium carbide manufacture from which acetylene is produced:

$$\text{Coke} \rightarrow \text{CaC}_2$$

$$\text{CaC}_2 + \text{H}_2\text{O} \rightarrow \text{HC}{\equiv}\text{CH}$$

2.5 Conclusions

In summary, a fuel is any material that stores potential energy in a form, which, upon burning in the presence of oxygen, liberates heat energy. The calorific value of fuel is the total quantity of heat liberated when a unit mass or volume of fuel is completely burnt. A higher-calorific value (also known as the gross calorific value (HCV)) is the total amount of heat produced when a unit mass/volume of fuel has been burnt completely and the products of combustion have been cooled to room temperature (15°C or 60°F). The lower-calorific value (also known as net calorific value (LCV)) is the heat produced when unit mass (volume) of the fuel is burnt completely and the products are permitted to escape.

$$\text{LCV} = \text{HCV} - \text{Latent heat of water formed}$$

Natural or primary fuels, such as wood, peat, coal, natural gas, crude oil, heavy crude oil, extra heavy oil, tar sand bitumen, coal and oil shale, can be found in nature. Artificial or secondary fuels are prepared from primary fuels. Fuels are further classified as follows: (i) gaseous fuels, (ii) liquid fuels, and (iii) solid fuels; they can be further described using the general (or desirable) properties (Table 2.7).

References

Absi-Halabi, M., Stanislaus, A., and Qabazard, H. 1997. Reformulated gasoline. *Hydrocarbon Process* 76(2): 45.

Amanda, D., Mangesh, G.K., Lekha, C.M., and Ajay, K.D. 2007. Synthesis of biodiesel from canola oil using heterogeneous base catalyst. 84: 937–943.

API. 2004. Heavy fuel oils. Report No. 201-15368B. American Petroleum Institute, Washington DC. June 15.

ASTM. 2019. Annual Book of Standards, ASTM International, West Conshocken, Pennsylvania.

Boenheim, A.F. and Pearson, A.J. 1973. In *Modern Petroleum Technology*, G.D. Hobson and W. Pohl (Editors). Applied Science Publishers Inc., Barking, Essex, United Kingdom, Chapter 19.

Donato, A.G., Rafael, T.P., Neyda, C.O., André, L.D.R., and Octavio, A.C.A. 2008. Acid-catalyzed homogeneous esterification reaction for biodiesel production from palm fatty acids. *Catalysis Letters* 122: 20–25.

Gary, J.H., Handwerk, G.E., and Kaiser, M.J. 2007. *Petroleum Refining: Technology and Economics*, 5th Edition. CRC Press, Taylor & Francis Group, Boca Raton, Florida.

Giavarini, C., Mastrofini, D., and Scarsella, M. 2000. Macrostructure and rheological properties of chemically modified residues and bitumens. *Energy & Fuels* 14: 495–502.

Gibbs, L.M. 1989. Gasoline. *Oil Gas J.* 87(17): 60.

Hadley, D.J. and Turner, L. 1973. In *Modern Petroleum Technology*, G.D. Hobson and W. Pohl (Editors). Applied Science Publishers Inc., Barking, Essex, United Kingdom, Chapter 12.

Hoffman, H.L. 1992. In *Petroleum Processing Handbook*, J.J. McKetta (Editor). Marcel Dekker Inc., New York, p. 2.

Hossein, N., Santosh, R.P.B., and Emily, A.G. 2009. A novel method for the production of biodiesel from the whole stillage-extracted corn oil. *J. Am. Oil Chem. Soc.* 86: 83–91.

Hsu, C.S. and Robinson, P.R. (Editors) 2017. *Handbook of Petroleum Technology*. Springer International Publishing AG, Chaim, Switzerland.

Kumar, S., Nautiyal, S.P., and Agrawal, K.M. 2007. Physical properties of petroleum waxes 1: Effect of oil content. *Petroleum Sci. Technol.* 25: 1531–1537.

Lappas, A.A., Iatridis, D.K., and Vasalos, I.A. 1999. Production of reformulated gasoline in the FCC unit: Effect of feedstock type on gasoline composition. *Catalysis Today* 50: 73–85.

Lu, Y., Zhai, Y., Liu, M., and Wu, Q. 2009. Biodiesel production from algal oil using cassava (Manihot esculenta Crantz) as feedstock. *J. Appl. Phycol.* 22: 573–578.

Mills, G.A. and Ecklund, E.E. 1987. Alcohols as components of transportation fuels. *Ann. Rev. Energy* 12: 47–80.

Nehal, S.A. 2008. Lubricants additives from maleate copolymers. *Petroleum Sci. Technol.* 26(3): 298–306.

Nielsen, K.E., Dittmer, J., Malmendal, A., and Nielsen, N.C. 2008. Quantitative analysis of constituents in heavy fuel oil by ^1H nuclear magnetic resonance (NMR) spectroscopy and multivariate data analysis. *Energy & Fuels* 22: 4070–4076.

Owen, K. 1973. In *Modern Petroleum Technology*, G.D. Hobson and W. Pohl (Editors). Applied Science Publishers Inc., Barking, Essex, United Kingdom, Chapter 15.

Pandey, S.C., Ralli, D.K., Saxena, A.K., and Alamkhan, W.K. 2004. Physicochemical characterization and application of naphtha. *J. Sci. Industr. Res.* 63(3): 276–282.

Parkash, S. 2003. *Refining Processes Handbook*. Gulf Professional Publishing, Elsevier, Amsterdam, Netherlands.

Pauli, A.T. and Branthaver, J.F. 1998. Relationship between asphaltenes, heithaus compatibility parameters, and asphalt viscosity. *Petroleum Sci. Technol.* 16(9&10): 1125–1147.

Rafael, C.R., Giandra, V., Keiko, W., and Marco, A.Z.A. 2008. Enzymatic synthesis of biodiesel from transesterification reactions of vegetable oils and short chain alcohols. *J. Am. Oil Chem. Soc.* 85: 925–930.

Ramon, P.R., Roger, S., and Sebastian, V. 2009. Thermal and kinetic evaluation of biodiesel derived from soybean oil and higuereta. *J. Therm. Anal. Calorim.* 96: 897–901.

Speight, J.G. 2008. *Synthetic Fuels Handbook: Properties, Processes, and Performance*. McGraw-Hill, New York.

Speight, J.G. 2011a. *The Refinery of the Future*. Gulf Professional Publishing, Elsevier, Oxford, United Kingdom.

Speight, J.G. (Editor). 2011b. *The Biofuels Handbook*. Royal Society of Chemistry, London, United Kingdom.

Speight, J.G. 2013. *The Chemistry and Technology of Coal*, 3rd Edition. CRC Press, Taylor & Francis Group, Boca Raton, Florida.

Speight, J.G. 2014. *The Chemistry and Technology of Petroleum*, 5th Edition. CRC Press, Taylor & Francis Group, Boca Raton, Florida.

Speight, J.G. 2015a. *Handbook of Petroleum Product Analysis*, 2nd Edition. John Wiley & Sons Inc., Hoboken, New Jersey.

Speight, J.G. 2015b. *Asphalt Materials Science and Technology*. Butterworth-Heinemann, Elsevier, Oxford, United Kingdom.

Speight, J.G. 2017. *Handbook of Petroleum Refining Processes*. CRC Press, Taylor & Francis Group, Boca Raton, Florida.

Speight, J.G. 2019. *Handbook of Petrochemical Processes*. CRC Press, Taylor & Francis Group, Boca Raton, Florida.

Walmsley, A.G. 1973. In *Modern Petroleum Technology*, G.D. Hobson and W. Pohl (Editors). Applied Science Publishers Inc., Barking, Essex, United Kingdom, Chapter 17.

Watanabe, K., Nagai, K., Aratani, N., Saka, Y., Chiyoda, N., and Mizutani, H. 2010. Techniques for octane number enhancement in FCC gasoline.

Proceedings of the 20th Annual Saudi-Japan Symposium on Catalysts in Petroleum Refining & Petrochemicals. Dhahran, Saudi Arabia, December.

Zhang, J., Wu, C., Li, W., Wang, Y., and Han, Z. 2003. Study on the performance mechanism of pour point depressants with differential scanning calorimeter and X-Ray diffraction methods. *Fuel* 82(11): 1419–1426.

Zhang, J., Wu, C., Li, W., Wang, Y., and Cao, H. 2004. DFT and MM calculation: The performance mechanism of pour point depressants. *Fuel* 83(3): 315–326.

Chapter 3

Alternate Fuels

3.1 Introduction

Biofuels is a generic term for those gaseous, liquid, or solid fuels that are not derived from fossil fuels or that contain a proportion of non-fossil fuel. For the purposes of this chapter, the term *alternate fuels* is used for fuels produced from plant sources and other sources such as the organic constituents (predominantly biological in nature) of municipal and industrial waste. Thus, biofuels are *biomaterials* produced from renewable biological sources which are now grown specifically for the purpose of providing useful heat upon combustion. Biofuels are produced from many sources, such as corn, soybeans, flaxseed, rapeseed, sugarcane, palm oil, sugar beet, raw sewage, food scraps, animal parts, and rice.

Biofuels are fuels derived from plant materials, they are entering the market, driven by factors such as oil price spikes and the need for increased energy security. Examples of solid biofuels include wood, sawdust, grass cuttings, domestic refuse, charcoal, agricultural waste, non-food energy crops, and dried manure.

Biofuels are also known as non-conventional fuels or alternative fuels. Alternative fuels can be classified as those fuels that are not derived from conventional sources, such as natural gas, crude oil, and coal.

Apart from biofuels, other examples of alternative fuels are solar, wind, and tidal power, hydrogen, air engine power and non-conventional oil. Even since mankind discovered fire, wood has been the first form of biofuel used for heating and cooking and has also been used to produce electricity. Liquid biofuels have been used in the automotive industry since its inception. Unlike fossil fuels, which are necessarily derived from long deceased and metamorphosed biological organisms, biofuels (also known as agrofuels) are obtained from either just recently deceased or from living biological organisms; in other words, biofuels are derived from biomass or biowaste.

When raw biomass is already in a suitable form (such as firewood), it can be combusted directly in a stove or a furnace to provide heat or (industrially) to raise steam. When raw biomass is in an inconvenient form (such as sawdust, wood chips, grass, urban waste wood, and agricultural residues), the typical procedure is to densify the biomass. This procedure includes grinding the raw biomass to an appropriate particulate size, ranging from 0.5 to 1.5 inches depending on the densification type, which is then followed by concentrating into a fuel product. The types of densification processes currently used are pellet, cube, and puck. The pellet process is most common in Europe and produces typically a pure wood product. The other types of densification yield larger size products compared to a pellet and are compatible with a wide range of input feedstocks. The resulting densified fuel is easier to transport and to feed into thermal generation systems such as boilers.

Bioprocessing routes have a number of compelling advantages over conventional petrochemical production; however, it is only in the last decade that a rapid progress in biotechnology has facilitated the commercialization of a number of plant-based chemical processes. It is widely recognized that more significant production of plant-based chemicals is economically viable only in highly integrated and efficient production complexes producing a diverse range of chemical products. Also, in contrast to fossil fuels, biofuels do not contribute to the stock of total carbon dioxide in the atmosphere, since the source of many biofuels is plants, which normally absorb carbon dioxide from the atmosphere, then give off the same amount when burned and are therefore considered to be carbon dioxide-neutral, although some observers would doubt that this balance is the result of a mathematical manipulation not taking into consideration other important and influential factors.

In any given location, the type of biofuel used will depend, among other minor factors, on the available natural resources and the local energy needed for processing requirements. Some of the main reasons for the shift of interest from fossil fuels to biofuels are the rising price of oil, the increasing emission of greenhouse gases, fewer barriers to the entry of biofuels, and the increasing government support. There is also a need for the utilization of biofuels for the reduction or total elimination of noxious emissions in an economical manner.

Biofuels can be classified as follows: (i) first-generation biofuels, (ii) second-generation biofuels, and (iii) third-generation biofuels, also known as advanced biofuels, and the composition varies according to the feedstock and conversion techniques used. First-generation biofuels are derived from simple materials, such as mono and disaccharides (simple sugars); amylose and amylopectin (starch); esters of glycerol and fatty acids, free fatty acids, and mono and diglycerides (fats and vegetable oils for biodiesel); methanol, ethanol, propanol, and butanol (bioalcohols); methyl tertiary-butyl ether (MTBE) and ethyl tertiary-butyl ether (ETBE), also known as bioether derivatives; methane, carbon dioxide, nitrogen, hydrogen, hydrogen sulfide, oxygen, carbon monoxide, and hydrogen (synthesis gas, often referred to as syngas); and cellulose, hemicellulose derivatives, and lignin (wood for solid biofuels). Second-generation biofuels are typically produced using the biomass-to-liquid (BTL) technology and these fuels include ethanol from lignocellulose, biohydrogen and wood diesel. The third-generation type of biofuel is a biofuel from algae, also known as oilgae. *Botryococcus braunii* and *Chlorella vulgaris*, as well as macroalgae (seaweed), are typically used to produce such biofuels.

In fact, the free fatty acid from non-edible oils (Table 3.1) can be mixed with crude oil-derived diesel fuel (up to 30% w/w of the non-edible oil with diesel may be used without having too much of an adverse effect on the properties and efficiency of the diesel fuel). However, it must be remembered that olefins in fuel (such as gasoline or diesel fuel) can often result in the formation of gums that have an adverse effect on fuel flow and efficiency (Speight, 2014, 2015, 2017). This tendency can be mitigated if the fatty acids are modified to produce more suitable fuels (i.e. fuels that do not form gum products) that can act as diesel substitutes. One method involves cracking at high temperatures on certain metal oxide catalysts (decarboxylation), which yields hydrocarbon products that can be hydrogenated to the saturated hydrocarbon derivative (*n*-heptadecane, $C_{17}H_{36}$). Using linoleic acid as an example, the pyrolysis product is heptadecadiene, which hydrogenates to heptadecane:

$$C_{18}H_{32}O_2 \rightarrow C_{17}H_{32} + CO_2 \qquad \text{(pyrolysis/decarboxylation)}$$

$$C_{17}H_{32} \rightarrow C_{17}H_{36} \qquad \text{(hydrogenation)}$$

Table 3.1: Fatty acid content of selected non-edible oils.

Fatty acid (w/w%)	No. of carbons[a]	Neem oil	Pongamia oil	Jatropha oil
Myrisitic	C14:0	0.2–2.6		0.5–1.4
Palmitic	C16:0	13.6–16.2	3.7–7.9	12.0–17.0
Stearic	C18:0	14.4–24	1 2.4–8.0	5.0–9.7
Oleic	Cl8:1	49.1–61.9	44.5–71.3	37.0–63.0
Linoleic	C18:2	2.3–15.8	10.8–18.3	19.0–41.0
Arachidic	C20:0	0.8–3.4	2.2–4.7	0.0–0.3
Eicosenoic	C20:1		9.5–12.4	
Behenic	C22:0		4.2–5.3	
Lignoceric	C24:0		1.1–3.5	

Notes: [a]The first number is the total number of carbon atoms in the molecule (including the carboxylic carbon atom) and the second number (0, 1, or 2) indicates the degree of unsaturation in terms of the number of double bonds. For example, linoleic acid ($C_{18}H_{32}O_2$) has 18 carbon atoms with two olefinic bonds on the C9 and C12 atoms, hence C18:2. Thus,

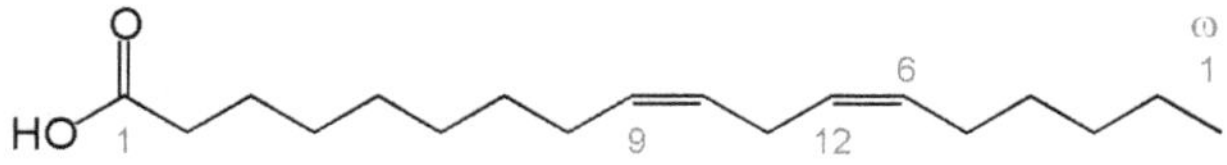

More generally, many food sources and plant crops are the main feedstocks for biofuels, and as such, one must take into consideration the deleterious effects of biofuel production on the natural resources of any particular region. These feedstocks may enter both animal and human food chains, and since biofuel production has increased, biofuel has been criticized for diverting food away from human consumption, which could lead to food shortage and increase in food prices. While the demand for biofuels is directly correlated with the price of oil, other issues also arise such as deforestation, soil erosion, water usage, carbon emissions, and biofuel prices, which complicate the food vs fuel debate with regard to energy balance and efficiency. Henceforth, when discussing the benefits of biofuels, the pitfalls must be made apparent, so that sustainable biofuel production can be realized.

This chapter provides an overview of different types of biomass and processes by which fuels can be produced from biomass. In addition, the variations in biofuel properties with the biomass feedstock are also presented.

3.2 Composition and Properties

Biomass is a term used to describe any material of recent biological origin, including plant materials, such as trees, grasses, agricultural crops, and even animal manure (Chapter 1). Other biomass components, which are generally present in minor amounts, include the following; (i) triglyceride derivatives, which are esters derived from glycerol and three fatty acids, (ii) sterol derivatives, also known as steroid alcohol derivatives, (iii) alkaloid derivatives, which are a class of naturally occurring organic compounds that mostly contain basic nitrogen atoms, (iv) terpene derivatives, which are the primary constituents of essential oils of many types of plants and flowers, (v) terpenoid derivatives (sometimes referred to as isoprenoids), which are a large and diverse class of naturally occurring organic chemicals derived from terpenes, most of which are multicyclic structures with oxygen-containing functional groups, and (vi) wax derivatives, which are a diverse class of organic compounds that are lipophilic, malleable solids near ambient temperature and include higher molecular weight alkane derivatives and lipids, typically with melting points above approximately 40°C (104°F) and melt to give low-viscosity liquids. Waxes are insoluble in water but soluble in organic, non-polar solvents. Natural waxes of different types are produced by plants and animals, and occur in crude oil.

The above list includes everything from the primary sources of crops and residues harvested/collected directly from the land, to secondary sources such as sawmill residuals, to tertiary sources of post-consumer residuals that often end up in landfills. A fourth source, although not usually categorized as such, includes the gases that result from anaerobic digestion of animal manures or organic materials in landfills (Wright *et al.*, 2006).

Thus, a knowledge of the composition of biomass feedstocks is critical to the selection of varieties with optimized properties for downstream conversion (De Jong, 2015). This can be partly achieved by selecting varieties with biomass compositions that are better suited to the conversion process. Lignocellulosic biomass displays considerable recalcitrance to biochemical conversion because of the inaccessibility of its polymer components to enzymatic digestion and the release or production of fermentation inhibitors during pretreatment. If the proportion of hemicellulose, cellulose, and lignin in a woody biomass feedstock is optimized for a specific biochemical conversion method, then the pretreatment methods could be reduced or avoided.

Plants use the light energy from the sun to convert water and carbon dioxide to sugar derivatives (photosynthesis) that can be stored within the plant system. Some plants, such as sugarcane and sugar beets, store energy as simple sugars, while other plants store energy as more complex starch derivatives and such plants include grains like corn, which is a prominent food source. Another type of plant matter, cellulosic biomass, is composed of complex sugar polymers and is not generally used as a food source. Cellulosic feedstocks under consideration for biofuels include the following: (i) agricultural residues, which are the leftover material from crops, such as the stalks, leaves, and husks of corn plants, (ii) forestry wastes, which include chips and sawdust from lumber mills, dead trees, and tree branches, (iii) municipal solid wastes, which include household garbage and paper products, (iv) food processing and other industrial wastes such as black liquor, which is a paper manufacturing by-product, and (v) energy crops, such as fast-growing trees and grasses, developed just for this purpose.

Composition of black liquor (Kavalov and Peteves, 2005)

Element	% w/w
Carbon	35.7
Hydrogen	3.7
Nitrogen	≥0.1
Oxygen	35.8
Sulfur	4.4
Chlorine	0.3
Potassium	1.1
Sodium	19.0

The main components of these types of biomass are as follows: (i) cellulose, which is the most common form of carbon in biomass, accounting for 40–60% w/w of the biomass, depending on the biomass source, and is a complex sugar polymer, or polysaccharide, made from a six-carbon sugar, glucose: The crystalline structure of cellulose renders it resistant to hydrolysis, a chemical reaction that releases simple, fermentable sugars from a polysaccharide; (ii) hemicellulose, which is also a major source of carbon in biomass, at levels between 20% and 40% w/w of the biomass, and is a complex polysaccharide made from a variety of five-carbon and six-carbon sugar derivatives: Hemicellulose is relatively easy to hydrolyze into simple sugars, but the sugars can be difficult to ferment; and (iii) lignin, which is a complex polymer making up 10% to 24% w/w of the biomass and provides structural integrity in plants: Lignin remains as the residual material after the sugars in the biomass have been converted.

In the simplest sense, biomass is carbon-based and is composed of a mixture of organic molecules containing hydrogen, usually including atoms of oxygen, often nitrogen, and also small quantities of other atoms, including alkali, alkaline earth metals and heavy metals. However, because of the wide variations in the character and properties of biomass, it is anticipated and realized that the character and properties of the biofuels produced form biomass are very much dependent on the initial biomass. The exception is the fuels produced by gasification and Fischer–Tropsch synthesis.

In addition to the chemical composition, three properties of biomass properties that are significant to the performance of biomass as a fuel are as follows: (i) the mineral matter content, manifested in thermal processes as mineral ash, (ii) susceptibility of the mineral matter to slagging and fouling, and (iii) the volatile matter content. The mineral matter content is the mass fraction of biomass composed of non-combustible inorganic material. Grasses, bark, and field crop residues typically have a much higher content of mineral matter than wood. Systems that are designed to combust wood can be overwhelmed by the volume of ash if other biofuels are used. Slagging and fouling are problems that occur if ash begins to melt during combustion, forming deposits on combustor surfaces (fouling) or leaving hard chunks of glassy material in the bottom of the combustion chamber (slag, often referred to as clinkers). Certain mineral components in biomass fuels, primarily silica, potassium, and chlorine, can cause these problems to occur at lower temperatures than normal. Dirt contamination also adds to the mineral content and associated slagging and fouling problems, so it is important that biomass feedstock be as clean (dirt-free) as possible. Slagging and fouling are minimized by keeping combustion temperatures low. Alternatively, some biomass combustion equipment is designed to encourage the formation of clinkers (often referred to in the singular form, clinker) but is able to dispose of the hardened ash in an effective manner.

The content of volatile constituents (or volatile matter) in a fuel is a lesser-known property that refers to the fraction of the fuel that will readily volatilize (turn into gas) when heated at a high temperature. Fuels with high volatile content will tend to vaporize before combusting whereas fuels with low volatiles will burn primarily as char. This affects the performance of the combustion chamber and should be taken into account when designing a combustor.

Other properties such as the particle size and density of biomass fuels are also important as they affect the thermal processing characteristics (especially the combustion characteristics) of biomass, especially the rate of heating and drying during the thermal process. The feedstock particle size also dictates the type of handling equipment required. For example, an incorrect fuel particle size will negatively impact the process efficiency and may cause jamming or damage to the handling equipment. Smaller-sized fuel is more common for commercial-scale systems because smaller-sized fuel is easier to use in automatic feed systems and allows for finer control of the processing rate by controlling the rate at which fuel is added to the reaction chamber. Fuel particle size and density are probably the most overlooked factors affecting fuel performance and should be given careful consideration when selecting a fuel type. Bulk density is the mass of a material divided by the volume it occupies. The bulk density of granular materials is dependent on the manner in which it is handled insofar as a freely settled material has a lower bulk density than tapped or compacted materials.

3.3 Classification of Biofuels

Put simply, a biofuel is a fuel produced using contemporary processes from a biological source (biomass) rather than that produced using the geological processes involved in the formation of fossil fuels. In addition, the word *biofuel* is usually reserved for liquid or gaseous fuels used for transportation. Biofuels can be produced from plants (such as energy crops) or from agricultural, commercial, domestic, and/or industrial wastes (if the waste has a biological origin).

Renewable biofuel sources generally involve carbon fixation, such as those that occur in plants or in micro-algae through the process of photosynthesis. Furthermore, some observers argue that biofuel can be carbon-neutral because all biomass crops sequester carbon to a certain extent and because all crops move carbon dioxide from above-ground circulation to below-ground storage in the roots and surrounding soil. However, the simple proposal that biofuel is carbon-neutral requires that the total carbon sequestered by the root system of the energy crop must compensate for all the above-ground emissions, which must include any emissions caused by direct or indirect land-use change. Many first-generation biofuel projects are not carbon-neutral, given these demands — in fact, some may even have higher total emission of greenhouse gases than some fossil-based alternatives.

Generally, the classification of biofuels is related to the method of preparation, which, in turn, is related to the starting material of the agent by which the biomass is converted into biofuels. For convenience, biofuels are generally classified as follows: (i) first-generation biofuels, (ii) second-generation biofuels, and (iii) third-generation biofuels.

3.3.1 *First-generation biofuels*

First-generation biofuels are those fuels directly prepared from biomass, which is often a direct part of the food chain. Such fuels, upon purification, can be used as fuel for various types of equipment. Examples of such fuels are alcohol fuels, such as methanol and ethanol, which are prepared from sugar, starches, and cellulose and which have properties that make the alcohols suitable for use as alternate fuels (Table 3.2). However, alcohols such as dry methanol and ethanol, can be corrosive under certain conditions and this must be given due consideration when either of these chemicals is used as fuel.

Before dealing with the character and properties of such fuels, an introduction to the structure and chemistry of the sugars is warranted.

3.3.1.1 *Sugars and starch*

Biomass is typically composed of 75–90% w/w sugar derivatives, and the other 10–25% w/w is predominantly lignin. Plants capture solar energy as fixed carbon, during which carbon dioxide is converted to water and sugars $(CH_2O)_x$:

$$CO_2 + H_2O \rightarrow (CH_2O)_x + O_2$$

The sugars produced are stored in three types of polymeric macromolecules: (i) starch, (ii) cellulose, and (iii) hemicellulose.

Table 3.2: Properties of methanol and ethanol compared to *iso*-octane.

Item	Methanol	Ethanol	*Iso*-octane
1. Formula	CH_3OH	C_2H_5OH	C_8H_{18}
2. Molecular weight	32.042	46.07	114.224
3. Carbon/hydrogen (atomic ratio)	3.0	4.0	5.25
4. Carbon (w/w%)	37.5	52.17	84.0
5. Hydrogen (w/w%)	12.5	13.4	16.0
6. Oxygen (w/w%)	50.0	34.78	0
7. Boiling point at 1 atm (°C)	64.5	78.40	99.239
8. Freezing point at 1 atm (°C)	–97.778	–80.00	–107.378
9. Density at 15.5 (°C lb/gal)	6.637	6.63	5.795
10. Viscosity at 20°C/1 atm, Centipoise	0.596	1.20	0.503
11. Specific heat at 25°C/1 atm (Btu/lb)	0.6	0.6	0.5
12. Heat of vaporization at boiling point/1 atm, (Btu/lb)	473.0	361.0	116.69
13. Heat of vaporization at 25°C/1 atm, (Btu/lb)	503.3		132.0
14. Heat of combustion at 25°C, (Btu/lb)			
(a) Higher heating value	9776	20555	12780
(b) Lower heating value	8593	19065	11550
15. Research octane number	106	105	100
16. Flash point temp. (°C)	11.112	12.778	–42.778
17. Auto-ignition temp. (°C)	463.889	422.778	257.23
18. Flammability limits			
(a) Lower	6.7	1.4	4.3
(b) Higher	36.0	7.6	19.0
19. Cetane number	5	8	—

Monosaccharides (the simple sugars) (Figure 3.1) are colorless, water-soluble crystalline solids. Because of the presence of numerous chiral centers, various isomeric forms of the different monosaccharides exist. However, since biofuels are necessarily obtained from natural sources, the forms which are abundant in nature predominate.

In general, sugar polymers such as cellulose and starch (Figure 3.2) can be readily broken down into their constituent monomers by hydrolysis, preparatory to conversion to ethanol or other chemicals. In contrast,

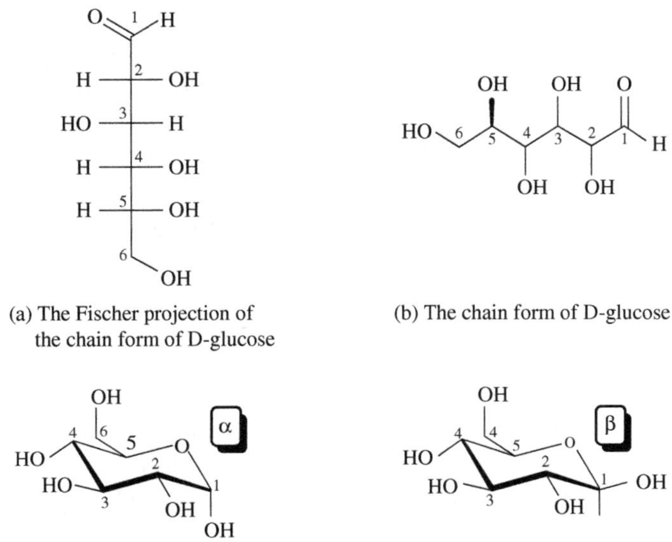

(a) The Fischer projection of the chain form of D-glucose

(b) The chain form of D-glucose

(c) α-D-glucopyranose

(d) β-D-glucopyranose

Figure 3.1: Various representations of the naturally occurring form of the simple monosaccharide, glucose.

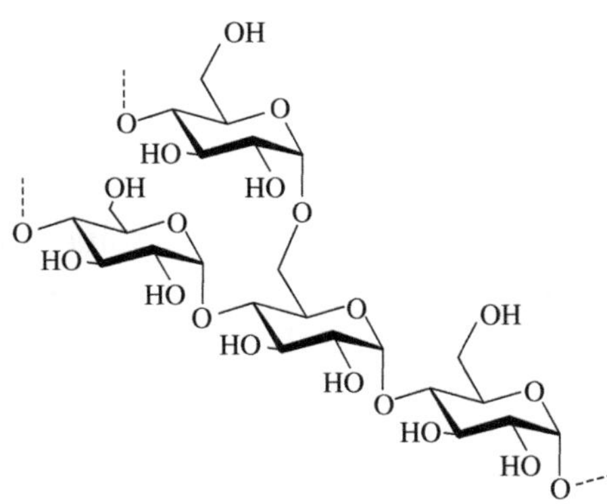

Figure 3.2: Portion of the polysaccharide starch used as a feedstock for the production of biofuel.

lignin is a complex structure containing aromatic groups and is less readily degraded. Although lignocellulose is one of the cheapest and most abundant forms of biomass, it is difficult to convert this relatively unreactive material into sugars. Among other factors, the walls of lignocellulose are composed of lignin, which must be broken down in order to render the cellulose and hemicellulose accessible to acid hydrolysis. It is for this reason that many programs focusing on ethanol production from biomass are based almost entirely on the fermentation of sugars derived from the starch present in corn grain.

Glucose (dextrose), fructose (laevulose), galactose, xylose, and ribose are common. Glucose is a simple sugar that can be readily fermented into biofuel if it can be accessed; the equation for its combustion is given as follows:

$$C_6H_6O_6(s) + 6O_2(g) \rightarrow 6CO_2(g) + 6H_2O(g)\ \Delta H = 2810\ \text{kJ mol}^{-1}$$

A large amount of heat energy is evolved in the complete combustion of one mole of pure glucose, and this exothermic reaction is the main source of metabolic energy (ATP) in cells. Glucose and other simple sugars are normally fermented into alcohols for direct use as fuel. Starch, also known as amylum, is the polysaccharide present in all green plants and is the form in which food is stored. Glucose monomers are joined by β1–4 glycosidic bonds in the linear helical portions of the polymer (amylose) and by β1–5 glycosidic bonds in the branched portions (amylopectin). Plant sugars or starches of biofuel crops such as sugarcane and maize are fermented to produce ethanol.

3.3.1.2 *Cellulose*

Cellulose is the main material, that is used as a solid biofuel, the biomass being obtained from corn, switchgrass, and sawdust. Cellulosic ethanol is obtained through fermentation of the monomers of the polysaccharide, as with starch, but a greater amount of processing is required to liberate the simpler sugars. Cellulose is the most abundant, naturally occurring organic polymer on the Earth and is a linear chain polymer made up of thousands of glucose residues linked through β1–4 glycosidic linkages. Its availability from switchgrass and miscanthus has encouraged biofuel production.

Chemically, cellulose is an organic compound with the formula $(C_6H_{10}O_5)_n$ and is a polysaccharide consisting of a linear chain of several hundreds to many thousands of β(1–4)-linked D-glucose units. It is an important structural component of the primary cell wall of green plants, many forms of algae, and oomycetes (fungus-like, filamentous, microscopic, absorptive organisms that reproduce both sexually and asexually). The cellulose content of cotton fiber is 90% w/w, that of wood is 40–50% w/w, and that of dried hemp (a strain of the *Cannabis sativa* plant species grown specifically for the industrial uses of its derived products; it is one of the fastest growing plants and is one of the first plants to be spun into usable fiber millennia ago) is of the order of 55–60% w/w.

3.3.1.3 *Lignin*

Lignin is a complex chemical compound that is most commonly derived from wood and that is an integral part of the cell wall of plants. The chemical structure of lignin is unknown and, at best, can be represented only by hypothetical formulas. Lignin makes up approximately one-quarter to one-third of the dry mass of wood and is generally considered to be a large, cross-linked hydrophobic moieties, aromatic macromolecule with a molecular mass that is estimated to be in excess of 10,000. Degradation studies indicate that the molecule consists of various types of substructures that appear to repeat in a random manner. Lignin fills the spaces in the cell wall between cellulose, hemicellulose, and pectin components and is covalently linked to hemicellulose. Lignin also forms covalent bonds to polysaccharides and thereby cross-links different plant polysaccharides.

3.3.2 *Second-generation biofuels*

Second-generation biofuels are produced from those plants that are not necessarily part of the food chain. Hence, their production relies heavily on biomass materials, such as waste from urban, forestry and agricultural sources as well as the non-reusable portion of municipal solid waste, which is generally landfilled. Since the main component of these biomass materials is cellulose, they are also known as *cellulosic fuels*. Residual forest biomass, such as bark, needles, limbs, etc., as well as the non-sugar portions of plants that are left in the fields after harvesting and processing, such as bagasse, rice hulls, and wheat straw, are utilized.

Apart from cellulosic ethanol, which is the principal second-generation biofuel, biomethanol, dimethyl ether, biohydrogen, synthetic diesel, and synthetic gasoline also emerge from the aforementioned biomass materials. These are the biomass-to-liquid projects which seek to further diversify the product range that can be derived from lignocellulose using advanced technologies, such as the Fischer–Tropsch process.

The drive toward second-generation biofuels arose due to earlier studies, which suggest that first-generation biofuel production may actually produce more carbon dioxide emission than they save, since in many instances, natural habitats are converted into cropland, thus releasing the carbon contained in trees and grasses and in the soil in which they grow. Next-generation fuel producers have adopted two main approaches: an enzymatic platform for homogeneous biomass or a thermochemical platform for a variety of feedstocks, including non-homogeneous residues.

The reduction of landfills, which are ultimately converted to methane (an important greenhouse gas), and reduction of crude oil dependence are among the benefits of second-generation biofuels. Ongoing research involves increasing the supply from available acreages and alteration of the composition of feedstock for reduction of processing cost and increasing the value of co-products.

3.3.3 *Third-generation biofuels*

Third-generation biofuels (sometimes called "oilgae") are derived from algae, which are either unicellular or multicellular autotrophic organisms that are able to photosynthesize, but do not contain typical plant organs. During photosynthesis, algae are able to convert carbon dioxide and the energy from sunlight into biomass and oxygen. The veg-oil algae produce can be harvested and converted into biodiesel, while the carbohydrate content of algae can be fermented to produce alcohol derivatives. Some studies have shown that algae can produce up to 60% of their biomass in the form of oil, which can be converted into biodiesel for use in automobiles; hence, the energy utilized for transesterification in the production of oil from alcohol and catalyst is eliminated. Algae can be grown on saline, arid, or drought-stricken land and has a harvesting cycle of 1–10 days. This minimizes the use of land for the production of food crops. Algae are also completely biodegradable and can be produced in oceans or fresh water — if there is a spill, this can be easily dealt with. A third-generation biofuel via direct cellulose fermentation involves a process called consolidated bioprocessing (CPB), in which cellulase production, substrate hydrolysis, and fermentation are accomplished in a single step process by microorganisms.

3.4 Types of Biofuels

As explained, in the simplest sense, a *biofuel* is any fuel that is derived from biomass, i.e. recently living organisms or their metabolic by-products. A *biofuel* has also been defined as any fuel with 80% minimum content (% v/v) of materials derived from living organisms harvested within 10 years preceding its manufacture. Biofuels are primarily used in fuel vehicles, but they can also be used in fuel engines or fuel cells for electricity generation.

The production of biofuels to replace oil and natural gas is in active development, focusing on the use of cheap organic matter (usually cellulose and agricultural and sewage waste) for the efficient production of liquid and gas biofuels that yield a high net energy gain. The carbon in biofuels was recently extracted from atmospheric carbon dioxide by the growing plants, so burning the carbon does not result in a net increase of the amount of carbon dioxide in the atmosphere of the Earth. As a result, biofuels are considered by many observers as a means to reduce the amount of carbon dioxide released into the atmosphere by using them to replace non-renewable sources of energy.

3.4.1 *Gaseous fuels*

Gaseous fuels are those fuels that are gaseous under ordinary conditions and, in the context of this book, are those produced from non-fossil fuel sources categorized as biomass.

Most biomass materials are easier to gasify than coal because they are more reactive with a higher ignition stability. This characteristic also makes it easier to process them thermochemically into higher value fuels such as methanol or hydrogen. The ash content is typically lower than for most coals, and the sulfur content is much lower than for many fossil fuels. Unlike coal ash, which may contain toxic metals and other trace contaminants, biomass ash may be used as a soil amendment to help replenish nutrients removed by harvest. A few biomass feedstocks stand out for their peculiar properties, such as high silicon or alkali metal contents — these may require special precautions for harvesting, processing, and combustion equipment. Note also that mineral content can vary as a function of the soil type and the timing of the feedstock harvest. In contrast to their fairly uniform physical properties, biomass fuels are rather heterogeneous with respect to their chemical elemental composition.

The gaseous products generated from the biomass feedstocks as defined in Chapter 5 are varied insofar as the gas composition varies with the system employed (Speight, 2013). It should be emphasized that the gas product must be first freed from any pollutants such as particulate matter and sulfur compounds before further use, particularly when the intended use is a water–gas shift or methanation.

A number of processes allow biomass to be transformed into gaseous fuels such as methane or hydrogen (Sørensen *et al.*, 2006). One pathway uses algae and bacteria that have been genetically modified to produce hydrogen directly instead of the conventional biological energy carriers. A second pathway uses plant materials such as agricultural residues in a fermentation process leading to biogas, from which the desired fuels can be isolated.

3.4.1.1 *Biogas*

Biogas, sometimes incorrectly known as swamp gas, typically refers to a biofuel gas produced by anaerobic digestion or fermentation of organic matter, including manure, sewage sludge, municipal solid waste, biodegradable waste, or any other biodegradable feedstock, under anaerobic conditions. Depending on where it is produced, biogas is also called swamp gas, marsh gas, landfill gas, and digester gas.

Biogas is produced by means of anaerobic digestion, which is a process whereby organic matter is broken down by microbiological activity and, as the name suggests, it is a process which takes place in the absence of air. It is a phenomenon that occurs naturally at the bottom of ponds and marshes and gives rise to marsh gas or methane, which is a combustible gas. There are two common man-made technologies for obtaining biogas: the first one (which is more widespread) is fermentation of human and/or animal waste in specially designed digesters. The second is a more recently developed technology for capturing methane from municipal waste landfill sites. The digestion of animal and human waste offers several benefits: (i) the production of methane for use as a fuel, (ii) the waste is reduced to slurry that has a high nutrient content, which makes an ideal fertilizer that, in some cases, is the main product from the digester while the biogas is merely a by-product, and (iii) during the digestion process, bacteria in the manure are killed, which is a great benefit to environmental health.

In the process, the waste is fed into the digester via the inlet pipe and undergoes digestion in the digestion chamber. The temperature of the process is quite critical, and methane-producing bacteria operate most efficiently at temperatures between 30°C and 40°C (86°F and 104°F) or 50°C and 60°C (122°F to 140°F) and in colder climates, heat may have to be added to the chamber to encourage the bacteria to carry out their function. The product is a combination of methane and carbon dioxide, typically in a ratio of the order of 6:4. Digestion time ranges from a couple of weeks to a couple of months depending on the feedstock and the digestion temperature. The residual slurry is removed at the outlet and can be used as a fertilizer.

The composition of biogas varies depending on the composition of the waste material in the landfill and the anaerobic digestion process. Landfill gas typically has methane concentrations of around 50%. Advanced waste treatment technologies can produce biogas with 55–75% v/v methane; often air is introduced (equivalent to 5% v/v) for microbiological desulfurization. For example, the constituents of biogas are generally methane (50–75% v/v), carbon dioxide (25–50% v/v), nitrogen (0–10% v/v), hydrogen (0–1% v/v), hydrogen sulfide (0–3% v/v), and oxygen (0–2% v/v).

If biogas is cleaned by any one of the variety of relevant processes (Chapter 9), the product gas has the same characteristics as natural gas. In this instance, the producer of the biogas can utilize the local gas distribution networks. The gas must be very clean to reach pipeline quality. Water (H_2O), hydrogen sulfide (H_2S), and particulates are removed if present at high levels or if the gas is to be completely cleaned. Carbon dioxide is less frequently removed, but it must also be separated to achieve pipeline quality gas. If the gas is to be used without extensive cleaning, it is sometimes co-fired with natural gas to improve combustion. Biogas cleaned up to pipeline quality is called *renewable natural gas.* In this form, the gas can be used in any application in which natural gas is used.

Biogas is almost identical to natural gas (methane, CH_4) (Table 3.3). The equation for the complete combustion of methane is given by:

$$CH_4\,(g) + 2O_2\,(g) \rightarrow CO_2\,(g) + 2H_2O \tag{1}$$

Natural gas, which is generally found associated with fossil fuels, coal, and gas hydrates (clathrates of methane in ice crystals), may also contain small amounts of ethane, propane, butane, and pentane, which are higher boiling hydrocarbon derivatives removed prior to use as a consumer fuel, as well as carbon dioxide, nitrogen, helium and hydrogen

Table 3.3: Composition of biogas from different sources.

Component	Agricultural biogas	Sewage gas	Landfill gas
Methane (v/v%)	55–75	55–65	40–45
Carbon dioxide (v/v%)	25–45	30–40	35–50
Nitrogen (v/v%)	0–10	0–10	0–20
Hydrogen sulfide (v/v%)	0–1.5	Up to 200 ppm	Approx. 200 ppm
Water (v/v%)	Saturated	Saturated	Saturated
Halogens (v/v%)	Trace amount	Up to 4 ppm	Amount dependent on the landfill
Higher hydrocarbons (v/v%)	Trace amount	Trace amount	Up to 200 ppm

sulfide. Biogas is produced by methanogenic organisms in landfills, swamps and marshes by the decomposition of organic (both animal and plant) matter. The anaerobic (low or no oxygen present in the system) decomposition produces methane, carbon dioxide, a small amount of hydrogen, and a small amount of heat. Acidity, the carbon-to-nitrogen ratio of the feedstock, temperature and percentage of solids all influence the quality and quantity of biogas produced. Normally, cow manure enriched with vegetable matter is the ideal feedstock. The other principal type of biogas is wood gas, which is created by gasification of wood or other biomass. This type of biogas is composed primarily of nitrogen, hydrogen, and carbon monoxide, with trace amounts of methane. Industrial anaerobic digesters and mechanical biological treatment systems are becoming increasingly popular. Methane is not only a potent greenhouse gas but also a very clean, easy burning fuel, and hence, harnessing it from natural sources such as landfills is beneficial to the natural environment.

Finally, high-temperature gasification supplies a crude gas, which may be transformed into hydrogen by a second reaction step. In addition to biogas, there is also the possibility of using the solid by-product as a biofuel. The technologies for gas production from biomass include the following: (i) fermentation, (ii) gasification, and (iii) direct biophotolysis.

3.4.1.2 *Low-Btu gas*

Low-Btu gas (low heat-content gas) is produced during the process by oxidation of the feedstock with air; the oxygen is not separated from the air and, as a result, the gas product invariably has a low heat content (150–300 Btu/ft^3). Low heat-content gas is also the usual product of *in situ* gasification of coal (Speight, 2013), which is essentially used as a method for obtaining energy from coal without the necessity of mining the coal, especially if the coal cannot be mined or if mining is uneconomical.

Several important chemical reactions and a host of side reactions are involved in the manufacture of low heat-content gas under the high temperature conditions employed (Speight, 2013). Low heat-content gas contains several components, four of which are always major components present at levels of at least several percent; a fifth component, methane, is marginally a major component.

The nitrogen content of low heat-content gas ranges from somewhat less than 33% v/v to slightly more than 50% v/v and cannot be removed by any reasonable means; the presence of nitrogen at these levels makes the product gas *low heat content* by definition. The nitrogen also strongly limits the applicability of the gas to chemical synthesis. Two other non-combustible components [water (H_2O) and carbon dioxide (CO)] further lower the heating value of the gas; water can be removed by condensation and carbon dioxide by relatively straightforward chemical means.

The two major combustible components of the gas are hydrogen and carbon monoxide; the H_2/CO ratio ranges from about 2:3 to about 3:2. Methane may also make an appreciable contribution to the heat-content of the gas. Of the minor components, hydrogen sulfide is the most significant, component and the amount produced is, in fact, proportional to the sulfur content of the feed coal. Any hydrogen sulfide present must be removed by only one, or more, of several procedures available (Mokhatab *et al.*, 2006; Speight, 2019b).

Low heat-content gas is of interest to industry as a fuel gas or even, on occasion, as a raw material from which ammonia, methanol, and other compounds may be synthesized.

3.4.1.3 *Medium-Btu gas*

Medium heat-content gas has a heating value in the range 300–550 Btu/ft^3 and the composition is much like that of low heat-content gas, except that there is virtually no nitrogen. The primary combustible gases in medium heat-content gas are hydrogen and carbon monoxide (Kasem, 1979). Medium heat-content gas is considerably more versatile than low heat-content gas; like low heat-content gas, medium heat-content gas may be used directly as fuel to raise steam or used through a combined power cycle to drive a gas turbine, with the hot exhaust gases employed to raise steam, but medium heat-content gas is especially amenable to synthesize methane (by methanation), higher hydrocarbon derivatives (by Fischer–Tropsch synthesis), methanol, and a variety of synthetic chemicals.

The reactions used to produce medium heat-content gas are the same as those employed for low heat-content gas synthesis, the major difference being the application of a nitrogen barrier (such as the use of pure oxygen) to keep diluent nitrogen out of the system.

In medium heat-content gas, the H_2/CO ratio varies from 2:3 to 3:1 and the increased heating value correlates with higher methane and hydrogen contents as well as with lower carbon dioxide content. Furthermore, the very nature of the gasification process used to produce the medium heat-content gas has a marked effect on the ease of subsequent processing. For example, the CO_2-acceptor product is quite amenable to use for methane production, because it has (i) the desired H_2/CO ratio just exceeding 3:1, (ii) an initially high methane content, and (iii) relatively low water and carbon dioxide contents. Other gases may require appreciable shift reaction and removal of large quantities of water and carbon dioxide prior to methanation.

3.4.1.4 *High heat-content gas*

High-Btu gas (high heat-content gas) is essentially pure methane and often referred to as *synthetic natural gas* or *substitute natural gas* (SNG). However, to qualify as substitute natural gas, a product must contain at least 95% v/v methane, giving an energy content (heat content) of synthetic natural gas of the order of 980–1080 Btu/ft^3.

The commonly accepted approach to the synthesis of high heat-content gas is the catalytic reaction of hydrogen and carbon monoxide:

$$3H_2 + CO \rightarrow CH_4 + H_2O$$

To avoid catalyst poisoning, the feed gases for this reaction must be quite pure and, therefore, impurities in the product are rare. The large quantities of water produced are removed by condensation and recirculated as very pure water through the gasification system. The hydrogen is usually present in slight excess to ensure that the toxic carbon monoxide is reacted; this small quantity of hydrogen will lower the heat-content to a small degree.

The carbon monoxide/hydrogen reaction is somewhat inefficient as a means of producing methane because the reaction liberates large quantities of heat. In addition, the methanation catalyst is troublesome and prone to poisoning by sulfur compounds and the decomposition of metals can destroy the catalyst. Hydrogasification may be thus employed to minimize the need for methanation:

$$[C]_{coal} + 2H_2 \rightarrow CH_4$$

The product of hydrogasification is far from pure methane and additional methanation is required after hydrogen sulfide and other impurities are removed.

3.4.1.5 *Landfill gas*

Landfill gas (LFG; sometimes categorized as biogas) is produced by the microbial decomposition of landfilled waste in an oxygen-free (anaerobic) atmosphere. This generally occurs under semi-controlled conditions in a landfill; its constituents depend on the composition and age of the waste (Chapter 7).

Landfill gas, a fuel gas, is, in fact, a by-product of current landfilling practices and hence occurs only after municipal solid waste has been sent away for disposal to a landfill site. Anaerobic digestion of the buried solid organic waste produces landfill gas naturally while bacterial decomposition of the organic matter continues over time. In the long term, as the use of landfills will necessarily dwindle, landfill gas will stop to exist as a resource. It is thus of inherently transient nature, but it will be covered here as many examples of landfill gas plants exist.

Landfill gas is approximately 65% v/v methane and 35% v/v carbon dioxide as well as traces of other organic vapors, but the composition does vary with age or the type of waste. The odor of landfill gas is associated with trace compounds, such as hydrogen sulphide (H_2S or H–SH), mercaptan derivatives (RSH), and ethylene (CH_2=CH_2). In addition, several physical conditions affect the production (the rate of production and the volume of the gas) of the gas, including the following: (i) the characteristics of the waste, namely, the composition and age of the refuse and (ii) environmental factors, such as the presence of oxygen in the landfill, the moisture content, and the temperature. Other factors can affect the migration of the gas.

3.4.1.6 *Synthesis gas*

Synthesis gas is a mixture composed of mainly hydrogen and carbon monoxide that is comparable in combustion efficiency to natural gas. This reduces the emissions of sulfur, nitrogen oxides, and mercury, resulting in a much cleaner fuel. The resulting hydrogen gas can be used for electricity generation or as a transport fuel. The gasification process also facilitates capture of carbon dioxide emissions from the combustion effluent (see the following discussion of carbon capture and storage).

Although synthesis gas can be used as a stand-alone fuel, the energy density of synthesis gas is approximately half that of natural gas and is therefore mostly suited for the production of transportation fuels and other chemical products. Synthesis gas is mainly used as an intermediary building block for the final production (synthesis) of various fuels such as synthetic natural gas, methanol, and synthetic crude oil fuel (dimethyl ether — synthesized gasoline and diesel fuel) (Chadeesingh, 2011; Speight, 2013).

The use of synthesis gas offers the opportunity to furnish a wide range of environmentally clean fuels and chemicals and there has been a steady growth in the number of traditional uses of synthesis gas. It is from synthesis gas that almost all hydrogen gas is manufactured and so there has been an increase in the demand for this basic chemical. In fact, the main use of synthesis gas is in the manufacture of hydrogen for a growing number of purposes, especially in crude oil refineries (Speight, 2014). Methanol not only remains the second largest consumer of synthesis gas but also has shown remarkable growth as one of the methyl ethers used as octane enhancers in automotive fuels.

The Fischer–Tropsch synthesis remains the third largest consumer of synthesis gas, mostly for transportation fuels and also as a growing feedstock source for the manufacture of chemicals, including polymers. The hydroformylation of olefins (the Oxo reaction), a completely chemical use of synthesis gas, is the fourth largest use of carbon monoxide and hydrogen mixtures. A direct application of synthesis gas as fuel (and eventually also for making chemicals) that shows promise to grow is its use for *integrated gasification combined cycle* (IGCC) units for the generation of electricity (and for making chemicals) from coal, crude oil coke or high-boiling crude oil resids (particularly vacuum resids). Finally, synthesis gas is the principal source of carbon monoxide, which is used in a growing number of carbonylation reactions that are of major industrial interest.

3.4.1.7 *Hydrogen*

Hydrogen is one of the basic raw materials for technological processes, including refinery processes, and will continue to play a major role in the foreseeable future (Parkash, 2003; Gary *et al.*, 2007; Speight, 2011a, 2014, 2016, 2017, 2019; Hsu and Robinson, 2017). In the refinery, hydrogen is produced by any one (or more) of several processes, not the least of which is the production of hydrogen from synthesis gas (Speight, 2014, 2017). Other methods, in the context of this book, include the following: (i) steam methane reforming, (ii) biomass pyrolysis, and (iii) biomass gasification (Mudhoo *et al.*, 2011; Speight, 2014, 2016, 2017; Sołowski, 2018).

Steam methane reforming (SMR) is a process requiring high temperature and pressure. During this process, hydrogen is generated in reactions with methane/CO and water participation under pressure:

$$CH_4 + H_2O \rightarrow CO + 3H_2$$

$$CO + H_2O \rightarrow CO_2 + H_2$$

The process requires catalysts from nickel or nickel alloy. The gas needs to be desulfurized to prevent catalyst poisoning.

The pyrolysis of biomass reaction is a thermal decomposition that depends on the properties of biomass, the use of catalysts, and the temperature: The process is achieved by rapid heating of organic wastes at high temperatures of 400–600°C (750–1,120°F) and pressure. The products of the process are hydrogen, carbon monoxide, methane, tar and oils. Optimally, hemicelluloses are degraded at temperatures of the order of 250–350°C (480–660°F), cellulose at temperatures of the order of 325–400°C (615–750°F), and lignin at temperatures of the order of 300–550°C (570–1,020°F).

Biomass gasification is the decomposition of biomass using heat and gasifying agents (such as steam). The waste biomass used for the process should have a dry matter content higher than 65%. The catalysts proposed to be used include dolomites minerals, alkali catalysts, and noble metals. The temperature of the process depends on the properties of the feedstock. The biomass feedstock is fragmented and dried.

3.4.2 *Liquid fuels*

Among liquid biomass fuels, biodiesel (vegetable oil ester) is worth noting for its similarity to crude oil-derived diesel fuel, apart from its negligible sulfur and ash content. Bioethanol has only approximately 70% of the heating value of crude oil distillates such as gasoline, but its sulfur and ash contents are also very low. Both of these liquid fuels have lower vapor pressure and flammability than their crude oil-based competitors — an advantage in some cases (e.g. they can be used in confined spaces such as mines) but a disadvantage in others (e.g. starting an engine at cold temperatures).

Currently, the production of ethanol by fermentation of corn-derived carbohydrates is the main technology used to produce liquid fuels from biomass resources. Furthermore, among different biofuels suitable for application in transport, bioethanol and biodiesel seem to be the most feasible ones at present. The key advantage of bioethanol and biodiesel is that they can be mixed with conventional gasoline and diesel, respectively, which allows using the same handling and distribution infrastructure. Another important strong point of bioethanol and biodiesel is that when they are mixed at low concentrations ($\leq$10% bioethanol in gasoline and $\leq$20% biodiesel in diesel), no engine modifications are necessary.

Alternatively, biomass can be converted into fuels and chemicals indirectly (by gasification to synthesis gas followed by catalytic conversion to liquid fuels) or directly (to a liquid product by thermochemical means). Direct thermochemical conversion processes include pyrolysis, liquefaction, and solvolysis (Kavalov and Peteves, 2005). Thermochemical treatment of biomass to produce hydrocarbon derivatives has been studied since the 1930s, and despite being relatively simple to perform, typically such processes are non-selective, producing a wide range of products.

The production of biodiesel from vegetable oil represents another means for producing liquid fuels from biomass and one which is growing rapidly in commercial importance. Commercially, biodiesel is produced from vegetable oils, including rapeseed, sunflower and soybean oils, and animal fats. These oils and fats are typically composed of C_{14}–C_{20} fatty acid triglycerides (constituting approximately 90–95%, by weight, of the oil). In order to produce a fuel that is suitable for use in diesel engines, these triglycerides are converted to the respective alkyl esters and glycerol by base-catalyzed transesterification with short chain alcohols (generally methanol).

Thus, for every 10 units of biodiesel produced, approximately 1 unit of glycerol ($HOCH_2CHOHCH_2OH$) is formed. Glycerol has found application in a wide range of industries, such as cosmetics, pharmaceuticals, and plasticizers.

3.4.2.1 *Alcohols*

Alcohols are organic compounds that contain at least one hydroxyl functional group (–OH) bound to a saturated carbon atom. The term *alcohol* originally referred to the primary alcohol ethanol (ethyl alcohol, CH_3CH_2OH, or C_2H_5OH), which is used as a drug or as a relaxant–intoxicant in the form of alcoholic beverages. One important class of alcohols, of which methanol (methyl alcohol, CH_3OH) and ethanol are the simplest members, includes all compounds for which the general formula is $C_nH_{2n}+_1OH$. Simple mono-alcohols (one hydroxyl group per molecule) include the primary alcohol derivatives (RCH_2OH), the secondary alcohol derivatives (R_2CHOH), and the tertiary alcohol derivatives (R_3COH).

3.4.2.1.1 Methanol

Methanol is a colorless, odorless and nearly tasteless alcohol that is also produced from crops and used as a fuel. Methanol, similar to ethanol, burns more completely but releases as much as or more carbon dioxide than its gasoline counterpart. The balance is often seen as the various bi-processes that draw carbon dioxide from the atmosphere, so there is no net modern release as there is for fossil fuels.

Methanol and other chemicals were routinely extracted from wood in the 19th century and in the early 20th century. However, the original route for recovery of methanol from biomass was quite different from current routes. Methanol was originally recovered from wood as a by-product of charcoal manufacture and was often called *wood alcohol*. Pyrolysis (heating of wood in the absence of air) to above 270°C in a retort causes thermal cracking or breakdown of the wood and allows much of the wood to be recovered as charcoal. The watery condensate leaving the retort contains methanol, among other compounds. In 1923, commercial production of methanol from synthesis gas by a catalytic process was commenced. Now, almost all of the methanol used worldwide comes from the processing of natural gas.

In general, methanol production from natural gas as feed consists of three steps: (i) the generation of synthesis gas — in the case of natural gas feed, synthesis gas production consists of converting methane (CH_4) into carbon monoxide (CO) and hydrogen (H_2) via steam reforming, (ii) synthesis gas upgrading — primarily removal of carbon dioxide plus any contaminants such as sulfur, and (iii) methanol synthesis and purification — reacting carbon monoxide, hydrogen, and steam over a catalyst in the presence of a small amount of carbon dioxide and at elevated temperature and pressure. The methanol synthesis is an equilibrium reaction and excess reactants must be recycled to optimize yields.

Modern methods proposed for the production of methanol from biomass involve the conversion of the biomass to a suitable synthesis gas, after which the processing steps are very similar to those developed for methanol from natural gas. However, the gasification techniques proposed are still at a relatively early stage of development using biomass as feed and the methods are based on similar techniques already used widely with natural gas as feed.

Before biomass can be gasified, it must be pre-treated to meet the processing constraints of the gasifier (Chapter 7). This typically involves size reduction and drying to keep moisture contents below specific levels. Thereafter, biomass gasification involves heating the biomass in the presence of low levels of oxygen (i.e. less than the level required for complete combustion to carbon dioxide and water). Above certain temperatures, the biomass will break down into a gas stream and a solid residue. The composition of the gas stream is influenced by the operating conditions of the gasifier, with some gasification processes being more suited than others to produce a gas for methanol production. In particular, simple gasification with air creates a synthesis gas stream that is diluted with large quantities of nitrogen. This nitrogen is detrimental to subsequent processing into methanol and so techniques using indirect gasification or an oxygen feed are preferred. For large-scale gasification, pressurized systems are considered to be more economical than atmospheric systems.

Once the economically optimum synthesis gas is available, the methanol synthesis takes place. This typically uses a copper–zinc catalyst at temperatures of 200–280°C and pressures of 50–100 atmospheres. The crude methanol from the synthesis loop contains the water produced during synthesis as well as other minor by-products. Purification is achieved in multistage distillation, with the complexity of distillation dictated by the final methanol purity required.

3.4.2.1.2 Ethanol

Ethanol is the predominant fuel produced from crops and has been used as fuel in the United States since at least 1908. Although early efforts to sustain an ethanol program failed, oil supply disruptions in the

Middle East and environmental concerns over the use of lead as a gasoline octane booster renewed interest in ethanol in the late 1970s. At present, extending the volume of conventional gasoline is a significant end use for ethanol as is its use as an oxygenate. To succeed in these markets, the cost of ethanol must be close to the wholesale price of gasoline, currently made possible by the federal ethanol subsidy. However, in order for ethanol to compete on its own merits the cost of producing it must be reduced substantially.

The production of ethanol from corn is a mature technology that holds much potential (Nichols *et al.*, 2006). Substantial cost reductions may be possible, however, if cellulose-based feedstocks are used instead of corn. Producers are experimenting with units equipped to convert cellulose-based feedstocks, using sulfuric acid to break down cellulose and hemicellulose into fermentable sugar. Although the process is expensive at present, advances in biotechnology could decrease conversion costs substantially. The feed for all ethanol fermentations is sugar — traditionally a hexose (a six-carbon or "C6" sugar) such as those present naturally in sugarcane, sugar beet, and molasses. Sugar for fermentation can also be recovered from starch, which is actually a polymer of hexose sugars (*polysaccharide*).

Biomass (Chapter 7), in the form of wood and agricultural residues such as wheat straw, is viewed as a low-cost alternative feed to sugar and starch. It is also potentially available in much greater quantities than sugar and starch feeds. As such, it has received significant attention as a feed material for ethanol production. Similar to starch, wood and agricultural residues contain polysaccharides. However, unlike starch, while the cellulose fraction of biomass is principally a polymer of easily fermented C6 sugars, the hemicellulose fraction is principally a polymer of C5 sugars, with quite different characteristics for recovery and fermentation. The cellulose and hemicellulose in biomass are bound together in a complex framework of crystalline organic material known as lignin.

These differences mean that recovery of these biomass sugars is more complex than recovery of sugars from a starch feed. Once recovered, fermentation is also more complex than a simple fermentation of C6 sugars. The current focus is on the issues of releasing the sugars (hydrolysis) and then fermenting as much of the C6 and C5 sugars as possible to produce ethanol.

There are several different methods of hydrolysis: (i) concentrated sulfuric acid, (ii) dilute sulfuric acid, (iii) nitric acid, and (iv) acid pretreatment followed by enzymatic hydrolysis.

Ethanol is produced from the fermentation of sugar by enzymes produced from specific varieties of yeast. The five major sugars are the five-carbon xylose and arabinose and the six-carbon glucose, galactose, and mannose. Traditional fermentation processes rely on yeasts that convert six-carbon sugars to ethanol. Glucose, the preferred form of sugar for fermentation, is contained in both carbohydrates and cellulose. Because carbohydrates are easier than cellulose to convert into glucose, the majority of ethanol currently produced in the United States is made from corn, which produces large quantities of carbohydrates. Also, the organisms and enzymes for carbohydrate conversion and glucose fermentation on a commercial scale are readily available.

The conversion of cellulosic biomass to ethanol parallels the corn conversion process. The cellulose must first be converted to sugars by hydrolysis and then fermented to produce ethanol. Cellulosic feedstocks (composed of cellulose and hemicellulose) are more difficult to convert into sugar than carbohydrates. Two common methods for converting cellulose into sugar are dilute acid hydrolysis and concentrated acid hydrolysis, both of which use sulfuric acid.

Dilute acid hydrolysis occurs in two stages to take advantage of the differences between hemicellulose and cellulose. The first stage is performed at low temperature to maximize the yield from hemicellulose, and the second, higher temperature stage is optimized for hydrolysis of the cellulose portion of the feedstock. Concentrated acid hydrolysis uses a dilute acid pretreatment to separate hemicellulose and cellulose. The biomass is then dried before the addition of the concentrated sulfuric acid. Water is added to dilute the acid and then heated to release the sugars, producing a gel that can be separated from residual solids. Column chromatography is used to separate the acid from sugars.

Both the dilute and concentrated acid processes have several drawbacks. Dilute acid hydrolysis of cellulose tends to yield a large amount of by-products. Concentrated acid hydrolysis forms fewer by-products, but for economic reasons the acid must be recycled. The separation and concentration of the sulfuric acid adds more complexity to the process. The concentrated and dilute sulfuric acid processes are performed at high temperatures (100°C and 220°C), which can degrade the sugars, reducing the carbon source and ultimately lowering the ethanol yield. Thus, the concentrated acid process has a smaller potential for cost reductions from process improvements such as acid recovery and the sugar yield for the concentrated acid process could provide higher efficiency for both technologies.

Another approach involves countercurrent hydrolysis, a two-stage process. In the first stage, cellulose feedstock is introduced into a horizontal co-current reactor with a conveyor. Steam is added to raise the temperature to 180°C (no acid is added at this point). After a residence time of approximately 8 minutes, during which about 60% of the hemicellulose is hydrolyzed, the feed exits the reactor. It then enters the second stage through a vertical reactor operated at 225°C. Very dilute sulfuric acid is added to the feed at this stage, where virtually all of the remaining hemicellulose and, depending on the residence time, anywhere from 60% to all of the cellulose is hydrolyzed. The countercurrent hydrolysis process offers higher efficiency (and, therefore, cost reductions) than the dilute sulfuric acid process. This process may allow an increase in glucose yields to 84%, an increase in fermentation temperature to 55°C, and an increase in fermentation yield of ethanol to 95%.

The greatest potential for ethanol production from biomass, however, lies in enzymatic hydrolysis of cellulose. The enzyme cellulase, now used in the textile industry to stone wash denim and in detergents, simply replaces the sulfuric acid in the hydrolysis step. The cellulase can be used at lower temperatures, 30–50°C (86–122°F), which reduces the degradation of the sugar. In addition, process improvements now allow simultaneous saccharification and fermentation (SSF). In the saccharification and fermentation process, cellulase and fermenting yeast are combined, so that as sugars are produced, the fermentative organisms convert them to ethanol in the same step. Once the hydrolysis of the cellulose is achieved, the resulting sugars must be fermented to produce ethanol. In addition to glucose, hydrolysis produces other six-carbon sugars from cellulose and five-carbon sugars from hemicellulose that are not readily fermented to ethanol by naturally occurring organisms. They can be converted to ethanol by genetically engineered yeasts that are currently available, but the ethanol yields are not sufficient to make the process economically attractive. It also remains to be seen whether the yeasts can be made hardy enough for production of ethanol on a commercial scale.

A large variety of feedstocks are currently available to produce ethanol from cellulosic biomass. The materials being considered can be categorized as agricultural waste, forest residue, and energy crops. Agricultural waste available for ethanol conversion includes crop residues such as wheat straw, corn stover (leaves, stalks, and cobs), rice straw, and bagasse (sugarcane waste). Forestry waste includes underutilized wood and logging residues; rough, rotten, and salvable dead wood; and excess saplings and small trees. Energy crops, developed and grown specifically for fuel, include fast-growing trees, shrubs, and grasses such as hybrid poplars, willows, and switchgrass.

Briefly, corn stover is the leaves and stalks of field crops, such as corn (maize), sorghum or soybean that are commonly left in a field after harvesting the grain. It is similar to straw, the residue left after any cereal grain or grass has been harvested at maturity for its seed. It can be directly grazed by cattle or dried for use as fodder; stover has attracted some attention as a potential fuel source, and as a biomass feedstock for fermentation or as a feedstock for ethanol production from cellulose.

Although the choice of feedstock for ethanol conversion is largely a cost issue, feedstock selection has also focused on environmental issues. Materials normally targeted for disposal include forest thinnings collected as part of an effort to improve forest health, and certain agricultural residues, such as rice straw. Also, forest residues offer an opportunity to decrease the fire hazard associated with the dead wood present in many forests. Small quantities of forest thinnings can be collected at relatively low cost, but collection costs rise rapidly as the quantities increase.

Agricultural residues, in particular, corn stover, represent a tremendous resource base for biomass ethanol production. Agricultural residues, in the long term, would be the sources of biomass that could support substantial growth of the ethanol industry. At conversion yields of around 60–100 gallons per dry ton, the available corn stover inventory would be sufficient to support 7–12 billion gallons of ethanol production per year.

Dedicated energy crops such as switchgrass, hybrid willow, and hybrid poplar are another long-term feedstock option. Switchgrass is grown on a 10-year crop rotation basis, and harvest can begin in year 1 in some locations and year 2 in others. Willows require a 22-year rotation, with the first harvest in year 4 and subsequent harvests every 3 years thereafter. Hybrid poplar requires 6 years to reach harvest age in the Pacific Northwest, 8 years in the Southeast, Southern Plains, and South Central regions, and 10 years in the Corn Belt, Lake States, Northeast, and Northern Plains regions. Thus, if it were planted in the spring of 2000, switchgrass could be harvested in 2000 or 2001, willow could be harvested in 2004, and poplars could be harvested in 2006, 2008, or 2010, depending on the region.

The use of cellulosic biomass in the production of ethanol has environmental benefits as well. Converting cellulose to ethanol increases the net energy balance of ethanol compared to converting corn to ethanol. The net energy balance is calculated by subtracting the energy required to produce a gallon of ethanol from the energy contained in a gallon of ethanol (approximately 76,000 Btu). Corn-based ethanol has a net energy balance of 20,000–25,000 Btu per gallon, whereas cellulosic ethanol has a net energy balance of more than 60,000 Btu per gallon. In addition, cellulosic ethanol use can reduce greenhouse gas emissions. Cellulosic ethanol can cause an 8–10% reduction in greenhouse gas emissions when used in E10 and a 68–91% reduction when used in E85.

3.4.2.1.3 Propanol and butanol

Propanol and butanol are considerably less toxic and less volatile than methanol. In particular, butanol has a high flashpoint (35°C; 95°F), which is an advantage for fire safety, but may be a disadvantage for starting engines in cold weather (Ezeji *et al.*, 2006).

The fermentation processes to produce propanol and butanol from cellulose are fairly tricky to execute, and the *Clostridium acetobutylicum* currently used to perform these conversions produces an extremely unpleasant smell, and this must be taken into consideration when designing and locating a fermentation plant. This organism also dies when the butanol content of whatever it is fermenting rises to 7%. For comparison, yeast dies when the ethanol content of its feedstock hits 14%. Specialized strains can tolerate even greater ethanol concentrations — the so-called turbo yeast can withstand up to 16% ethanol. However, if ordinary *Saccharomyces* yeast can be modified to improve its ethanol resistance, scientists may one day produce a strain of the Weizmann organism with a butanol resistance higher than the natural boundary of 7%. This would be useful because butanol has a higher energy density than ethanol, and because waste fiber leftover from the sugar crops used to make ethanol could be turned into butanol, raising the alcohol yield of fuel crops without the need for more crops to be planted.

3.4.2.2 *Biodiesel*

Biodiesel is a diesel-equivalent fuel derived from biological sources (such as vegetable oil) and can be used in unmodified diesel-engine vehicles. It is thus distinguished from the unchanged vegetable oil or waste vegetable oil used as fuel in some diesel vehicles. In the current context, biodiesel refers to alkyl esters that are produced by the transesterification of vegetable oil or animal fat:

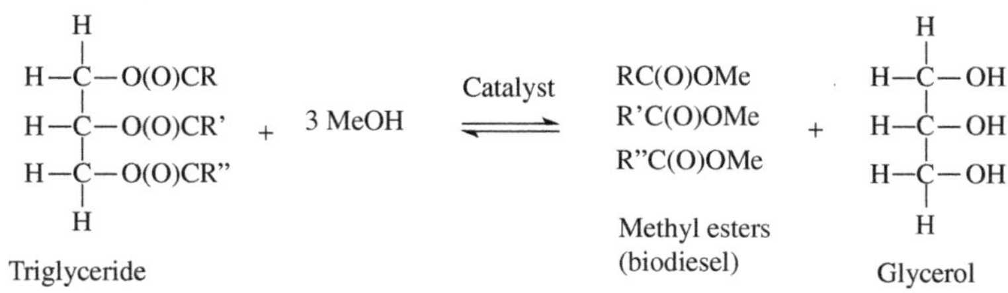

3.4.2.2.1 Feedstocks

Biodiesel fuel is a fuel made from the oil of certain oilseed crops such as soybean, canola, palm kernel, coconut, sunflower, safflower, corn, and hundreds of other oil-producing crops (Bourne, 2007). The oil is extracted by the use of a press and then mixed in specific proportions with other agents, which causes a chemical reaction. The results of this reaction are two products, biodiesel and soap. After a final filtration, the biodiesel is ready for use. After curing, the glycerin soap which is produced as a by-product can be used as is or can have scented oils added before use.

Biodiesel is made through a chemical process (*transesterification*) whereby the glycerin is separated from the fat or vegetable oil. The process leaves behind two products: (i) methyl esters (the chemical name for biodiesel) and (ii) glycerin (a valuable by-product, usually sold for use in soaps and other products).

Biodiesel (fatty acid methyl esters; FAME) is a notable alternative to the widely used crude oil-derived diesel fuel since it can be generated by domestic natural sources such as soybeans, rapeseeds, coconuts, and even recycled cooking oil and thus reduces dependence on the diminishing crude oil fuel from foreign sources. In addition, because biodiesel is largely made from vegetable oils, it reduces the lifecycle of greenhouse gas emissions by as much as 78% (Ban-Weiss *et al.*, 2007).

Vegetable oils and animal fats belong to an extensive family of chemicals called lipids. Lipids are bioproducts from the metabolism of living creatures. As a result, they can be found widely distributed in nature. Their biofunctions are diverse, but they are most known for their energy storage capacity. Most lipids can easily dissolve in common organic solvents, meaning that they are hydrophobic. If a lipid is a solid at 25°C (77°F), it is classified as a fat; otherwise, it is oil. Typically, fats are produced by animals and oils are produced by plants, but both are mainly made of triglyceride (TG) molecules, which are tri-esters of glycerol (a triol) and free fatty acids (long alkyl chain carboxylic acids). Other glyceride species, such as diglycerides and monoglycerides, are obtained from triglycerides by the substitution of one and two fatty acid moieties, respectively, with hydroxyl groups (Lotero *et al.*, 2006).

Biodiesel production is a very modern and technological area for researchers due to the relevance that it is gaining every day because of the increase in crude oil price and the environmental advantages (Marchetti *et al.*, 2005). The successful introduction and commercialization of biodiesel in many countries around the world has been accompanied by the development of standards to ensure high product quality and user confidence (Knothe, 2005). In general, biodiesel compares well to crude oil-based diesel (Lotero et al., 2006). Pure biodiesel fuel (100% esters of fatty acids) is called B100. When blended with diesel fuel the designation indicates the amount of B100 in the blend, e.g. B20 is 20% B100 and 80% diesel, and B5 used in Europe contains 5% B100 in diesel (Pinto *et al.*, 2005).

Biodiesel is biodegradable and non-toxic and typically produces approximately 60% less net carbon dioxide emission than crude oil-based diesel, as it is itself produced from atmospheric carbon dioxide by way of the photosynthesis process in plants. However, this figure can actually differ widely between fuels, depending on the production and processing methods employed for their creation. Pure biodiesel is available at many gas stations in Germany.

Soybeans are widely used as a source of biodiesel. However, a variety of oils can be used to produce biodiesel. These include the following: (i) virgin oil feedstock such as rapeseed oil and soybean oil as well as field pennycress, Jatropha and other crops such as mustard, flax, sunflower, canola, palm oil, hemp, and algae; (ii) waste vegetable oil; and (iii) animal fat, including tallow, lard, yellow grease, chicken fat, and the by-products of the production of omega-3-fatty acids from fish oil.

3.4.2.2.2 Processes

Biodiesel is produced from vegetable oils, yellow grease, used cooking oils, or animal fats. The fuel is produced by transesterification, a process that converts fats and oils into biodiesel and glycerin (a co-product), which is commonly used in the manufacture of pharmaceuticals and cosmetics.

3.4.2.2.2.1 Transesterification

Transesterification (alcoholysis) is the conversion of triacylglycerol lipids by alcohols to alkyl esters without first isolating the free fatty acids (May, 2004). The purpose of transesterification of vegetable oils to their methyl esters (biodiesel) is to lower the viscosity of the oil. The transesterification reaction is affected by the alcohol type, molar ratio of glycerides to alcohol, type and amount of catalyst, reaction temperature, reaction time, and free fatty acids and water content of vegetable oils or animal fats. The transesterification reaction proceeds with or without a catalyst by using primary or secondary monohydric aliphatic alcohols having 1–8 carbon atoms as follows (Demirbas, 2006a, 2006b):

Triglyceride derivatives + monohydric alcohol ↔
glycerin + mono-alkyl ester derivatives

Generally, a reaction temperature close to the boiling point of the alcohol is recommended (Çanakçı and Özsezen, 2005). The reactions take place at low temperature (approximately 65°C, 160°F) and at modest pressure (2 atmospheres, 1 atmosphere = 14.7 psi). Biodiesel is further purified by washing and evaporation to remove any remaining methanol. The oil (87%), alcohol (9%), and catalyst (1%) are the inputs in the production of biodiesel (86%), the main output (Lucia *et al.*, 2006). Pretreatment is not required if the reaction is carried out under high pressure (9,000 kPa) and high temperature (240°C, 465°F), where simultaneous esterification and transesterification take place, with maximum yield obtained at temperatures ranging from 60°C to 80°C (140 to 175°F) at a molar ratio of 6:1 (Barnwal and Sharma, 2005). The alcohols employed in the transesterification are generally short-chain alcohols such as methanol, ethanol, propanol, and butanol. It has been reported that when the transesterification of soybean oil using methanol, ethanol and butanol was performed, 96–98% of ester could be obtained after 1 h (Dmytryshyn *et al.*, 2004).

3.4.2.2.2.2 Catalytic transesterification

Transesterification reactions can be catalyzed by *alkalis* (Korytkowska *et al.*, 2001; Dmytryshyn *et al.*, 2004; Stavarache *et al.*, 2005; Varghaa and Truterb, 2005; Meher *et al.*, 2006a), *acids* (Lee *et al.*, 2000; Goff *et al.*, 2004; Lopez *et al.*, 2005; Liu *et al.*, 2006) or *enzymes*

(Watanabe *et al.*, 2000, 2002; Ghanem, 2003; Reyes-Duarte *et al.*, 2005; Royon *et al.*, 2007; Shah and Gupta, 2007; Bernardes *et al.*, 2007). The catalytic transesterification of vegetable oils with methanol is an important industrial method used in biodiesel synthesis. Also known as methanolysis, this reaction has been well studied and established using acids or alkalis, such as sulfuric acid or sodium hydroxide, as catalysts. However, these catalytic systems are less active or completely inactive for long-chain alcohols. Usually, industries use sodium or potassium hydroxide or sodium or potassium methoxide as catalyst, since they are relatively cheap and quite active for this reaction (Macedo *et al.*, 2006). Enzyme-catalyzed procedures, using lipase as catalyst, do not produce side reactions, but the lipases are very expensive for industrial scale production and a three-step process was required to achieve a 95% conversion. The acid-catalyzed process is useful when a range amount of free acids is present in the vegetable oil, but the reaction time is very long (48–96 hours), even at the boiling point of the alcohol, and a high molar ratio of alcohol was needed (20:1 w/w to the oil) (Stavarache *et al.*, 2005).

The transesterification process is catalyzed by alkaline metal alkoxides and hydroxides as well as sodium or potassium carbonates. Alkali-catalyzed transesterification with short-chain alcohols, for example, generates high yields of methyl esters in short reaction times (Jeong and Park, 1996). The alkaline catalysts show high performance for obtaining vegetable oils with high quality, but a question often arises: do the oils contain significant amounts of free fatty acids which cannot be converted to biodiesels but to a lot of soap? (Furuta *et al.*, 2004). These free fatty acids react with the alkaline catalyst to produce soaps that inhibit the separation of the biodiesel, glycerin, and wash water (Çanakcı and Van Gerpen, 2003). Triglycerides are readily transesterified in a batch operation in the presence of alkaline catalyst at atmospheric pressure and at a temperature of approximately 60–70°C with an excess of methanol (Srivastava and Prasad, 2000). It often takes at least several hours to ensure the alkali (NaOH or KOH) catalytic transesterification reaction is complete. Moreover, removal of these catalysts is technically difficult and brings extra cost to the final product (Demirbas, 2002, 2003). Nevertheless, they are a good alternative since they can give the same high conversions of vegetable oils just by increasing the catalyst concentration to 1 or 2 mol%. Alkaline metal alkoxides (such as CH_3ONa for the methanolysis) are the most active catalysts, since they give very high yields (>98%) in short reaction times (30 min) even if they are applied at low molar concentrations (0.5 mol%) (Schuchardta *et al.*, 1998).

The transesterification process is catalyzed by sulfuric, hydrochloric, and organic sulfonic acids. In general, acid-catalyzed reactions are performed at high alcohol-to-oil molar ratios, low-to-moderate temperatures and pressures, and high acid catalyst concentrations (Lotero *et al.*, 2006). These catalysts give very high yields of alkyl esters, but these reactions are slow, requiring typically temperatures above 100°C (212°F) and more than 3 h to complete the conversion (Meher *et al.*, 2006b). Studies of the acid-catalyzed system have been very limited in number. Despite the relatively slow reaction rate, the acid-catalyzed process offers benefits with respect to its lack of dependence on free fatty acid content and the consequent absence of a pretreatment step. These advantages favor the use of the acid-catalyzed process when using waste cooking oil as the raw material (Zhang *et al.*, 2003).

Enzyme (such as lipase)-catalyzed reactions have some advantages over traditional chemical-catalyzed reactions: no by-products generated, easy product recovery, mild reaction conditions, and catalyst recycling. Also, enzymatic reactions are insensitive to the free fatty acid and water contents in waste cooking oil (Kulkarni and Dalai, 2006). As for the enzyme-catalyzed system, it requires a much longer reaction time than the other two systems (Zhang *et al.*, 2003). The enzyme reactions are highly specific and chemically clean. Because alcohol can be inhibitory to the enzyme, a typical strategy is to feed the alcohol into the reactor in three steps of 1:1 mole ratio each. The reactions are very slow, with a three-step sequence requiring 4–40 h or more. The reaction conditions are modest from 35°C to 45°C (Van Gerpen *et al.*, 2004). The main problem with the enzyme-catalyzed process is the high cost of lipases used as catalyst (Royon *et al.*, 2007).

Synthesis of biodiesel using enzymes such as *Candida antarctica*, *Candida rugasa*, *Pseudomonas cepacia*, immobilized lipase (Lipozyme RMIM), *Pseudomonas sp*, and *Rhizomucor miehei* has been well reported in the literature. In the work by Shah and Gupta (2007), the best yield of 98% (w/w) was obtained using *Pseudomonas cepacia* lipase immobilized on celite at 50°C in the presence of 4–5% (w/w) water in 8 h.

3.4.2.2.2.3 Supercritical methanol transesterification

The transesterification of triglycerides by supercritical methanol (SCM), ethanol, propanol, and butanol has proved to be the most promising process. Recently, a catalysts-free method was developed for biodiesel production by employing supercritical methanol (Saka and Kusdiana, 2001). The supercritical treatment at 350°C, 43 MPa and 240 s with a molar ratio of 42 in methanol is the optimum condition for the transesterification of rapeseed oil to biodiesel fuel (Kusdiana and Saka, 2004a).

To achieve more moderate reaction conditions, further effort was made through the two-step preparation (Kusdiana and Saka, 2004b). In this method, oils/fats are, first, treated in subcritical water for a hydrolysis reaction to produce fatty acids. After hydrolysis, the reaction mixture is separated into the oil phase and the water phase by decantation. The oil phase (upper portion) is mainly fatty acids, while the water phase (lower portion) contains glycerol in water. The separated oil phase is then mixed with methanol and treated under supercritical conditions to produce FAME through methyl esterification. After removing the unreacted methanol and water produced in the reaction, FAME can be obtained as a biodiesel. Therefore, in this process, methyl esterification is the main reaction for FAME formation, while in the one-step method, transesterification is the main one (Saka and Minami, 2006).

Reaction with supercritical methanol has some advantages: (i) glycerides and free fatty acids are reacted at equivalent rates, (ii) the homogeneous phase eliminates diffusive problems, (iii) the process tolerates great percentages of water in the feedstock, and the catalytic process requires the periodical removal of water in the feedstock or in the intermediate stage to prevent catalyst deactivation, (iv) the catalyst removal step is eliminated, and (v) if high methanol-to-oil ratios are used, the total conversion of the oil can be achieved in a few minutes (Vera *et al.*, 2005). Some disadvantages of the one-stage supercritical method are obvious: (i) the method operates at very high pressures (25–40 MPa), (ii) the high temperatures bring along proportionally high heating and cooling costs, (iii) high methanol-to-oil ratio (usually set at 42) involves high costs for the evaporation of the unreacted methanol, and (iv) the process as proposed to date does not explain how to reduce free glycerol to less than 0.02% as established in the ASTM D6584 or other equivalent international standards.

3.4.2.2.2.4 Effect of reaction parameters on yield

The main factors affecting transesterification are the molar ratio of glycerides to alcohol, catalyst, reaction temperature and pressure, reaction time, and the contents of free fatty acids and water in oils.

The free fatty acid and moisture contents are key parameters for determining the viability of the vegetable oil transesterification process (Meher *et al.*, 2006b). In the transesterification, free fatty acids and water always produce negative effects, since the presence of free fatty acids and water causes soap formation, consumes catalyst, and reduces catalyst effectiveness, all of which result in a low conversion (Demirbas and Karslıoğlu, 2007). To carry the base catalyzed reaction to completion, a free fatty acid value lower than 3% is needed (Meher *et al.*, 2006b).

The presence of water has a greater negative effect on transesterification than the free fatty acids. In the transesterification of beef tallow catalyzed by sodium hydroxide (NaOH) in the presence of free fatty acids and water, the water and free fatty acid contents must be maintained at specified levels (Ma *et al.*, 1998; Kusdiana and Saka, 2004a; Bala, 2005).

Transesterification can occur in different temperatures, depending on the type of oil employed (Ma and Hanna, 1999). A few works have reported the reaction at room temperature (Marinetti, 1962, 1966; Graboski and McCormick, 1998; Encinar *et al.*, 2002).

The effect of reaction temperature on production of propyl oleate was examined at the temperature range of 40–70°C with free *P. fluorescens* lipase (Iso *et al.*, 2001). The conversion ratio to propyl oleate was observed to be highest at 60°C (140°F), whereas the activity was highly decreased at 70°C (158°F).

The conversion rate increases with reaction time. The transesterification of rice bran oil with methanol was studied at molar ratios of 4:1, 5:1, and 6:1 (Gupta *et al.*, 2007). At molar ratios of 4:1 and 5:1, there was a significant increase in yield when the reaction time was increased from 4 to 6 h. Among the three molar ratios studied, the ratio 6:1 gave the best results.

One of the most important factors that affect the yield of ester is the molar ratio of alcohol to triglyceride. Although the stoichiometric molar ratio of methanol to triglyceride for transesterification is 3:1, higher molar ratios are used to enhance the solubility and to increase the contact between the triglyceride and alcohol molecules (Noureddini *et al.*, 1998). In addition, an investigation of the effect of the molar ratio on the transesterification of sunflower oil with methanol showed that the molar ratio varied from 6:1 to 1:1 and concluded that 98% conversion to ester was obtained at a molar ratio of 6.1 (Freedman *et al.*, 1986). Another important variable affecting the yield of methyl ester is the type of alcohol to triglyceride. In general, short-chain alcohols such as methanol, ethanol, propanol, and butanol can be used in the transesterification reaction to obtain high methyl ester yields (Çanakcı and Van Gerpen, 2003).

The catalysts used in the transesterification of triglycerides are classified as alkali, acid, and enzyme. Alkali-catalyzed transesterification is much faster than acid-catalyzed transesterification and is more often used commercially (Ma and Hanna, 1999) and, quite often, for the base-catalyzed transesterification, the best yields are obtained when the catalyst is used at a small concentration, i.e. 0.5% w/w of oil (Stavarache *et al.*, 2005). On the other hand, during the production of free and bound ethyl ester (FAEE) from castor oil, hydrochloric acid is much more effective than sodium hydroxide.

3.4.2.3 *Hydrocarbons*

There are certain species of flowering plants belonging to different families which convert a substantial amount of photosynthetic products into latex. The latex of such plants contains liquid hydrocarbon derivatives of high molecular weight (10,000). These hydrocarbon derivatives can be converted into high-grade transportation fuel (i.e. crude oil). Therefore, hydrocarbon-producing plants are called crude oil plants or petroplants and their crop as petrocrop. Natural gas is one of the products obtained from hydrocarbon derivatives. Thus, crude oil plants can be an alternative source for obtaining crude oil to be used in diesel engines. Normally, some of the latex-producing plants of the families Euphorbiaceae, Apocynaceae, Asclepiadaceae, Sapotaceae, Moraceae, Dipterocarpaceae, etc. are petroplants. Similarly, sunflower (family *Composiae*) and *Hardwickia pinnata* (family Leguminosae) are also petroplants. Some algae also produce hydrocarbon derivatives.

Different species of *Euphorbia* of the family Euphorbiaceae serve as petroplants that have the potential to be a renewable substitute for the conventional crude oil sources. Latex of *Euphorbia lathyrus* contains a fairly high percentage of terpenoids. These can be converted into high-grade transportation fuel. Similarly, the carbohydrates (hexoses) from such plants can be used for ethanol formation.

Sugarcane (*Saccharum officinarum*, family *Gramineae*) is the main source of raw material for the sugar industry. The wastes from the sugar industry include bagasse, molasses, and press mud. After extracting the cane juice for sugar production, the cellulosic fibrous residue that remains is called bagasse. It is used as the raw material (biomass) and processed variously for the production of fuel, alcohols, and single-cell protein as well as in paper mills. Molasses is an important by-product of sugar mills and contains 50–55% fermentable sugars. One ton of molasses can produce approximately 280 L of ethanol. Molasses is used for the production of animal feed, liquid fuel, and alcoholic beverages.

Sugar beet (*Beta vulgaris*, family *Chenopodiaceae*) is yet another plant which contains a high percentage of sugars stored in fleshy storage roots. It is also an important source for the production of sugar as well as ethanol. Raising crops like sugarcane, sugar beet, tapioca, potato, maize, etc. purely for production of ethanol is described as energy cropping, and the crops are called energy crops. Cultivation of plants (trees) for obtaining fuel/fire wood is called as energy plantation.

Other plants, such as *Euphorbiaceae* plants like *Euphorbia lathyris*, milkweed (*Asclepia speciosa*) and the tree legume *Copaifera multijuga*, produce hydrocarbon derivatives, which can be converted into and used as a diesel called biodiesel. In addition, some freshwater and marine algae are also known to accumulate hydrocarbon derivatives. Some algae like Chlamydomonas and anaerobic bacteria like Clostridium produce hydrogen gas, which can be used as a pollution-free fuel. Long-term research efforts are needed to develop these organisms as sources of bioenergy. The Euphorbias and milkweeds can be grown in relatively dry environments on lands not suited for crop production; this makes them highly attractive sources of biofuels. The Euphorbias are relatives of the plants used to produce rubber; they produce a latex, which has approximately 30% hydrocarbon derivatives emulsified in water.

However, hydrocarbon derivatives, as such, are not usually produced from crops, as the amount of hydrocarbon derivatives present in plant tissue is not sufficient to make the process economical. On the other hand, biodiesel is produced from crops and thereby represents an excellent renewable fuel for diesel engines.

3.4.2.4 *Bio-oil*

Bio-oil (sometimes called *biocrude*) is the liquid condensate produced from biomass (forestry residues, crop residues, waste paper, and organic waste) by means of processes such as pyrolysis and thermochemical conversion. Depending on the composition of the feedstock and the process, the yield of bio-oil from wood, paper, and other biomass ranges from 60% to 95% w/w. The yield of bio-oil from wood can range from 72% to

80% w/w, depending on the relative amounts of cellulose and lignin in the material. Materials with a high lignin content, such as bark, tend to give lower liquid yields (typically, 60–65% w/w), but these are offset somewhat by a higher energy density in the liquid. Materials such as paper, which are high in cellulose content, give bio-oil yields in the range of 75–93% — yields in excess of 75 wt.% are typical of a mixed paper stream.

Bio-oil is typically a liquid, almost black through dark red-brown. The density of the liquid is approximately 1200 kg/m^3, which is higher than that of fuel oil and significantly higher than that of the original biomass. The viscosity of the oil varies from as low as 25 cP up to 1,000 cP depending on the water content and the original feedstock. Due to the large amounts of oxygenated components present (Table 3.4), bio-oil has a polar nature and does not mix readily with hydrocarbon derivatives. The degradation products from the biomass constituents include organic acids (like formic and acetic acids), giving the oil its low pH of approximately 2–4. Typically, bio-oil is a low-sulfur, non-viscous, renewable liquid fuel — it is an oxygenated and, as such, not a hydrocarbon fossil fuel. The oil has a fuel value of at least 7,500 BTU/lb, which is approximately the same as air-dried wood and is between the fuel value of methanol and crude oil-derived fuel oil. Bio-oil is a clean-burning fuel that can be fired

Table 3.4: Typical properties and composition of bio-oil[a].

Property	
Water content (w/w%)	15–30
pH	2.8–4.0
Density (kg/L)	1.1–1.25
Elemental analysis (moisture free)	
Carbon (w/w%)	55–64
Hydrogen (w/w%)	5–8
Nitrogen (w/w%)	0.05–1
Sulfur (w/w%)	0–0.05
Oxygen (w/w%)	100 — (C + H + N + S)
Ash (w/w%)	0.03–0.3
Viscosity (42°C, 108°F, cP)	25–1000

Note: [a]Also called *pyrolysis oil*.

in existing fossil fuel boilers with minimal modifications, although the tendency for any signs of incompatibility with crude oil-derived fuels oils should be investigated before blending.

Water is an integral part of the single-phase chemical solution. The (hydrophilic) bio-oils have a water content of typically 15–35 wt.%. Typically, phase separation does occur if water content is higher than approximately 30–45%. The heating values are as follows: the higher heating value (HHV) is below 19 MJ/kg compared to 42–44 MJ/kg for conventional fuel oils.

A process entirely different from the one used for biodiesel production can be used to convert biomass into a type of fuel similar to diesel, namely bio-oil. This process (called *fast pyrolysis* or *flash pyrolysis*) occurs when solid fuels are heated at temperatures between 350°C and 500°C (660°F and 930°F) for a very short period of time (<2 s). The bio-oils currently produced are suitable for use in boilers for electricity generation.

In another process, the feedstock is fed into a fluidized bed (at 450–500°C, 840–930°F), and the feedstock flashes and vaporizes. The resulting vapors pass into a cyclone where solid particles, char, are extracted. The gas from the cyclone enters a quench tower, where they are quickly cooled by heat transfer using the bio-oil already made in the process. The bio-oil condenses in a product receiver and any non-condensable gases are returned to the reactor to maintain process heating. The entire reaction from injection to quenching takes only 2 s.

In the hydrothermal upgrading process (HTU process), biomass is treated with water at high temperature (300–350°C, 570–660°C) and pressure 1,770–2650 psi) to produce biocrude. This can be separated by flashing or extraction into heavy (high-density) crude oil (suitable for co-combustion in coal power stations) and light (low-density) crude oil, which can be upgraded by hydrodeoxygenation (HDO) to biofuels.

The *rapid thermal processing* process (RTP™ process) occurs at a temperature of approximately 500°C (930°F), when a turbulent stream of hot sand flashes the biomass into a vapor. The vapor is then rapidly condensed into a liquid. This process occurs in less than 2 s, yielding high quantities of bio-oil (typically, 65–75% w/w yield of pyrolysis oil from dried lignocellulosic biomass).

Bio-oil is a dark brownish viscous liquid that bears some resemblance to fossil crude oil. However, bio-oil is a complex oxygenated compound composed of water, water-soluble compounds, such as acids, esters, etc., and water-insoluble compounds, usually called pyrolytic lignin (because it comes from the lignin fraction of the biomass). The elemental composition of bio-oil is similar to that of the parent biomass. Because of its high oxygen content, the heating value (Btu per gallon) of bio-oil is lower than that of fossil fuel, typically only approximately half the heating value of fossil crude such as high-boiling fuel oil. However, it contains less nitrogen and only traces of metals or sulfur.

Bio-oil is acidic with a pH in the 2–4 range, making it highly unstable and corrosive — the acidity can be lessened by the addition of readily available base compounds. It, therefore, presents transportation/piping and storage challenges including the tendency to corrode most metals. Hence, it is usually transported and stored in stainless steel containers.

Like other biofuels, the properties of bio-oil are variable and dependent on the feedstock. The specific gravity of the liquid is approximately 1.10–1.25, which means it is slightly heavier than water, heavier than fuel oil, and significantly heavier than the bulk density of the original biomass. The viscosity of the oil varies from as low as 25 cP up to 1,000 cP, depending on the water content and the original feedstock.

3.4.2.5 *Bioethers*

Ethers are a class of carbon compounds that contain an ether group (C–O–C) in which either carbon may be attached to other carbon atoms as well (such as the commonly used diethyl ether, $CH_3CH_2OCH_2CH_3$). The most commonly used fuel additive ethers are methyl-tertiary-butyl-ether (MTBE) and ethyl-tertiary-butyl-ether (ETBE).

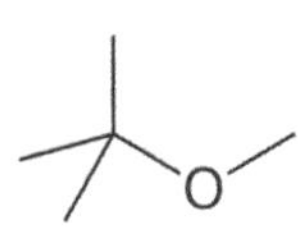

MTBE

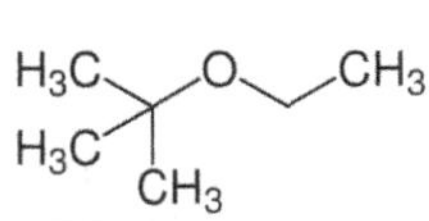

ETBE

Bioethers (Table 3.5) are produced by the reaction of reactive *iso*-olefin derivatives, such as *iso*-butylene, with bioethanol, which is ethanol produced from biosources. When added to gasoline, a bioether can make

Table 3.5: Examples of bioether derivatives used in fuels.

Ether	Use in fuels
Dimethyl ether (DME)	Alternative fuel for gasoline engines
Diethyl ether (DEE)	Used as an ignition improver for gasoline
	Possible alternative fuel for gasoline engines
Methyl tertiary-butyl ether (MTBE)	Additive for gasoline
Ethyl tert-butyl ether (ETBE)	Additive for gasoline
Tert-amyl methyl ether (TAME)	Additive for gasoline
Tert-amyl ethyl ether (TAEE)	Additive for gasoline
	Increases the solubility of ethanol in diesel

the gasoline burn cleanly and completely and enhance engine performance, while reducing engine wear and toxic exhaust emissions, as well as the amount of ground-level ozone.

3.4.2.6 *Liquid fuels from synthesis gas*

Biomass can be gasified to produce synthesis gas composed primarily of hydrogen and carbon monoxide, also called *biosynthesis gas* or *biosyngas* (Cobb, 2007) The production of high-quality synthesis gas from biomass, which is later used as a feedstock for biomass-to-liquids (BTL) production (Figure 3.3), requires particular attention. This is due to the fact that the production of synthesis gas from biomass is indeed the novel component in the gas-to-liquids concept — obtaining synthesis gas from fossil raw materials (natural gas and coal) is a relatively mature technology.

In the current context, gasification is the thermal degradation in the presence of an externally supplied oxidizing (oxygen-containing) agent, e.g. air, steam, and oxygen. Various gasification concepts have been developed over the years, mainly for the purposes of power generation. However, efficient BTL production imposes completely different requirements for the composition of the gas. The reason is that in power generation the gas is used as a fuel, while in BTL processing it is used as a chemical feedstock to obtain other products. This difference has implications with respect to the purity and composition of the gas.

The calorific value of the gas is the prime factor for power generation — the higher the value, the better. Hence, the availability in the gas

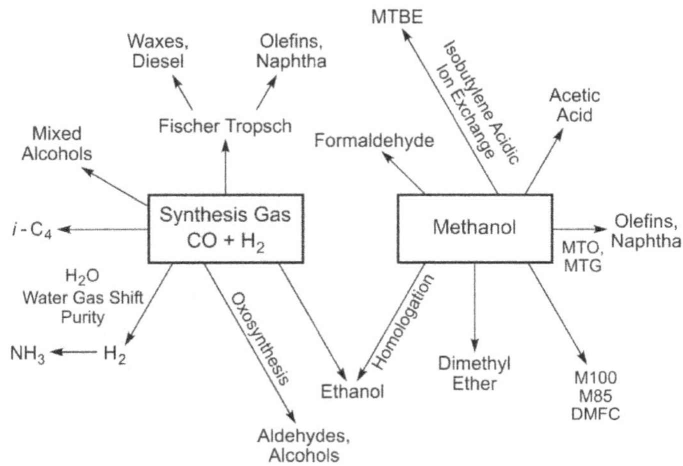

Figure 3.3: Products from synthesis gas.

of any compounds that increase the calorific value is generally welcome e.g. product gas, which contains carbon monoxide (CO), hydrogen (H_2), and various hydrocarbon derivatives (methane (CH_4), ethylene (C_2H_4), ethane (C_2H_6), tars and chars). The presence of inert components (water (H_2O), carbon dioxide (CO_2), and nitrogen (N_2)) is also acceptable, provided it is kept within certain limits. In contrast, for BTL production, the amount of carbon monoxide and hydrogen is only important (the larger the amount, the better), while the calorific value is irrelevant. The presence of other hydrocarbon derivatives and inert components should be avoided or at least kept as low as possible.

The amount of various components can be minimized via a more complete decomposition of biomass, thereby preventing the formation of undesirable components at the gasification step. This approach seems to be more appropriate from an energy efficiency and cost point of view. The minimization of the content of various hydrocarbon derivatives is achieved by increasing the temperatures in the gasifier along with shortening the residence time of feedstocks inside the reactor. Because of this short residence time, the particle size of feedstocks should be small enough (in any case, smaller than in gasification for power generation) for a complete and efficient gasification to occur.

In gasification for power generation, typically air is employed as an oxidizing agent, being the cheapest among all possible oxidizing agents. However, the application of air results in large amounts of nitrogen in the product gas, since nitrogen is the main constituent of air. The presence of such large quantities of nitrogen in the product gas does not hamper (very much) the power generation, but it does hamper BTL production. Removing this nitrogen via liquefaction under cryogenic temperatures is extremely energy intensive, reduces substantially the overall BTL energy efficiency, and increases costs. Among other potential options (steam, carbon dioxide, and oxygen), from a technical and economical point of view oxygen appears to be the most suitable oxidizing agent for BTL manufacturing. It is true that the oxygen-blown gasification implies additional costs compared to the air-blown gasification, because of the oxygen production. Nevertheless, the energy and financial cost of producing oxygen seems to be far lower than the alternative energy and financial cost of cleaning up the product gas from air-blown gasification from nitrogen. This is partly due to the fact that the production of high-purity oxygen (above 95% v/v oxygen) is a mature technology.

In principle, the larger the carbon and hydrogen content in raw materials employed in GTL processing is, the easier and more efficient the carbon monoxide and hydrogen. Hence, the natural gas pathway is the most convenient one, since natural gas is gaseous and contains virtually carbon and hydrogen only. Solid raw materials (biomass, coal) involve more processing because they have to be gasified first and then the obtained product gas should be cleaned up from other components such as nitrogen oxides (NO_x), sulfur oxides (SO_x), and particulate matter to the extent of achieving the best possible purity of synthesis gas. Two basic types of biomass raw material are distinguished; woody material and herbaceous material. Currently, woody material accounts for approximately 50% of the total bioenergy potential of the world. Another 20% is the straw-like feedstock obtained as a by-product from agriculture. The dedicated cultivation of straw-like energy crops could increase the herbaceous share to 40% (Boerrigter and Van der Drift, 2004; Van der Drift *et al.*, 2004).

There are three main types of gasifiers: fixed bed, fluidized bed, and entrained flow. The air-blown direct gasifiers operated at atmospheric pressure and used in power generation — fixed-bed updraft and downdraft gasifiers, and fluidized-bed bubbling and circulating gasifiers — are not suitable for BTL production. In addition, downdraft fixed-bed gasifiers face severe limitations in scaling and are fuel-inflexible, being able to process only fuels with well-defined properties. Fixed-bed updraft gasifiers have fewer restrictions in scaling, but the produced gas contains a lot of tars and methane.

Fluidized-bed gasifiers generally do not encounter limitations in scaling and are more flexible concerning the particle size of fuels. Nevertheless, they still have limited fuel flexibility due to a risk of slagging and fouling, agglomeration of bed material, and corrosion. The operating temperatures of air-blown fluidized bed gasifiers are therefore kept relatively low (800–1000°C), which implies incomplete decomposition of feedstocks, unless long residence times are used. Fluidized-bed gasifiers (especially the bubbling bed ones) tend to contaminate the product gas with dust. The oxygen-blown atmospheric or pressurized circulating fluidized-bed gasifiers and the steam-blown gas or char indirect gasifiers are better solutions for BTL production. Both gasifying concepts reduce significantly the amount of nitrogen in the product gas. In the first case it is achieved via substituting air with oxygen. In the second case, nitrogen ends up in the flue gas, but not in the product gas, because gasification and combustion are separate — the energy for the gasification is obtained by burning the chars from the first gasifier in a second reactor.

Typically, synthesis gas is used directly to generate heat and power, but several types of biofuels may be derived from synthesis gas. Hydrogen can be recovered from synthesis gas, or it can be catalytically converted to methanol. The gas can also be run through a biological reactor to produce ethanol or can also be converted using Fischer–Tropsch catalyst into a liquid stream with properties similar to diesel fuel. However, all of these fuels can also be produced from natural gas using a similar process.

3.4.3 *Solid fuels*

The term *solid fuel* generally refers to various forms of solid material that can release energy (in the form of heat and/or light) through the process of combustion. Solid fuels can be contrasted with gaseous fuels and liquid fuels. Common examples of solid fuels include wood, dried cow dung, charcoal, peat, and coal, which have been used throughout human history to create fire.

One widespread use of such fuels is in home cooking and heating in which the solid fuel may be burned on an open fireplace or in a special stove. The efficiency of this process may vary widely, from 10% for a well-made fire (even less if the fire is not made carefully) up to 40% for a custom designed charcoal stove. Inefficient use of fuel is a cause of deforestation (though this is negligible compared to the deliberate destruction to clear land for agricultural use) but more importantly it means that more effort has to be put into gathering fuel, thus the quality of cooking stoves has a direct influence on the viability of biofuels.

Solid fuels, compared to gaseous fuels and liquid fuels, are often cheaper, easier to extract, stabler to transport, and in many places, more readily available. Coal, in particular, is utilized for the generation of electricity because it is less expensive and more powerful than its gaseous or liquid fuel counterparts. However, solid fuels are also heavier to transport, require more destructive methods to extract/burn, and often have higher carbon, nitrate and sulphate emission. With the exception of sustainable wood/biomass, solid fuel (such as peat or coal) is typically considered to be non-renewable because it requires thousands of years to form.

3.5 Property Variation with Source

The quality and composition of a biofuel depends on the source of the biomass feedstock as well as the types of processing and conversion techniques utilized in its manufacture. Biomass feedstock composition ultimately decides the yield from the chemical or biochemical conversion processes, which, in turn, affects the economics involved. There are many plant varieties which are used as biofuel sources — the geography, weather conditions, soil composition, and legislation of a location normally dictate what types are grown specifically for biofuel production. Ethanol, biodiesel, and butanol are the main types of commercially produced biofuels.

3.5.1 *General observations*

The soil organic matter content contributes greatly to the grain and stover and hence carbohydrate content of maize plants. Lignin and cell-wall cross-linking also affects the ethanol production. Selection for reduced lignin content and increased cellulose content in stover can potentially be expected to increase mechanical strength as well as ethanol yield. Although pretreatment and enzyme hydrolysis constitute two of the more costly steps in cellulosic ethanol production, stover with reduced lignin may still need to be treated before being subjected to enzyme hydrolysis.

Genetics as well as environmental factors affect the chemical composition of the various parts of the plant, and it was found that husk, followed by rind and pith, has the highest sugar (glucan + xylan) content. The term "glucan" represents a diverse group of glucose polymers that differ in the position of glycosidic bonds, which can be short or long, branched or unbranched, alpha or beta isomers, and soluble or insoluble. On the other hand, the term *xylan* represents a group of hemicellulose derivatives.

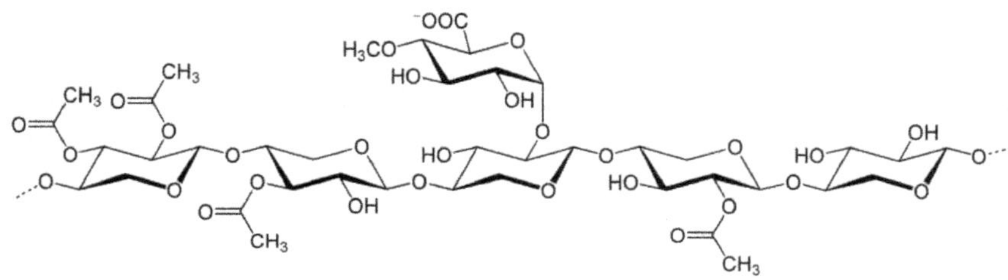

Structure of xylan from hardwood

The variation in the structure of the glucan derivatives and the xylan derivatives is due to differences in the amounts of the main chemical constituents of biomass (cellulose, hemicelluloses and lignin, all of which have different uses) being present in different proportions in the various parts of the plant.

The production of fuels from ligno-cellulosic feedstocks can be achieved through two very different processing routes: (i) the biochemical route, whereby enzymes and other micro-organisms are used to convert cellulose and hemicelluloses components of the feedstocks to sugars prior to their fermentation to produce ethanol and (ii) the thermochemical route, where pyrolysis and gasification technologies produce a synthesis gas (carbon monoxide and hydrogen) from which a wide range of long-chain biofuels, such as synthetic diesel or aviation fuel, can be reformed. In terms of the biochemical route, much remains to be done related to (i) improving feedstock characteristics, (ii) improving the efficiency of enzymes, and (iii) improving overall process integration. One key difference between the biochemical and thermochemical routes is that the lignin component is a residue of the enzymatic hydrolysis process and can be used for heat and power generation.

The production of biogas is essentially achieved by wastewater treatment facilities, which use anaerobic digestion to reduce the organic content of sewage sludge and animal wastes. The variation in the composition of biogas will depend on the organic content of the biomass as well as the pretreatment and processing (Chapter 9). In addition to the main components (methane and carbon dioxide), biogas can also contain a variety of contaminants and impurities, such as sulfur compounds (hydrogen sulfide, mercaptans, etc.), halogens, ammonia, and dust particles.

Concerning biodiesel production, transesterification optimization is desired, which would depend on the chemical composition of alkyl esters of vegetable oils, animal fats and cooking oil. Most common feedstocks possess fatty acid profiles consisting mainly of five C_{16} and C_{18} fatty acids, namely, palmitic (hexadecanoic), stearic (octadecanoic), oleic (9(*Z*)-octadecenoic), linoleic (9(*Z*),12(*Z*)-octadecadienoic) and linolenic (9(*Z*),12(*Z*),15(*Z*)-octadecatrienoic) acids, with the exception of a few oils such as coconut oil, which contains large amounts of saturated acids in the C_{12}–C_{16} range or others. Changing the fatty acid profile can be achieved by physical means, genetic modification of the feedstock or use of alternative feedstocks with different fatty acid profiles, such as wood-derived fatty-acids and related compounds.

Tall oil is a dark, viscous, and odorous liquid that phase-separates from the used pulping liquor (alkaline black liquor) that remains after the pulping of wood chips, and contains sodium soaps of rosin and fatty acids. Fish oil is also a potentially good source of fatty acids for the production of bio-oil. In a particular study, waste fish oil was converted into bio-oil by a fast pyrolysis process at 525°C (975°F) in a continuous pilot plant reactor with a yield of the order of 72% yield. Many different chemical groups can be produced during the pyrolysis reaction and the liquid product (bio-oil) obtained from triglyceride pyrolysis has a very complex composition. This bio-oil can be used directly as fuel or can be fractionated to obtain purified hydrocarbon derivatives in the range of gasoline and diesel.

Green algae (aquatic and unicellular) biomass is also used in biodiesel production, and is an attractive source of triglycerides, as under good conditions, green algae can double its biomass in less than 24 h. Factors such as carbon dioxide availability, sunlight, water, and space affects algal density, and the annual productivity and oil content of algae is far greater than that of seed crops.

One of the largest issues seems to be overall greenhouse gas emissions from the various biofuels when compared with crude oil fuels. To estimate the impacts of increases in renewable and alternative fuels on greenhouse gas emissions, the entire fuel lifecycle including fossil fuel extraction or feedstock growth, fuel production, distribution, and combustion should be taken into consideration to provide a comparison of the carbon dioxide emissions from different fuels.

It is generally accepted that biofuels have the potential to more drastically lower carbon dioxide emissions than oil-derived fuels, but in many instances, this is not the case. For example, ethanol made from corn requires a substantial amount of energy in terms of fertilization, irrigation, harvesting, and fermentation processes and most of this energy comes from fossil fuels. As a result, some ethanol production scenarios emit more lifecycle carbon dioxide emissions than gasoline. Cellulose-based ethanol, however, allows for more efficient and cost-effective fuel production, and the carbon footprint is decreased. Conversely, biodiesel blended into diesel at low levels reduces emissions of volatile organic compounds, carbon monoxide, particulate matter, and sulfur oxides during combustion. Furthermore, over the full lifecycle, biodiesel blends reduce carbon monoxide, particulate matter, and sulfur oxides compared with diesel.

The bulk density (related to energy density) of most solid biomass feedstocks is generally low, even after compression, and is approximately between 10% and 40% of the bulk density of most fossil fuels; however, liquid biofuels have comparable values. Since biomass materials are more reactive, with higher ignition stability, they are generally easier to gasify and thermochemically process into methanol and hydrogen (higher value fuels). Coal ash may contain toxic metals and small amounts of other trace contaminants, while biomass ash may be used as a soil amendment to assist in replacing nutrients removed by harvest. Bioethanol possesses approximately 70% of the heating value of crude oil distillates, but its sulfur and ash contents are appreciably lower.

Ethanol has a lower energy density than gasoline, so for a given volume of gasoline, a larger volume of ethanol is needed to produce an equivalent amount of energy. Unfortunately, if ethanol is used to power vehicles, it leads to lower gas mileage, since energy density is correlated with gas mileage. Ethanol, which has approximately 30% lower energy density than gasoline, contributes to a reduction in fuel mileage when it is mixed with gasoline, as is now commonly practised in the United States. Another problem with ethanol as an auto fuel is that it is extremely hygroscopic. Ethanol is difficult to separate from water and readily absorbs water from its surroundings. This is a problem in the transport and storage of ethanol, as well as the purification of ethanol from water. Ethanol-containing fuels can easily absorb water, and decontamination (i.e. removal of water from fuel) is difficult. The difference between the properties of gasoline and ethanol also causes compatibility problems in engines. Biodiesel is the only alternative fuel to have fully completed the health effects testing requirements of the Clean Air Act.

The use of biodiesel in a conventional diesel engine results in substantial reduction of unburned hydrocarbon derivatives, carbon monoxide, and particulate matter compared to emissions from diesel fuel. In addition, the exhaust emissions of sulfur oxides and sulfates (major components of acid rain) from biodiesel are essentially eliminated compared to diesel. Of the major exhaust pollutants, both unburned hydrocarbon derivatives and nitrogen oxides are ozone or smog forming precursors. The use of biodiesel results in a substantial reduction of unburned hydrocarbon derivatives. Emissions of nitrogen oxides are either slightly reduced or slightly increased depending on the duty cycle of the engine and the testing methods used. Based on engine testing, using the most stringent emission testing protocols required by EPA for certification of fuels or fuel additives in the USA, the overall ozone-forming potential of the speciated hydrocarbon emissions from biodiesel was nearly 50% lower than that measured for diesel fuel. In a review published in 2009, the environmental indicators related to biofuel production were examined.

The postulated superior properties of agrofuels when compared with fossil fuels, as we have seen, must be weighed very carefully among the various factors of cost, environmental impact, energy density, chemical composition, and availability and life cycle of food crops. No doubt, increasing advances in technology will tip the scales in favor of biofuels.

In conclusion, biofuels are classed according to the source and type. They are derived from forest, agricultural or fishery products or municipal wastes, as well as from agro-industry, food industry and food service by-products and wastes. They may be solid, such as fuelwood, charcoal and wood pellets; liquid, such as ethanol, biodiesel, and pyrolysis oils; or gaseous, such as biogas.

3.5.2 *Fuel specifications and performance*

Biodiesel has promising lubricating properties and cetane ratings compared to low-sulfur diesel fuels. Fuels with higher lubricity may increase the usable life of high-pressure fuel injection equipment that relies on the fuel for its lubrication. Depending on the engine, this might include high-pressure injection pumps, pump injectors (also called *unit injectors*), and fuel injectors.

Generally, the heat content (calorific value) of biodiesel is lower than that of the regular crude oil-derived diesel, and variation in biodiesel energy density is dependent more on the feedstock used than on the production process. Biodiesel can provide better lubricity and more complete combustion, thus increasing the engine energy output and partially compensating for the higher energy density of petrodiesel. The color of biodiesel ranges from golden to dark brown, depending on the production method. It is slightly miscible with water, and has a high boiling point and, consequently a low vapor pressure. The flash point of biodiesel exceeds 130°C (266°F), which is significantly higher than that of crude oil-derived diesel, which may be as low as 52°C (126°F). Biodiesel has a density of the order of 0.88 g/cm^3, which is higher than that of petrodiesel (which is approximately 0.85 g/cm^3). Typically, biodiesel does not contain sulfur, and it is often used as an additive to ultra-low sulfur diesel (ULSD) fuel to aid in lubrication, as the sulfur compounds in petrodiesel provide much of the lubricity.

ASTM International (ASTM), previously known as the American Society for Testing Materials, is an international organization which develops and publishes information on the technical standards of various products, materials, systems, and services. It is one of the largest and most highly regarded standards development organizations in the world. The available literature on the performance of biofuels, compared with traditional fossil fuels, normally uses ASTM and International Standards Organization (ISO) specifications and parameters. The specifications provide details of requirements for fuel characteristics as well as the relevant standard test methods to use for each.

Because of the focus on alternate fuel standards (particularly the standards for biodiesel) over the past two decades, the development of biodiesel standards started in the 1990s in order to support the increasing use of alkyl ester-based biodiesel and its blends as automotive fuels. The first ASTM standard (ASTM D6751) was adopted in 2002, while in Europe, EN 14214 biodiesel standard (based on the former DIN 51606) was finalized in October 2003. The US and EU standards have international significance; they are usually the starting point for biodiesel specifications developed in other countries. However, in the United States, the standard (ASTM D6751) establishes specifications for a *biodiesel blend stock* for middle distillate fuels and, while the specification was written for B100, it is not intended for neat biodiesel used as automotive fuel. Rather, it is for the biodiesel component that is to be blended to produce biodiesel/diesel fuel blends. Since 2012, the ASTM D6751 standard has defined two grades of biodiesel: (i) grade 2-B, which is identical to biodiesel defined by earlier versions of the standard, and (ii) grade 1-B,

Table 3.6: Examples of ASTM biodiesel blend specifications.

Test method	Title and comment
ASTM D975	Specification for diesel fuel oils. Used for on- and off-road diesel applications. Revised to include requirements for up to 5% v/v biodiesel.
ASTM D396	Specification for fuel oils. Used for home heating and boiler applications. Revised to include requirements for up to 5% biodiesel.
ASTM D7467	Specification for diesel fuel oil, biodiesel blend (B6 to 20). Covers finished fuel blends of between 6% (B6) and 20% (B20) biodiesel for on- and off-road diesel engine use.

which has stricter controls on monoglycerides and cold soak filterability. In addition, the ASTM Standard Specification for Diesel Oil (ASTM D975) was modified in 2008 to allow up to 5% biodiesel to be blended into the fuel, while the standard for biodiesel blends (ASTM D7467) is a specification for biodiesel blends from B6 to B20 (Table 3.6).

Most countries of the world use a system known as the "B" factor to state the amount of biodiesel in any fuel mix, in contrast to the "BA" or "E" system used for bioalcohol. Pure biodiesel is referred to as B100, while a fuel containing 20% biodiesel is labeled "B20". The common international standard for biodiesel is EN 14214, while the ASTM 6751 standard is most referenced in the USA. In Germany, the requirements for biodiesel are fixed in the DIN EN 14214 standard.

One of the most important fuel properties of biodiesel and conventional diesel fuel derived from crude oil is viscosity, which is also an important property of lubricants. The ranges of kinematic viscosity acceptable are specified in various biodiesel and crude oil standards. Reducing the viscosity is one of the main purposes of the transesterification of vegetable oils or fats to biodiesel because high viscosity of neat vegetable oils or fats ultimately will lead to operational problems such as engine deposits.

The viscosity of biodiesel is slightly higher than that of petrodiesel but approximately one order of magnitude lower than that of the parent starting material (vegetable oil or fat). Biodiesel and its blends with petrodiesel display temperature-dependant viscosity similar to that of neat petrodiesel. Influencing factors are the chain length, position, number and nature of double bonds, and the nature of the oxygenated moieties.

Generally, biodiesel has a higher cloud point (the temperature at which a fuel becomes hazy or cloudy and starts to gel) than petrodiesel. This makes its use impractical in cooler climates and limits its potential market. Other important chemical and physical properties described in ASTM standards for biodiesel are acid number (total acid number, TAN, indicates the presence of free fatty acids and carboxylic acids), corrosion (describes the potential for copper corrosion, measured using ASTM method D130), low temperature performance (describes pour points (PP) and cloud points (CP) using ASTM D5949 and ASTM D5773 methods), and oxidation stability (normally evaluated using Differential Scanning Calorimetry and Oxidation Stability Index). In addition, according to the EMA (Engine Manufacturers Association), a blend of crude oil diesel fuel meeting ASTM D975 and 100% (neat) biodiesel fuel meeting either ASTM 6751 or EN 14214, where the biodiesel content of the blended fuel is no more than 20% biodiesel by volume (B20), shall meet the requirements identified at the point of delivery of the fuel to the end user (Table 3.7).

Table 3.7: Examples of final blend fuel requirements (at the point of delivery).

Item	Performance characteristic	Requirements		Test procedure[a]
		D1 blends	D2 blends	
1	Flash point (°C, min)	38	52	ASTM D93
2	Water and Sediment (v/v%, max)	0.05	0.05	ASTM D2709 or D1796
3	Physical distillation (°C, max)	343	343	ASTM D86
4	Kinematic Viscosity, cSt at 40°C	1.3–4.1	1.9–4.1	ASTM D445
5	Ash (w/w%, max)	0.01	0.01	ASTM D482
6	Sulfur (w/w%, max)	Per regulation	Per regulation	
7	Copper strip corrosion rating (max)	No. 3	No. 3	ASTM D130
8	Cetane number (min)	43	43	ASTM D613
9	Cloud point	Ambient temperature		ASTM D2500
10	Ramsbottom carbon residue	0.15	0.35	ASTM D524
11	Lubricity, 60°C (micron, max)	460	460	ASTM D6079
12	Acid number, mg KOH/g (max)	0.3	0.3	ASTM D664
13	Phosphorous (w/w%, max)	0.001	0.001	ASTM D4951
14	Thermo-oxidative stability	10	10	ASTM D2274

Note: [a]Annual Book of Standards, 2019. ASTM International, West Conshohocken, Pennsylvania.

In terms of the effects of biodiesel on fuel filters, the use of biodiesel blends which do not meet the designated specifications has been shown to drastically reduce filter life. Blends greater than B20 may have enough of a solvent to break down the varnish deposits on the walls of existing fuel storage tanks or fuel systems. The breakdown of these varnish deposits will contaminate the fuel with particulates, which can cause fuel filters to plug rapidly.

Another disadvantage of biodiesel is that it tends to reduce fuel economy. Energy efficiency is the percentage of the thermal energy of the fuel that is delivered as engine output, and biodiesel has shown no significant effect on the energy efficiency of any test engine. Volumetric efficiency, a measure that is more familiar to most vehicle users, is usually expressed as miles traveled per gallon of fuel (or kilometers per liter of fuel). The energy content per gallon of biodiesel is approximately 11% lower than that of crude oil diesel. Vehicles running on B20 are therefore expected to achieve 2.2% (20% × 11%) fewer miles per gallon of fuel.

Areas of concern and interest are for the biofuel industry to have in place a good quality-control protocol for the measurement of bioalcohols, to avoid metal corrosion from water and acid corrosion (due to weak and strong acids and inorganic chlorides in solution). Also of importance are the limits set on phosphorus content (less than 5.0 mg/L in ethanol) to prevent engine catalyst deterioration, and copper content (less than 0.1 mg/kg), along with sulfur content (less than 10 mg/kg).

Up to a 10% blend level, the performance of bioethanol-blended gasoline is very similar to ordinary gasoline. At higher levels, however, some engines may begin to exhibit problems, for example, stumbling under slight acceleration. The fuel also has more aggressive properties at higher concentrations of bioethanol, which increases the possibility of deterioration of some components. Gasoline must be volatile enough to move from the carburetor or injectors into the cylinders and to vaporize prior to combustion. However, gasoline cannot have such a volatility that allows it to vaporize and boil in the injectors, carburetor, fuel lines or fuel pump, which could prevent it from being metered correctly. Also, if the gasoline is too volatile, more gasoline evaporates into the air, adding to environmental problems. There are a number of volatility specifications to ensure that suppliers get this balancing act right. Adding bioethanol to gasoline as low-level blends increases the volatility of the blended fuel. The Engine Fuel Specifications Regulations specify volatility measures for bioethanol-blended petrol. The limits for blends are similar to those for gasoline so as to ensure that no changes in vehicles are required. Bioethanol introduces more oxygen into the fuel. In vehicles with simple fuel metering systems such as carburetors, this causes the mixture to become a little leaner. Leaning is good for fuel economy and is generally good for lowering some types of exhaust emissions. However, it may cause some engines to stumble if they are already tuned reasonably lean. If a vehicle stumbles on bioethanol-blended gasoline, re-tuning should solve the problem. A vehicle tuned correctly for use on ordinary gasoline would normally not exhibit problems when using bioethanol blends.

References

Bala, B.K. 2005. Studies on biodiesels from transformation of vegetable oils for diesel engines. *Energy Edu. Sci. Technol.* 15: 1–43.

Ban-Weiss, G.A., Chen, J.Y., Buchholz, B.A., and Dibble, R.W. 2007. A numerical investigation into the anomalous slight NO_x increase when burning biodiesel; A new (old) theory. *Fuel Process. Technol.* 88: 659–667.

Barnwal, B.K. and Sharma, M.P. 2005. Prospects of biodiesel production from vegetable oils in India. *Renew. Sust. Energy Rev.* 9: 363–378.

Bernardes, O.L., Bevilaqua, J.V., Leal, M.C.M.R., Freire, D.M.G., and Langone, M.A.P. 2007. Biodiesel fuel production by the transesterification reaction of soybean oil using immobilized lipase. *Appl. Biochem. Biotechnol.* 137–140: 105–114.

Boerrigter H. and Van Der Drift A. 2004. Biosyngas: Description of R&D Trajectory Necessary to Reach Large-scale Implementation of Renewable Syngas from Biomass. Energy Research Center of the Netherlands, Petten, The Netherlands.

Çanakçı, M. and Van Gerpen. J. 2003. A pilot plant to produce biodiesel from high free fatty acid feedstocks. *Trans. ASAE* 46: 945–955.

Çanakçı, M. and Özsezen, A.N. 2005. Evaluating waste cooking oils as alternative diesel fuel. G.U. *J. Sci.* 18: 81–91.

Chadeesingh, R. 2011. The Fischer–Tropsch Process. In *The Biofuels Handbook*, J.G. Speight (Editor). The Royal Society of Chemistry, London, United Kingdom. Part 3, Chapter 5, pp. 476–517.

Cobb, J.T. Jr. 2007. Production of Synthesis Gas by Biomass Gasification. Proceedings. Spring National Meeting AIChE. Houston, Texas, April 22–26, 2007.

De Jong, W. 2015. Biomass Composition, Properties, and Characterization. In *Biomass as a Sustainable Energy* Source *for the Future: Fundamentals of*

Conversion Processes, W. De Jong and J.R. Van Ommen (Editors). American Institute of Chemical Engineers Inc., Washington, DC.

Demirbaş, A. 2002. Biodiesel from vegetable oils via transesterification in supercritical methanol. *Energy Convers. Mgmt.* 43: 2349–2356.

Demirbaş, A. 2003. Biodiesel fuels from vegetable oils via catalytic and non-catalytic supercritical alcohol transesterifications and other methods: A survey. *Energy Convers. Mgmt.* 44: 2093–2109.

Demirbaş, A. 2006a. Biodiesel from sunflower oil in supercritical methanol with calcium oxide. *Energy Convers Mgmt.* 48: 937–941.

Demirbaş, A. 2006b. Biodiesel production via non-catalytic SCF method and biodiesel fuel characteristics. *Energy Convers Mgmt.* 47: 2271–2282.

Demirbaş, A. and Karslıoğlu, S. 2007. Biodiesel production facilities from vegetable oils and animal fats. *Energy* Sources, *Part A* 29: 133–141.

Dmytryshyn, S.L., Dalai, A.K., Chaudhari, S.T., Mishra, H.K., and Reaney, M.J. 2004. Synthesis and characterization of vegetable oil derived esters: Evaluation for their diesel additive properties. *Biores. Technol.* 92: 55–64.

Encinar, J.M., Gonzalez, J.F. Rodriguez, J.J., and Tejedor, A. 2002. Biodiesel fuels from vegetable oils: Transesterification of *Cynara cardunculus L.* oils with ethanol. *Energy Fuel* 16: 443–450.

Freedman, B., Butterfield, R.O., and Pryde, E.H. 1986. Transesterification kinetics of soybean oil. *J. Am. Oil. Chem. Soc.* 63: 1375–1380.

Furuta, S., Matsuhasbi, H., and Arata, K. 2004. Biodiesel fuel production with solid superacid catalysis in fixed bed reactor under atmospheric pressure. *Catalysis Commun.* 5: 721–723.

Gary, J.H., Handwerk, G.E., and Kaiser, M.J. 2007. *Petroleum Refining: Technology and Economics*, 5th Edition. CRC Press, Taylor & Francis Group, Boca Raton, Florida.

Ghanem, A. 2003. The utility of cyclodextrins in lipase-catalyzed transesterification in organic solvents: Enhanced reaction rate and enantioselectivity. *Org. Biomol. Chem.* 1: 1282–1291.

Goff, M.J., Bauer, N.S., Lopes, S., Sutterlin, W.R., and Suppes, G.J. 2004. Acid-catalyzed Alcoholysis of Soybean Oil. *J. Am. Oil. Chem. Soc.* 81: 415–420.

Graboski, M.S. and McCormick, R.L. 1998. Combustion of fat and vegetable oil derived fuels in diesel engines. *Prog. Energy Combust. Sci.* 24: 125–164.

Gupta, P.K., Kumar, R., Panesar, P.S., and Thapar, V.K. 2007. Parametric studies on bio-diesel prepared from rice bran oil. *Agricultural Engineering International*: *The CIGR Ejournal*, Manuscript EE 06 007, Vol. IX.

Hsu, C.S. and Robinson, P.R. (Editors) 2017. *Handbook of Petroleum Technology*. Springer International Publishing AG, Chaim, Switzerland.

Iso, M., Chen, B., Eguchi, M., Kudo, T., and Shrestha, S. 2001. Production of biodiesel fuel from triglycerides and alcohol using immobilized lipase. *J. Molecular Catal. B: Enzymatic* 16: 53–58.

Jeong, G.T. and Park, D.H. 1996. Batch (one- and two-stage) production of biodiesel fuel from rapeseed oil. *Appl. Biochem. Biotechnol.* 131: 668–679.

Kavalov, B. and Peteves, S.D. 2005. Status and perspectives of Biomass-to-Liquid Fuels in the European Union. European Commission. Directorate General Joint Research Centre (DG JRC). Institute for Energy, Petten, Netherlands.

Knothe, G. 2005. Dependence of biodiesel fuel properties on the structure of fatty acid alkyl esters. *Fuel Process. Technol.* 86: 1059–1070.

Korytkowska, A., Barszczewska-Rybarek, I., and Gibas, M. 2001. Side-reactions in the transesterification of oligoethylene glycols by methacrylates. *Designed Monomers Polym.* 4: 27–37.

Kulkarni, M.G. and Dalai, A.K. 2006. Waste cooking oils: An economical source for biodiesel: A Review. *Ind. Eng. Chem. Res.* 45: 2901–2913.

Kusdiana, D. and Saka, S. 2004a. Effects of water on biodiesel fuel production by supercritical methanol treatment. *Biores. Technol.* 91: 289–295.

Kusdiana, D. and Saka, S. 2004b. Two-step preparation for catalyst-free biodiesel fuel production: Hydrolysis and methyl esterification. *Appl. Biochem. Biotechnol.* 115: 781–791.

Lee, Y., Park, S.H., Lim, I.T., Han, K., and Lee, S.Y. 2000. Preparation of alkyl (R)-(2)-3-hydroxybutyrate by acidic alcoholysis of poly-(R)-(2)-3-hydroxybutyrate. *Enzyme Microbial Technol.* 27: 33–36.

Liu, Y., Lotero, E., and Goodwin, J.G. Jr. 2006. Effect of water on sulfuric acid catalyzed esterification. *J. Molecular Catalysis A: Chem.* 245: 132–140.

Lopez, D.E., Goodwin, J.G. Jr, Bruce, D.A., and Lotero, E. 2005. Transesterification of triacetin with methanol on solid acid and base catalysts. *Appl. Catalysis A: General* 295: 97–105.

Lotero, E., Goodwin, J.G. Jr., Bruce, D.A., Suwannakarn, K., Liu, Y., and Lopez, D.E. 2006. The catalysis of biodiesel synthesis. *Catalysis* 19: 41–83.

Lucia, L.A., Argyropoulos, D.S., Adamopoulos, L., and Gaspar, A.R. 2006. Chemicals and energy from biomass. *Can. J. Chem.* 8: 960–970.

Ma, F., Clements, L.D., and Hanna, M.A. 1998. The effect of catalyst, free fatty acids, and water on transesterification of beef tallow. *Trans. ASAE* 41: 1261–1264.

Ma, F. and Hanna, M.A. 1999. Biodiesel production: A review. *Biores. Technol.* 70: 1–15.

Macedo, C.C.S., Abreu, F.R., Tavares, A.P., Alves, M.P., Zara, L.F., Rubim, J.C., and Suarez, P.A.Z. 2006. New heterogeneous metal-oxides based catalyst for vegetable oil transesterification. *J. Braz. Chem. Soc.* 17: 1291–1296.

Marchetti, J.M., Miguel, V.U., and Errazu, A.F. 2005. Possible methods for biodiesel production. *Renew. Sust. Energy Rev.* 11: 1300–1311.

Marinetti, G.V. 1962. Hydrolysis of lecithin with sodium methoxide. *Biochemistry* 1: 350–353.

Marinetti, G.V. 1966. Low temperature partial alcoholysis of triglycerides. *J. Lipid Res.* 7: 786–788.

Meher, L.C., Kulkarni, M.G., Dalai, A.K., and Naik, S.N. 2006a. Transesterification of Karanja (*Pongamia pinnata*) oil by solid basic catalysts. *Eur. J. Lipid Sci. Technol.* 108: 389–397.

Meher, L.C., Sagar, D.V., and Naik, S.N. 2006b. Technical aspects of biodiesel production by transesterification — A review. *Renew. Sust. Energ. Rev.* 10: 248–268.

Mudhoo, A., Forster-Carneiro, T., and Sánchez, A. 2011. Biohydrogen production and bioprocess enhancement: A review. *Crit. Rev. Biotechnol.* 31(3): 250–263.

Nichols, N.N., Dien, B.S., Bothast, R.J., and Cotta, M.A. 2006. The corn ethanol industry. In *Alcoholic Fuels*, S. Minteer (Editor). CRC-Taylor & Francis, Boca Raton, Florida. Chapter 4.

Noureddini, H., Harkey, D., and Medikonduru, V. 1998. A continuous process for the conversion of vegetable oils into methyl esters of fatty acids. *JAOCS* 75: 1775–1783.

Parkash, S. 2003. *Refining Processes Handbook.* Gulf Professional Publishing, Elsevier, Amsterdam, Netherlands.

Pinto, A.C., Guarieiro, L.N.N., Rezende, M.J.C., Ribeiro, N.M., Torres, E.A., Lopes, W.A., Pereira, P.A.P., and De Andrade, J.B. 2005. Biodiesel: An overview. *J. Braz. Chem. Soc.* 16: 1313–1330.

Reyes-Duarte, D., Lopez-Cortes, N., Ferrer, M., Plou, F., and Ballesteros, A. 2005. Parameters affecting productivity in the lipase-catalysed synthesis of sucrose palmitate. *Biocatalysis Biotransform.* 23: 19–27.

Royon, D., Daz, M., Ellenrieder, G., and Locatelli, S. 2007. Enzymatic production of biodiesel from cotton seed oil using *t*-butanol as a solvent. *Biores. Technol.* 98: 648–653.

Saka, S. and Kusdiana, D. 2001. Biodiesel fuel from rapeseed oil as prepared in supercritical methanol. *Fuel* 80: 225–231.

Saka, S. and Minami, E. 2006. A novel non-catalytic biodiesel production process by supercritical methanol as NEDO high efficiency bioenergy conversion project. In *The 2nd Joint International Conference on "Sustainable Energy and Environment (SEE 2006)"*, Bangkok, Thailand, November 21–23.

Sørensen, B.E., Njakou, S., and Blumberga, D. 2006. Gaseous Fuels Biomass. In *Proceedings of the World Renewable Energy Congress IX*. WREN, London, United Kingdom.

Schuchardta, U., Serchelia, R., and Vargas, R.M. 1998. Transesterification of vegetable oils: A review. *J. Braz. Chem. Soc.* 9: 199–210.

Shah, S. and Gupta, M.N. 2007. Lipase catalyzed preparation of biodiesel from Jatropha oil in a solvent free system. *Process Biochemis.* 42: 409–414.

Sołowski, G. 2018. Biohydrogen Production — Sources and methods: A review. *Int. J. Bioprocess. Biotech.* IJBBT-101. doi: 10.20911/IJBBT-101. 100001, https://www.researchgate.net/publication/323869568_Biohydrogen_Production_-_Sources_and_Methods_A_Review.

Speight, J.G. 2011. *The Refinery of the Future*. Gulf Professional Publishing, Elsevier, Oxford, United Kingdom.

Speight, J.G. 2013. *The Chemistry and Technology of Coal*, 3rd Edition. CRC Press, Taylor & Francis Group, Boca Raton, Florida.

Speight, J.G. 2014. *The Chemistry and Technology of Petroleum*, 5th Edition. CRC Press, Taylor & Francis Group, Boca Raton, Florida.

Speight, J.G. 2015. *Handbook of Petroleum Product Analysis*, 2nd Edition. John Wiley & Sons Inc., Hoboken, New Jersey.

Speight, J.G. 2016. Hydrogen in Refineries. In *Hydrogen Science and Engineering: Materials, Processes, Systems, and Technology*, D. Stolten and B. Emonts (Editors). Wiley-VCH Verlag GmbH & Co., Weinheim, Germany. Chapter 1. pp. 3–18.

Speight, J.G. 2017. *Handbook of Petroleum Refining*. CRC Press, Taylor & Francis Group, Boca Raton, Florida.

Speight, J.G. 2019. *Handbook of Petrochemical Processes*. CRC Press, Taylor & Francis Group, Boca Raton, Florida.

Srivastava. A. and Prasad, R. 2000. Triglycerides-based diesel fuels. *Renew. Sust. Energy Rev.* 4: 111–133.

Stavarache, C., Vinatoru, M., Nishimura, R., and Maed, Y. 2005. Fatty acids methyl esters from vegetable oil by means of ultrasonic energy. *Ultrasonics Sonochem.* 12: 367–372.

Van Der Drift, A., Van Ree, R., Boerrigter, H., and Hemmes, K. 2004. Bio-Syngas: Key Intermediate for Large Scale Production of Green Fuels and

Chemicals, In *Proceedings of the 2nd World Conference and Technology Exhibition "Biomass for Energy*, Industry and Climate Protection", Rome, http://www.conference-biomass.com/Biomass2004/conference_Welcome.asp.

Van Gerpen, J., Shanks, B., Pruszko, R., Clements, D., and Knothe, G. 2004. Biodiesel Analytical Methods: August 2002–January 2004. National Renewable Energy Laboratory, NREL/SR-510-36240, Colorado, July.

Varghaa, V., and Truterb, P. 2005. Biodegradable polymers by reactive blending trans-esterification of thermoplastic starch with poly(vinyl acetate) and poly(vinyl acetate-co-butyl acrylate). *Eur. Polym. J.* 41: 715–726.

Vera, C.R., D'Ippolito, S.A., Pieck, C.L., and Parera, J.M. 2005. Production of biodiesel by a two-step supercritical reaction process with adsorption refining. In *2nd Mercosur Congress on Chemical Engineering and 4th Mercosur Congress on Process Systems Engineering* (*ENPROMER-2005*), Rio de Janeiro, Brasil, August 14–18.

Watanabe, Y., Shimada, Y., and Sugihara, A. 2000. Continuous production of biodiesel fuel from vegetable oil using immobilized *Candida antarctica lipase*. *J. Am. Oil. Chem. Soc.* 77: 355–360.

Watanabe, Y., Shimada, Y., Sugihara, A., and Tominaga, T. 2002. Conversion of degummed soybean oil to biodiesel fuel with immobilized *Candida antarctica lipase*. *J. Mol. Catal. B: Enzym.* 17: 151–155.

Wright, L., Boundy, R, Perlack, R., Davis, S., and Saulsbury, B. 2006. *Biomass Energy Data Book*: *Edition 1*. Office of Planning, Budget and Analysis, Energy Efficiency and Renewable Energy, United States Department of Energy. Contract No. DE-AC05-00OR22725. Oak Ridge National Laboratory, Oak Ridge, Tennessee.

Zhang, Y., Dube, M.A., McLean, D.D., and Kates, M. 2003. Biodiesel production from waste cooking oil: 1. Process design and technological assessment. *Biores. Technol.* 89: 1–16.

Part 2

Feedstocks

Chapter 4

Feedstocks for Fossil Fuels

4.1 Introduction

Alternate fuels (also called synthetic fuels or synfuels), such as gases, liquid fuels, and solid fuels, are making inroads into the fuel market. Currently, the production of alternate fuels from non-conventional sources (i.e. non-fossil fuel sources, such as the various types of biomass) is starting to become active at a level of prominence not seen in the latter parts of the 20th century. Also, the production of crude oil is costlier and some types of the crude oil family (Chapter 1) have a larger environmental impact. In addition, the non-conventional fossil fuel sources of fuels may be increasingly relied upon as fuel for transportation as the price of conventional crude oil increases and supplies dwindle.

While conventional sources of liquid fuels from crude oil are currently preferred, voices are being raised for the establishment of an industry that produces and develops liquid fuels from non-conventional sources but there is still a long way to go. Incentives are still needed to develop such resources but the dynamics are now coming into place for the establishment of a synthetic fuels industry and it is up to various levels of government not just to promote the establishment of such an industry but lead the way recognizing that it is not only supply and demand but also the available and variable technology.

Natural gas and crude oil from shale and tight formations offer additional energy-producing resources can be recovered from shale formations that typically function as both reservoir rocks and source rocks.

In terms of chemical makeup, shale gas is typically a dry gas composed primarily of methane, but some formations do produce wet gas while the crude oil from tight formations is typically more volatile than many of the crude oils from conventional reservoirs. The shale formations that yield natural gas and crude oil are organic-rich shale formations but were previously regarded only as source rocks and seals for the gas accumulating in the strata near sandstone and carbonate reservoirs of traditional onshore gas development. On the other hand, the crude oil from shale formations is a light, highly volatile crude oil that contains more of the volatile hydrocarbon derivatives (C1–C4) than many conventional crude oils. In addition, the gas and oil storage properties of shale are quite different from conventional reservoirs. In the case of shale formations, the natural gas and crude oil present in the matrix system of pores — similar to those found in conventional reservoir rocks — are accompanied by a gas or oil that is bound to, or adsorbed on, the surface of inorganic minerals in the shale. The relative contributions and combinations of free gas and oil from matrix porosities and from desorption of adsorbed constituents are a key determinant of the production profile of the well.

Thus, in any book on the production of alternate fuels, a chapter on fossil fuels is necessary in order that meaningful comparisons can be made between the conventional fossil fuel industry (natural gas and the crude oil family) and the alternate fuel industry (extra heavy crude oil, tar sand bitumen, coal, and oil shale).

This chapter provides an overview of the fuels and other products derived from crude oil refining in order for the reader to place each process in the correct context of the production of conventional fuels and to understand the process modifications often necessary to convert other feedstocks to fuels that meet the necessary composition and specification.

4.2 Natural Gas

Natural gas (also called marsh gas and swamp gas in the older literature and more recently landfill gas) is a naturally occurring gaseous fossil fuel found in conjunction with crude oil in coal seams as well as in reservoirs that contain natural gas without any crude oil. Briefly, for the sake of clarity, natural gas is not the same as *town gas*, which is manufactured from coal, and *coal gas*, *manufactured gas*, *producer gas*, and *synthetic natural gas* (SNG) that are also produced from coal. Depending on the production process, a gas from a carbonaceous feedstock is a mixture of hydrogen, carbon monoxide, carbon dioxide, methane, and volatile hydrocarbon derivatives in varying amounts as well as small amounts of nitrogen.

Prior to the development of natural gas resources, virtually all fuel gases and lighting gases were manufactured from coal (Speight, 2013), and the history of natural gas cleaning has its roots in town gas cleaning (Mokhatab *et al.*, 2006; Speight, 2019a). The by-product coal tar produced during the manufacture of gas from coal was an important feedstock for the chemical industry. The development of manufactured gas parallels that of the Industrial Revolution and urbanization.

4.2.1 *History*

The uses of natural gas did not necessarily parallel its discovery. In fact, the discovery of natural gas dates from ancient times in the Middle East. During recorded historical time, there was little or no understanding of natural gas; it posed somewhat of a mystery to man. Sometimes, such things as lightning strikes would ignite natural gas escaping from the crust of the Earth. This would cause a fire to emerge from the earth burning the natural gas seeping out of the ground. These fires puzzled most early civilizations and were the root of much myth and superstition. One of the most famous of these types of flames was found in Ancient Greece, on Mount Parnassus, approximately 1,000 BC and appeared as a flame rising from a fissure in the rock. These types of springs became prominent in the religions of Greece, India, and Persia because of their (supposedly) divine or supernatural origin. However, the energy value of natural gas was not recognized until approximately 900 BC in China, when the first known natural gas well was drilled in 211 BC. Pipelines were formed from bamboo to transport the gas to a collection point where it was used to boil sea water, thereby separating the salt and making the water drinkable.

On the other side of the world, natural gas was discovered and identified in North America as early as 1626, when French explorers discovered natives igniting gases that were seeping into and around Lake Erie. In Europe, natural gas was unknown until it was discovered in Great Britain in 1659 and Britain was the first country to commercialize the use of gas in 1785, when gas produced from coal was used to light houses as well as provide street lighting. Manufactured gas of this type (as opposed to naturally occurring gas) was first brought to the United States in 1816, when it was used to light the streets of Baltimore, Maryland. This manufactured gas was much less efficient, and less environmentally friendly, than modern natural gas that comes from the ground.

In 1821, in Fredonia, United States, residents observed gas bubbles rising to the surface from a creek. This led to the first natural gas well in North America and the beginnings of the natural gas industry (Mokhatab *et al.*, 2006; Speight, 2019). In 1859, Colonel Edwin Drake (a former railroad conductor who seems to have adopted the title "Colonel") dug the first well. Drake hit a reservoir of crude oil and natural gas at 69 feet below the surface of the earth. In these early days, natural gas was often discovered as a consequence of prospecting for crude oil but it was an unwelcome by-product since, as natural gas reservoirs were not tapped in the drilling process, workers were forced to stop drilling to let the gas vent freely into the air.

Throughout the 19th century, natural gas was used almost exclusively as a source of light and its use remained localized because of the lack of transport structures, making it difficult to transport large quantities of natural gas over long-distances. There was an important change in 1890 with the invention of leak proof pipeline couplings but transportation of natural gas to long-distance customers did not become practical until the 1920s as a result of technological advances in pipelines. Moreover, it was only after World War II that the use of natural gas grew rapidly because of the development of efficient pipeline networks and storage systems.

4.2.2 *Formation and occurrence*

Natural gas is found occurring along with crude oil (where it is termed *associated gas*) or alone (where it is termed *non-associated gas*) in reservoirs and coal seams (also called coal beds with the gas being known as coalbed methane) (Mokhatab *et al.*, 2006; Speight, 2013, 2019). Natural gas often occurs in conjunction with crude oil, although natural gas reservoirs that contain condensate (higher molecular weight hydrocarbon derivatives up to approximately C10) but no crude oil are equally well known.

Similar to crude oil, natural gas has been generated over geological time (i.e. millions of years) by the decay of animal remains and plant remains (organic debris). In such situations, the mud and soil that covered the organic debris metamorphosed into impermeable rock and trapped the debris beneath the newly-formed rock sediments. Pressure and, to some extent, heat (the geothermal gradient) changed some of the organic material into coal, some into crude oil, and some into natural gas. Whether or not the debris formed coal, crude oil, or gas depended on the nature of the debris and the localized conditions under which the changes occurred.

Although the geothermal gradient varies from place to place, it is generally of the order of 25–30°C/km (15°F/1000 ft or 120°C/1000 ft, i.e. 0.015°C per foot of depth or 0.012°C per foot of depth), i.e. approximately 1° for every 100 feet below the surface. Thus, on increasing the depth of the reservoir, there is a tendency for crude oil to become lower boiling (lower density) insofar as it now contains increasing amounts of low-molecular-weight hydrocarbon derivatives and decreasing amounts of higher molecular weight constituents.

However, there is considerable discussion on the heat to which the organic precursors have been subjected. It is by no means certain that cracking temperatures (≥300°C, ≥572°F) have played a role in natural gas formation. Maturation of the organic debris through temperature effects occurred over geological time and shortening the time to laboratory time (hours, days, or months, which cannot simulate geological time) and increasing the temperature to above and beyond the cracking temperature (at which the chemistry changes) do not offer conclusive proof that high temperatures have played a role in the formation of natural gas or crude oil (Speight, 2014a).

Nevertheless, at some point during or after the maturation process, the gas and crude oil once formed migrated from the source rock either upward or sideways or in both directions (subject to the structure of the accompanying and overlying geological formations). Eventually, the gas and crude oil became trapped in reservoirs that may be many miles from the source rock. It is rarely that the source rock and the reservoir were one and the same. Thus, a natural gas field may have a series of layers of crude oil/gas and gas reservoirs in the subsurface. In some instances, the natural gas and crude oil parted company, leading to the occurrence of reservoirs containing only gas (non-associated gas) or crude oil reservoirs containing very little gas.

Typically, a reservoir is generally a geological formation that is made up of layers of porous, sedimentary rock, such as sandstone, in which the gas and/or the crude oil can collect. However, for retention of the gas each trap must have an impermeable base rock and an impermeable cap rock to prevent further movement of the gas. Such formations, known as *reservoirs* or *traps* (i.e. naturally occurring storage areas), vary in size and can retain varying amounts of natural gas and/or crude oil.

There are a number of different types of these formations, but the most common is, characteristically, a folded rock formation such as an anticline as occurs in many crude oil reservoirs (Parkash, 2003; Gary *et al.*, 2007; Speight, 2014a, 2017a; Hsu and Robinson, 2017), which traps and holds natural gas and/or crude oil. On the other hand, a reservoir may be formed by a geological fault that occurs when the normal sedimentary layers sort of split vertically, so that impermeable rock shifts down to trap natural gas in the more permeable limestone or sandstone layers. Essentially, the geological formation which layers impermeable rock over more porous formations has the potential to form a reservoir.

Reservoirs vary in size from a few hundred feet to several miles across and tens to hundreds of feet thick, with the natural gas and crude oil trapped against a lower and an upper impermeable formation (Speight, 2014a, 2019). Some reservoirs may be only hundreds of feet below the surface, while other reservoirs are thousands or even tens of thousands of feet underground. In the United States, several reservoirs have been discovered at depths greater than 30,000 feet. Many offshore wells are drilled in thousands of feet of water and penetrate tens of thousands of feet into the sediments below the sea floor.

Natural gas reservoirs, similar to crude oil reservoirs, exist in many forms such as the dome (syncline–anticline) structure (Parkash, 2003; Gary *et al.*, 2007; Speight, 2014a, 2017a; Hsu and Robinson, 2017) with water below or a dome of gas with a crude oil rim and water below the oil. When the water is in direct contact with the gas, pressure effects may dictate that a considerable portion of the gas (20% or more) is dissolved in the crude oil as well as in the water. As gas is produced (or recovered) from the reservoir, the reservoir pressure declines, allowing the dissolved gas to emerge from the liquid and enter the gas phase. In addition, and because of the variability of reservoir structure, gas does not always flow equally to wells placed throughout the length, breadth, and depth of the reservoir and at equal pressure. Recovery wells must be distributed throughout the reservoir to recover as much of the gas as efficiently as possible. Furthermore, as the gas pressure in the reservoir declines, the reservoir energy (i.e. reservoir pressure) declines, and the gas requires stimulation for continued production. In addition, reduction in the gas pressure may allow compaction of the reservoir rock by the weight of rock above, eventually resulting in subsidence of the surface above the reservoir. This can be a gradual process or a sudden catastrophic process, depending on the structures of the geological formation above the reservoir.

A reservoir containing *wet* gas with a large amount of valuable *natural gas liquids* (e.g. ethane, propane, and butane — any hydrocarbon derivatives other than methane) and even crude oil containing low-boiling hydrocarbon constituents and condensate. When the reservoir pressure drops below the critical point for the mixture, the liquids may condense out and remain in the reservoir. In such cases, it may be necessary to implement a *cycling* process in which the wet gas is brought to the surface and the natural gas liquids are condensed as a separate stream and the gas is compressed and injected back into the reservoir to maintain the pressure.

Once brought out of the ground (recovered from the reservoir), natural gas is refined to remove impurities like water, other gases, sand, and other compounds (Mokhatab *et al.*, 2006; Speight, 2014a, 2019). Some hydrocarbon derivatives are removed and sold separately — such as ethane (for use in the production of petrochemicals) as well as propane and butane (for use as fuel gases) (Speight, 2014a, 2019b). Other impurities such as hydrogen sulfide are also removed — hydrogen sulfide can be used to produce sulfur, which is then also sold separately. After refining (Chapter 9), the clean natural gas is transmitted through a network of pipelines that deliver the gas to its point of use.

In terms of reserves, natural gas is produced on all the continents except Antarctica (BP, 2019). It should also be remembered that the total gas resource base (like any fossil fuel or mineral resource base) is dictated by economics. Therefore, when resource data are quoted, some attention must be given to the cost of recovering those resources. Most important, the economics must include a cost factor that reflects the willingness to secure total, or a specific degree of, energy independence. Nevertheless, natural gas is produced by many countries around the world and most of those countries produce both oil and natural gas; a few produce solely natural gas. Because transportation costs add to the cost of natural gas, in most countries, natural gas is consumed within the country or exported to a neighboring country by pipeline. Technology for liquefying natural gas so that it can be transported in tankers is improving (Mokhatab *et al.*, 2006). As technology continues to expand the methods for gas transportation, the demand for natural gas is expected to increase.

A common misconception related to natural gas is that resources are being depleted at an alarming rate and the supplies are quickly running out. In fact, a vast amount of natural gas remains to be retrieved from a variety of reservoirs, especially the low-permeability reservoirs — the so-called tight formation reservoirs (Speight, 2017a, 2020). However, many proponents of the rapid-depletion theory believe that price spikes indicate that natural gas resources are depleted beyond the point of no return, which is not always true. Price spikes of any commodity are not always caused by waning resources but can be the outcome of other forces at play in the market (Speight and Islam, 2016).

4.2.2.1 *Associated gas*

Associated or *dissolved* natural gas occurs either as free gas in a crude oil reservoir or as gas in solution in the crude oil. Gas that occurs in solution in crude oil is *dissolved* gas whereas the gas that exists in contact with the crude oil (*gas cap*) is *associated* gas (Parkash, 2003; Gary *et al.*, 2007; Speight, 2014a, 2017a; Hsu and Robinson, 2017). Crude oil cannot be produced without producing some of the associated gas, which comes out of the solution as the pressure is reduced on the way to and on the surface. Crude oil in the reservoir with minimal or no dissolved associated gas is rare and *dead crude oil* is often difficult to produce as there is little energy to drive it.

After the production fluids are brought to the surface, they are separated at a tank battery at or near the production lease into a hydrocarbon liquid stream (crude oil or *gas condensate*), a produced water stream (brine or salty water), and a gaseous stream. The gaseous stream is traditionally very rich (*rich gas*) in *natural gas liquids* (NGLs). Natural gas liquids include ethane, propane, butanes, and pentane (C_5H_{12}) and higher molecular weight hydrocarbon derivatives. The higher molecular weight hydrocarbon product, which may also contain some pentane and hexane (as well as heptane and octane), is commonly referred to as *natural gasoline.*

Rich gas has a high heating value and a high hydrocarbon dew point — the temperature to which a gas must be cooled to become saturated with vapor. However, the terms *rich gas* and *lean gas*, as used in the gas processing industry, are not precise indicators of gas quality but only indicate the relative amount of natural gas liquids in the gas stream. When referring to natural gas liquids in the natural gas stream, *gallons per thousand cubic feet* of gas is used as a measure of hydrocarbon richness.

Thus, in the case of associated gas, crude oil may be assisted up the wellbore by gas lift (Mokhatab *et al.*, 2006; Speight, 2019a). Thus, gas is compressed into the annulus of the well and then injected by means of a gas lift valve near the bottom of the well into the crude oil column in the tubing. At the top of the well, the crude oil and gas mixture passes into a separation plant that drops the pressure down to nearly atmospheric in two stages. The crude oil and water exits the bottom of the lower pressure separator, from where it is pumped to tanks for separation of the crude oil and water. The gas produced in the separators is recompressed and the gas that comes out of the solution with the produced crude oil (surplus gas) is then treated to separate out the natural gas liquids (NGLs) that are treated in a gas plant to provide propane and butane or a mixture of the two (liquefied petroleum gas, LPG). The higher boiling residue, after the propane and butane are removed, is condensate, which is mixed with the crude oil or exported as a separate product. The gas itself is then *dry* and, after compression, is suitable for injection into the natural gas system, where it substitutes for natural gas from the non-associated gas reservoir.

Another use for the gas is as fuel for the gas turbines on site but the gas must be treated in a fuel gas plant to ensure that it is clean and at the correct pressure. The startup fuel gas supply will be from the main gas system, but facilities exist to collect and treat low-pressure gas from the various other plants as a more economical fuel source.

4.2.2.2 *Non-associated gas*

Non-associated gas (sometimes called *gas well gas*) is produced from geological formations that typically do not contain much, if any, crude oil, or higher boiling hydrocarbon derivatives (*gas liquids*) than methane. However, non-associated gas can contain non-hydrocarbon gases such as carbon dioxide and hydrogen sulfide.

The non-associated gas recovery system is somewhat simpler than the associated gas recovery system. The gas flows up the well under its own energy, through the wellhead control valves and along the flow line to the treatment plant. Treatment requires the temperature of the gas to be reduced to a point dependent on the pressure in the pipeline so that all liquids that would exist at pipeline temperature and pressure condense and are removed (Mokhatab *et al.*, 2006; Speight, 2014a, 2019a). The water in the gas must also be dealt with to stop the formation of *gas hydrates* that may block the pipes. One method is to inject ethylene glycol (*glycol*) that combines with the water and is later recovered by a glycol plant. The treated gas then flows from the top of the treatment vessel into the pipeline. The water is treated in a glycol plant to recover the glycol. Any natural gas liquids are pumped as additional feedstock to the liquefied petroleum gas plant.

When natural gas is cooled to a temperature of approximately –160°C (approximately –260°F) at atmospheric pressure, it condenses to a liquid (*liquefied natural gas*, LNG). One volume of this liquid takes up approximately 1/600th the volume of natural gas. Liquefied natural gas weighs less than one-half that of water and is odorless, colorless, non-corrosive, and non-toxic. When vaporized, it burns (sometimes explosively) in concentrations of 5–15% v/v when mixed with air, but neither liquefied natural gas (LNG) nor the vapor therefrom can explode in an unconfined environment. Since liquefied natural gas takes less volume and weight, it represents a more convenient option for storage and transportation.

4.2.3 *Unconventional gas*

The boundary between conventional gas and unconventional gas resources is not well defined, because they result from a continuum of geological conditions. Coal seam gas, more frequently called coalbed methane (or coal-bed methane, CBM), is typically referred to as unconventional gas. Tight shale gas and gas hydrates are also placed into the category of *unconventional gas*.

4.2.3.1 *Coalbed methane*

Coalbed methane and *gas hydrates* are two relatively new resources that are recognized as having plentiful supplies of methane and other lower boiling hydrocarbon derivatives (Berecz and Balla-Achs, 1983; Sloan, 1997; Gudmundsson *et al.*, 1998; Max, 2000; Sloan, 2000). Coalbed methane is the generic term given to methane gas held in underground coal seams and released or produced when the water pressure within the seam is reduced by pumping from either vertical or inclined to horizontal surface holes. The methane is predominantly formed during the coalification process whereby organic matter is slowly transformed into coal by increasing temperature and pressure as the organic matter is buried deeper and deeper by additional deposits of organic and inorganic matter over long periods of geological time. This is referred to as *thermogenic coalbed methane*. Alternatively, and more often in (but not limited to) lower rank and thermally immature coals, recent bacterial processes (involving naturally occurring bacteria associated with meteoric water recharge at outcrop or sub-crop) can dominate the generation of coalbed methane. This is referred to as late-stage *biogenic coalbed methane*.

During the coalification process, a range of chemical reactions take place which produce substantial quantities of gas. While much of this gas escapes into the overlying or underlying rock, a large amount is retained within the forming coal seams. However, unlike conventional natural gas reservoirs, where gas is trapped in the pore or void spaces of a rock such as sandstone, methane formed and trapped in coal is actually adsorbed onto the coal grain surfaces or micropores and held in place by reservoir (water) pressure. Therefore, because the micropore surface area is very large, coal can potentially hold significantly more methane per unit volume than most sandstone reservoirs.

The amount of methane stored in coal is closely related to the rank and depth of the coal — the higher the coal rank and the deeper the coal seam is presently buried (causing pressure on coal), the greater the capacity of coal to produce and retain methane. Because coal has a very large internal surface area of over one billion square feet per ton of coal, coal can hold on average three times as much gas in place as the same volume of a conventional sandstone reservoir at equal depth and pressure. In order to allow the absorbed gas to be released from the coal, it is often necessary to lower the pressure on coal. This generally involves removing the water contained in the coal seam. After the gas is released from the internal surfaces of coal it moves through the internal matrix of coal until it reaches natural fracture networks in coal (*cleats*). The gas then flows through these cleats or fractures until it reaches the wellbore.

Gas derived from coal is generally pure and requires little or no processing because it is solely methane and not mixed with higher boiling hydrocarbon derivatives, such as ethane, which is often present in conventional natural gas. Coalbed methane has a slightly higher energy value than some natural gases. A coal seam gas well's productivity depends mostly on reservoir pressure and water saturation.

To recover coalbed methane, multi-well patterns are necessary to dewater coal and to establish a favorable pressure gradient. Since the gas is adsorbed on the surface of coal and trapped by reservoir pressure, initially there is low gas production and high water production. Therefore, an additional expense relates to the disposal of coalbed water, which may be saline, acidic, or alkaline. As production continues, water production declines and gas production increases, before eventually beginning a long decline. In general, however, coal seam gas recovery rates have been low and unpredictable. Average per-well conventional gas production in a mature gas-rich basin is approximately five times higher than average per-well coal seam gas production. Thus, several times as many wells have to be drilled in coal seams than in conventional gas accumulations to achieve similar gas recovery levels.

4.2.3.2 *Gas hydrates*

In addition to coalbed methane (Section 4.2.3.1), another relatively new, and possibly large source of methane that can be expected to extend the availability of natural gas is *methane hydrate* (also called *gas hydrate* and *methane ice*) (Berecz and Balla-Achs, 1983; Sloan, 1997, 2000; Gudmundsson *et al.*, 1998; Max, 2000). Their production technologies have only recently been developed and these sources are now becoming economically competitive. A gas hydrate is a molecule consisting of an ice lattice or cage in which low molecular-weight hydrocarbon derivatives, such as methane, are embedded. The two major conditions that promote hydrate formation are thus: (i) high gas pressure and low gas temperature and (ii) the gas at or below its water dew point with free water present.

Gas hydrates are common constituents of the shallow marine geosphere and occur both in deep sedimentary structures and as outcrops on the ocean floor. Methane hydrates are believed to form by migration of gas from depths along geological faults, followed by precipitation, or crystallization, on contact of the rising gas stream with cold sea water. At high pressures methane hydrates remain stable at temperatures up to 18°C (64°F) and the typical methane hydrate contains one molecule of methane for every 6 molecules of water that forms the ice cage, but this ratio is dependent on the number of methane molecules that fit into the various cage structures of the water lattice. One volume of solid methane hydrate can contain up to 168 volumes of methane gas.

Methane hydrates are restricted to the shallow lithosphere (i.e. less than approximately 6,500 feet depth). Furthermore, necessary conditions are found only either in polar continental sedimentary strata where the surface temperature is less than 0°C (32°F) or in oceanic sediment at water depths greater than 1,000 feet where the bottom-water temperature is of the order of 2°C (35°F).

The methane in gas hydrates is dominantly generated by bacterial degradation of organic matter in low-oxygen environments. Organic matter in the uppermost few centimeters of sediments is first attacked by aerobic bacteria, generating carbon dioxide, which escapes from the sediments into the water column. In this region of aerobic bacterial activity sulfates are reduced to sulfides. If the sedimentation rate is low (<1 cm of sediment formed per 1,000 years), the organic carbon content is low (<1%), and oxygen is abundant, then aerobic bacteria use up all the organic matter present in the sediments. But where sedimentation rates and the organic carbon content are high, the pore waters in the sediments are anoxic at depths of only a few inches, and methane is produced by anaerobic bacteria. The presence of clathrates at a given site can often be determined by the observation of a bottom simulating reflector (BSR), which is a seismic reflection at the sediment to clathrate stability zone interface caused by the different densities of normal sediments and sediments laced with clathrates.

The size of the oceanic methane hydrate reservoir is not well defined and estimates of its size have varied considerably over a wide range. However, improvements in understanding the nature of the gas hydrate resource have revealed that hydrates only form in a narrow range of depths (such as in the area of continental shelves) and typically are found at low concentrations (0.9–1.5% v/v) at sites where they do occur.

The potential for hydrate development is high and commercial development of natural gas hydrates could cause a shift in energy supply geopolitics because of the amount of natural gas sequestered in hydrate form. If only a small part of the gas can be produced, especially adjacent to countries that are presently energy importers, there will be major economic and geopolitical consequences that would dramatically change the present energy supply system.

4.2.3.3 *Shale gas*

Large continuous gas accumulations are sometimes present in low permeability shale, (tight) sandstones, siltstones, sandy carbonates, limestone, dolomite, and chalk (Speight, 2017b, 2020). Such gas deposits are commonly classified as unconventional because their reservoir characteristics differ from conventional reservoirs and they require stimulation to be produced economically.

The tight gas is contained in lenticular or blanket reservoirs that are relatively impermeable, which occur downdip from water-saturated rocks and cut across lithologic boundaries. They often contain a large amount of in-place gas, but exhibit low recovery rates. Gas can be economically recovered from the better quality continuous tight reservoirs by creating downhole fractures with explosives or hydraulic pumping. The nearly vertical fractures provide a pressure sink and channel for the gas, creating a larger collecting area so that the gas recovery is at a faster rate. Sometimes massive hydraulic fracturing is required, using a half million gallons of gelled fluid and a million pounds of sand to keep the fractures open after the fluid has been drained away.

In the United States, unconventional gas accumulations account for approximately 2 trillion cubic feet (tcf) of gas production per year, about 10% of the total gas output. In the rest of the world, however, gas is predominantly recovered from conventional accumulations.

4.2.4 *Composition and properties*

Natural gas (also called marsh gas and swamp gas in the older literature and more recently landfill gas) is a naturally occurring gaseous fossil fuel found in crude oil reservoirs, natural gas reservoirs, and coal seams (in which case it is referred to as coalbed methane). For the sake of clarity, natural gas is not the same as *town gas*, which is manufactured from coal, and *coal gas*, *manufactured gas*, *producer gas*, and *synthetic natural gas* (SNG) that are also produced from fossil fuel sources.

4.2.4.1 *Composition*

Natural gas is a combustible mixture of hydrocarbon gases (Mokhatab *et al.*, 2006). The principal constituents of natural gas, in varying amounts depending upon the source (Table 4.1), are methane (CH_4), ethane (C_2H_6), propane (C_3H_8), butanes (C_4H_{10}), pentanes (C_5H_{12}), hexane (C_6H_{14}), heptane (C_7H_{16}), and sometimes trace amounts of octane (C_8H_{18}) and higher molecular weight hydrocarbon derivatives. Some aromatics (BTX — benzene (C_6H_6), toluene ($C_6H_5CH_3$), and the xylenes ($CH_3C_6H_4CH_3$)) can also be present, raising safety issues due to their toxicity. The non-hydrocarbon gas portion of natural gas contains nitrogen (N_2), carbon dioxide (CO_2), helium (He), hydrogen sulfide (H_2S), water vapor (H_2O), and other sulfur compounds (such as carbonyl sulfide (COS) and mercaptans (e.g. CH_3SH)) and trace amounts of other gases (Mokhatab *et al.*,

Table 4.1: Typical range of compositions of natural gas.

Name	Formula	% v/v
Methane	CH_4	85–95
Ethane	C_2H_6	3–8
Propane	C_3H_8	1–5
Butane	C_4H_{10}	1–2
Pentane+	C_5H_{12}	1–5
Carbon dioxide	CO_2	1–2
Hydrogen sulfide	H_2S	1–2
Nitrogen	N_2	1–5
Helium	He	<0.5

Notes: Pentane+: pentane and higher molecular weight hydrocarbon derivatives up to octane as well as benzene (C_6H_6), toluene ($C_6H_5CH_3$), and in some cases, the xylene isomers (1,2-, 1,3- and 1,4-$CH_3C_6H_4CH_3$).

2006; Speight, 2019). Carbon dioxide and hydrogen sulfide are commonly referred to as *acid gases* since they form corrosive compounds in the presence of water. Nitrogen, helium, and carbon dioxide are also referred to as *diluents* since none of these burn, and thus they have no heating value. Mercury can also be present either as a metal in vapor phase or as an organo-metallic compound in liquid fractions. Concentration levels are generally very small, but even at very small concentration levels, mercury can be detrimental due to its toxicity and its corrosive properties (reaction with aluminum alloys).

However, in its commercial form, the natural gas that is delivered to the consumer is almost pure methane and the remaining hydrocarbon derivatives and non-hydrocarbon derivatives have been removed through refining (Mokhatab *et al.*, 2006; Speight, 2014a, 2019a). The non-hydrocarbon constituents include, but are not limited to, carbon dioxide (CO_2), hydrogen sulfide (H_2S), nitrogen (N_2), and helium (He). Because natural gas is colorless, shapeless, odorless, and tasteless in its pure form, it is not possible to see or smell natural gas. Therefore, an odorant (a mercaptan, also called a thiol, i.e. a sulfur-containing compound having the general formula R-SH) is added to natural gas for safety reasons so that it can be smelled if there is a gas leak. A mercaptan is a chemical odorant that smells a little like rotten eggs or skunk spray.

Several general definitions have been given to natural gas based on its composition. For example, *lean* gas is a gas in which methane is the major constituent. On the other hand, w*et* gas contains considerable amounts of the higher molecular weight hydrocarbon derivatives. To further define the terms *dry* and *wet* in quantitative terms, the term *dry* natural gas indicates that there is less than 0.1 gallon (1 US gallon = 264.2 m^3) of gasoline vapor (higher molecular weight paraffins) per 1,000 ft^3 (1 ft^3 = 0.028 m^3). The term *wet natural gas* indicates that such paraffins are present in the gas, in fact, more than 0.1 gal per 1,000 ft^3. Natural gas is considered *dry* if it is almost pure methane with most of the other commonly associated hydrocarbon derivatives removed. If other hydrocarbon derivatives are present, natural gas is *wet.*

S*our* gas contains hydrogen sulfide whereas *sweet* gas contains very little, if any, hydrogen sulfide. *Residue gas* is natural gas from which the higher molecular weight hydrocarbon derivatives have been extracted and *casing head gas* is derived from crude oil but is separated at the separation facility at the well-head.

4.2.4.2 *Properties*

The properties of unrefined natural gas are variable because the composition of natural gas is never constant. Therefore, the properties and behavior of natural gas are best understood by investigating the properties and behavior of the constituents (Table 4.2). Thus, assuming that natural gas has been cleaned (i.e. any constituents such as carbon dioxide and hydrogen sulfide have been removed and the only constituents remaining are hydrocarbon derivatives), the properties and behavior of natural gas become clear from a study of the properties and behavior of the relevant hydrocarbon derivatives.

The composition of natural gas varies depending on the field, the formation, or the reservoir from which it is extracted and these are an artifact of its formation. The different hydrocarbon derivatives that form natural gas can be separated using their different physical properties, such as weight, boiling point, or vapor pressure. Depending on its content of higher molecular weight hydrocarbon components, natural gas can be considered as rich (five or six gallons or more of recoverable hydrocarbon components per cubic feet) or lean (less than one gallon of recoverable hydrocarbon components per cubic feet).

Table 4.2: General properties of the constituents of natural gas up to and including *n*-octane (C_8H_{18}) as well as toluene, ethyl benzene, and the xylene isomers.

Constituent	Molecular weight	Specific gravity	Vapor density air = 1	Boiling point (°C)	Ignition temperature (°C)	Flash point (°C)
Methane	16	0.553	0.56	–160	537	–221
Ethane	30	0.572	1.04	–89	515	–135
Propane	44	0.504	1.50	–42	468	–104
Butane	58	0.601	2.11	–1	405	–60
Pentane	72	0.626	2.48	36	260	–40
Hexane	86	0.659	3.00	69	225	–23
Benzene	78	0.879	2.80	80	560	–11
Heptane	100	0.668	3.50	98	215	–4
Octane	114	0.707	3.90	126	220	13
Toluene	92	0.867	3.20	161	533	4
Ethyl benzene	106	0.867	3.70	136	432	15
o-Xylene	106	0.88	3.7	144	463	17
m-Xylene	106	0.86	3.7	139	527	25
p-Xylene	106	0.861	3.7	138	528	25

4.2.4.2.1 Density

Density is a physical property of matter and is a measure of the relative *heaviness* of hydrocarbon derivatives and other chemicals at a constant volume, and each constituent of natural gas has a unique density associated with it. Another term, *specific gravity*, is commonly used in relation to the properties of hydrocarbon derivatives. The specific gravity of a substance is a comparison of its density to that of water. For most chemical compounds (i.e. those that are solid or liquid), the density is measured relative to water (1.00). For gases, the density is more likely to be compared to the density of air (also given the number 1.00, but this is arbitrary and bears no relationship to the density of water). As a comparison, the density of liquefied natural gas (LNG) is approximately 0.41–0.5 kg/L, depending on temperature, pressure, and composition; in comparison, the density of water is 1.0 kg/L.

Table 4.3: Boiling point and density of methane relative to air and water.

- Boiling Point (760 mmHg): –161.5°C (–258.7°F)
- Vapor density: 0.55 to 0.64 (air = 1.00)
- Gas density (varies slightly): 0.0438 lb/scf
- Specific gravity of liquefied natural gas: 0.42–0.46 (water = 1.00)

Density values including those of natural gas hydrocarbon derivatives (Table 4.3) are given at room temperature unless otherwise indicated by a superscript numeral. For example, 2.487^{15} indicates a density of 2.487 g/cm at 15°C. The density of any gas compared to the density of air is *vapor density*, a very important characteristic of the constituents of natural gas. Furthermore, a statement is often made that *Natural gas is lighter than air*. This statement often arises because of the continued insistence by engineers and scientists that the properties of a mixture are determined by the mathematical average of the properties of individual constituents of the mixture. Such mathematical bravado and inconsistency of thought are *detrimental to safety* and need to be qualified.

Relative to air, methane is less dense but the other hydrocarbon constituents of unrefined natural gas (e.g. ethane, propane, butane, etc.) are denser than air. Therefore, should a natural gas leak occur in field operations, especially where the natural gas contains constituents other than methane, only methane dissipates readily into the air and the higher molecular weight hydrocarbon constituents pose a considerable risk if the ambient conditions are such that they accumulate at ground level.

4.2.4.2.2 Heat of combustion

All of the hydrocarbon constituents of natural gas are combustible, but non-flammable non-hydrocarbon components (carbon dioxide, hydrogen sulfide, nitrogen, and helium) detract slightly from the heating value of natural gas.

The *heat of combustion* (*energy content*) of natural gas is the amount of energy obtained from the burning of a volume of natural gas and is measured in British thermal units (Btu). The value of natural gas is calculated from its Btu content. One Btu is the quantity of heat required to raise the temperature of one pound of water of 1 degree Fahrenheit at atmospheric pressure. A cubic foot of natural gas has an energy content of approximately 1031 Btu, but it ranges between 500 and 1,500 Btu depending on the composition of the gas.

Thus, the energy content of natural gas is variable, because natural gas has variations in the amounts and types of energy gases (methane, ethane, propane, and butane) it contains: the higher the amount of non-combustible gases present in natural gas, the lower the energy (Btu). In addition, the volume and mass of energy gases present in a natural gas accumulation also influence the Btu value of natural gas. The larger the number of carbon atoms in a hydrocarbon gas, the higher its Btu value. It is necessary to conduct a Btu analysis of natural gas at each stage of the supply chain. Gas chromatographic process analyzers are used to conduct a fractional analysis of the natural gas streams, separating natural gas into identifiable components. The components and their concentrations are converted into a gross heating value in Btu-cubic foot.

In USA, at retail, natural gas is often sold in units of therms (th) (1 therm = 100,000 Btu). Wholesale transactions are generally done in decatherms (Dth), thousand decatherms (MDth), or million decatherms (MMDth). A million decatherms is roughly a billion cubic feet of natural gas.

The gross heats of combustion of crude oil and its products are given with fair accuracy by the equation:

$$Q = 12{,}400 - 2100d^2$$

where d is the 60/60°F specific gravity. Deviation from the formula is generally less than 1%.

4.2.4.2.3 Volatility and flammability

The boiling point (boiling temperature) of a substance is the temperature at which the vapor pressure of the substance is equal to atmospheric pressure.

At the boiling point, a substance changes its state from liquid to gas. A stricter definition of boiling point is the temperature at which the liquid and vapor (gas) phases of a substance can exist in equilibrium. When heat is applied to a liquid, the temperature of the liquid rises until the *vapor pressure* of the liquid equals the pressure of the surrounding atmosphere (gases). At this point, there is no further rise in temperature, and the additional heat energy supplied is absorbed as *latent heat* of vaporization to transform the liquid into gas. This transformation occurs not only at the surface of the liquid (as in the case of *evaporation*) but also throughout the volume of the liquid, where bubbles of gas are formed. The boiling point of a liquid is lowered if the pressure of the surrounding atmosphere (gases) is decreased. On the other hand, if the pressure of the surrounding atmosphere (gases) is increased, the boiling point is raised. For this reason, it is customary, when the boiling point of a substance is mentioned, to include the pressure at which it is observed in case that pressure is other than standard, namely 760 mm of mercury or 1 atmosphere standard temperature and pressure (STP).

The boiling points of crude oil fractions are rarely, if ever, distinct temperatures. It is, in fact, more correct to refer to the boiling ranges of the various fractions; the same is true of natural gas. To determine these ranges, the material in question is tested by various methods of distillation, either at atmospheric pressure or at reduced pressure. Thus, the boiling points of the hydrocarbon constituents of natural gas increase with molecular weight and the initial boiling point of natural gas corresponds to the boiling point of the most volatile constituent (i.e. methane).

Purified natural gas is neither corrosive nor toxic, its ignition temperature is high, and it has a narrow flammability range, making it an apparently safe fossil fuel compared to other fuel sources. In addition, purified natural gas (i.e. methane) having a specific gravity (0.60) lower than that of air (1.00) rises if escaping and dissipates from the site of any leak.

However, methane is highly flammable, burns easily and almost completely. Therefore, natural gas can also be hazardous to life and property through an explosion. When natural gas is confined, such as within a house or in a coal mine, the concentration of the gas can reach explosive mixtures that, if ignited, result in blasts that could destroy buildings. As defined by a standard test method, the *fire point* is the temperature to which the gas must be heated under the prescribed conditions of the method to burn continuously when the mixture of vapor and air is ignited by a specified flame (ASTM D92, 2019).

The *flash point* of crude oil or a crude oil product, including natural gas, is the temperature to which the product must be heated under specified conditions to give off sufficient vapor to form a mixture with air that can be ignited momentarily by a specified flame (ASTM D56, D92, and D93, 2019). As with other properties, the flash point is dependent on the composition of the gas and the presence of other hydrocarbon constituents.

From the viewpoint of safety, information related to the flash point is of most significance at or slightly above the maximum temperatures (30–60°C, 86–140°) that may be encountered during the storage, transportation, and use of liquid crude oil products in either closed or open containers. In this temperature range the relative fire and explosion hazard can be estimated from the flash point. For products with flash point below 40°C (104°F) special precautions are necessary for safe handling. Flash points above 60°C (140°F) gradually lose their safety significance until they become indirect measures of some other quality. The flash point of a crude oil product is also used to detect contamination. A substantially lower flash point than expected for a product is a reliable indicator that a product has become contaminated with a more volatile product, such as gasoline. The flash point is also an aid in establishing the identity of a particular crude oil product.

A further aspect of volatility that has received considerable attention is the vapor pressure of crude oil and its constituent fractions. The *vapor pressure* is the force exerted on the walls of a closed container by the vaporized portion of a liquid. Conversely, it is the force that must be exerted on the liquid to prevent it from vaporizing further (ASTM D323, 2019). The vapor pressure increases with temperature for any given gasoline, liquefied petroleum gas, or other product. The temperature at which the vapor pressure of a liquid, either a pure compound or a mixture of many compounds, equals 1 atmosphere (14.7 psi, absolute) is designated as the boiling point of the liquid.

The flammable range is expressed by the lower explosive limit (LEL) and the upper explosive limit (UEL). The LEL is the concentration of natural gas in air below which the propagation of a flame will not occur on contact with an ignition source. The LEL of natural gas is 5% by volume in air and, in most cases, the smell of gas will be detected well before combustion conditions are met. The UEL is the concentration of natural gas in air above which the propagation of a flame will not occur on contact with an ignition source. The natural gas UEL is 15% by volume in air.

In any form, a minute amount of an odorant with an obvious smell is added to the otherwise colorless and odorless gas so that leaks can be detected before any fire or explosion occurs. Odorants are considered nontoxic at the extremely low concentrations occurring in the natural gas delivered to the end user.

4.3 The Crude Oil Family

The crude oil family consists of various types of crude oil as well as heavy crude oil (Chapter 1) from which the typical bulk fractions that have commercial value as a bulk product are produced (Table 4.4) (Parkash, 2003; Gary *et al.*, 2007; Speight, 2014a, 2017a; Hsu and Robinson, 2017). By comparison and in the strictest sense, petrochemicals are also crude oil products but they are individual chemicals that are used as the basic building blocks of the chemical industry (Parkash, 2003; Gary *et al.*, 2007; Speight, 2014a, 2017a, 2019b; Hsu and Robinson, 2017). Thus petrochemicals are not usually classified as crude oil products.

The use of crude oil and its products was established in pre-Christian times, as is known largely through documentation by many of the older civilizations (Speight, 2014a) and, thus, the use of crude oil and the development of related technology is not as much a modern subject as is often assumed in the active literature. However, there have been many changes in product demand since crude oil first came into use some five to six millennia ago (Speight, 2014a). It is these changes in product demand that have been largely responsible for the evolution of the industry from asphalt used in ancient times to gasoline and other liquid fuels of the modern crude oil era.

Table 4.4: Separation of crude oil into different generic boiling fractions.

Fraction	Boiling range ([a] °C)	°F
Low-boiling naphtha	−1–150	30–300
High-boiling naphtha	150–205	300–400
Kerosene	205–260	400–500
Atmospheric gas oil[b]	260–400	400–750
Vacuum gas oil[b]	400–600	750–1100
Residuum	>510	>950

Notes: [a]The boiling ranges are approximate and are refinery dependent and, for convenience, boiling ranges are converted to the nearest 5°.

[b]Atmospheric gas oil and vacuum gas oil (or parts thereof) are often used as fuels under the general name *fuel oil*.

4.3.1 *Conventional crude oil*

Crude oil (often called conventional crude oil) is an extremely complex mixture of hydrocarbon compounds, usually with minor amounts of nitrogen-, oxygen-, and sulfur-containing compounds as well as trace amounts of metal-containing compounds. In addition, the properties of crude oil vary widely and, as a result, crude oil is not used in its raw state. A variety of processing steps are required to convert crude oil from its raw state to products that have well-defined properties.

Product complexity becomes even more meaningful when various fractions from different types of crude oil as well as fractions from synthetic crude oil are blended with the corresponding crude oil stock. The implications for refining the fractions to salable products have increased. However, for the main part, the crude oil industry was inspired by the development of the automobile and the continued demand for gasoline and other fuels. Such a demand has been accompanied by the demand for other products: diesel fuel for engines, lubricants for engine and machinery parts, fuel oil to provide power for industrial complexes, and asphalt for roadways.

Unlike processes, products are more difficult to place on an individual evolutionary scale. Processes have changed and evolved to accommodate the demand for, say, higher-octane fuels, longer-lasting asphalt, or lower sulfur coke. In this section, a general overview of some crude oil products is provided to show the *raison d'être* of the industry. Another factor that must be acknowledged is the change in the character and composition of the original crude oil feedstock. In the early days of the crude oil industry, several products were obtained by distillation and could be used without any further treatment. The different character and composition of crude oil in the modern refinery dictates that any liquids obtained by distillation must go through one or more of the several available product improvement processes (Speight, 2014a, 2017a). Such changes in feedstock character and composition have caused the refining industry to evolve in such a way that the changes in crude oil can be accommodated.

The demand for fuels derived from crude oil — particularly transportation fuels (gasoline and diesel) and petrochemical feedstocks (such as aromatics and olefins) — is increasing throughout the world. Traditional markets such as North America and Europe are experiencing a moderate increase in demand, whereas emerging Asian markets such as India and China are witnessing a rapid surge. This has prompted a fresh technological approach to optimize efficiency and throughput. Major oil companies and technology suppliers/licensors are investing heavily in revamping their refining technologies in an effort to cater to the growing needs of customers.

The steady evolution of the specifications for fuel products caused by an endless waves of fresh environmental regulations has played a major role in the development of crude oil refining technologies. In the United States and Europe, gasoline and diesel specifications have changed radically in the past few decades and will continue to do so in the future. Currently, reducing the sulfur levels of finished products is the dominant objective. Sulfur is ubiquitous in crude oil and refiners are seeking technologies that can help to achieve the mandated levels of sulfur in crude oil products.

Refineries are eager to adapt to the changing circumstances and are willing to try new technologies radically different in character. This is evident from the increasing use of ultrasonic technology and novel separation methods. Currently, refineries are also looking to exploit heavy (more viscous) crude oils, provided there are refinery technologies capable of handling them. Heavy crude oil, relative to low-density or conventional crude oil, is rich in higher boiling constituents and transforming the heavier components into low-boiling fractions is thus a necessity. However, this requires technological changes, including more effective use of hydrogen within the refinery, which increases the operating costs.

Heavier crude oil could also be contaminated with sulfur and metal particles that must be removed to meet quality standards. A deeper understanding of how catalysts work, both chemically and physically, will provide greater scope for technological improvements. Nanotechnology and combinatorial chemistry are among the techniques that are likely to help push forward the frontiers of efficiency and selectivity. Even though the nature of crude oil is changing, refineries are here to stay in the foreseeable future, since crude oil products satisfy wide-ranging energy requirements/demands that are not fully catered to by natural gas, liquefied petroleum gas (LPG), or coal. At present, alternative energy schemes lack technological sophistication or economic sufficiency to be considered a substitute for crude oil products.

A major group of fuel products from crude oil is the basis of a major industry (Speight, 2014a, 2019b). They are, in the strictest sense, different from crude oil products insofar as petrochemicals are the basic building blocks of the chemical industry. They are not considered here but are dealt with separately in this book. Crude oil products (in contrast to petrochemicals) are those bulk fractions that are derived from crude oil and have commercial value as a bulk product. In the strictest sense, petrochemicals (Speight, 2014a, 2019b) are also crude oil products, but they are individual chemicals that are used as the basic building blocks of the chemical industry as well as the liquid fuels industry.

There have been many changes in product demand since crude oil first came into use some five to six millennia ago (Speight, 2014a). These changes in product demand have been largely responsible for the evolution of the industry from asphalt used in ancient times to gasoline and other liquid fuels in current use. The modern crude oil industry began in 1859 with the discovery and subsequent commercialization of crude oil for the production of products in Pennsylvania (Bell, 1945). After the completion of the first well (by Edwin Drake), the surrounding areas were immediately leased and extensive drilling took place. It is from this time that modern crude oil products have evolved.

The constant demand for products, such as liquid fuels, is the main driving force behind the crude oil industry (Guthrie, 1960; Hobson and Pohl, 1973). Other products, such as lubricating oils, waxes, and asphalt, have also added to the popularity of crude oil as a national resource. Indeed, such fuel products derived from crude oil supply more than half of the total global supply of energy. Gasoline, kerosene, and diesel oil provide fuel for automobiles, tractors, trucks, aircraft, and ships. Fuel oil and natural gas are used to heat homes and commercial buildings, as well as to generate electricity. Crude oil products are the basic materials used for the manufacture of synthetic fibers for clothing and in plastics, paints, fertilizers, insecticides, soaps, and synthetic rubber. The use of crude oil as a source of raw material in manufacturing is central to the functioning of modern industry.

There are a myriad of other products that have evolved throughout the short life of the crude oil industry. The complexities of product composition have matched the evolution of the products (Hoffman, 1992). In fact, it is the complexity of product composition that has served the industry well and, at the same time, had an adverse effect on product use. Product complexity has made the industry unique among industries. Indeed, current analytical techniques that are accepted as standard methods for, as an example, the aromatics content of fuels, as well as proton and carbon nuclear magnetic resonance methods, yield different information (Speight, 2014a, 2015).

Product complexity, and the means by which the product is evaluated, has made the industry unique among industries. But product complexity has also brought to the fore issues such as instability and incompatibility. In order to understand the evolution of the products it is essential to have an understanding of the composition of the various products. Product complexity becomes even more meaningful when various fractions from different types of crude oil as well as fractions from synthetic crude oil are blended with the corresponding crude oil stock. The implications for refining the fractions to salable products have increased (Dooley *et al.*, 1979). However, for the main part, the crude oil industry was inspired by the development of the automobile and the continued demand for gasoline and other fuels (Hoffman, 1992). Such a demand has been accompanied by the demand for other products: diesel fuel for engines, lubricants for engines and machinery parts, fuel oil to provide power for industrial complexes, and asphalt for roadways.

Unlike processes, products are more difficult to place on an individual evolutionary scale. Processes have changed and evolved to accommodate the demand for, say, higher-octane fuels, longer-lasting asphalt, or lower sulfur coke. In this section, a general overview of some crude oil products is provided to show the *raison d'être* of the industry. Another factor that must be acknowledged is the change in character and composition of the original crude oil feedstock (Long and Speight, 1998). In the early days of the crude oil industry several products were obtained by distillation and could be used without any further treatment. In the modern refinery, the different character and composition of the crude oil-dictates that any liquids obtained by distillation must go through one or more of the several available product improvement processes (Speight, 2014a, 2017a). Such changes in feedstock character and composition have caused the refining industry to evolve in such a way that changes in the crude oil can be accommodated.

4.3.2 *Opportunity crude oil*

Opportunity crude oils are either new crude oils with unknown or poorly understood properties relating to processing issues or existing crude oils with well-known properties and processing concerns (Ohmes, 2014; Speight, 2014a, 2014b; Yeung, 2014). Opportunity crude oils are often, but not always, heavy crude oils but in either case are more difficult to process due to high levels of solids (and other contaminants) produced with the oil, high levels of acidity, and high viscosity. These crude oils may also be incompatible with other oils in the refinery feedstock blend and cause excessive equipment fouling when processed either in a blend or separately (Speight, 2015). There is also the need for a refinery to be configured to accommodate *opportunity crude oils* and/or *high acid crude oils*, which, for many purposes, are often included with heavy feedstocks.

4.3.3 *High acid crude oil*

High acid crude oils are crude oils that contain considerable proportions of naphthenic acids, which is a term commonly used in the crude oil industry to refer collectively to all of the organic acids present in the crude oil (Shalaby, 2005; Ghoshal and Sainik, 2013; Speight, 2014b). By the original definition, a naphthenic acid is a monobasic carboxyl group attached to a saturated cycloaliphatic structure. However, it is a convention accepted in the oil industry that all organic acids in crude oil are called naphthenic acids. Naphthenic acids in crude oils are now known to be mixtures of low to high molecular-weight acids and a naphthenic acid fraction also contains other acidic species. Naphthenic acids, which are not *user friendly* in terms of refining (Kane and Cayard, 2002; Ghoshal and Sainik, 2013; Speight, 2014c), can be either water-soluble or oil-soluble (or both), depending on their molecular weight, process temperatures, salinity of waters, and fluid pressures.

4.3.4 *Foamy oil*

Foamy oil is oil-continuous foam that contains dispersed gas bubbles produced at the well head from heavy oil reservoirs under solution gas drive. The nature of the gas dispersions in oil distinguishes foamy oil behavior from conventional heavy oil. The gas that comes out of the solution in the reservoir does not coalesce into large gas bubbles nor into a continuous flowing gas phase. Instead it remains as small bubbles entrained in the crude oil, keeping the effective oil viscosity low while providing expansive energy that helps drive the oil toward the production well. Foamy oil accounts for the unusually high production in heavy oil reservoirs under solution-gas drive (Sun *et al.*, 2013).

Foamy oil behavior is a unique phenomenon associated with production of heavy oils. It is believed that this mechanism contributes significantly to the abnormally high production rate of heavy oils observed in the Orinoco Belt. During production of heavy oil from solution gas drive reservoirs, the oil is pushed into the production wells by energy supplied by the dissolved gas. As fluid is withdrawn from the production wells, the pressure in the reservoir declines and the gas that was dissolved in the oil at high pressure starts to come out of the solution (hence, foamy oil). As pressure declines further with continued removal of fluids from the production wells, more gas is released from the solution and the gas already released expands in volume. The expanding gas, which at this point is in the form of isolated bubbles, pushes the oil out of the pores and provides energy for the flow of oil into the production well. This process is very efficient until the isolated gas bubbles link up and the gas itself starts flowing into the production well. Once the gas flow starts, the oil has to compete with the gas for available flow energy. Thus, in some heavy oil reservoirs, due to the properties of the oil and the sand and also due to the production methods, the released gas forms foam with the oil and remains subdivided in the form of dispersed bubbles much longer.

4.3.5 *Tight oil*

Tight oil is a low-viscosity crude oil (sometimes erroneously referred to as shale oil — by way of definition, shale oil is the liquid product produced by the decomposition of the kerogen component of oil shale) is crude oil that is confined in the pore spaces of these impermeable shale formations. On the other hand, oil shale is a kerogen-rich petroleum source rock that was not buried under the correct maturation conditions to experience the temperatures required to generate oil and gas (Speight, 2014a). Economic production from tight oil formations requires the same hydraulic fracturing and often uses the same horizontal technology as used in the production of tight gas.

Tight formations (such as shale formations) are heterogeneous and vary widely over relatively short distances. Tight oil reservoirs subjected to fracking can be divided into four different groups: (i) Type I has low matrix porosity and permeability, leading to fractures dominating both storage capacity and fluid-flow pathways, (ii) Type II also has low matrix porosity and permeability, but here the matrix provides storage capacity while the fractures provide fluid-flow paths, (iii) Type III are microporous reservoirs with high matrix porosity but low matrix permeability, thus giving induced fracture dominance in fluid-flow paths, and (iv) Type IV are macroporous reservoirs with high matrix porosity and permeability, thus the matrix provides both storage capacity and flow paths while the fractures only enhance permeability.

Even in a single horizontal drill hole, the amount recovered may vary, as may recovery within a field or even between adjacent wells. This makes evaluation of plays and decisions regarding the profitability of wells on a particular lease difficult. Production of oil from tight formations requires a gas cap representing at least 15–20% natural gas in the reservoir pore space to drive the oil toward the borehole; tight reservoirs which contain only oil cannot be economically produced, but such reserves may be limited (Wachtmeister *et al.*, 2017).

4.3.6 *Heavy crude oil*

The name *heavy oil* can often be misleading as it has also been used in reference to (i) fuel oil that contains the residuum left over from distillation, i.e. heavy fuel oil or residual fuel oil, (ii) coal tar creosote, or (iii) viscous crude oil. Thus, for the purposes of this text, the term is used to mean *viscous crude oil*. Extra heavy oil has been included in many of these categories. On the other hand, tar sand bitumen (often called simply *bitumen*) is often confused with manufactured asphalt (confusingly referred to as bitumen in many countries). To add further to this confusion, in some countries, tar sand bitumen is referred to as natural asphalt!

Heavy oil is a viscous type of crude oil that contains higher level of sulfur than conventional crude oil and occurs in locations similar to that of crude oil (IEA, 2005; Ancheyta and Speight, 2007; Speight, 2016b). The nature of heavy oil is a problem for recovery operations and for refining — the viscosity of the oil may be too high, thereby leading to difficulties in recovery and/or difficulties in refining the oil (Speight, 2016a, 2017). However, tar sand bitumen, in terms of properties and behavior, is far apart from conventional crude oil (Table 4.5). Success with this material and with extra heavy oil depends as much on understanding the fluid (or non-fluid) properties of the material and the behavior of the fluids in the deposit in which they occur as it does on knowing the geology of the deposit. The reason is that the chemical and physical differences between heavy oil, extra heavy oil, and tar sand bitumen oil ultimately

Table 4.5: Simplified differentiation between conventional crude oil, tight oil, heavy crude oil, extra heavy crude oil, and tar sand bitumen.[a]

Conventional Crude Oil
Mobile in the reservoir; API gravity: >25°
High-permeability reservoir
Primary recovery
Secondary recovery
Tight Oil
Properties similar to those of conventional crude oil; API gravity: >25°
Immobile in the reservoir
Low-permeability reservoir
Horizontal drilling into reservoir
Fracturing (typically multi-fracturing) to release fluids/gases
Heavy Crude Oil
More viscous than conventional crude oil; API gravity: 10–20°
Mobile in the reservoir
High-permeability reservoir
Secondary recovery
Tertiary recovery [enhanced oil recovery (EOR); e.g. steam stimulation]
Extra Heavy Crude Oil
Similar properties to those of tar sand bitumen; API gravity: <10°
Mobile in the reservoir
High-permeability reservoir
Secondary recovery
Tertiary recovery (EOR; e.g. steam stimulation)
Tar Sand Bitumen
Immobile in the deposit; API gravity: <10°
High-permeability reservoir
Mining (often preceded by explosive fracturing)
Steam-assisted gravity draining (SAGD)
Solvent methods (VAPEX)
Extreme heating methods
Innovative methods[b]

Notes: [a]This list is not intended for use as a means of classification.

[b]Innovative methods exclude tertiary recovery methods and others methods such as steam-assisted gravity drainage (SAGD) and vapor-assisted extraction (VAPEX) but include variants or hybrids thereof.

affect the viscosity and other relevant properties that, in turn, influence the individual aspects of recovery and refining operations.

Heavy oil has a much higher viscosity (and lower API gravity) than conventional crude oil and recovery of heavy oil usually requires thermal stimulation of the reservoir. The generic term *heavy oil* is often applied to a crude oil that has less than 20° API and usually, but not always, sulfur content higher than 2% w/w (Ancheyta and Speight, 2007). Furthermore, in contrast to conventional crude oils, heavy oils are darker in color and may even be black.

4.4 Extra Heavy Crude Oil and Tar Sand Bitumen

Fuels from extra heavy oil and tar sand bitumen have also attracted attention (Ancheyta *et al.*, 2001, 2002a, 2002b; Speight, 2014a, 2017a, 2019c). *Extra heavy oil* is a recently-evolved term (related to viscosity) of little scientific meaning. While this type of oil may resemble tar sand bitumen and does not flow easily, extra heavy oil is generally recognized as having mobility in the reservoir compared to tar sand bitumen, which is typically incapable of mobility (free flow) under reservoir conditions. For example, the tar sand bitumen located in Alberta Canada is not mobile in the deposit and requires extreme methods of recovery to recover the bitumen. On the other hand, much of the extra heavy oil located in the Orinoco belt of Venezuela requires recovery methods that are less extreme because of the mobility of the material in the reservoir (Total SA, 2007; Schenk *et al.*, 2009). Whether the mobility of extra heavy oil is due to a high reservoir temperature (i.e. higher than the pour point of the extra heavy oil) or other factors is uncertain and is subject to localized conditions in the reservoir.

Tar sand, also called *oil sand* (in Canada) or the more geologically correct term *bituminous sand*, is commonly used to describe a sandstone reservoir that is impregnated with a heavy, viscous bituminous material. Tar sand is actually a mixture of sand, water, and bitumen but many of the tar sand deposits in countries other than Canada lack the water layer that is believed to facilitate the hot water recovery process. The heavy bituminous material has a high viscosity under reservoir conditions and cannot be retrieved through a well by conventional production techniques.

In the context of this book, extra heavy oil and tar sand bitumen typically have relatively low proportions of volatile compounds with low molecular weights and quite high proportions of high-molecular-weight compounds of lower volatility. The high molecular-weight fraction of extra heavy oil and tar sand bitumen is composed of a complex assortment of different molecular and chemical types — a complex mixture of compounds and not necessarily just paraffin derivatives or asphaltene constituents — with high melting points and high pour points that greatly contribute to the poor fluid properties of the material, thereby contributing to low mobility (compared to conventional crude oil).

4.4.1 *Extra heavy crude oil*

Extra heavy crude oil is a material that occurs in the solid or near-solid state and, because of the high temperature in the reservoir, generally has mobility under reservoir conditions (Chapter 1). While this type of oil resembles tar sand bitumen at ambient temperatures and is not easily mobile, extra heavy crude oil is generally recognized as having mobility in the reservoir compared to tar sand bitumen, which is typically incapable of mobility (free flow) under reservoir conditions (Tables 4.5 and 4.6). Whether the mobility of extra heavy crude oil is due to a high reservoir temperature (i.e. higher than the pour point of the extra heavy crude oil) or other factors is uncertain and is subject to localized conditions in the reservoir.

4.4.2 *Tar sand Bitumen*

Tar sand, also called *oil sand* (in Canada) or the more geologically correct term *bituminous sand*, is another source of liquid fuels that is distinctly

Table 4.6: Comparison of selected properties of Athabasca tar sand bitumen (Alberta, Canada) and Zuata extra heavy crude oil (Orinoco, Venezuela).

		Athabasca bitumen	Zuata extra heavy oil
Whole bitumen	API gravity	8	8
	Sulfur, % w/w	4.8	4.2
650°F+	% v/v	85	86
	Sulfur, % w/w	5.4	4.6
	Ni + V, ppm	420	600
	CCR, % w/w[a]	14	15

Note: [a]Conradson carbon residue.

different from conventional crude oil (US Congress, 1976; Speight, 2005, 2007). The term *tar sand* is commonly used to describe a sandstone reservoir that is impregnated with a high-boiling, viscous bituminous material. Tar sand is actually a mixture of sand, water, and bitumen but many of the tar sand deposits in countries other than Canada lack the water layer that is believed to facilitate the hot water recovery process, which can then lead to a variety of products (Table 4.7). The high-boiling bituminous material has a high viscosity under reservoir conditions and cannot be retrieved through a well by conventional production techniques.

Geologically, the term *tar sand* is commonly used to describe a sandstone reservoir that is impregnated with bitumen, a naturally occurring material that is solid or near-solid and is substantially immobile under reservoir conditions. The bitumen cannot be retrieved through a well by conventional production techniques, including currently used enhanced recovery techniques. In fact, tar sand is defined (FE-76-4) in the United States as follows:

Table 4.7: Simplified flow chart for processing tar sand bitumen.

Atmospheric distillation			
	C1–C4 gases		
	Synthetic crude oil blend stock		
Vacuum distillation			
	C1–C4 gases		
	Synthetic crude oil blend stock		
	Solvent deasphalting		
		Deasphalted oil	Further processing
		Deasphalter bottoms	Gasification
Delayed coking/ fluid coking	C1–C4 gases Synthetic crude oil blend stock		
	Coke	Gasification	C1–C4 gases
			Synthesis gas

> *The several rock types that contain an extremely viscous hydrocarbon which is not recoverable in its natural state by conventional oil well production methods including currently used enhanced recovery techniques. The hydrocarbon-bearing rocks are variously known as bitumen-rocks oil, impregnated rocks, tar sands, and rock asphalt.*

In addition to this definition, there are several tests that must be carried out to determine whether or not, in the first instance, a resource is a tar sand deposit (Speight, 2014a). Most of all, a core taken from a tar sand deposit, and the bitumen isolated therefrom, are certainly not identifiable by the preliminary inspections (sight and touch) alone. In the United States, the final determinant is whether or not the material contained therein can be recovered by primary, secondary, or tertiary (enhanced) recovery methods (US Congress, 1976).

The relevant position of tar sand bitumen in Nature is best illustrated by comparing its position to crude oil and heavy crude oil. Thus, crude oil is referred to generically as a *fossil energy resource* (Speight, 2014a) and is further classified as a hydrocarbon resource and, for illustrative (or comparative) purposes, in this work, coal and oil shale kerogen have also been included in this classification. However, the inclusion of coal and oil shale under the broad classification of hydrocarbon resources has required (incorrectly) that the term hydrocarbon be expanded to include these resources. It is essential to recognize that resources such as coal, oil shale kerogen, and tar sand bitumen contain large proportions of heteroatomic species. Heteroatomic species are those organic constituents that contain atoms other than carbon and hydrogen, e.g. nitrogen, oxygen, sulfur, and metals (nickel and vanadium).

4.5 Coal

Coal, an often-maligned fossil fuel, should again be seriously considered as an alternate fuel source to crude oil not so much for the generation of power and as a source of chemicals as for the production of liquid fuel. The reaffirmation by the United States and many other countries of the world for a clean environment through the passages of the various environmental laws as well as the serious consideration of the effects of combustion products (carbon dioxide) on the atmosphere are all an inter to the consideration of coal as a source of fuels.

Coal occurs in different forms or *types* (Table 4.8) (Speight, 2013). Variations in the nature of the source material and local or regional

Table 4.8: Coal types and selected properties.

Types	
Lignite	Increases in maturity by becoming darker and harder and is then classified as sub-bituminous coal. After a continuous process of burial and alteration, chemical and physical changes occur until the coal is classified as bituminous — dark and hard coal.
Sub-bituminous coal	Also called black lignite. Generally dark brown to black coal that is intermediate in rank between lignite and bituminous coal. The woody structure characteristic of most lignite is absent from sub-bituminous coal.
Bituminous coal	Ignites easily and burns long with a relatively long flame. If improperly fired, bituminous coal is characterized by excess smoke and soot.
Anthracite coal	Very hard and shiny. Anthracite coal creates a steady and clean flame and is preferred for domestic heating. Furthermore, it burns longer with more heat than the other types.

Properties

Typical sulfur content in coal

- Anthracite coal: 0.6–0.77 wt.%
- Bituminous coal: 0.7–4.0 wt.%
- Lignite coal: 0.4 wt.%

Typical moisture content in coal

- Anthracite coal: 2.8–16.3 wt.%
- Bituminous coal: 2.2–15.9 wt.%
- Lignite coal: 39 wt.%

Typical fixed carbon content in coal

- Anthracite coal: 80.5–85.7 wt.%
- Bituminous coal: 44.9–78.2 wt.%
- Lignite coal: 31.4 wt.%

Typical bulk density of coal

- Anthracite coal: 50–58 (lb/ft^3), 800–929 (kg/m^3)
- Bituminous coal: 42–57 (lb/ft^3), 673–913 (kg/m^3)
- Lignite coal: 40–54 (lb/ft^3), 641–865 (kg/m^3)

Typical mineral matter ash content in coal (manifested as mineral ash)

- Anthracite coal: 9.7–20.2 wt.%
- Bituminous coal: 3.3–11.7 wt.%
- Lignite coal: 4.2 wt.%

variations in the coalification processes cause the vegetal matter to evolve differently. Thus, various classification systems exist to define the different types of coal. Thus, as geological processes increase their effect over time, the coal precursors are transformed into the following: (i) lignite, (ii) sub-bituminous coal, (iii) bituminous coal, and (iv) anthracite.

Lignite — also referred to as brown coal — is the lowest rank of coal and is used almost exclusively as fuel for steam-electric power generation. Jet is a compact form of lignite that is sometimes polished and has been used as an ornamental stone since the Iron Age (a period in human history that started between 1200 BC and 600 BC, depending on the region, and that was preceded by the Stone Age and Bronze Age). Sub-bituminous coal is a next to lowest rank coal in which the properties range from those of lignite to those of bituminous coal and is used primarily as fuel for steam-electric power generation. Bituminous coal is a dense coal, usually black and sometimes dark brown, often with well-defined bands of bright and dull material, and is used primarily as fuel in steam-electric power generation, with substantial quantities also used for heat and power applications in manufacturing and to produce coke. Anthracite is the highest rank coal and is a harder, glossy, black coal used primarily for residential and commercial space heating.

Coal classification systems are based on the degree to which coals have undergone coalification. Such varying degrees of coalification are generally called coal *ranks* (or *classes*). The determination of coal rank has a number of practical applications such as the definition of the coal properties. The properties include the amount of heat produced during combustion, the amount of gaseous products released upon heating, and the suitability of the coals for producing coke.

The *rank* of a coal indicates the progressive changes in carbon, volatile matter, and probably ash and sulfur that take place as coalification progresses from the lower-rank lignite through the higher ranks of sub-bituminous, high-volatile bituminous, low-volatile bituminous, and anthracite. The rank of a coal should not be confused with its grade. A high-rank coal (e.g. anthracite) represents coal from a deposit that has undergone the greatest degree of metamorphosis and contains very little mineral matter, ash, and moisture. On the other hand, any rank of coal, when cleaned of impurities through coal preparation, will be of a higher grade.

The most commonly employed systems of classification are those based on analyses that can be performed relatively easily in the laboratory

(ASTM, 2019) on the basis of fixed carbon content, volatile matter content, and calorific value. In addition to the major ranks (lignite, sub-bituminous, bituminous, and anthracite), each rank may be subdivided into coal groups, such as high-volatile A bituminous coal. Other designations, such as coking coal and steam coal, have been applied to coals, but they tend to differ from country to country.

The term *coal type* is also employed to distinguish between banded coals and non-banded coals. Banded coals contain varying amounts of vitrinite and opaque material. They include bright coal, which contains more than 80% vitrinite, and splint coal, which contains more than 30% opaque matter. The non-banded varieties include boghead coal, which has a high percentage of algal remains, and cannel coal with a high percentage of spores. The usage of all the above terms is quite subjective.

By analogy to the term *mineral*, which is applied to inorganic material, the term *maceral* is used to describe organic constituents present in coals. Maceral nomenclature has been applied differently by some European coal petrologists who studied polished blocks of coal using reflected light microscopy (their terminology is based on morphology, botanical affinity, and mode of occurrence) and by some North American petrologists who studied very thin slices (thin sections) of coal using transmitted light microscopy.

Three major maceral groups are generally recognized: *vitrinite*, *exinite*, and *inertinite*. The vitrinite group is the most abundant and is derived primarily from cell walls and woody tissues. Several varieties are recognized, e.g. telinite (the brighter parts of vitrinite that make up cell walls) and collinite (clear vitrinite that occupies the spaces between cell walls).

Coal is also classified on the basis of its macroscopic appearance (generally referred to as coal of rock type, litho-type, or *kohlen-type*) and there are four main types: (i) vitamin (*Glanzkohle* or *charbon brillant*), which is dominated by vitrinite group macerals and appears glassy to the unaided eye, (ii) clarain (*Glanzstreifenkohle* or *charbon semi-brillant*), which is composed of both vitrinite and exinite and has an appearance between that of vitrain and durain, (iii) durain (*Mattkohle* or *charbon mat*), which is generally composed of fine-grained inertinites and exinites and has a dull, mat-like luster, and (iv) fusion (*Faserkohle* or *charbon fibreux*), which is composed mainly of fusinite and resembles wood charcoal (because it soils the hands just as charcoal would).

Coal analysis may be presented in the form of *proximate* and *ultimate* analyses, whose analytic conditions are prescribed by organizations such as the American Society for Testing and Materials. A typical proximate analysis includes the moisture content, ash yield (that can be converted to mineral matter content), volatile matter content, and fixed carbon content.

It is important to know the moisture and ash contents of a coal because they do not contribute to the heating value of a coal. In most cases, ash becomes an undesirable residue and a source of pollution, but for some purposes (e.g. use as a chemical feedstock or for liquefaction), the presence of mineral matter may be desirable. Most of the heat value of a coal comes from its volatile matter, excluding moisture, and fixed carbon content. For most coals, it is necessary to measure the actual amount of heat released upon combustion, which is expressed in Btu per pound.

The fixed carbon data are derived using a thermal technique that converts the non-volatile constituents of the feedstock to a carbon-like mass and it is, therefore, the material other than the mineral matter (manifested as mineral ash) that does not vaporize when heated in the absence of air. It is determined by subtracting the sum of the moisture, ash, and volatile matter in weight percent from 100%.

Ultimate analyses are used to determine the carbon, hydrogen, sulfur, nitrogen, ash, oxygen, and moisture contents of a coal. For specific applications, other chemical analyses may be employed. These may involve, for example, identifying the forms of sulfur present; sulfur may occur in the form of sulfide minerals (pyrite and marcasite, FeS_2), sulfate minerals (gypsum, Na_2SO_4), or organically bound sulfur. In other cases, the analyzes may involve determining the trace elements present (e.g. mercury and chlorine), which may influence the suitability of a coal for a particular purpose or help establish methods for reducing environmental pollution.

4.6 Oil Shale

By way of introduction, shale oil is produced from a special class of kerogen-containing rocks (sometimes referred to as bituminous rocks, but this term is more appropriate for application to tar sand formations and oil sand formations in Canada). Bituminous rocks have gained some importance for the resource called *oil shale*, which is an argillaceous, laminated sediment of generally high organic content that can be thermally decomposed to yield appreciable amounts of oil, commonly referred to as *shale oil*. Oil shale does not yield shale oil without the application of high temperatures and the resulting thermal decomposition that is necessary to decompose the organic material (*kerogen*) in the shale. Kerogen produces a liquid product (shale oil) upon thermal decomposition at high temperature (>500°C, >930°F). Shale oil also contains heteroatom species (nitrogen-, oxygen-, and sulfur-containing organic compounds). Refined (hydrotreated) shale oil (Chapter 16), in which the heteroatom content has been reduced to acceptable levels, may also be referred to as *synthetic crude oil* that is sent to a refinery for further processing into various products. The raw oil shale can even be used directly as a fuel akin to a low-quality coal. Indeed, oil shale deposits have been exploited as such for several centuries and shale oil has been produced from oil shale since the 19th century (Speight, 2012, 2020).

Geologically, oil shale is a sedimentary marlstone rock that is embedded with rich concentrations of an organic material known as kerogen. The oil shale deposits in the Western United States contain approximately 15% w/w organic material (Scouten, 1990; Speight, 2012, 2020; Lee *et al.*, 2014). When the oil shale is heated to high temperatures (typically temperatures in excess of 500°C, 930°F), the kerogen is converted (in part) to gases and a distillable liquid that, once upgraded, can be refined into a variety of liquid fuels, gases, and high-value chemical and mineral by-products. The remainder of the kerogen is converted to a char that remains on the shale.

The United States has vast known oil shale resources and they are concentrated in the Green River Formation in the states of Colorado, Wyoming, and Utah, which account for nearly three-quarters of this potential. Because of the abundance and geographic concentration of the known resource, oil shale has been recognized as a potentially valuable United States energy resource since as early as 1859, the year Colonel Drake completed his first oil well in Titusville, Pennsylvania (Speight, 2014a). Early products derived from shale oil included kerosene and lamp oil, paraffin, fuel oil, lubricating oil and grease, naphtha, illuminating gas, and ammonium sulfate fertilizer.

Since the beginning of the 20th century, when the United States Navy converted its ships from coal to fuel oil and the national economy was transformed by gasoline-fueled automobiles and diesel-fueled trucks and trains, concerns have been raised related to assuring adequate supply of liquid fuels at affordable prices to meet the growing needs of the nation and its consumers. The abundant resources of oil shale in the

United States were initially viewed as a major source for these fuels. Numerous commercial entities sought to develop the oil shale resources. The Mineral Leasing Act of 1920 rendered crude oil and oil shale resources on federal lands available for development under the terms of federal mineral leases. Soon, however, discoveries of more economically producible and refinable liquid crude oil in commercial quantities caused the interest in oil shales to decline.

Interest in oil shales resumed after World War II, when military fuel demand and domestic fuel rationing and the rising fuel prices made the economic and strategic importance of the oil shale resource more apparent. After the war, the booming post-war economy drove the demand for fuels ever higher. Public and private research and development efforts were commenced, including the 1946 United States Bureau of Mines Anvil Point, Colorado oil shale demonstration project. Significant investments were made to define and develop the resource and to develop commercially viable technologies and processes to mine, produce, retort, and upgrade oil shale into viable refinery feedstocks and by-products. Once again, however, major crude oil discoveries in the lower-48 United States, offshore, and in Alaska, as well as other parts of the world reduced the foreseeable need for shale oil and the interest and associated activities again diminished.

By 1970, oil discoveries were slowing, the demand was rising, and crude oil imports, largely from Middle Eastern countries, were rising to meet the demand. Global oil prices, although still relatively low, were also rising, reflecting the changing market conditions. Ongoing oil shale research and testing projects were reenergized and new projects were envisioned by numerous energy companies seeking alternative feedstocks for the production of fuels. These efforts were significantly amplified by the impact of the 1973 Arab Oil Embargo, which demonstrated the vulnerability of the United States to oil import supply disruptions, and were underscored by a new supply disruption associated with the 1979 Iranian Revolution.

By 1982, however, technological advances and discovery of offshore oil resources in the North Sea and elsewhere provided new and diverse sources for oil import into the United States and dampened the global energy prices. Global political shifts promised to open previously restricted provinces to oil and gas exploration and led economists and other experts to predict a long future of relatively low and stable oil prices. Despite significant investments by energy companies, and numerous variations and advances in mining, restoration, retorting, and *in situ* processes, the costs of oil shale production relative to foreseeable oil prices made continuation of most commercial efforts impractical. During that time, numerous projects were initiated but then terminated primarily because of their economic infeasibility relative to expected world oil prices or project design issues. Since that time, the recovery of crude oil from tight formations has turned the United States from an overall importer of crude oil to an overall exporter of crude oil and its products, but the issue of peak oil production remains to be a matter for serious discussion (Speight and Islam, 2016).

Oil shale is a complex and intimate mixture of organic and inorganic materials that vary widely in composition and properties. In general terms, oil shale is a fine-grained sedimentary rock that is rich in organic matter and yields oil when heated. Some oil shales are genuine shales, but others have been misclassified and are actually siltstones, impure limestone, or even impure coal. An oil shale does not contain oil and it produces oil only when heated to approximately 500°C (approximately 930°F), when some of the organic material is transformed into a distillate similar to crude oil.

There is no scientific definition of *oil shale* and the current definition is based on economics. Generally, an oil shale is a mixture of carbonaceous molecules dispersed in an inorganic (mineral) matrix. It is called a shale because it is found in a layered structure typical of sedimentary rocks, but the mineral composition can vary from true aluminosilicate shale to carbonate minerals. Thus, an *oil shale* is a compact, laminated rock of sedimentary origin that yields over 33% ash and contains insoluble organic matter that yields oil when distilled. *Kerogen* is the name given to the naturally occurring insoluble organic matter found in shale deposits. *Shale oil* is the synthetic fuel produced by thermal decomposition of *kerogen* at high temperature (>500°C, >930°F). Shale oil is referred to as *synthetic crude oil* after hydrotreating. The amount of kerogen in the shales varies with depth, with the richer portions appearing much darker. For example, in Colorado (USA), the richest layers are termed *the Mahogany Zone* after the rich dark brown color.

Despite being a huge energy resource, oil shale is an underutilized energy resource. In fact, one of the issues that arise when dealing with fuels from oil shale is the start–stop–start episodic nature of the various projects. The projects differ in time and economic investment and viability, the reasons being competition from cheaper energy sources, relatively high front-end investments and, of late, an unfavorable environmental record. Oil shale has, however, a definite potential to meet energy demand in an environmentally acceptable manner (Bartis *et al.*, 2005; Andrews, 2006).

Just like *oil sand* (tar sand in the United States), the term *oil shale* is a misnomer since the mineral does not contain oil nor is it always a *shale*. The organic material is chiefly *kerogen* and the *shale* is usually a relatively hard rock called marl. When properly processed, kerogen can be converted into a substance somewhat similar to crude oil, which is often better than the lowest grade oil produced from conventional oil reservoirs but of lower quality than conventional light (low-density, non-viscous) crude oil.

Retorting is the process of heating oil shale in order to recover the organic material, predominantly as a liquid. To achieve economically attractive recovery of product, temperatures of 400–600°C (750–1100°F) are required. A retort is simply a vessel in which the oil shale is heated from which the product gases and vapors can escape to a collector. Thus, *shale oil*, sometimes termed *retort oil*, is the liquid oil condensed from the effluent in oil shale retorting and typically contains appreciable amounts of water and solids, as well as having an irrepressible tendency to form sediments because of the various constituents (Tables 4.9 and 4.10). However, shale oils are sufficiently different from crude oil, and processing shale oil presents some unusual problems.

Table 4.9: Challenges for oil shale processing.

Constituents	Effects
Particulate matter	Plugging on processing
	Product quality
Arsenic content of shale oil	Toxicity
	Catalyst poison
High pour point of shale oil	Must be upgraded to pipeline quality
Nitrogen content of shale oil	Catalyst poison
	Contributes to instability
	Toxicity
Diolefin derivatives	Contribute to instability
	Plugging on processing

Table 4.10: Major compound types in shale oil.

Saturates	Heteroatom systems
Paraffins	Benzothiophenes
Cycloparaffins	Dibenzothiophenes
Olefins	Phenols
Aromatics	Carbazoles
Benzenes	Pyridines
Indans	Quinolines
Tetralins	Nitriles
Naphthalenes	Ketones
Biphenyls	Pyrroles
Phenanthrenes	
Chysenes	

The oil production capacity of an oil shale is measured using a laboratory pyrolysis method called the Fischer assay (Speight, 2017b, 2020) and expressed in barrel (42 gallons) of oil per ton of oil shale. Rich zones can yield more than 40 gallons per ton, while most shales fall in the range of 10–25 gallons per ton. Oil shale ore, which yields more than 25 United States gallon per ton, is generally considered as the most economically attractive ore and hence the most favorable ore for development.

References

Ancheyta, J., Betancourt, G., Marroquin, G., Perez, A.M., Maity, S.K., Cortez, M.T., and Del Rio, R. 2001. An exploratory study for obtaining synthetic crudes from heavy crude oils via hydrotreating. *Energy & Fuels* 15: 120–127.

Ancheyta, J., Betancourt, G., Centeno, G., Marroquin, G., Alonso, F., and Garciafigueroa. 2002a. Catalyst deactivation during hydroprocessing of Maya heavy crude oil. 1. Evaluation at constant operating conditions. *Energy & Fuels* 16: 1438–1443.

Ancheyta, J., Betancourt, G., Marroquin, G., Centeno, G., Castaneda, L.C., Alonso, F., Munoz, J.A., Gomez, M.T., and Rayo, P. 2002b. Hydroprocessing of Maya heavy crude oil in two reaction stages. *Appl. Catal. A: Gen.* 233: 159–170.

Ancheyta, J. and Speight, J.G. 2007. *Hydroprocessing of Heavy Oils and Residua.* CRC-Taylor and Francis Group, Boca Raton, Florida.

ASTM. 2019. *Annual Book of Standards*. ASTM International, West Conshohocken, Pennsylvania.

ASTM D56. 2019. Standard Test Method for Flash Point by Tag Closed Cup Tester. *Annual Book of Standards*. ASTM International, West Conshohocken, Pennsylvania.

ASTM D92. 2019. Standard Test Method for Flash and Fire Points by Cleveland Open Cup. *Annual Book of Standards*. ASTM International, West Conshohocken, Pennsylvania.

ASTM D93. 2019. Standard Test Method for Flash Point by Pensky-Martens Closed Cup. *Annual Book of Standards*. ASTM International, West Conshohocken, Pennsylvania.

ASTM D323. 2019. Standard Test Method for the Vapor Pressure of Petroleum Products (Reid Method). *Annual Book of Standards*. ASTM International, West Conshohocken, Pennsylvania.

Berecz, E. and Balla-Achs, M. 1983. *Gas Hydrates*. Elsevier BV, Amsterdam, Netherlands.

BP. 2019. *Statistical Review of World Energy*. British Petroleum, London, united Kingdom, http://www.bp.com/centres/energy2019.

Gary, J.G., Handwerk, G.E., and Kaiser, M.J. 2007. *Petroleum Refining: Technology and Economics*, 5th Edition. CRC Press, Taylor & Francis Group, Boca Raton, Florida.

Ghoshal, S. and Sainik, V. 2013. Monitor and minimize corrosion in high-TAN crude processing. *Hydrocarbon Process* 92(3): 35–38.

Gudmundsson, J.S., Andersson, V., Levik, O.I., and Parlaktuna, M. 1998. Hydrate Concept for Capturing Associated Gas. In *Proceedings of the SPE European Petroleum Conference*, The Hague, The Netherlands, October 20–22.

Hsu, C.S. and Robinson, P.R. (Editors). 2017. *Handbook of Petroleum Technology*. Springer International Publishing AG, Chaim, Switzerland.

IEA. 2005. *Resources to Reserves: Oil & Gas Technologies for the Energy Markets of the Future*. International Energy Agency, Paris, France.

Lee, S., Speight, J.G., and Loyalka, S. 2014. *Handbook of Alternative Fuel Technologies*, 2nd Edition. CRC Press, Taylor & Francis Group, Boca Raton, Florida.

Max, M.D. (Editor). 2000. *Natural Gas in Oceanic and Permafrost Environments*. Kluwer Academic Publishers, Dordrecht, The Netherlands.

Mokhatab, S., Poe, W.A., and Speight, J.G. 2006. *Handbook of Natural Gas Transmission and Processing*. Elsevier, Amsterdam, The Netherlands.

Ohmes, R. 2014. Characterizing and Tracking Contaminants in Opportunity Crudes. Digital Refining, http://www.digitalrefining.com/article/1000893,

Characterising_and_tracking_contaminants_in_opportunity_crudes_.html#.VJhFjV4AA; accessed November 1, 2014.

Parkash, S. 2003. *Refining Processes Handbook*. Gulf Professional Publishing, Elsevier, Amsterdam, Netherlands.

Scouten, C. 1990. In *Fuel Science and Technology Handbook*, J.G. Speight (Editor). Marcel Dekker Inc., New York.

Shalaby, H.M. 2005. Refining of Kuwait's Heavy Crude Oil: Materials Challenges. In *Proceedings of the Workshop on Corrosion and Protection of Metals*. Arab School for Science and Technology, Kuwait, December 3–7.

Sloan, E.D. 1997. *Clathrates of Hydrates of Natural Gas*. Marcel Dekker Inc., New York.

Sloan, E.D. 2000. Clathrates hydrates: The other common water phase. *Ind. Eng. Chem. Res*. 39: 31123–3129.

Speight, J.G. 2012. *Shale Oil Production Processes*. Gulf Professional Publishing Company, Elsevier, Oxford, United Kingdom.

Speight, J.G. 2013. *The Chemistry and Technology of Coal*, 3rd Edition. CRC-Taylor and Francis Group, Boca Raton, Florida.

Speight, J.G. 2014a. *The Chemistry and Technology of Petroleum*, 5th Edition. CRC-Taylor and Francis Group, Boca Raton, Florida.

Speight, J.G. 2014b. *High Acid Crudes*. Gulf Professional Publishing, Elsevier, Oxford, United Kingdom.

Speight, J.G. 2014c. *Oil and Gas Corrosion Prevention*. Gulf Professional Publishing, Elsevier, Oxford, United Kingdom.

Speight, J.G. 2015. *Handbook of Petroleum Product Analysis*, 2nd Edition. John Wiley & Sons Inc., Hoboken, New Jersey.

Speight, J.G. 2016a. *Introduction to Enhanced Recovery Methods for Heavy Oil and Tar Sands*, 2nd Edition. Gulf Professional Publishing, Elsevier, Oxford, United Kingdom.

Speight, J.G. 2016b. Hydrogen in Refineries. In *Hydrogen Science and Engineering*: *Materials, Processes, Systems, and Technology*, D. Stolten and B. Emonts (Editors). Wiley-VCH Verlag GmbH & Co., Weinheim, Germany, Chapter 1, pp. 3–18.

Speight, J.G. and Islam, M.R. 2016. *Peak Energy — Myth or Reality*. Scrivener Publishing, Beverly, Massachusetts.

Speight, J.G. 2017a. *Handbook of Petroleum Refining*. CRC Press, Taylor & Francis Group, Boca Raton, Florida.

Speight, J.G. 2017b. *Deep Shale Oil and Gas*. Gulf Professional Publishing Company, Elsevier, Oxford, United Kingdom.

Speight, J.G. 2019a. *Natural Gas: A Basic Handbook*, 2nd Edition. Gulf Publishing Company, Elsevier, Cambridge, Massachusetts.

Speight, J.G. 2019b. *Handbook of Petrochemical Processes*. CRC Press, Taylor & Francis Group, Boca Raton, Florida.

Speight, J.G. 2019c. *Heavy Oil Recovery and Upgrading*. Gulf Publishing Company, Elsevier, Cambridge, Massachusetts.

Speight, J.G. 2020. *Shale Oil and Gas Production Processes*. Gulf Publishing Company, Elsevier, Cambridge, Massachusetts.

Sun, X., Zhang, Y., Li, X., Cui, G., and Gu, J. 2013. A case study on foamy oil characteristics of the Orinoco belt, Venezuela. *Adv. Petroleum Exp. Develop.* 5(1): 37–41.

US Congress. 1976. Public Law FEA-76-4. United States Library of Congress, Washington, DC.

Wachtmeister, H., Linnea Lund, L., Aleklett, K., and Höök, M.M. 2017. Production decline curves of tight oil wells in eagle ford shale. *Nat. Res. Res.* 26(3): 365–377.

Yeung, T.W. 2014. Evaluating Opportunity Crude Processing. *Digital Refining*, http://www.digitalrefining.com/article/1000644; accessed October 25, 2014.

Chapter 5

Feedstocks for Alternate Fuels

5.1 Introduction

The widespread use of fossil fuels within the current energy infrastructure is considered by some observers as the largest source of anthropogenic emission of carbon dioxide (CO_2), which is largely blamed for global warming and climate change (Zanganeh and Shafeen, 2007) although other sources are evident but are often ignored (Speight, 2020). Nevertheless, having been identified as *one* of the causative agents of climate change, it is necessary to attempt to mitigate the emissions of carbon dioxide from fossil fuel combustion and to offset the depletion of fossil fuels such as the commonly used natural gas, and crude oil, and coal (Ragland *et al.*, 1991). Oil shale (Chapter 1) is of lesser importance, having received on-again and off-again popularity as a source of fuels but never really been recognized as a major source of fuels (Baughman, 1978; Allred, 1982; Scouten, 1990; Lee, 1991; US DOE, 2004a, 2004b, 2004c; Speight and Islam, 2016; Speight, 2012, 2013, 2014, 2019b).

The combustion of natural gas, crude oil, and coal (or, for that matter any carbonaceous feedstock) for power generation inevitably produces carbon dioxide, a key greenhouse gas, as well as other pollutants, including the gases that cause acid rain. These resources (natural gas, crude oil, and coal) are currently on a depletion curve with estimates of the longevity of these resources varying up to 50 years (Speight and Islam, 2016). However, reducing national dependence of any country on fossil fuels — in particular, the dependence on imported natural gas and crude oil — is of critical importance for its long-term security and continued economic growth. Supplementing crude oil consumption with renewable biomass resources is the first step towards this goal. The realignment of the chemical industry from a petrochemical refining to a biorefinery concept is, given time, feasible and has become a national goal of many oil-importing countries. However, clearly defined goals are necessary to increase the use of biomass-derived feedstocks in industrial chemical production and it is important to keep this goal in perspective. In this context, the increased use of biofuels should be viewed as one of a range of possible measures for achieving self-sufficiency in energy, rather than as a stand-alone cure-all solution (Demirbaş, 2009).

Alternate fuels (also known as alternative fuels, non-conventional fuels, or advanced fuels) are defined as any materials that can be used as fuels except conventional fuels (defined here as fuels derived from natural gas, crude oil, coal, or oil shale. Nuclear fuels such as uranium and thorium and artificial radio-isotope fuels made in nuclear reactors are not included in this definition. For example, common alternate fuels include biohydrogen, biogas, bioalcohol, biodiesel, refuse-derived fuel, non-fossil fuel methane, and related hydrocarbon derivatives. As explained in Chapter 1, the prefix *bio* used in this text indicates that the fuel was obtained from a biological source and did not necessarily exist, or occur naturally, in fuel form.

Biomass (Chapter 1) is a collective term for many types of biological materials but the focus of this book is on three types: (i) energy crops, (ii) wood, and (iii) organic waste. Biomass as a renewable resource has received considerable attention due to environmental impact considerations, which are focused somewhat negatively the use of on fossil fuels and the increasing demand for energy worldwide. Biomass is the oldest form of energy used by humans and, traditionally, biomass was utilized through direct combustion, and this process is still widely used in many parts of the world (Ragland *et al.*, 1991). Since the energy crises of the 1970s, many countries have been interested in biomass as a fuel source to expand the development of domestic and renewable energy sources and to reduce the environmental impacts of energy production. Biomass energy (bioenergy) can be an important alternative in the future for a more sustainable energy supply. Biomass is material derived from plants and there are many types of biomass resources currently used and potentially available for use. Biomass energy has the potential to be produced and used efficiently and cost-competitively, generally in the more convenient forms of gases, liquids, or electricity (Larson and Kartha, 2000).

Examples of modern biomass use are ethanol production from sugarcane, combined heat and power (CHP) district heating programs, and the co-combustion of biomass in conventional coal-based power plants (Hoogwijk *et al.*, 2005). In industrialized countries, the main biomass processes to be used in the future are expected to be the following: direct combustion of residues and wastes for electricity generation, use of bioethanol and biodiesel as liquid fuels, and combined heat and power production from energy crops. In fact, biomass can be considered as the best option and has the largest potential to meet these requirements and could ensure fuel supply in the future (Demirbas, 2007, 2008).

To elaborate further on the term *biomass*, it is used to describe any material of recent biological origin, including plant materials, such as trees, grasses, agricultural crops, and even animal manure. Other biomass components, which are generally present in minor amounts, include triglyceride derivatives, sterol derivatives, alkaloid derivatives, terpene derivatives, terpenoid derivatives, and waxes. This includes everything from *primary sources* of crops and residues harvested/collected directly from the land, to *secondary sources* such as sawmill residuals, to *tertiary sources* of post-consumer residuals that often end up in landfills. However, for the purpose of this chapter, three distinct sources of biomass energy exist: (i) agricultural crops, (ii) wood, and (iii) industrial waste and domestic waste (Chapter 6). Landfill gas is included in this book (Chapter 7) and, although not usually categorized as such, is the fourth source of biomass energy that includes the gases that result from anaerobic digestion of organic materials in landfills (Chapter 7) (Demirbaş and Ozturk, 2004).

Primary biomass is produced directly by photosynthesis and includes all terrestrial plants now used for food, feed, fiber, and fuel wood. All plants in natural and conservation areas (as well as algae and other aquatic plants growing in ponds, lakes, oceans, or artificial ponds and bioreactors) are also considered primary biomass. However, only a small portion of the primary biomass produced will ever be harvested as feedstock material for the production of bioenergy and by-products.

Biomass feedstocks and fuels exhibit a wide range of physical, chemical, and agricultural/process engineering properties. Despite their wide range of possible sources, biomass feedstocks are remarkably uniform in many of their fuel properties, compared with competing feedstocks such as coal or crude oil. For example, there are many types of coal and the gross heating value of these types varies from 8,600 to 12,900 Btu/lb

(Chapter 4) (Speight, 2013). However, nearly all kinds of biomass feedstocks destined for combustion fall in the range 6,450–8,200 Btu/lb. For most agricultural residues, the heating values are even more uniform — typically of the order of 6,450–7,300 Btu/lb while the values for most woody materials are of the order of 7,750–8,200 Btu/lb.

Moisture content is probably the most important determinant of the heating value of a feedstock or fuel. Air-dried biomass typically has approximately 15–20% w/w moisture, whereas the moisture content for oven-dried biomass is typically at 0% w/w. Moisture content is also an important characteristic of coals, varying in the range of 2–30% w/w (Chapter 4) (Speight, 2013). However, the bulk density (and, hence, the energy density) of most biomass feedstocks is generally low, even after densification, and is typically of the order of 10–40% of the bulk density of most fossil fuels. Liquid biofuels have comparable bulk densities to fossil fuels.

In addition, in the case of many staple food crops, a potentially large economic fuel resource is effectively being discarded. For example, the straw associated with the wheat crop is often plowed back into the soil, even though only a small proportion is needed to maintain the level of organic matter. Thus, a huge renewable resource is not being usefully exploited since wheat straw contains a range of potentially useful chemicals. These include: (i) cellulose and related compounds, which can be used for the production of paper and/or bioethanol, and (ii) long-chain lipid derivatives, which can be used in cosmetics or other specialty chemicals.

Cellulose is an organic compound with the empirical formula ($C_6H_{10}O_5$) and is a polysaccharide consisting of a linear chain of several hundreds to many thousands of linked glucose units — it is an important structural component of the primary cell wall of green plants.

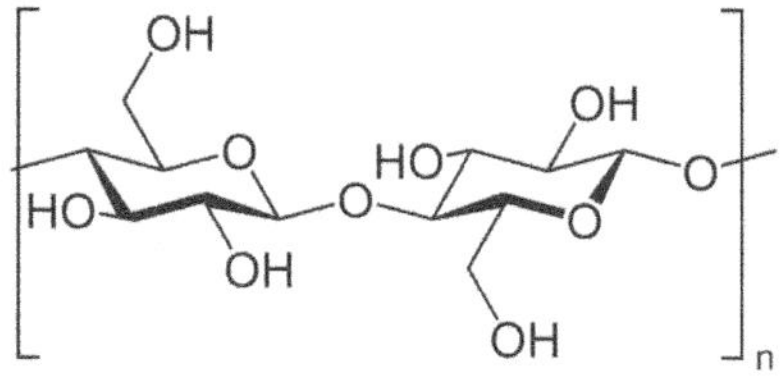

On the other hand, lipid derivatives are very diverse in both their respective structures and functions. These diverse compounds that make up the lipid family are so grouped because they are insoluble in water (hydrophobic). They are also soluble in other organic solvents such as ether, acetone, and other lipids. Lipids serve a variety of important functions in living organisms and also act as chemical messengers, serve as valuable energy sources, provide insulation, and are the main components of membranes. Major lipid derivatives include the phospholipid derivatives.

Briefly, the phospholipids are a class of lipids that are a major component of all cell membranes. They can form lipid bilayers because of their amphiphilic characteristic. The structure of the phospholipid molecule generally consists of two hydrophobic fatty acid tails and a hydrophilic head consisting of a phosphate group. The two components are usually joined together by a glycerol ($HOCH_2CHOHCH_2OH$) molecule.

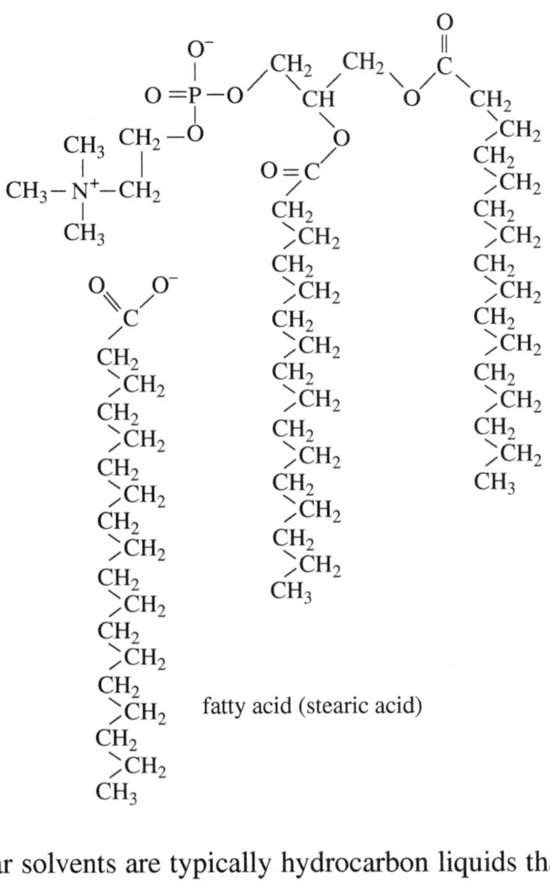

fatty acid (stearic acid)

Non-polar solvents are typically hydrocarbon liquids that are used to dissolve other naturally occurring hydrocarbon lipid molecules that do not (or do not easily) dissolve in water, including fatty acids, waxes, sterols, fat-soluble vitamins (such as vitamins A, D, E, and K), monoglyceride derivatives, diglyceride derivatives, and triglyceride derivatives. Thus:

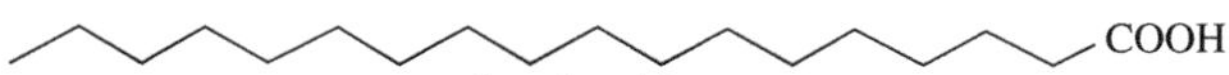

Stearic acid

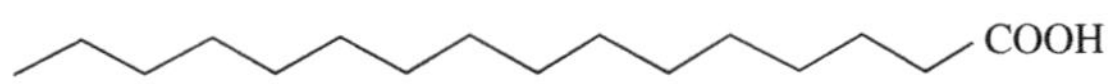

Palmitic acid

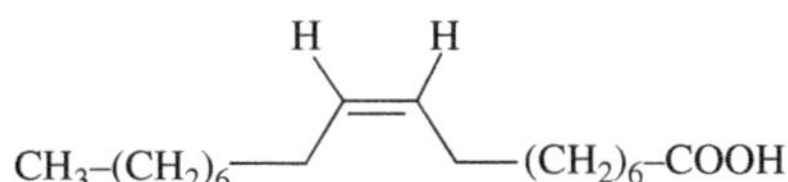

Oleic acid

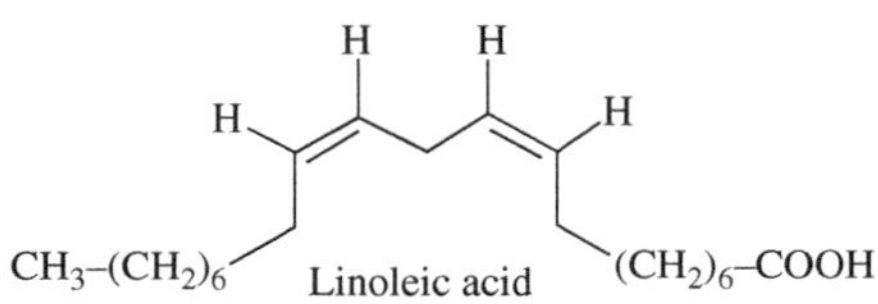

Linoleic acid

Examples of fatty acids

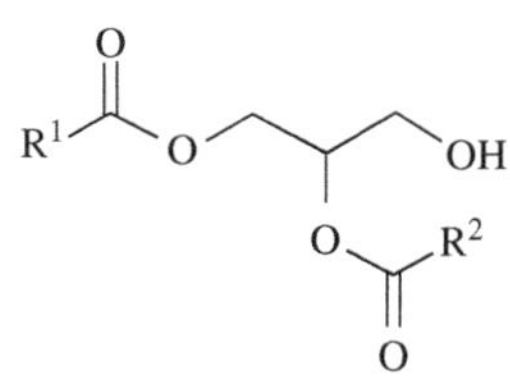

An example of a diglyceride

The production of fuels from alternate sources (i.e. non-fossil energy sources) offers many potential environmental and economic benefits. The production of the raw biomass material and its subsequent conversion to fuels also offers many environmental benefits including reduction of carbon dioxide emissions associated with global climate change and improved waste utilization. The chemical composition of many biofuels also leads to improved engine performance and reduces the generation of unwanted pollutants such as carbon monoxide and unburned hydrocarbon derivatives.

By way of introduction, and in contrast to crude oil, biomass (because of its heterogeneous nature) is subdivided into various types, such as woody biomass, grass biomass and the like. Each different type is then subdivided into the various chemical constituents that make up the biomass.

Lignocellulosic feedstocks (which are three-dimensional polymeric composites formed by plants as structural material) are composed primarily of carbohydrate polymers (cellulose and hemicellulose) and polymers (lignin) of phenol derivatives. Lower concentrations of various other compounds, such as proteins, acids, salts, and minerals, are also present. Cellulose and hemicellulose, which typically make up two-thirds of cell wall dry matter (the portion of biomass that is not water), are polysaccharide derivatives that can be hydrolyzed to sugars and then fermented to ethanol. Process performance, in this case the ethanol yield from biomass, is directly related to the cellulose, hemicellulose, and individual sugar concentration in the feedstock. Lignin cannot be used in fermentation processes, but it may be useful for other purposes.

Cellulose (30–50% w/w of total feedstock dry matter) is a glucose polymer linked by β-1,4 glycosidic bonds. The basic building block of this linear polymer is cellobiose, which is a glucose–glucose dimer — the hydrolysis of cellulose (saccharification) produces its glucose monomers.

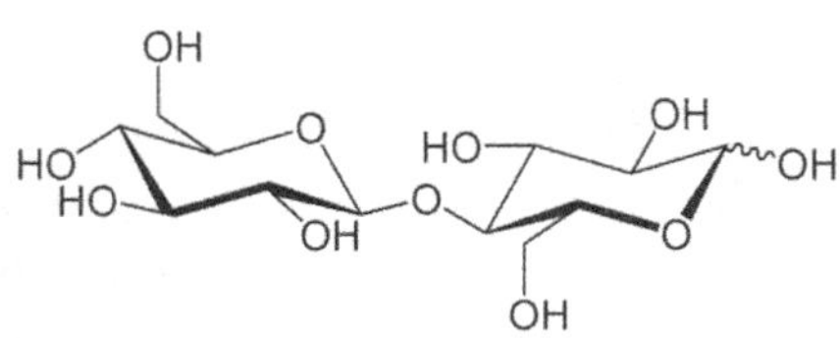

Cellobiose

On the other hand, hemicellulose (20–40% w/w of total feedstock dry matter) is a short, highly branched polymer of five- (C_5) and six-carbon (C_6) sugar derivatives. Specifically, hemicellulose contains xylose and arabinose (C_5 sugars) and galactose, glucose, and mannose (C_6 sugars). Hemicellulose is more readily hydrolyzed than cellulose because of its branched, amorphous nature and a major product of hemicellulose hydrolysis is xylose (a C_5 sugar).

This chapter presents an overview of feedstocks that are commonly used for the production of alternate fuels in order for the reader to understand the chemical and physical parameters of each type of feedstock, which are an indication of the methods that are necessary for the production of alternate fuels. Examples of the common biomass-type feedstocks presented here are as follows: (i) crops, (ii) wood, and (iii) organic waste.

5.2 Crops

Biomass currently provides varying proportions (depending on the country) of the primary energy supply. Projections are that this proportion will need to increase dramatically to meet the growing energy demand. Given a relatively fixed land base, the production of energy crops will compete with traditional agricultural and forestry uses of land. It is essential that agriculture and forestry work together to create integrated biomass production systems that landowners can use to help meet the growing energy demands of the United States. Producing fast-growing short rotation woody crops on agricultural lands is one such approach that shows considerable promise. When produced in integrated systems, short rotation woody crops combine the goals of sustainable forestry with those of sustainable agriculture at multiple temporal and spatial scales. In addition to using woody biomass for energy and for power and heat generation via cofiring and gasification, woody crops with their high hemicellulose and cellulose content are well suited for biorefining to yield liquid fuels such as ethanol, methanol, distillable oil (sometimes referred to as pyrolysis oil or bio-oil), and other products, such as biodegradable plastics and specialty chemicals (Tables 5.1 and 5.2).

Crops are the annual or seasonal yield of any plant that is grown in significant quantities to be harvested as food, as livestock fodder, or as fuel, or for any other economic purpose. This category includes crop species as well as agricultural techniques related to cropping. There are many types of crops that are used for industrial purposes. For example, crops are grown and harvested for the sole purpose of making profit and feeding people, as they are grown in large amounts in a certain area suitable for growing crops. Crops have essentially many functions and benefits, particularly for the *human animal*! Products derived from crops serve not only as a primary source of human foods and animal feed but also as a source of timber, fibers, and biomass energy. In addition, crops also have an essential function: to maintain ecological systems and the natural environment. Most of the crop production is used as foods, but in the last century (the 20th century), crops were also cultivated for non-food purposes — examples include pharmaceutical and nutritional products, chemical derivative products such as adhesives, paints, polymers, plastics, and industrial oils in the form of biodiesel, transmission fluids, and lubricants.

Many efforts have been made to grow crops for food production in order to strengthen food security and to alleviate the hunger and poverty

Table 5.1: Types of pyrolysis.

Type	Parameters
Slow pyrolysis	Several hours to complete
	Char is the main product
Flash pyrolysis	High heating rate of the feedstock
	High temperature: 450–1,000°C
	Rapid devolatilization in an inert atmosphere
	Up to 75% w/w yield of bio-oil
	Tendency for oil to be thermally unstable
Fast pyrolysis	Takes seconds to complete
	Up to 60% yield of bio-oil
	Also yields synthesis gas and char

Table 5.2: Biomass liquefaction by pyrolysis.

Feedstock	Process	Products	
Biomass	Pyrolysis, 550°C (1,120°F), no air	Gases	H_2, CO, CO_2, $C_nH_{2n}+_2$
		Liquids	$C_nH_{2n}+_2$
		Char	
	Combustion	Gases	CO_2, H_2O

particularly seen in developing countries. However, serious attention should also be directed toward cultivation of crops for non-food use.

Thus, by definition, energy crops are plants grown specifically for use as a fuel. Although growing crops for fuel dates back to medieval times, in their modern form, energy crops are the most recent and innovative renewable energy option. Energy crops are important as a renewable energy technology because their use will produce a variety of economic, environmental, and energy benefits (Lago *et al.*, 2019). Commercial energy crops are typically densely planted, high-yielding crop species where the energy crops will be burnt to generate power. Woody crops such as willow and poplar are widely utilized, as well as tropical grasses such as miscanthus and elephant grass (*Pennisetum purpureum*).

Grasses are usually herbaceous plants with narrow leaves growing from the base. They include the true grasses of the Poaceae (or Gramineae) family, as well as the sedges (Cyperaceae) and the rushes (Juncaceae). The true grasses include cereals, bamboo and the grasses of lawns (turf) and grassland. Sedges include many wild marsh and grassland plants, and some cultivated ones such as water chestnut (*Eleocharis dulcis*) and papyrus sedge (*Cyperus papyrus*). Most of the interest in grass biomass tends to focus on its economics aspects, but there is a list of traits that should be considered and valued when evaluating a potential solid biomass energy source. These traits are beneficial to society in general or impact the suitability of biomass for a farm operation.

If the carbohydrate content is desired for the production of biogas, whole crops such as maize, Sudan grass, millet, white sweet clover and many others can be made into silage after which they can be converted into biogas. On the other hand, crop residues are the residues remaining after crops have been harvested. Crop residues typically contain 40% w/w of the nitrogen (N), 80% w/w of the potassium (K), and 10% w/w of the phosphorus (P) applied to the soil in the form of fertilizer. If these residues are subjected to direct combustion for energy, only a small percentage of the nutrients is left in the ash.

The valuable portion of sugar and starch crops (in terms of biofuel production) is the stalks and leaves, which are composed mainly of cellulose. The individual six-carbon sugar units in cellulose are linked together in extremely long chains by a stronger chemical bond than exists in starch. In *starch crops*, most of the six-carbon sugar units are linked together in long, branched chains (*starch*). Yeast cannot use these chains to produce ethanol. The starch chains must be broken down into individual six-carbon units or groups of two units. The starch conversion process is relatively simple because the bonds in the starch chain can be broken in an inexpensive manner by the use of heat and enzymes or by a mild acid solution.

Sugar crops include a variety of plants such as fodder beets, fruit crops, Jerusalem artichokes, sugar beets, sugarcane, and sweet sorghum. Interest in ethanol production from such agricultural crops has prompted the development of sugar crops that have not been cultivated on a widespread commercial basis in many countries. Preparation is basically crushing and extraction of the sugars, which the yeast can immediately use. But sugar crops must be dealt with fairly quickly before their high sugar and water content causes spoilage. Because of the risk of such spoilage, storage of sugar crops is not practical.

As with starch, cellulose must be broken down into sugar units before it can be used by yeast to make ethanol. However, the breaking of the cellulose bonds is much more complex and costly than the breaking of the starch bonds. Breaking the cellulose into individual sugar units is complicated by the presence of lignin, a complex compound surrounding cellulose, which is even more resistant than cellulose to enzymatic or acidic pretreatment (Hwang and Obst, 2003). Because of the high cost of converting liquefied cellulose into fermentable sugars, agricultural residues (as well as other crops having a high percentage of cellulose) are not yet a practical feedstock source for small ethanol plants.

Crop residues (cobs, stems, leaves, in particular straw, and other plant matter) left in agricultural fields after harvest could potentially be used for solid biofuel production. Due to its high energy content, straw is one of the best crop residues for solid biofuels. However, straw has several disadvantages — it has a higher ash content, which results in lower calorific value. In order to improve its bulk density, the straw is generally baled before transportation. Straw burning requires a specific technology and there are four basic types of straw burners: (i) the burners that accept shredded, loose straw, (ii) the burners that use densified straw products such as pellets, briquettes, or cubes and straw logs, (iii) the burners for small, square bales of feedstock, and (iv) the burners for round bales of feedstock. To be suitable for heat and electricity production straw should not have a large content of moisture, preferably not more than 20%, as the moisture reduces the boiler efficiency. Also straw color as well as straw chemistry should be considered before burning as it indicates the quality of the straw. Most crop residues are returned to the soil, and the humus resulting from their decomposition helps maintain soil nutrients, soil porosity, water infiltration and storage, as well as reducing soil erosion.

Wood is the oldest fuel known to man. Burning wood rather than fossil fuels can reduce the carbon dioxide emissions responsible for global climate change. Wood fuel is carbon dioxide (CO_2) neutral. It gives off only as much carbon dioxide when burned as it stores during its lifetime.

In addition, wood fuel has very low levels of sulphur, a chemical that contributes to acid rain.

Regularly coppiced plantations will actually absorb more carbon dioxide than mature trees, since carbon dioxide absorption slows once a tree has grown. Growing crops for fuel, particularly wood coppice, offers very promising developments for the future. Short rotation arable coppicing, using fast-growing willows, is currently seen as an important source of fuel for electricity generation. The overall process involves several stages: growing over two or three years, cutting and converting to wood chip, storage and drying, and transport to a power plant for combustion. The combustion process can be very efficient, given the development of advanced cogeneration techniques.

Moreover, the security of energy supply will be significantly enhanced for many crop-growing countries. Energy crop production and use could reduce dependence on imported oil, although complete energy independence will remain a possible future event.

5.2.1 *Energy crops*

Energy crops are low-cost and low-maintenance crops grown solely for energy production (not for food). The crops are processed into solid, liquid, or gaseous fuels, such as pellets, bioethanol, or biogas. The fuels are burned to generate power or heat. Furthermore, the term *energy crops* can be used both for biomass crops that simply provide high output of biomass per hectare for low inputs and for those that provide specific products that can be converted into other biofuels such as sugar or starch for bioethanol by fermentation or into vegetable oil for biodiesel by transesterification. Energy crops such as grasses, miscanthus oilseed crops, short-rotation woody crops, residual herbaceous biomass, starch crops, sugar crops, and switchgrass can be converted to liquid biofuels by biochemical and thermochemical conversion process (Table 5.3).

In the biochemical process, bacteria, yeasts, and enzymes also break down carbohydrates. For example, the fermentation process used to make wine changes biomass liquids into alcohol, a combustible fuel. A similar process is used to turn corn into ethanol, which is mixed with gasoline to make gasohol. Also, when bacteria break down biomass, methane and carbon dioxide are produced. This methane can be captured in sewage

Table 5.3: Summary of the methods for the conversion of biomass to fuels.

Feedstock	Stage 1	Stage 2	Products
Biomass	Extraction	Transesterification	Biodiesel
	Hydrolysis	Fermentation	Biogas Ethanol
	Gasification	Synthesis gas	Biogas Hydrogen Methanol Ethanol
	Pyrolysis		Hydrogen Bio-oil
	Hydrotreating		Diesel

treatment plants and landfills, for example, and burned for heat and power. Also, biomass oils, such as soybean oil and canola oil, can be chemically converted into a liquid fuel similar to diesel fuel and into gasoline additives. Used cooking oil has been used as a source to make biodiesel.

In the thermochemical process, heating plant matter breaks down the feedstock into gaseous products, liquid products, and a carbonaceous solid (commonly referred to as char). These products can then be processed further and refined into useful fuels such as methane and alcohol. Another approach is to take these fuels and run them through fuel cells, converting the hydrogen-rich fuels into electricity and water, with few or no emissions. However, the direct conversion thermal processes, such as combustion, may encounter the same problems as those encountered when coal is the feedstock (Speight, 2013). The conversion of biomass into other useful forms such as gaseous fuels or liquid fuels is considered an alternative route to make use of biomass energy. The thermochemical conversion is an effective and well-known technology that has been applied to coal. However, the technology developed for coal may not be suitable for biomass utilization because of the differences in the physical and chemical properties of coal and biomass.

In terms of the feedstock for the biochemical processes and thermal processes (often referred to as the biochemical platform and the thermochemical platform, respectively) — which give rise to the concepts of the biorefinery (Chapter 12) — energy crops can be divided into short-rotation energy crops that are harvested on a cycle of anything from 2 to 20 years, depending on the crop and the system, and herbaceous energy crops that are harvested annually. Herbaceous energy crops such as the woody, perennial, rhizomatous grasses like miscanthus offer the advantage of an annual harvest, although typically establishment costs are high. All crops, including some biennial, permanent, and multiannual crops, are produced for energy purposes. Energy is generated by burning plants grown for the purpose, often after the dry matter is pelletized. Energy crops are used for firing power plants, either alone or cofired with other fuels. Alternatively, they may be used for heat or combined heat and power (CHP) production.

Energy crop fuel contains almost no sulfur and has significantly less nitrogen than fossil fuels, therefore reductions in the pollutants causing acid rain (SO_2) and smog (NO_x) may be realized. For example, the use of energy crops will greatly reduce greenhouse gas emissions. Burning fossil fuels removes carbon that is stored underground and transfers it to the atmosphere. Burning energy crops, on the other hand, releases carbon dioxide but as their growth requires carbon dioxide there is no net release of carbon into the atmosphere, i.e. it creates a closed carbon cycle. Furthermore, where energy crops are gasified there is a net *reduction* of carbon dioxide. In addition, substantial quantities of carbon can be captured in the soil through energy crop root structures, creating a net carbon sink.

An additional environmental benefit is in water quality, as energy crop fuel contains less mercury than coal. Also, energy crop farms using environmentally pro-active designs will create water quality filtration zones as well as the uptake and sequestering of pollutants such as phosphorus from soils that leach into water bodies. Also, growing energy crops on agricultural land that might otherwise be converted to residential or industrial use will reduce erosion/chemical runoff and enhance wildlife habitat. This will give energy producers and consumers a renewable energy option with uniquely desirable characteristics. For example, energy crops differ from other sources of renewable energy in virtue of the fact that they can be grown to meet the needs of the market whereas other renewable resources (for example, wind and wave power) must be harnessed where and when they occur.

Thus, perennial crops that regenerate annually from buds at the base of the plant offer the greatest potential for energy-efficient production These include the following: (i) cordgrass and switchgrass, (ii) Jerusalem artichoke, (iii) miscanthus, (iv) reed plants, (v) residual herbaceous biomass, (vi) short rotation coppice, and (vii) sorghum.

5.2.1.1 *Cordgrass and switchgrass*

Cordgrass (genus *Spartina*), also called marsh grass or salt grass, is a genus of 16 species of perennial grasses in the family Poaceae, which is found on marshes and tidal mud flats of North American, Europe, and Africa and often forms dense colonies. Some species are planted as soil binders to prevent erosion and a few are considered invasive species in areas outside their native range. Prairie cordgrass (*Spartina pectinata*) and gulf cordgrass (*Spartina spartinae*) are the most widely distributed North American species.

Cordgrasses are erect, tough, long-leaved plants that range from 1 to 10 feet in height. Most species grow in clumps, with short flower spikes alternating along and often adherent to the upper portion of the stems. Many spread through rhizomes (underground stems) that send up new plants. One of the variations of cordgrass — referred to as smooth cordgrass — has smooth, blade-like leaves that taper to a point. The leaves grow 12–20 inches in length and one-half an inch wide and have round, hollow stems and a strong, interconnected root system. Smooth cordgrass grows in two forms: a short form that grows to 2 feet tall, and a tall form that can reach 7 feet tall.

Switchgrass (*Panicum virgatum*) is a perennial sod-forming grass with thick strong stems. It is a perennial warm season bunchgrass that is native to North America, where it occurs naturally from Canada southwards to the United States and Mexico. Switchgrass is one of the dominant species of the central North American tall grass prairie and can be found in prairie remnants (grassland areas in the Western and Midwestern United States and Canada that remain, to some extent, undisturbed), in native grass pastures, and naturalized along roadsides.

The advantages of switchgrass as an energy crop are that it is fast-growing, remarkably adaptable, and high-yielding. Further advantages of switchgrass are that it can be harvested, using conventional equipment, either annually or semi-annually for 10 years or more before replanting is needed and that it is able to reach deep into the soil for water and to use water very efficiently. Besides showing great promise as an energy crop for energy production, switchgrass also restores vital organic nutrients to farmed-out soils and with its extensive network of stems and roots (the plants extend nearly as far below ground as above), it is also a valuable soil stabilization plant.

Switchgrass has the potential to be a versatile bioenergy feedstock since the energy content is comparable to that of wood with significantly lower initial moisture content. Switchgrass is a very suitable substrate and produces high ethanol yield using the current simultaneous saccharification and fermentation technology. Extensive analysis of ash and alkali content of switchgrass indicates that it typically has relatively low alkali content and should have low slagging potential in coal-fired combustion systems. As an agro-fiber source for pulping, switchgrass has a relatively high cellulose content, low ash content, and good fiber length to width ratios. Switchgrass reaches its full yield potential after the third year planted, producing approximately 6–8 tons per acre, that is, 500 gallons of ethanol per acre.

The utilization of energy crops such as switchgrass (*Panicum virgatum*, L., Poaceae) is a concept with great relevance to current ecological and economic issues on a global scale. The development of a significant national capacity to utilize perennial forage crops such as switchgrass as biofuels could offer an important new source of energy from the perennial cropping systems, which are compatible with conventional farming practices, could help reduce degradation of agricultural soils, and lower national dependence on foreign oil

5.2.1.2 *Jerusalem artichoke*

The *Jerusalem artichoke* (*Helianthus tuberosus*; also called sunroot, sunchoke, or earth apple) is a species of sunflower native to central North America. It is a herbaceous perennial plant that grows up to 5–10 feet tall and the leaves are opposite on the lower part of the stem but are alternate above. The leaves have a rough, hairy texture. The larger leaves found on the lower stem are broad–ovoid–acute and can be up to 12 inches long while the leaves on the higher stem are smaller and narrower. The tubers are often elongated and uneven, typically 3–4 inches long and 1–2.2 inches thick with a crisp and crunchy texture when raw. They vary in color from pale brown to white, red, or purple.

The *Jerusalem artichoke* has shown excellent potential as an alternative sugar crop. A member of the sunflower family, this crop is native to North America and is well adapted to northern climates. Similar to sugar beet, the Jerusalem artichoke produces sugar in top growth and stores it in the roots and tuber. It can grow in a variety of soils and is not demanding of soil fertility. The *Jerusalem artichoke* is a perennial; small tubers left in the field will produce the next seasonal crop, so no plowing or seeding is necessary. The high-fructose syrups that can be derived from the tubers produced by the *Jerusalem artichoke* may be used for the production of ethanol and other industrial raw materials. *Jerusalem artichokes* also produce a large amount of top growth, which may also prove to be a useful source of biomass for energy purposes.

5.2.1.3 *Miscanthus*

Miscanthus (also called silvergrass or silver grass) is a hardy perennial grass that produces a crop of bamboo-like cane that is up to 15 feet tall. Miscanthus is high in lignin and lignocellulose fiber and uses the C4 pathway. It can be grown in a cool climate and on many types of arable land. Miscanthus does not require a large input of fertilizers due to its ability to recycle large amounts of nutrients. Miscanthus has a calorific value per unit weight similar to that of wood and therefore they could possibly be used in the same power plant or those designed for agricultural residues

Miscanthus is well equipped for high productivity under relatively cool temperatures and may require substantial amounts of water for maximal growth (its growth could therefore also have valuable environmental benefits by acting as absorbing and disposal areas for waste water and some industrial effluents). Furthermore, miscanthus seems to grow well under most soil conditions (barring thin droughty soils) but appears to thrive within areas currently best suited to maize production. The advantages of miscanthus as an energy crop are that it multiplies very rapidly, has a high yield which is relatively dry and can be harvested annually (from its second season onward) as compared with every 2–4 years for short-rotation coppice. Further advantages are that miscanthus can be grown and harvested with existing farm machinery, it requires little or no pesticide/fertilizer input after establishment, and the harvest can use the same infrastructure for storage and transport as short-rotation coppice.

5.2.1.4 *Reed plants*

Reed is a common name for several tall, grass-like plants that are commonly found in wetlands. *Reed plants* are a potentially prolific producer of biomass, capable of yielding 20–25 tons per hectare (2.47 acres) of dry matter annually for a number of years. They can grow up to 6 m, are spread by means of stout rhizomes (continuously growing horizontal underground stems, which put out lateral shoots and adventitious roots at intervals) and stolons (also known as runners, which are horizontal connections between organisms), and are commonly found in swampy ground and shallow water throughout temperate and subtropical areas.

Reed canary grass is a robust perennial grass widely distributed across temperate regions of Europe, Asia, and also North America. It occurs in wet places along the margins of rivers, streams, lakes, and pools. The species spreads naturally by creeping rhizomes, but plants can also be raised from seed. The advantages of reed canary grass as an energy crop are its good adaptation to cool temperate climates and poor wet soil conditions and, conversely, its ability to withstand drought. Crucially, for the purposes of biomass production, reed canary grass is also able to attain high dry matter content earlier than miscanthus. The crop responds well to nitrogen and phosphate and it may be used in a bed system to remove nutrients from waste water, as well as to stabilize areas at risk of soil erosion.

5.2.1.5 *Residual herbaceous biomass*

Residual herbaceous biomass (straw) is the main residual herbaceous material for energy application. As it is a residual product, its availability for energy purposes is driven by the cereal markets and does not have autonomous market behavior. In addition, farms consume significant quantities of straw internally — as bed material for livestock, grain drying, etc. Some straw is also chaffed and returned back to the field as a soil ameliorator. The net straw yield per hectare for energy application also depends on the crop, the grain yield per hectare, climate and cultivation conditions, etc. Nevertheless, one can roughly estimate that the average straw yield per hectare is approximately 50–65% of the grain yield per hectare from cereals and oilseeds.

Similar to herbaceous crops, straw usually has lower moisture content than woody biomass. Conversely, it has a lower calorific value, bulk density, and ash melting point and higher content of ash, and problematic inorganic components such as chlorine, potassium, and sulfur, which cause corrosion and pollution. The last two drawbacks can be relatively easily overcome by leaving straw on the field for a while. In such a way, rainfall provides a natural leaching process and separates a large part of the potassium and chlorine. Alternatively, fresh straw can be directly shipped to the gasification plant, where it is washed by dedicated facilities at moderate temperatures (50–60°C; 120–140°F). Due to washing, the initially low moisture content of straw becomes higher in both cases and hence a mandatory drying is applied afterwards. In both cases, also the content of corrosive components is reduced, but not completely taken out. In order to decrease handling costs, straw and dedicated herbaceous energy crops are usually baled before being shipped to the gasification plant. The weight and the size of bales depend on the baling equipment and on the requirements of the gasification plant (Luque and Speight, 2015).

There are many opportunities for producing fuel from agricultural crops and crop residues. Possible end products include ethanol, vegetable oil, and solid cellulosic fuels. Ethanol can be produced from any grain, root, fruit, or juice crop containing fermentable carbohydrates. It can also be made from crops, residues, or wood that contain cellulose or other long-chain carbohydrates, which can be hydrolyzed to fermentable sugars.

Vegetable oils are produced from numerous oil seed crops. Some of these oils have been evaluated in other laboratories as substitutes for diesel fuel. While all vegetable oils have high energy content, most require some processing to assure safe use in internal combustion engines.

The simplest form of agricultural biomass energy use involves direct combustion of cellulosic crops or residues, such as hay, straw, or corn fodder, to heat space or produce steam. Such fuels are useful for heating farm buildings and small commercial buildings in rural areas and for drying crops. Ideally, energy crops should be produced on land not needed for food production. This use should not increase the erosion hazard or cause other environmental damage. On the other hand, a variety of crops can be grown specifically to provide sources of energy and, once established, a stand of perennial biomass energy crop may remain productive for 6 years or more.

5.2.1.6 *Short-rotation coppice*

Short-rotation woody crops (short-rotation coppice, SRC) refers to fast-growing deciduous trees that are grown as energy crops, such as willow and poplar trees. The species of SRCs that are most suitable, and therefore most popular, for use as energy crops are poplar and willow (and possibly also birch) because they both require deep, moisture-retentive soils for proper growth. Willow, in particular, is able to endure periods of water logging and is therefore better suited to wetter soils.

SRC is harvested during winter when the dry matter percentage of the coppice is at its highest and it is then bundled or immediately chipped. It may then be stored for a few weeks in order to reduce its moisture content to a satisfactory level for use in energy production. Dry SRC can then be burnt under controlled conditions to produce other fuels, gas or liquid, which are then used for electricity generation.

5.2.1.7 *Sorghum*

Sorghum is an annual tropical grass with large genetic variation and represents a crop with the potential for energy production. It is a genus of flowering plants in the grass family Poaceae. Seventeen of the 25 species are native to Australia with the range of some extending to Africa, Asia, and Central America as well as to islands in the Indian Ocean and the Pacific Ocean. One species is grown for grain, while many others are used as fodder plants, either cultivated in warm climates throughout the world or naturalized in pasture lands.

Sweet sorghum has been selected for its sugar content and is normally grown for molasses production. Forage sorghum has been selected for high yields of reasonably good quality animal feed. Sorghum varieties producing tall plants with large stems make the best candidates for biomass production. Both sweet and forage sorghum have a high potential for lodging. Lodging can result in harvest problems with ensuing loss of yield from both initial and ratoon crops.

This crop has been cultivated on a small scale in the past for production of table syrup, but other varieties can be grown for production of sugar. The most common types of sorghum species are those used for production of grain. Sweet sorghum can be considered as an energy crop, because it can be grown in all continents, in tropical, sub-tropical and temperate regions, as well as in poor quality soils. Sweet sorghum is a warm-season crop that matures earlier under high temperatures and short days. Sweet sorghum is an extraordinarily promising multifunctional crop not only for its high economic value but also for its capacity to provide a very wide range of renewable energy products, industrial commodities, and food and animal feed products. Sweet sorghum biomass is rich in readily fermentable sugars and thus it can be considered as an excellent raw material for fermentative hydrogen production — hydrogen is an important commodity for the refining industry and new sources are continually sought (Parkash, 2003; Gary *et al.*, 2007; Speight, 2014, 2017; Hsu and Robinson, 2017). Integrated production of several crops (sweet sorghum/sugarcane/sweet sorghum corn/sweet sorghum/sweet potatoes) and simultaneous processing of the full crop components (starch, sugar, and lignocellulosic) can improve the global economics of bioenergy schemes considerably in view of a viable production of bioethanol.

Sweet sorghum crops produce sugar syrups, which could form the basis of fermentation processes for methane or ethanol production, and some of the forage types of the plant may be suitable for biomass production.

5.3 Wood

Wood is a porous and fibrous structural tissue found in the stems and roots of trees and other woody plants. It is a natural composite of cellulose fibers that are strong in tension and embedded in a matrix of lignin that resists compression. Wood is sometimes defined as only the secondary xylem in the stems of trees, or it is defined more broadly to include the same type of tissue elsewhere such as in the roots of trees or shrubs. In a living tree it performs a support function, enabling woody plants to grow large or to stand up by themselves. Wood also conveys water and nutrients between the leaves, other growing tissues, and the roots. As a result of this structure, wood has lent itself to a variety of uses throughout recorded history.

5.3.1 *History*

The use of wood by humans from biblical times to modern computer applications is reviewed here. In peace and in war, in the Old and the New World, man first turned to wood for his basic needs and later learned to use advanced science to employ wood as his most sophisticated raw material, being infinitely versatile and an easily renewable source.

The manner in which wood was used by early cultures is difficult to determine, as wood artifacts have largely disappeared. Certainly, the use of wood for fire is one of the first and most significant contributions of this resource to the development of society. No doubt, man built early pole structures from the small trees growing along the rivers and later he would build more solid structures from planks, turf, mud, and adobe. The Scandinavians developed the basic principles of timber framing which were probably known in Europe in the Bronze Age, and framing eventually became the pre-eminent method of wood building in the Western world, reflecting developments in structural engineering that had been worked out with wood mostly through trial and error.

One of the first uses of wood for water transport was probably a hollowed-out log. Around 4000 BC, the Egyptians were making ships from bundles of reeds and their earliest wooden boats copied the hull frame of the reed boats. For larger vessels, the Egyptians imported cedar from Lebanon. One reason for the northward expansion of the influence of Egypt was to ensure Egypt's cedar supply. Records show that Egyptian shipbuilders could use wood on a grand scale. The barge of Queen Hatshepsut, built in 1500 BC to transport granite obelisks from Aswan to Thebes, had a displacement of about 7,500 tons, and 30 oar-powered tugs were needed to tow it. According to Theophrastus, a pupil of Aristotle, the ship-building woods in Ancient Greece were silver fir, fir, and cedar. Silver fir was used for its lightness; for merchant ships, fir was used because of its resistance to rot. In Syria and Phoenicia, cedar was used because there was no fir.

Technological improvement of land transport was slower than that of water transport. From 7000 BC onward, wood sledges were used for heavy loads such as stones, and archeologists reason that the massive stones in the great monument at Stonehenge on Salisbury Plain, England (Hawkins and White, 1993), must have been moved on sledges placed on rollers, which may have inspired the discovery of the wheel.

Another significant contribution of wood to the ancient world was for war devices. Examples include the catapult, which enabled a man to attack his enemy from a safe distance, the battering ram and scaling ladder, the tortoise, and the siege tower. Although the choice of materials for these purposes was quite limited, the properties of wood made it eminently suitable. High strength and low weight were highly valued characteristics of wood then, just as they are currently. These siege engines were integral to the expansion of both Greek and Roman civilizations and of the science, technology, and philosophy that developed under the tutelage of the great thinkers and teachers of the times.

Ancient man was using wood to conquer his world as well as build it and explore it. Then, some unknown woodman in Ancient Greece invented a primitive wooden lathe, and man found himself on the threshold of the age of machines. When he entered that age, he looked for ways to make wood serve him to unprecedented degrees. Using the basic concept of the lathe and the ability to shape wood to circular symmetry, new concepts of both materials use and machine development were developed.

In Europe, the water-and-wood phase reached a high plateau around the 16th century with the work of Leonardo da Vinci and his talented contemporaries. At this time, the availability of timber diminished, particularly in many European countries. The scar-city was caused by the expansion of agriculture, the increasing use of wood as a structural material and fuel, and the increasing demand for wood to supply the smelting furnaces. To smelt one cannon it took several tons of wood. By the 17th century, Europeans were turning to coal for the domestic hearth, and when the secret of smelting metal with coal was discovered, coal became the unique basis for industrial technology until late in the 19th century.

In early 19th century America, a seemingly inexhaustible supply of timber existed. The technology here was geared to exploit the use of all natural resources to make up for the scarcity in capital and labor. But the technological advances of the 19th century, along with the increasing population, would have a major impact on American forests. Railroads, telegraph lines, charcoal-fueled steel mills, and other industries were consuming immense quantities of wood. The Civil War also caused a heavy demand. One gun factory alone used 28,000 walnut trees for gunstocks. During the latter half of the 19th century, the volume of lumber produced each year increased from 4000 million board feet to approximately 35,000 million.

As with many other industries of this time, lumbering was a highly competitive business. Quick profits were the name of the game. This encouraged careless and extravagantly wasteful harvesting and manufacturing methods. The visible devastation that resulted stimulated a new concern for North American forests. Theories were published that purported to prove that the fall of ancient empires, radical changes in climate, and the spread of epidemics could be attributed to deforestation.

But the wood-and-water phase in the United States reached its own plateau in the 1850s, approximately 200 years after that phase had peaked in Europe. Some of the traditional uses of wood — for fuel, pavement, sailing ships, charcoal, and iron smelting — were taken over by coal, steel, and stone. However, the demand for timber was maintained as many new uses of wood, for paper, plywood, telegraph and telephone poles, railroad ties, and chemicals, entered the picture. The selection from among competing materials was done partly based on cost and availability and partly based on properties and performance. It is also worth noting that such a range of choices coincided with the rapid mechanization and the increasing technical complexity of modern society. Nevertheless, in the late 19th century the use of wood products had begun to level off — coal was abundant, and iron and steel could be manufactured and a scarcity of timber was no cause for concern. The most significant decline in wood product use since then has been in fuel wood. Around 100 years ago, the USA consumed four exajoules of energy a year, and three were provided by wood.

Up to the latter part of the 19th century, no appreciable systematic research on wood occurred. Wood had been used by early experimenters to make instruments and other research equipment, and early engineers had used it as a construction material and a material with which to work out engineering problems and designs. Methods for pulping wood to make paper had been worked out by the paper industry, too. Furthermore, both cotton and wood had been used by chemists as a source of cellulose for man-made fibers. This led to work on cellulose acetate reactions with solvents that led to the ability to produce that compound as both film and fiber. These advances provided a base for the subsequent technology of nylon and established the principles by which countless numbers and kinds of linear high polymers can be synthesized. The carriage business provided an early milestone for a new era of wood research. In 1889 the Carriage Builders Association had concerns related to the scarcity of northern oak, a species long preferred for their craft.

During World War II, wood research covered the whole gamut of possible wartime uses of wood but after the war the importance of timber products declined, on a relative scale, as the importance of minerals increased, due in part to abundant low-cost energy in the form of coal first and then crude oil. It is worth noting, however, that tonnage of timber products produced in the USA then exceeded that of all metals and plastics combined, just as it does currently. So, while timber declined in relative importance and public awareness, it remained the major product of American manufacture.

5.3.2 *Composition and properties*

Wood is the hard, fibrous substance found beneath the bark in the stems and branches of trees and shrubs. Practically, all commercial wood, however, comes from trees. It is plentiful and replaceable. Since a new tree can be grown in the same location where one has been cut, wood is considered the only renewable natural resource of the world. Wood consists of cellulose ($C_6H_{19}O_5$), resins, lignin, various inorganic salts, and water, as is reflected in the ultimate analysis of wood (Table 5.4). The quantity of water present has a great effect on the heating value and it ranges from 25% to 50% w/w in green wood and 10–20% w/w in air-dried wood (Table 5.5).

Aliphatic extractives contain alkanes, fatty alcohols, fatty acids, fats, and waxes. Only minor amounts of alkane derivatives, alcohol derivatives, and free fatty acid derivatives are present in wood. The most common fatty acid constituents belong to both the saturated and unsaturated compounds. More than 30 fatty acids (or fatty acid moieties) have been identified in softwoods and hardwoods. The majority of the fatty acid derivatives present in wood are esterified with glycerol (to form fats) or with higher fatty alcohols and terpenoids (to form waxes). In softwoods, the resin is mainly composed of fats whereas in hardwoods, the resin contains a significant proportion of waxes as well as fats. Fats and waxes are hydrolyzed during wood storage.

In woods from temperate zones, elements other than carbon, hydrogen, oxygen, and nitrogen make up between 0.1% and 0.5% w/w of the

Table 5.4: Ultimate analysis of wood (% w/w, dry-ash-free).

Element	Average of 11 hardwoods	Average of 9 softwoods	Oak bark	Pine bark
C	50.2	52.7	52.6	54.9
H	6.2	6.3	5.7	5.8
O	43.5	40.8	41.5	39.0
N	0.1	0.2	0.1	0.2
S	—	0.0	0.1	0.1

Table 5.5: Properties of various woody feedstocks compared to coal and natural gas.

	Bituminous coal	Natural gas	Wood	Bark	Willow	Forest residues
Ash, % w/w	8.5–10.9	0	0.4–0.5	3.5–8	1.1–4.0	1–3
Moisture, % w/w	5–10	0	5–60	45–65	50–60	50–60
Volatile matter, % w/w	25–40	100	>70	70–77	>70	>70
Ash melting point, °C	1100–1400		1400–1700	1300–1700	n.a.	n.a.”
C, % w/w	76–87	75	46–52	46–52	47–51	48–52
H, % w/w	3.5–5	24	6.2–604	4.6–6.8	5.8–6.7	6.0–6.2
N, % w/w	0.8–1.5	0.9	0.1–0.5	0.3–0.8	0.2–0.8	0.3–0.5
O, % w/w	2.8–11.3	0.9	36–42	24.3–42.4	40–46	40–44
S, % w/w	0.5–3.1	0	<0.05	<0.05	0.02–0.10	<0.05
Cl, % w/w	<0.1		0.01–0.03	0.01–0.03	0.02–0.05	0.01–0.04
K, % w/w	0.003	—	0.02–0.05	0.1–0.4	0.2–0.5	0.1–004
Ca, % w/w	4–12		0.1–1.5	0.02–0.08	0.2–0.7	0.2–0.9

dry solids in wood, whereas in woods from tropical and subtropical regions, they make up even 5% w/w. In practice, the total amount of wood inorganics is measured from ash, the residue obtained after proper combustion of the organic matter of a wood sample. Ash contains mainly different metal oxides, and the average values for the ash content of commercial softwoods and hardwoods are generally in the range of 0.3–1.5% w/w of wood dry solids. There is also seen a significant dependence of the ash content and composition on the environmental conditions under which the tree has grown and, on the other hand, on the location within the tree. Some of the inorganic elements present in wood are essential to wood growth — in many cases, alkali and alkali earth elements such as potassium, calcium, and magnesium make up about 80% w/w of the total inorganic elemental constituents of softwoods and hardwoods.

Wood cut in spring and summer contains more water than that cut in the early part of winter. A cord (8 feet long by 4 feet wide by 4 feet high) of hardwood, such as ash or maple, is equal in heating value to approximately 1 ton of bituminous coal; softwoods, such as pine and poplar, have less than half this amount. Wood burns with a long flame and when it is kindled, the fire quickly reaches its maximum intensity, and a relatively small quantity of ash is formed. Wood is too expensive for industrial use, except in a few special cases where a fuel free from dirt and smoke is necessary. Of other cellulose materials, shavings, sawdust, and straw are used as fuel in some places. They are bulky and difficult to handle, while their heat value, which depends on the amount of moisture they contain, is seldom more than one-third to one-half that of good coal. Waste matter such as spent tan-bark and *bagasse* (crushed sugar cane), and the pulp from sugar beets is sometimes used as fuel for evaporation and for steam, but owing to the large amount of moisture they contain, the heat value is very low.

All woods dried to the same moisture content contain approximately the same heat value per pound: from 6,200 to 7,500 Btu for fully dried wood and 5,500 to 8,500 Btu for air-dried seasoned wood (Table 5.6).

Table 5.6: The gross calorific value of wood as one of the variety of fuels.

Fuel	Gross Calorific Value (Btu/lb)[a]
Alcohol, 96%	
Anthracite	14,000–14,500
Bituminous coal	7,300–10,000
Butane	20,900
Charcoal	12,800
Coal	8,000–14,000
Diesel	19,300
Ethanol	12,800
Gasoline	20,400
Gasoline	7,000
Kerosene	154,000
Wood (dry)	5,500–8,500

Notes: [a]The Gross Calorific Value (GCV) assumes that the water of combustion is entirely condensed and that the heat contained in the water vapor is recovered,

- 1 kJ/kg = 1 J/g = 0.4299 Btu/ lb_m = 0.23884 kcal/kg,
- 1 Btu/lb = 2.326 kJ/kg = 0.55 kcal/kg,
- 1 kcal/kg = 4.1868 kJ/kg = 1.8 Btu/lb_m.

However, the heat content of any fire depends on wood density, resin, ash, and moisture. A general rule for estimating the heat value of firewood is as follows: one cord of well-seasoned hardwood (weighing approximately 2 tons) burned in an airtight, draft-controlled wood stove with a 55–65% efficiency is equivalent to approximately 175 gallons of No. 2 fuel oil or 225 therms of natural gas consumed in normal furnaces having 65–75% efficiencies. There are also differences depending on the type of wood. Softwoods usually have a lot of resin content that in turn has high energy content, so their total energy content is usually higher than that of hardwoods (often by approximately 5%). The softwoods tend to burn up faster than hardwoods and have other characteristics that reduce their attractiveness as a fuel. The fact that the density of softwood is typically lower than that of hardwood implies that there is less weight of wood in one cord and the extra 5% of volatile fuel will not make up for the reduction in weight.

The two most important properties of any papermaking cellulosic raw material are as follows: (i) the amount of cellulose fiber and (ii) the length of the fibers. The amount of cellulose fiber present in wood determines the pulp yield, the ease of pulping, and the cost of pulp produced. The maximum average fiber length that pulp will have is that of wood because whatever the pulping method, full chemical or full mechanical, fiber is going to be damaged. In mechanical pulping, the damage is physical (cutting, bruising, etc.) whereas in chemical pulping, it is chemical degradation (lower degree of polymerization).

5.3.2.1 *Chemical composition*

The chemical composition of wood varies from species to species, but wood is made up of approximately 50% w/w carbon, 42% w/w oxygen, 6% w/w hydrogen, 1% w/w nitrogen, and 1% w/w other elements (mainly calcium, potassium, sodium, magnesium, iron, and manganese).

5.3.2.1.1 Cellulose

Cellulose, the major chemical constituent of wood, is in many respects the most important and also the easiest to define and describe. Wood cellulose is chemically defined as $(C_6H_{10}O_5)_n$. Cellulose is a high molecular-weight, stereoregular, linear polymer of repeating β-D-glucopyranose units. Put simply, it is the chief structural element and a major constituent of the cell wall of trees and plants.

Cellulose is derived from D-glucose units, which condense through β(1→4)-glycosidic bonds. This linkage motif contrasts with that of α(1→4)-glycosidic bonds present in starch and glycogen. The empirical formula of cellulose is $(C_6H_{10}O_5)_n$, where *n* is the degree of polymerization (DP). Cellulose is a straight-chain polymer. Unlike starch, no coiling or branching occurs and the molecule adopts an extended and rather stiff rod-like conformation, aided by the equatorial conformation of the glucose residues.

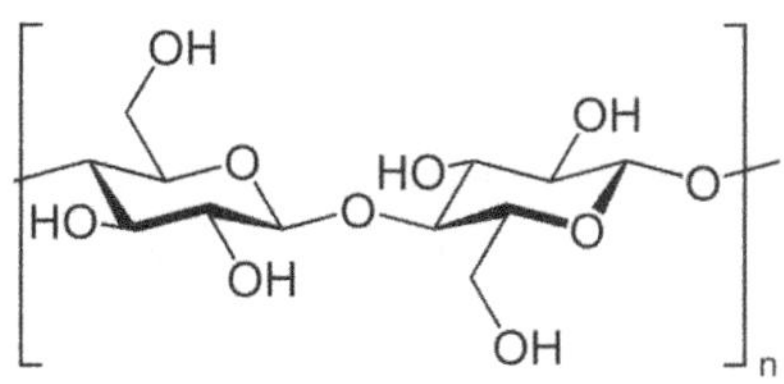

The multiple hydroxyl groups on the glucose from one chain form hydrogen bonds with oxygen atoms on the same or on a neighboring chain, holding the chains firmly together side by side and forming microfibrils with high tensile strength. The high tensile strength of plant stems and of the tree wood also arise from the arrangement of cellulose fibers intimately distributed into the lignin matrix.

5.3.2.1.2 Hemicellulose

Hemicellulose (or hemi-cellulose) is a constituent of woods that is, like cellulose, a polysaccharide, but less complex and easily hydrolyzable. Hemicellulose derivatives are polysaccharides that are often associated with cellulose, but have a very different composition compared to the composition of cellulose. Hemicellulose derivatives, which make up 20–35% of the dry weight of wood, are the second important constituent of wood, are also sugar polymers and unlike cellulose, which is made only from glucose, hemicelluloses consist of glucose and several other water-soluble sugar derivatives that are produced during photosynthesis.

Hemicellulose derivatives are usually isolated by successive extractions with dimethyl sulfoxide and aqueous alkali. However, degradation caused by the alkali cannot be completely avoided. Galactoglucomannan derivatives (water-soluble hemicellulose derivatives consisting of galactose, glucose, and mannose) can be depolymerized by acid treatment and especially the bond between galactose and the main chain. The acetyl groups are much more easily cleaved by alkali than by acid. For arabinoglucuronoxylan (hemicellulose polysaccharide derivatives in the plant cell wall which contain glucuronic acid and xylose with arabinose as the main constituents), the arabinose side chains can be hydrolyzed by acids due to their furanosidic structure, which is less resistant to hydrolysis. The reliable analysis of hemicelluloses is based on the separate determination of the polysaccharide constituents using the chromatographic method. The extractive-free wood sample is subjected to an acid hydrolysis after which the liberated monosaccharides are separated and quantified.

In the hemicellulose family, the degree of polymerization is lower than that cellulose and hemicelluloses are composed of shorter molecular chains than those found in cellulose. There exist many varieties of hemicelluloses, and hemicelluloses, in softwoods and hardwoods: markedly differ in composition and structure (Figure 5.1) generally, hemicellulose derivatives are present in a relatively greater proportion in hardwoods than in softwoods. Unlike cellulose, hemicellulose consists of 500–3,000 sugar units whereas cellulose consists of 7,000–15,000 glucose molecules per polymer.

Hemicellulose derivatives are classified according to the main sugar residue in the backbone as xylan derivatives, mannan derivatives, and glucan derivatives. Depending on the plant species, the developmental stage, and the tissue type, various subclasses of hemicellulose may exist and may be grouped into two general categories based on the hydration of the fibers: (i) *low-hydration polysaccharide derivatives* function primarily to stabilize the cell wall through hydrogen-bonding interactions with cellulose and covalent interaction with lignin. They are water-soluble due to their branched structure. (ii) *Hemicellulose derivatives composed mainly of hydrocolloids* (often called gums; hydrocolloids are hydrophilic polymers, of vegetable, animal, microbial or synthetic origin, that generally contain many hydroxyl groups and may be polyelectrolytes) function primarily as an extracellular energy and raw material storage system and as a water-retention mechanism in seeds.

Hemicellulose derivatives contain many different sugar monomers, whereas cellulose only contains anhydrous glucose. For example, in addition to glucose, sugar monomers in hemicelluloses may include the five-carbon sugars xylose and arabinose, the six-carbon sugars mannose and

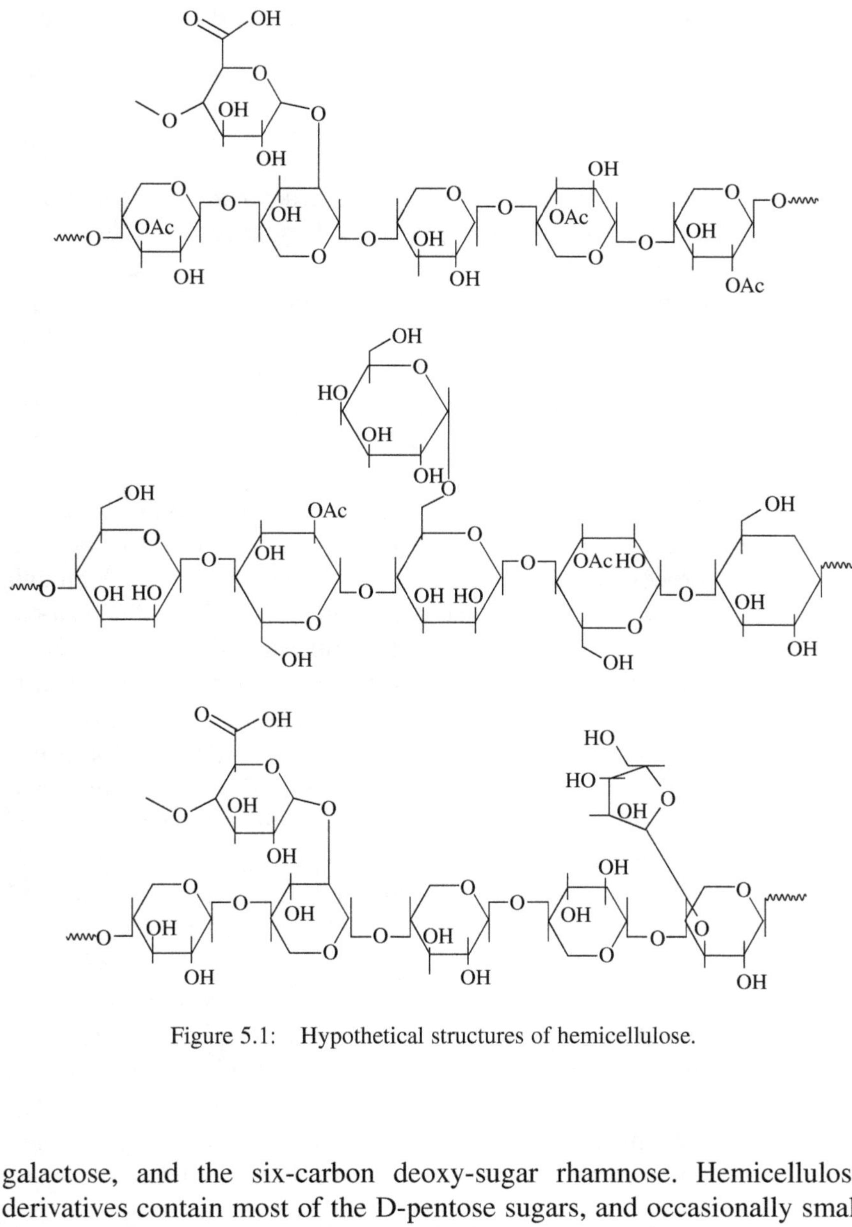

Figure 5.1: Hypothetical structures of hemicellulose.

galactose, and the six-carbon deoxy-sugar rhamnose. Hemicellulose derivatives contain most of the D-pentose sugars, and occasionally small amounts of L-sugars as well. Xylose is, in most cases, the sugar monomer present in the largest amount, although in softwoods, mannose can be the most abundant sugar, leading to the production of ethanol (Keller, 1996; Galbe and Zacchi, 2002). Not only regular sugars but also their acidified form, for instance, glucuronic acid and galacturonic acid can be present in hemicellulose.

5.3.2.1.3 Lignin

Lignin is a complex constituent of wood that cements the cellulose fibers together. Lignin is brown in color. Lignin is largely responsible for the strength and rigidity of plants. Lignin is a class of complex organic polymers that form key structural materials in the support tissues of vascular plants and some algae. Lignins are particularly important in the formation of cell walls, especially in wood and bark, because they do not rot easily. Chemically, lignins are cross-linked phenolic polymers.

A rough classification of lignin derivatives includes the following: (i) softwood lignin, (ii) hardwood lignin, and (iii) grass lignin. Several methods can be used to isolate lignin and therefore various forms of lignin, such as milled wood lignin, dioxane lignin or enzymically liberated lignin, Kraft process lignin, and alkali lignin, are available. Although native lignin derivatives behave like an insoluble and three-dimensional chemical network (Figure 5.2), isolated lignin derivatives often exhibit maximum solubility in solvents including dioxane, acetone, methyl cellosolve, tetrahydrofuran, dimethylformamide, and dimethyl sulfoxide.

The composition of lignin varies from species to species — for example, the composition of an aspen sample is carbon 63.4% w/w, hydrogen 5.9%, and oxygen 30% (by difference), and the mineral ash 0.7% corresponding to the approximate formula $(C_{31}H_{34}O_{11})_n$. As a biopolymer, lignin is unusual because of its heterogeneity and lack of a defined primary structure. Its most commonly noted function is its support in strengthening wood.

Structurally, lignin is a cross-linked polymer with a molecular mass in excess of 10,000 (Figure 5.2). It is relatively hydrophobic and rich in aromatic subunits. The actual degree of polymerization is difficult to measure, since the material is heterogeneous. The lignol derivatives that crosslink are of three main types, all derived from phenylpropane: 4-hydroxy-3-methoxy phenylpropane, 3,5-dimethoxy-4-hydroxy phenylpropane, and 4-hydroxy phenylpropane. The former tends to be more prevalent in conifers and the latter in hardwoods.

Different types of lignin have been described, depending on the means of isolation. The three common monolignols are (i) *trans*-coniferyl

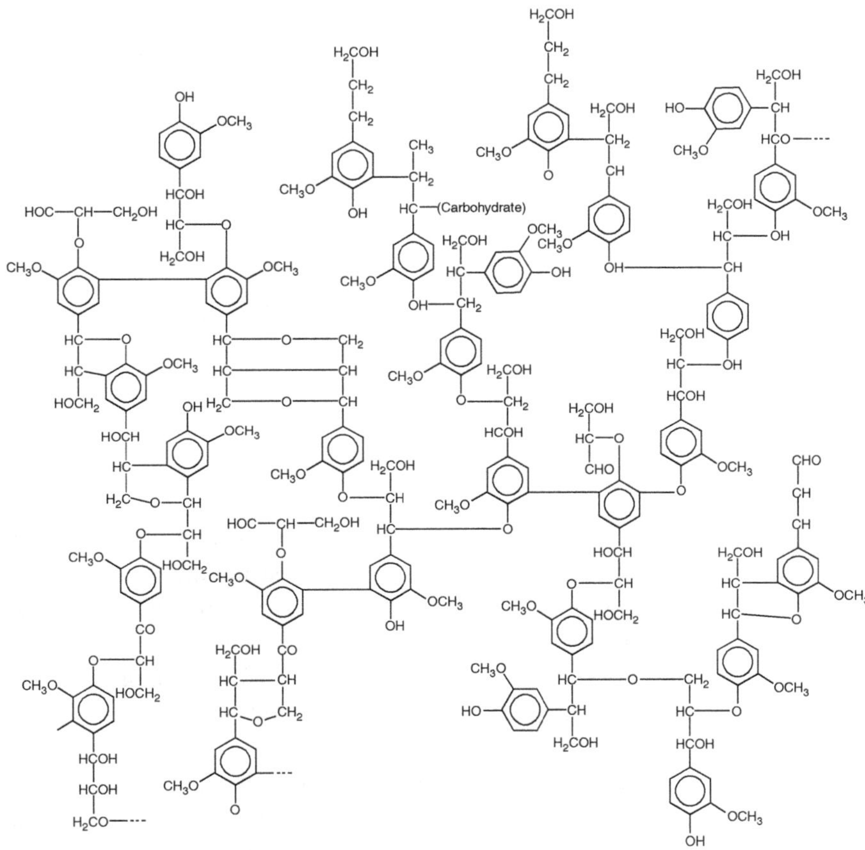

Figure 5.2: Hypothetical structures of lignin.

alcohol, (ii) *trans*-sinapyl alcohol, and (iii) *trans*-p-coumaryl alcohol (Figure 5.3). Thus, lignin can be defined as a polyphenolic material arising primarily from enzymic dehydrogenative polymerization of three phenylpropanoid units (p-hydroxy-cinnamyl alcohols). The proportions of the precursors in lignins vary with their botanical origin. Normal structural elements of softwood lignins are derived principally from *trans*-coniferyl alcohol (90%) with the remainder consisting mainly of *trans*-p-coumaryl alcohol. In contrast, hardwood lignins are mainly composed of *trans*-coniferyl alcohol and *trans*-sinapyl alcohol in varying ratios (about 50% for each alcohol).

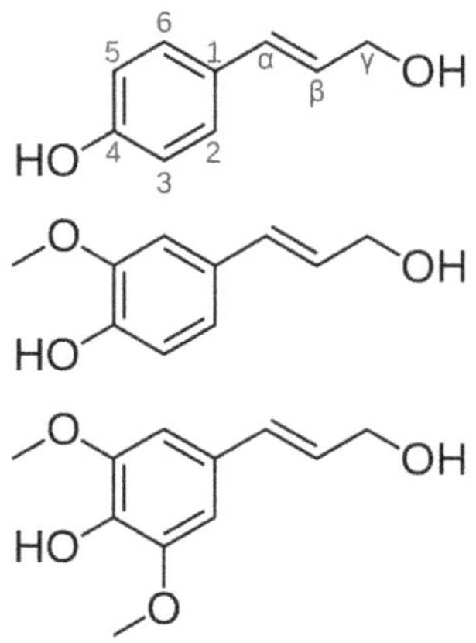

Figure 5.3: Structures of the monolignol derivatives.

Structural issues aside, lignin fills the spaces in the cell wall between cellulose, hemicellulose, and pectin components (structural acidic heteropolysaccharide derivatives contained in the primary cell walls of terrestrial plants), especially in vascular and support tissues. It is covalently linked to hemicellulose and therefore cross-links different plant polysaccharides, conferring mechanical strength on the cell wall and, by inference, on the whole plant.

Lignin plays a crucial role in conducting water in plant stems — the polysaccharide constituents of the plant cell are hydrophilic and thus permeable to water, whereas lignin is more hydrophobic and less permeable to water. The crosslinking of polysaccharides by lignin is an obstacle for water absorption to the cell wall and, thus, the presence of lignin makes it possible for the vascular tissue of the plant to conduct water efficiently.

5.3.2.1.4 Solvent extractives

The structure created by hydrogen bonds results in the typical material properties of the chemical constituents of wood and confers insolubility in most solvents. For the isolation of, for example, cellulose from wood, a direct nitration of wood yields undegraded cellulose trinitrate, which is soluble in organic solvents. On the other hand, the glycosidic linkages are easily cleaved by strong mineral acids and therefore cellulose can be hydrolyzed into simple sugars. However, for a complete hydrolysis of cellulose, concentrated acid solutions must be used in order to bring about the necessary swelling and at least a partial destruction of the ordered regions. Furthermore, although native lignin derivatives behave like an insoluble and three-dimensional network, isolated lignin derivatives exhibit maximum solubility in a variety of solvents including dioxane, acetone, methyl cellosolve, tetrahydrofuran, dimethylformamide, and dimethyl sulfoxide.

More generally, soluble materials or extractives in wood consist of those components that are soluble in neutral organic solvents and that are not only suitable for conversion to alternate fuels but also able to continue and expand (if there are minimal economic constraints) the petrochemical industry (Table 5.7) (Speight, 2019c). The di-chloromethane extractable content of wood is a measure of substances such as waxes, fats, resins, phyto-sterols, and non-volatile hydrocarbon derivatives. The amount of extractives is highly dependent on the seasoning or drying of wood. The ethanol–benzene extractable content of the wood consists of certain other di-chloromethane insoluble components such as low molecular-weight carbohydrates, salts, and other water-soluble substances. Most water-soluble and volatile compounds are removed during pulping. The extractives reduce the pulp yield, increase pulping and bleaching chemical consumption, and create problems such as foaming during papermaking if not removed.

For isolation of the extractives from wood, different methods can be used. Volatile extractives are represented by highly volatile compounds, which can be separated by water distillation. They are mainly composed of monoterpene derivatives and other volatile terpene derivatives, terpenoid derivatives as well as of many different low molecular-weight compounds. The term *terpenes* typically refers to pure hydrocarbon derivatives whereas the compounds collectively referred to as *terpenoids* have one or more oxygen-containing functional groups, such as hydroxyl, carbonyl, and carboxylic acid groups in the molecule. The basic structural unit of terpenes is isoprene and terpenes can be divided into subgroups based on the number of isoprene units linked in a terpene.

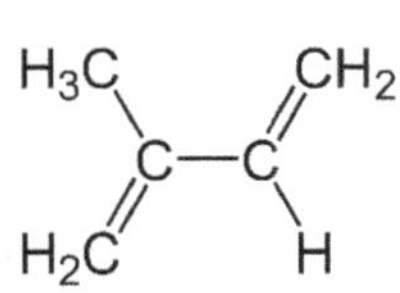

Isoprene (2-methyl-1,3-butadiene)

Table 5.7: Examples of chemicals that are extractable from wood.

Aliphatic and alicyclic compounds
Terpenes and terpenoids (including resin acids and steroids)
Esters of fatty acids (fats and waxes)
Fatty acids and alcohols
Alkanes
Phenolic derivative
Simple phenols
Stilbenes
Lignans
Isoflavones
Condensed tannins
Flavonoids
Hydrolyzable tannins
Gums (polysaccharides)
Linear structure
Branched structure
Branch-on-branch structure
Other compounds
Sugars
Cyclitols
Tropolones
Amino acids
Alkaloids
Coumarins
Quinones

Mono-, sesqui-, di-, tri-, and polyterpenoids (Figure 5.4) are the most abundant terpenes in wood. In addition to this classification, terpenes can be generally classified according to the number of rings within a structure, for example, into acyclic, monocyclic, bicyclic, tricyclic, and tetracyclic terpenes. Various polypropylene derivatives are also present in certain wood species in the form of rubber and gutta percha and, in these macromolecules, the number of isoprene units is high.

MONOTERPENES AND MONOTERPENOIDS

(-)-Limonene α-Pinene 3-Carene Camphene Borneol

SESQUITERPENES AND SESQUITERPENOIDS

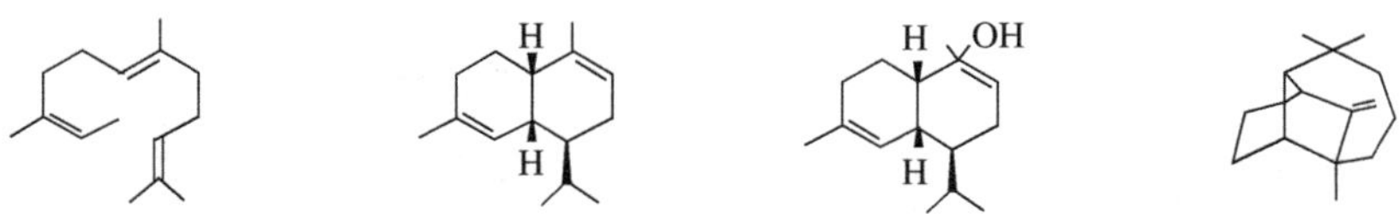

Farnesene α-Muorolene α-Cadinol Longifolene

DITERENES AND DITERPENOIDS

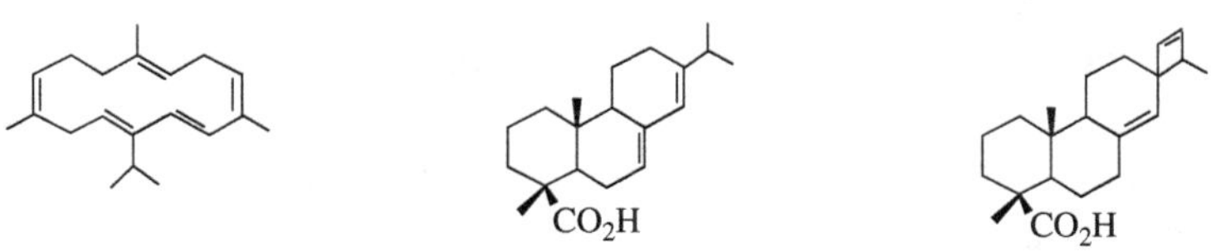

Thunbergene Abietic acid Pimaric acid

TRITERPENES AND STEROIDS

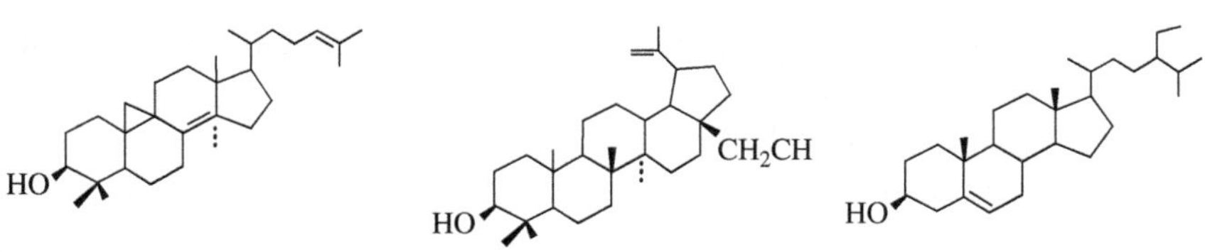

Cycloartenol Betulinol Sitosterol

Figure 5.4: Structures of some common terpene derivatives and terpenoid derivatives.

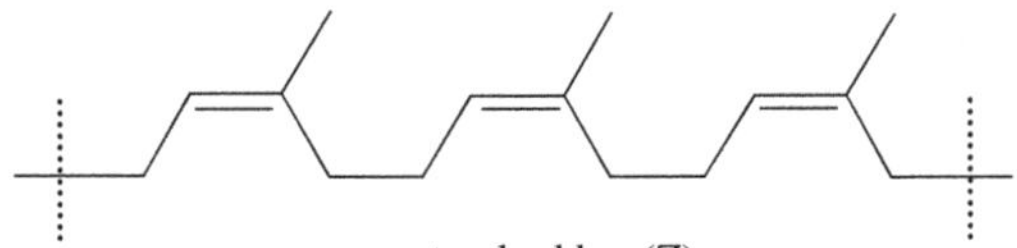

natural rubber (Z)
(*cis*-1,4-polyisoprene)

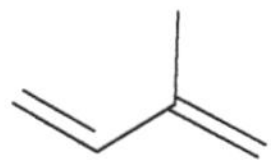

isoprene
(2-methy-1,3-butadiene)

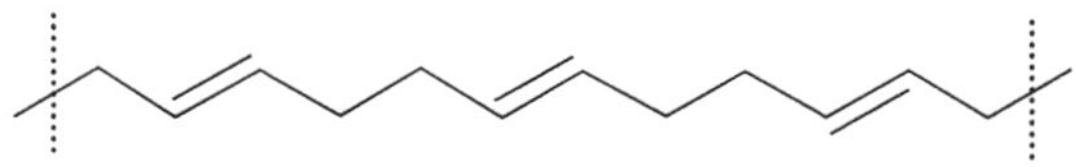

gutta-percha (E)
(*trans*-1,4-polyisoprene)

Structure of natural rubber and gutta percha

Monoterpenes and monoterpenoids are volatile compounds and contribute substantially to the odor of wood. Monoterpenoids occur mainly in softwood oleoresin, either as hydrocarbons or their derivatives. Monoterpenes represent one of the most important constituents of the resin canal extractives and exudates of softwoods. Although monoterpenes and monoterpenoids are rare in common hardwood species, some of these compounds are minor constituents of the oleoresins of tropical hardwoods.

Sesquiterpenes and sesquiterpenoids represent a wide variety of compounds, and they can be found as components of canal resin and of the heartwood deposits of softwoods. Sesquiterpenes and sesquiterpenoids are found in many tropical hardwoods but are rare components of hardwoods from temperate zones. Since these compounds are found only in small amounts, usually they are commercially less important.

Diterpenes and diterpenoids constitute a major part of the canal extractives and are of great industrial importance. Diterpenoids can be divided into acyclic, bicyclic, tricyclic, tetracyclic, and macrocyclic structure types. Diterpenoids are present either as hydrocarbons or as derivatives with hydroxyl, carbonyl, or carboxyl groups. They seem to be restricted to softwood species mainly in the form of resin acids and only some diterpenoids have been found in tropical hardwoods.

Triterpenes and triterpenoids comprise mainly oxygenated derivatives and are traditionally treated as two classes of compounds, i.e. triterpenoids and steroids. Triterpenoids can be roughly divided into three subgroups: tetracyclic lanostane, pentacyclic lupine, and pentacyclic oleanane derivatives. Steroids are similar compound groups compared with triterpenoids, but they differ from some of the tetracyclic terpenoids only by the post-cyclization loss of methyl groups in the biosynthesis process from the acyclic squalene precursor. Triterpenoids and steroids occur mainly as fatty acid esters and as glycosides, but also in the free form. Triterpenoids and steroids are common in softwoods and are generally present in relatively small amounts. A large variety of triterpenes and steroids are also present in many hardwoods of tropical and temperate zones. In both softwood and hardwood, the most prominent compound is sitosterol, which is a potential raw material for making wood-based chemicals.

Resin is the collective name for lipophilic extractives (with the exception of phenolic substances). Resin extractives can be extracted with organic solvents. Water-soluble compounds consist of various phenol derivatives, carbohydrates, glycoside derivatives, and soluble salts, which can be extracted with cold or hot water (Figure 5.5).

Resins are divided into free acids (e.g. resin acid and fatty acid) and neutral compounds (such as fats and waxes). A resin fraction is soluble in organic solvents but insoluble in water and therefore can be extracted with organic solvents, such as hexane, dichloromethane, diethyl ether, acetone or ethanol. Different non-polar and polar solvents can be selected for the isolation of a certain type of extractive from wood.

Analysis of the extractive material can be done at different levels: (i) gravimetric determination of total extractives, (ii) determination of different component groups, and (iii) analysis of individual components. The component groups in extracts can be determined by several chromatographic techniques: gas chromatography (GC), high-performance liquid chromatography (HPLC), size-exclusion chromatography (SEC), supercritical fluid chromatography (SFC), and thin-layer chromatography (TLC).

5.3.3 *Types of wood*

The amount and type of wood fuel used vary considerably across regions, mainly due to the different local situations and conditions. However, some of the use of wood is derived from the use of black liquor (Kavalov and Peteves, 2005) in the pulp and paper industries.

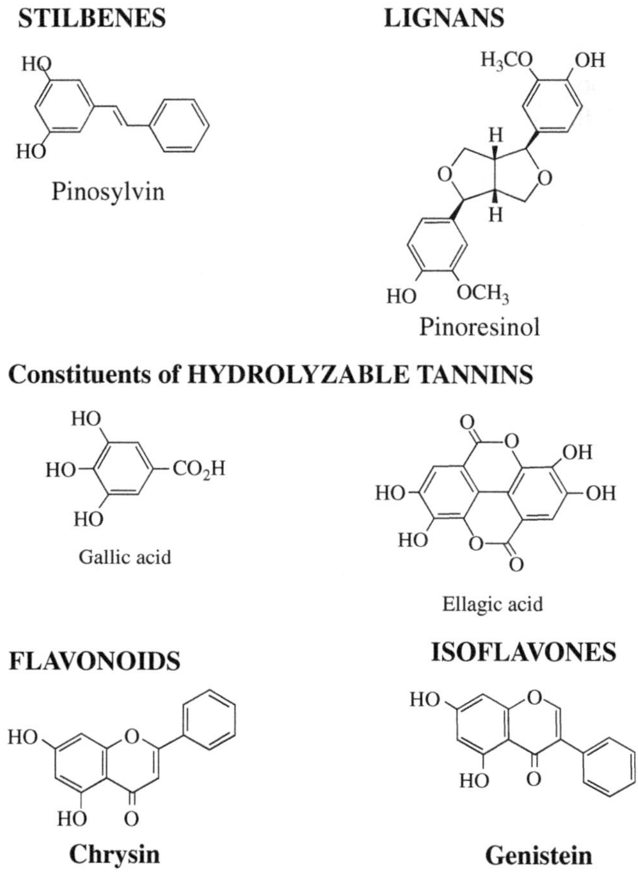

Figure 5.5: Examples of the chemical structures of some phenolic extractable constituents.

Composition of black liquor.

Element	w/w %
Carbon	35.7
Hydrogen	3.7
Nitrogen	≥0.1
Oxygen	35.8
Sulfur	4.4
Chlorine	0.3
Potassium	1.1
Sodium	19.0

The quality of wood fuels is determined for fuel types by choosing for each delivery batch the limit values for the energy density, moisture content, and particle size of the fuel as received from the quality. The energy density is dependent on net calorific value (Table 5.6), moisture content, bulk density, and particle size of the fuel concerned. When choosing limit values for energy density and moisture, the interdependence of different characteristics should be considered by using characteristic values typical of different wood fuels. In deliveries of different wood fuel blends or mixtures (e.g. bark/sawdust, cutter chips/grinding dust/other wood residues), the parties should agree upon the application of quality classification and the quality determination of the fuel considering the use and safety issues. Quality limits for other characteristics of the fuel can be specified case by case for mechanical properties of the fuel (i.e. oversize particles) and other properties. When preparing delivery agreements, differences in properties due to possible seasonal variations should also be considered and agreed upon by parties separately.

The components of wood include cellulose, hemicellulose, lignin, extractives, lipids, proteins, simple sugars, starches, water, hydrocarbon derivatives, ash, and other compounds. The proportion of these wood constituents varies across species, and there are distinct differences between hardwoods and softwoods.

5.3.3.1 *Hardwood*

As the name suggests, the wood from hardwoods is generally harder than that from softwoods, but there are significant exceptions. In both groups, there is enormous variation in the actual wood hardness, with the range of density of hardwoods completely encompassing that of softwoods; some hardwoods (such as balsa) are softer than most softwoods, while yew is an example of a hard softwood. Trees grown in tropical climates are generally hardwood. Hardwood grows faster than softwood but has shorter fibers compared to softwood. Hardwoods or deciduous woods have a higher proportion of cellulose, hemicelluloses, and extractives than that of softwoods, but softwoods have a higher proportion of lignin (Table 5.8). Generally, hardwoods, which produce long-burning fires, have the highest total heating value per unit volume.

Hardwoods have a more complex structure than that of softwoods and are often much slower growing as a result. The dominant feature separating "hardwoods" from softwoods is the presence of pores, also known as

Table 5.8: Composition of different biomass types (% w/w, dry basis).

Type	Cellulose	Hemicellulose	Lignin	Others	Ash
Softwood	41	24	28	2	0.4
Hardwood	39	35	20	3	0.3
Pine bark	34	16	34	14	2
Straw (wheat)	40	28	17	11	7
Rice husks	30	25	12	18	16
Peat	10	32	44	11	6

Source: Prakash and Karunanithi (2008).

vessels. The vessels may show considerable variation in the size, shape of perforation plates (simple, scalariform, reticulate, or foraminate), and structure of the cell wall, such as spiral thickenings.

Hardwoods are employed in a large range of applications, including fuel, tools, construction, and the manufacture of charcoal. Solid hardwood joinery tends to be expensive compared to softwood. In the past, tropical hardwoods were easily available, but the supply of some species, such as teak and mahogany, is now becoming scarce due to over-exploitation. Hardwoods may be used in a variety of objects, but are most frequently seen in furniture or musical instruments because of their density, which adds to the durability, appearance, and performance. Different species of hardwoods lend themselves to different end uses or construction processes due to the variety of characteristics apparent in different timbers, including density, grain, pore size, growth and fiber pattern, flexibility, and ability to be steam-bent.

5.3.3.2 *Softwood*

Softwood is usually wood from gymnosperm trees such as pine trees and spruce trees, which often reproduce using cones and occasionally nuts. The trees classified as softwoods have needle-like or scale-like leaves that, with a few exceptions, remain on the tree all through the year. Hence softwood trees are sometimes called evergreens. Botanically, they are known as gymnosperms and instead of bearing seeds from flowers, gymnosperms have exposed seeds in cones.

Within the softwood and hardwood groups, there is considerable variation in the actual wood hardness, with the range of density of hardwoods completely encompassing that of softwoods. Some hardwoods (such as balsa) are softer than most softwoods, while the hardest hardwoods are much harder than any softwood. In short, the terms *softwood* and *hardwood* are archaic with questionable meaning and often belie the properties of the wood.

Softwoods are generally most used by the construction industry and are also used to produce paper, pulp, and card products. In many of these applications, there is a constant need for density and thickness monitoring and γ-ray sensors have shown good performance in this case. Certain species of softwoods are more resistant to insect attack from woodworm, as certain insects prefer damp hardwood. Softwoods, which produce a fast-burning, cracking blaze, are less dense and have a lower total heating value per unit volume.

5.3.4 *Uses*

The past abundance of timber and the decline of the industry have worked against advances in technology for the efficient production, conversion, and use of wood products. Fortunately, and despite its relatively recent origin as a recognized field of study, wood science has had an appreciable effect on wood technology as well as science in general. The study of wood chemistry has contributed to our understanding of the principal components of wood — cellulose and lignin — and their relationship with each other.

For example, the moisture content of a living tree varies seasonally and the average values are in the range of 40–50% w/w of the total wood mass. The chemical constituents of dry wood species are the so-called structural and non-structural substances. Structural substances are cellulose, hemicelluloses, and lignin. Non-structural substances are mostly low-molecular-mass compounds, such as extractives, some water-soluble organics, and inorganics (Table 5.9). In addition, softwood differs typically from hardwood with regard to the chemical composition. Furthermore, the chemistry of the hemicellulose derivatives and lignin also differs across the various types of trees. On the other hand, the cellulose, hemicelluloses, and lignin are not uniformly distributed in wood cells, and their relative mass proportions can vary widely, depending on the morphological region and age of the wood.

Early research on the hydrolysis of cellulose was prompted by fuel needs in World War I, but have contributed much to our knowledge of this

Table 5.9: Interrelationships of the various constituents of wood.

Feedstock	General separation	Primary products	Secondary products
Dry wood	High molecular-weight materials	Polysaccharides	Cellulose Hemicellulose
		Lignin	
	Low molecular-weight materials	Organic substances	Extractable constituents
		Inorganic substances	

form of chemical reaction. As an example of an early process, consider the Madison Process as it was described in the 1940s; it was developed to hydrolyze softwood species, which are particularly valuable for the production of fuels such as ethanol (Keller, 1996; Galbe and Zacchi, 2002). The hemicellulose sugars were recovered in the form of furfural, and only one stage was required. In case hardwood species are used, a two-stage process is more desirable for maximizing the recovery of both hemicellulose and cellulose. High yields of hemicellulose products can be obtained under the milder pre-hydrolysis conditions, whereas a higher temperature is required to maximize glucose yields from cellulose in the second stage.

Several acid hydrolysis processes are now being promoted to produce ethanol from wood, but they do so in different types of equipment under slightly different acid–temperature–time conditions. The stake process uses a horizontal screw reactor. The Iotech process uses a high-pressure, short-time hydrolysis followed by rapid release of pressure. The twin-screw extruder process developed by New York University uses a high-pressure, special reactor design. The plug-flow reactor under study at Dartmouth and another developed by American Can use different methods to pump the wood and acid into the reactor. The New Zealand process under license to Ultra Systems is probably a modern version of the Madison Process.

After the hydrolysis of the cellulose, the processes could be identical if the same products were to be recovered. Some of the two-stage processes differ in that the residue from the first stage is delignified with a lignin solvent. This dissolves the lignin to leave only cellulose to be hydrolyzed in the second stage. The lignin is then recovered by distilling off the solvent. This is in contrast to hydrolyzing the first stage residue to solubilize the cellulose and leave the lignin.

Wood is bulky, has less than half the heat of combustion of fuel oil, and is heavy to ship in its green state. Furthermore, the cost of a wood-burning system may be three to four times that of a gas-burning installation because of fuel storage, handling, and air quality control systems. These drawbacks have kindled an interest in the production of liquid and gaseous fuels from wood. Much research is devoted to improving existing technology and devising new approaches, but wood fuels are still expensive compared with crude oil-based fuels.

Finally, a topic closely related to the conversion of wood into liquid or gaseous fuel is the use of the chemical storehouse wood to produce a wide range of *silvichemicals.* Many processes of this type have already formed the basis of chemical production on a commercial scale. But the potential to use wood as a chemical feedstock is much greater than it has so far been realized. Whole wood can be gasified, liquefied, or pyrolyzed by methods similar to those used for coal to yield a wide variety of chemicals. Cellulose, as a glucose polymer, can be hydrolyzed to the glucose monomer by acid or enzymes, and the glucose is then fermented to ethanol. The ethanol can be used as a fuel or as a source of other important chemicals such as ethylene or butadiene. As an alternative, the use of glucose as a substrate for fermentation would make possible production of antibiotics, vitamins, and enzymes. Hemicelluloses can easily be converted to simple sugars, which can be used to produce ethanol or furfural, a potential raw material for nylon or other synthetics.

Lignin can be pyrolyzed, hydrogenated, and hydrolyzed to yield phenols, which can be further processed into benzene. Once the technology and economics become feasible, future plants will manufacture a variety of these very significant chemicals from wood, which are now derived from crude oil or other resources.

Charcoal continues to be used as an important industrial source of energy. For example, in Brazil, about 6 million tonnes of charcoal are produced every year for use in heavy industries, such as steel and alloy production. The industrial demand for charcoal in the last few years has led to new, more efficient, large-scale technologies, mainly aimed at improving charcoal yield and quality. Furthermore, although fuel wood is mainly a local source of energy, there are signs of an international trade in wood fuel developing between European and North American countries.

The dynamics of wood fuel flow are complex and very site-specific. The development of sustainable wood energy systems remains one of the most critical issues to be addressed by policy-makers and community planners. With society giving increasing attention to sustainability issues, in the case of wood energy in both developing and developed countries, economic, environmental, and social issues deserve particular attention.

Most of the uses of wood are accounted for by combustion in intermediate or large-scale units outside the forest industries (e.g. in schools, hospitals, barracks, or district heating plants), with minor volumes going to the production of charcoal. Very small volumes were used in a few European countries to generate electricity or to manufacture solid fuels (e.g. briquettes). No wood is used at present in the region to make synthetic liquid or gaseous fuels. The use of energy wood by the forest industries and other users has grown faster than its use by households.

5.4 Waste

Waste is a result of human activities or a by-product of a process (or processes), for which no future use is planned or foreseen. The words *domestic* and *industrial* are qualifiers of the source of the waste and, to some extent, are also descriptive of the contents of the waste. Once a material has been designated as waste, it remains waste until it has been fully recovered and no longer poses a potential threat to the environment or a use is found for the waste. The term *waste* incorporates a number of streams in the modern society. Relevant to gasification, pyrolysis, and incineration processes are streams such as chemical waste, medical waste, waste from auto-shredders, paper, plastics, textiles, metals, glass, and sundry other components which may form part of the industrial and domestic waste. Like feedstocks for any process, these raw materials require different processes for optimal operation.

Thus, *domestic waste* (also known as *rubbish*, *garbage*, *trash*, or *junk*) is unwanted or undesired material (although the old adage "One man's waste is another man's treasure" sometimes applies) (Table 5.10). *Waste* is the general term; although the other terms are used loosely as synonyms, they have more specific meanings. Thus, (i) rubbish or trash is mixed household waste, including paper and packaging; (ii) food waste or garbage (North America) is kitchen and table waste; and (iii) junk or scrap is metallic or industrial material. There are other categories of waste as well: sewage, ash, manure, and plant materials from garden operations,

Table 5.10: Major sources of waste.

Dredging and irrigation	Waste consists of soil and sediments removed from waterways, harbors, estuaries, and irrigation canals. The quantities may be considerable and contain hazardous materials discharged from industrial and farming activities. It is not suited for energy production.
Farming, livestock rearing, and dairy activities	Waste is composed largely of spoilt food, manure, crop waste, and waste from chemical or pesticide use. Not all of the agricultural waste generated can be put to energy production; some of this waste is returned to the land as part of good agricultural practices.
Industrial	Some of the major industrial sources of waste are construction and demolition, fabrication, light and heavy manufacturing, refineries, chemical plants and non-nuclear power plants. Some large industrial facilities have their own recycling initiatives and operate their own landfills; it is difficult to determine the amount of waste discarded.
Mining and quarrying	Waste consists of mine tailings (silts, fine sands, or other aggregate materials) and may pose problems to the environment due to the large quantities of waste produced and, in some cases, its hazardous nature. Not suited for energy production.
Nuclear power and nuclear defense	The civilian nuclear power industry, defense facilities and nuclear research projects generate dangerous radioactive waste. The toxicity due to exposure and concerns over long-range health and environmental effects make nuclear waste extraordinarily difficult to dispose of safely.
Residential, commercial, and institutional	Homes, commercial, or institutional businesses, construction and demolition activities, municipal services and treatment plants (inclusive of waste incinerators) are major contributors to waste streams.

including grass cuttings, fallen leaves, and pruned branches. On the other hand, the term *industrial waste* refers to waste produced by industrial operations such as factories, mills, and mines, and has existed since the onset of the Industrial Revolution. *Chemical waste* and *toxic waste* are two additional (but specific) designations of industrial waste (Table 5.10).

Municipal solid waste (MSW) is a waste type that includes predominantly household waste (domestic waste) with sometimes the addition of commercial waste collected by a municipality within a given area. It exists in either solid or semisolid form and generally excludes industrial hazardous waste. The term *residual waste* refers to the waste leftover from household sources containing materials that have not been separated out or sent for reprocessing.

There are five broad categories of municipal solid waste: (i) biodegradable waste, such as food and kitchen waste and green waste, (ii) recyclable materials such as paper, glass, cans metals, and certain types of plastic, (iii) inert waste such as construction and demolition waste, dirt, rocks, and debris, (iv) composite waste, which includes waste clothing, and waste plastics, and (v) domestic hazardous waste (also called *household hazardous waste*) and toxic waste such as discarded medications, paints, chemicals, light bulbs, fluorescent tubes, spray cans, fertilizer containers, pesticide containers, batteries, and shoe polish.

Some components of waste, such as plastic bottles, metals, glass, or paper, can be recycled once recovered from the waste stream. The biodegradable components of wastes (such as paper and food waste), including agricultural waste (Tables 5.10 and 5.11), can be composted or anaerobically digested to produce soil improvers and alternate (renewable) fuels though a variety of reactions (Table 5.12). If it is not dealt with in a sustainable manner biodegradable waste can contribute to greenhouse gas emissions and, by implication, climate change.

On the other hand, there is also electronic waste, which is a waste consisting of any broken or unwanted electrical or electronic appliance. Although cases, electronic waste consists of electronic products that were used for data processing, telecommunications, or entertainment in private households and businesses but are now considered obsolete, broken, or irreparable. Despite being commonly classified as a waste, disposed electronic products are a considerable secondary resource due to their significant suitability for direct reuse (for example, many fully functional computers and components are discarded during upgrades), for refurbishing, and for material recycling of its constituent raw materials. It is a point of concern considering that many components of such equipment are considered toxic and are not biodegradable but they are not precursors to fuels and other than recognition, though the above paragraph will not be considered in the context of the present text.

Table 5.11: Amounts (% w/w) of cellulose, hemicellulose and lignin in common agricultural residues and wastes.

Agricultural residue	Cellulose	Hemicellulose	Lignin
Bamboo	41–49	24–28	24–26
Coastal Bermuda grass	25	35.7	6.4
Corn cobs	45	35	15
Corn stover	35	28	16–21
Cotton seed hairs	80–90	5–20	0
Grasses	25–40	35–50	10–30
Hardwood stem	40–50	24–40	18–25
Leaves	15–20	80–85	0
Newspaper	40–55	25–40	18–30
Nut shells	25–30	25–30	30–40
Paper	85–99	0	0–15
Primary wastewater solids	8–15	NA	24–29
Rice straw	40	18	5.5
Softwood stem	45–50	25–35	25–35
Solid cattle manure	1.6–4.7	1.4–3.3	2.7–5.7
Sorted refuse	50–60	10–20	15–20
Sugar cane bagasse	32–48	19–24	23–32
Sweet sorghum	27	25	11
Swine waste	6.0	28	—
Switch grass	30–51	10–50	5–20
Waste papers from chemical pulps	60–70	10–20	5–10
Wheat straw	33–40	20–25	15–20

In nature, there is no waste but, on the other hand, waste is a by-product of the human economic system and the technology that drives it. Since the Industrial Revolution, human society has developed economies that are largely unrelated to nature and the natural order of events. It has created public and private wealth, material well-being, a plentiful food supply, comfortable living conditions, personal freedom, and enhanced culture, but at a price.

Table 5.12: Common reactions in the anaerobic digestion process.

Substrate	Reactions
Alcohol derivatives	$4CH_3OH \rightarrow 3CH_4 + CO_2 + 2H_2O$ $CH_3OH + H_2 \rightarrow CH_4 + H_2O$ $4HCOO^- + 2H^+ \rightarrow CH_4 + 2H_2O$
Monosaccharide derivatives	$C_6H_{12}O_6 + 2H_2O \rightarrow 2H_2 + \text{butyrate} + 2HCO_3^- + 3H^+$ $C_6H_{12}O_6 + 4H_2O \rightarrow 4H_2 + 2\ \text{acetate} + 2HCO_3^- + 4H^+$
Organic acid derivatives	$\text{Butyrate} + 2H_2O \rightarrow 2H_2 + 2\ \text{acetate} + H^+$ $\text{Propionate} + 3H_2O \rightarrow 3H_2 + \text{acetate} + HCO_3^- + H^+$ $HCOO^- + 3H_2 + H^+ \rightarrow CH_4 + 2H_2O$ $CH_3COO^- + H_2O \rightarrow CH_4 + HCO_3^-$ $4CH_2NH_2 + 2H_2O + 4H^+ \rightarrow 3CH_4 + CO_2 + 4NH_4^+$
Methanogenic substances	$2(CH_3)_2NH + 2H_2O + 2H^+ \rightarrow 3CH_4 + CO_2 + 2NH_4\, 4(CH_3)_3N$ $+ 6H_2O + 4H^+ \rightarrow 9CH_4 + 3CO_2 + 4NH_4\, 2CH_3CH_2\text{-}N(CH_3)_2$ $+ 2H_2O \rightarrow 3CH_4 + CO_2 + 2CH_3CH_2NH_2$ $4H_2 + HCO_3^- + H^+ \rightarrow CH_4 + 3H_2O$ $4H_2 + 2HCO_3^- + H^+ \rightarrow \text{acetate} + 4H_2O$
Sulfate derivatives	$4H_2 + SO_4^{2-} \rightarrow HS^- + 3H_2O + OH^-$

In modern society, waste comes in a variety of forms and must be dealt with on a day-to-day basis. Industry produces huge amounts of *industrial waste* and *domestic waste* that make a large contribution to the general waste problem. In spite of the recognition, many insidious waste products escape (inadvertently or deliberately) into the surrounding environment. Thus, there are numerous pollution incidents. On the other hand, there are technologies available for the treatment of most of the wastes that are produced. The level of treatment is largely a matter of cost, but conversion of waste into new products is a concept that has long been ready to hatch. Now is the time.

5.4.1 *Domestic and industrial waste*

Much of what human society discards contains usable material, much of it in the form of recoverable energy. Paper, wood, cloth, food waste, and plastics are the main potential energy sources present in waste. The remainder of the waste consists of glass, metals, and miscellaneous rubble. Domestic waste is typically disposed of by tipping it into large holes in the ground–landfill sites. Sometimes the waste is incinerated first and only the remaining ash and non-combustible material is sent to a landfill. Increasingly, a proportion of the waste is separated for recycling at some stage along the way.

It is possible to recover the energy from landfill sites and waste incinerators. Domestic waste could also provide the feedstock for a number of other conversion systems, all of which could recover useful energy while reducing the need for landfill sites. However, whatever the energy technology used, domestic waste provides a low-grade fuel. Its consistency is variable and not well suited to mechanical handling systems; the proportions of various constituents will vary from load to load; the moisture content and heating value will vary; and the amount of non-combustible material will keep the heating value low. All of these can lead to an inefficient combustion if the process is not well controlled, making it more difficult to control the toxic emissions from plastics and other materials. There is also a potential conflict between the recycling of materials and the recovery of energy from those materials. The main benefit of domestic waste as a fuel source is that, similar to many other waste streams, energy technology can reduce the waste disposal problem.

Recently, there was a surge in interest in the burning of garbage/domestic waste to generate electricity. This is not a new idea, although in the past, when waste was burned, it created pollution that could even be toxic. Technology does exist to remove almost all the pollutants from the off-gases produced during this energy production cycle (Mokhatab *et al.*, 2016; Speight, 2019a). Special filters remove the dangerous chemicals and particles that would normally be found in off-gases. The domestic waste is sorted out usually by hand to remove those materials that can be recycled. Steel is removed using electromagnets and stored until there is enough quantity for recycling to be economically viable. Aluminum, in the form of cans, is removed by hand. The waste is then fed into the hopper of a furnace. When the doors slide open, it falls into a burning chamber. Gas is normally used to start a fire, which burns at a high temperature to destroy the domestic waste. While the waste burns, it heats a water tank that, in turn, produces steam. The steam is used to turn turbines that produce electricity. Once steam has been produced, the production method of electrical power is no different from that used in any other power station. The high-pressure steam is used to turn electrical turbines that produce electricity. The advantage of this way of producing electricity is that domestic waste that would normally be buried in landfill sites or even dumped far out at sea is burned instead. This implies that vast areas of land that would have been used for landfill are now free for agriculture or building purposes.

Finally, although not presented in the current context, electronic waste is also worthy of mention. Electronic waste [*e-waste* or WEEE (waste electrical and electronic equipment)] is a type of waste that consists of any broken or unwanted electrical or electronic appliance. It is a point of concern considering that many components of such equipment are considered toxic and are not biodegradable. Electronic waste includes computers, entertainment electronics, mobile phones, and other items that have been discarded by their original users. Reconceptualization of electronic waste as a resource thus preempts its potentially hazardous qualities.

However, electronic waste is a valuable source of secondary raw materials, if treated properly; however, if not treated properly, it is a major source of toxins. Rapid technology change, low initial cost, and even planned obsolescence have resulted in a fast-growing problem around the globe. Technical solutions are available but in most cases a legal framework, a collection system, logistics, and other services need to be implemented before a technical solution can be applied.

5.4.2 *Effects of waste*

Wastes generated from the domestic and industrial sources increase continuously with rising population. In general, the lack of facilities for disposal of waste caused overuse of landfill sites, resulting in hazards for the environment and for public health. These effects include (i) air pollution, (ii) pollution of surface waters, (iii) changes in soil fertility, and (iv) changes in the landscape and visual discomfort.

Air pollution by unpleasant odors and wind-carried suspensions is obvious in areas located near urban waste landfills, since these are currently run in complete ignorance of practices such as cell operation and inert matter lagging. The leakage from the slopes of landfills situated near surface water bodies adds to the pollution of such waters by organic substances and suspensions. The urban waste landfills that are not waterproof often become sources of groundwater pollution by nitrates and nitrites, as well as other pollutants. Water leakage adversely affects the quality of the adjoining soils, which brings about consequential effects on their utilization.

Withdrawing land from the natural or economic circuit in order to set up waste landfills is a process that may be regarded as temporary; yet, in terms of sustainable development, it extends over at least two generations when adding up the time periods required for planning (1–3 years), running (15–30 years), ecological reconstruction, and subsequent monitoring (15–20 years).

In terms of biodiversity, any domestic or industrial waste landfill means the subsequent elimination of 30–300 species (microbiological population of the soil not included) on each hectare of the area intended to host a landfill. Moreover, changes are likely to occur in areas close to the landfill, such as: (i) ruderal species, which are plant species that are the first to colonize disturbed lands that are specific to polluted areas, would become dominant in the vegetal assemblies, and (ii) some mammals, birds, and insects would desert the area, in favor of species that feed on refuse (rats and crows).

Although the effects on flora and fauna are, in theory, limited to the period over which the landfill is operated, the ecological reconstruction performed after the area has been relieved from its technological use will not be able to retrieve the initial biological balance, as the evolution of that biosystem has been irreversibly modified. The practices used currently in the collection, transport, and storage of urban waste facilitate the multiplication and dissemination of the pathogenic agents and their accompanying breed: insects, rats, crows, and stray dogs. Thus, waste in any form (domestic or industrial) represents a health hazard, due to its disposal and composition. In terms of composition, the focus is often on toxic substances such as heavy metals (lead and cadmium), pesticides, solvents, and used oils that form an integral part of the waste.

Indeed, the challenge for waste disposal arises from the joint storage of hazardous materials (including toxic sludge, oil products, dyeing residues, and metallurgical slag) and solid domestic waste. This situation is likely to generate inflammable, explosive, or corrosive mixtures and combinations thereof. On the other hand, the presence of easily degradable household waste may facilitate the decomposition of complex hazardous components and thus diminish environmental pollution. Another negative aspect is the fact that several recyclable and useful materials are stored in the same place as materials that cannot be recycled; consequently, these materials blend together and become chemically and biologically contaminated, which renders their retrieval rather difficult.

Thus, the problems faced by waste management activities may be summarized as follows: (i) storage in open grounds is the most used method to remove waste ultimately, (ii) existing landfills may be located in sensitive places, such as those places in close proximity to lodgings, surface or ground water, and leisure areas), (iii) existing waste landfills may be improperly designed from an environmental protection point of view, thus allowing for water and soil pollution in those areas, (iv) currently, waste landfills may require a review of waste handling practices insofar as waste layers are not compacted and there is no strict control of the quality and quantity of waste that is dumped on the landfill, leading to the potential for fire and/or the emanation of unpleasant odors.

All of the above lead to the conclusion that specific measures need to be taken with regard to waste management, which would be adequate in each phase of the waste dumping process. Environmental monitoring activities should comprise the observance of these measures. However, one answer to these issues is to convert the waste to usable products either through (i) the production of gaseous fuels or (ii) the production of liquid fuels or (iii) the production of solid fuels. Such efforts may not only solve the depletion of fuels from fossil sources but also assist in the disposal of waste materials and the ensuing environmental issues. However, before entering the process descriptions of waste conversion, it is necessary to understand better the composition of domestic and industrial wastes.

Generating waste at current levels is incompatible with a sustainable future. While the problem of waste is serious, a variety of initiatives are being taken to address the various threats. These include moves toward waste minimization, waste segregation and recycling, cleaner production with regular waste audits, green chemistry, renewable energy, and energy efficiency, and developing the concept of industrial ecosystems. The issues involved are much more than technical problems.

The historic approach to solid waste has been to bury it in landfills. This is becoming increasingly problematic because the diminishing availability of suitable landfill sites and the increasingly stringent conditions being applied to landfills mean that costs have increased and will continue to increase. Among the major problems associated with landfills are that the leachate-contains toxic heavy metals and methane gas is produced.

This is a major issue because of the variability of industrial solid, liquid and gaseous wastes and the capacity of modern processing industries to produce huge quantities of waste. Fortunately, regulatory processes are now such that the numerous damages caused previously will not be repeated. However, there remain problem sites that represent long-term hazards.

Thus, in order to reduce the amount of landfill, the amount of waste must be reduced. This involves either (i) cutting back on the use of many materials or (ii) making use of the waste by converting if into useful products. Either option would reduce the amount of waste sent to landfill sites. The first option certainly reduces the amount of landfill material but is often more difficult to achieve. However, the second option offers an attractive proposition of the production of fuel products. Thus, waste conversion has become an attractive option to landfill disposal and the end result is the generation of a usable product in the form of a gaseous, liquid, or solid fuel. However, a disadvantage of the second option is the heterogeneity of waste materials. In fact, it is obvious that many waste streams are not subject to direct processing but require special measures in the form of specific pretreatments of the waste prior to processing.

One form of pretreatment is the separation and recycling of waste components, thereby setting aside a portion of the waste stream for recycling and other uses. The result of this separation at the source is that just the remaining residual waste (i.e. the waste stream from which recyclable materials have been removed), not destined for any use other than landfill, is sent to the conversion reactor.

References

Allred, V.D. (Editor). 1982. *Oil Shale Processing Technology*. Center for Professional Advancement, East Brunswick, New Jersey.

Baughman, G.L. 1978. *Synthetic Fuels Data Handbook*, 2nd Edition, Cameron Engineers, Inc., Denver, Colorado.

Demirbaş, A. and Ozturk, T. 2004. Anaerobic digestion of agricultural solid residues. *Int. J. Green Energy*. 1: 483–494.

Demirbaş, A. 2007. Modernization of biomass energy conversion facilities. *Energy Sources Part B*, 2: 227–235.

Demirbaş, A. 2008. Biofuels sources, biofuel policy, biofuel economy and global biofuel projections. *Energy Convers. Mgmt*. 49: 2106–2116.

Demirbaş, A. 2009. *Biofuels: Securing the Planets Future Energy Needs*. Springer Verlag, London, United Kingdom.

Galbe, M. and Zacchi, G. 2002. A review of the production of ethanol from softwood. *Appl. Microbiol. Biotechnol*. 59(6): 618–628.

Gary, J.H., Handwerk, G.E., and Kaiser, M.J. 2007. *Petroleum Refining: Technology and Economics*, 5th Edition. CRC Press, Taylor & Francis Group, Boca Raton, Florida.

Hawkins, G.B. and White, J.B. 1993. *Stonehenge Decoded*, 2nd Edition. Hippocrene Books Inc., New York.

Hoogwijk, M., Faaij, A., Eickhout, B., de Vries, B., and Turkenburg, W. 2005. Potential of Biomass Energy out to 2100, for Four IPCC SRES Land-use Scenarios. *Biomass and Bioenergy* 29: 225–257.

Hsu, C.S. and Robinson, P.R. (Editors). 2017. *Handbook of Petroleum Technology*. Springer International Publishing AG, Chaim, Switzerland.

Hwang, B. and Obst, J.R. 2003. Basic studies on the pyrolysis of lignin compounds, Proceedings of the IAWPS 2003 International Conference on Forest Products: Better Utilization of Wood for Human, Earth and Future. The Korean Society of Wood Science and Technology, International Association of Wood Products Societies, Daejeon, Korea, April 21–24, 2: 1165–1170.

Kavalov, B. and Peteves, S.D. 2005. *Status and Perspectives of Biomass-to-Liquid Fuels in the European Union*. European Commission. Directorate General Joint Research Centre (DG JRC). Institute for Energy, Petten, Netherlands.

Keller, F.A. 1996. Integrated Bioprocess Development for Bioethanol Production. In *Handbook on Bioethanol*: *Production and Utilization*, Wyman, C.E. (Editor) Taylor & Francis, Washington, DC, pp. 351–379.

Lago, C., Caldés, N., and Lechón, Y. 2019. *The Role of Bioenergy in the Emerging Bioeconomy*: *Resources, Technologies, Sustainability and Policy*. Elsevier BV, Amsterdam, Netherlands.

Larson, E.D. and Kartha, S. 2000. Expanding roles for modernized biomass energy. *Energy Sustain. Develop*. 4: 15–25.

Lee, S. 1991. *Oil Shale Technology*. CRC Press, Taylor & Francis Group, Boca Raton, Florida.

Lee, S. 1996. *Alternative Fuels*. CRC Press, Taylor & Francis Group, Boca Raton, Florida.

Luque, R. and Speight, J.G. (Editors). 2015. *Gasification for Synthetic Fuel Production*: *Fundamentals, Processes, and Applications*. Woodhead Publishing, Elsevier, Cambridge, United Kingdom.

Mokhatab, S., Poe, W.A., and Speight, J.G. 2006. *Handbook of Natural Gas Transmission and Processing*. Elsevier, Amsterdam, Netherlands.

Parkash, S. 2003. *Refining Processes Handbook*. Gulf Professional Publishing, Elsevier, Amsterdam, Netherlands.

Ragland, K.W., Aerts, D.J., and Baker, A.J. 1991. Properties of wood for combustion analysis. *Bioresource Technology* 37: 161–168.

Scouten, C.S. 1990. Oil Shale. In *Fuel Science and Technology Handbook*. Marcel Dekker Inc., New York, Chapters 25–31, pp. 795–1053.

Speight, J.G. 2013. *The Chemistry and Technology of Coal*, 3rd Edition. CRC-Taylor and Francis Group, Boca Raton, Florida.

Speight, J.G. 2014. *The Chemistry and Technology of Petroleum* 5th Edition. CRC-Taylor and Francis Group, Boca Raton, Florida.

Speight, J.G. and Islam, M.R. 2016. *Peak Energy — Myth or Reality*. Scrivener Publishing, Beverly, Massachusetts.

Speight, J.G. 2017. *Handbook of Petroleum Refining*. CRC Press, Taylor & Francis Group, Boca Raton, Florida.

Speight, J.G. 2019a. *Natural Gas: A Basic Handbook*, 2nd Edition. Gulf Publishing Company, Elsevier, Cambridge, Massachusetts.

Speight, J.G. 2019b. *Synthetic Fuels Handbook: Properties, Processes, and Performance*. McGraw-Hill, New York.

Speight, J.G. 2019c. *Handbook of Petrochemical Processes*. CRC Press, Taylor & Francis Group, Boca Raton, Florida.

Speight, J.G. 2020. *Global Climate Change Demystified*. Scrivener Publishing, Beverly, Massachusetts.

US DOE 2004a. Strategic Significance of America's Oil Shale Reserves, I. Assessment of Strategic Issues, March, http://www.fe.doe.gov/programs/reserves/publications.

US DOE 2004b. Strategic Significance of America's Oil Shale Reserves, II. Oil Shale Resources, Technology, and Economics; March, http://www.fe.doe.gov/programs/reserves/publications.

US DOE 2004c. America's Oil Shale: A Roadmap for Federal Decision Making; USDOE Office of US Naval Petroleum and Oil Shale Reserves, http://www.fe.doe.gov/programs/reserves/publications.

Chapter 6

Production of Alternate Fuels

6.1 Introduction

Conversion of non-fossil fuel sources into alternate fuels and higher value products is not a new concept. For example, sugarcane has been used in the production of ethanol since 6000 BC. The production of biofuels from lignocellulosic feedstocks can be achieved through two very different processing routes: (i) the *thermochemical platform* and (ii) the *bioconversion platform*. There is no clear candidate for "best technology pathway" between the competing biochemical and thermochemical routes. Technical barriers for enzymatic hydrolysis include the low specific activity of current commercial enzymes, the high cost of enzyme production, and a lack of understanding of enzyme biochemistry and mechanistic fundamentals.

The bioconversion platform typically uses a combination of physical or chemical pretreatment and enzymatic hydrolysis to convert lignocellulose into its component monomers. The bioconversion platform (e.g. anaerobic digestion (AD) and fermentation) uses biological agents to carry out a structured deconstruction of lignocellulose components and combines process elements of pretreatment with enzymatic hydrolysis to release carbohydrates and lignin from wood. An advantage of the bioconversion platform is that it provides a range of intermediate products, including glucose, galactose, mannose, xylose, and arabinose, which can be relatively easily processed into value-added bioproducts. The bioconversion platform also generates a quantity of lignin or lignin components; depending on the pretreatment, lignin components may be found in the hydrolysate after enzymatic hydrolysis, or in the wash from the pretreatment stage.

Once hydrolyzed, six-carbon sugars can be fermented to ethanol using the age-old yeasts and processes. Five-carbon sugars, however, are more difficult to ferment; new yeast strains that can process these sugars are now being developed, but issues remain with the process efficiency and the length of fermentation. Other types of fermentations, such as bacterial fermentation under aerobic and anaerobic conditions, can produce a variety of other products from the sugar stream, including lactic acid. Bioconversion proceeds at lower temperatures and lower reaction rates and can offer high selectivity for products. Ethanol production is a biochemical conversion technology used to produce energy from biomass, depending on the type of biomass (Tables 6.1–6.3) (Prakash and Karunanithi, 2008). For ethanol production, biochemical conversion researchers have focused on a process model of dilute acid hydrolysis of hemicelluloses followed by enzymatic hydrolysis of cellulose. Biodiesel production is a biochemical conversion technology used to produce energy from oilseed crops.

Cellulosic materials can be used to produce ethanol, which represents an important, renewable liquid fuel for motor vehicles. Production of ethanol from biomass is one way of reducing both the consumption of crude oil and environmental pollution. In order to produce ethanol from cellulosic biomass, a pretreatment process is performed to reduce the sample size, break down the hemicelluloses to sugars, and open up the structure of the cellulose component. The cellulose portion is hydrolyzed by acids or enzymes into glucose sugar that is fermented to ethanol. The sugars from the hemicelluloses are also fermented to ethanol. The use of

Table 6.1: The biomass resource base.

Resource	Products
Cereals, grains, sugar crops	Can be fermented to produce ethanol
Cellulosic materials	Can be converted into alcohol or synthesis gas
Oil-seed crops	Can be converted into methyl-esters
Organic waste material	Can be converted into biodiesel Mature and organic household waste might be converted into biogas

Table 6.2: Elemental compositions, ash yield, and calorific heating value (dry basis).

Fuel	C (wt.%)	H (wt.%)	O (wt.%)	N (wt.%)	S (wt.%)	Ash (wt.%)wt	Heating value (kJ/g)
Birch wood	48.8	6.0	44.2	0.5	0.01	0.5	20.0
Pine wood	49.3	6.0	44.2	0.5	0.01	0.5	20.1
Bark	47.2	5.6	46.9	0.3	0.07	3.9	20.9
Wheat straw	49.6	6.2	43.6	0.6	n.a	4.7	18.6
Miscanthus	49.5	6.2	43.7	0.6	n.a	3.3	18.5
Sugarcane	49.5	6.2	43.8	0.5	n.a	3.7	18.5
Reed grass	49.4	6.3	42.7	1.6	n.a	8.8	18.8
Peat	53.1	5.5	38.1	1.3	0.2	5.6	20.5
Coal	80.4	5.0	6.7	1.3	0.53	7.0	30.4

Table 6.3: Composition of different biomass types (% w/w, dry basis).

Type	Cellulose	Hemicellulose	Lignin	Others	Ash
Softwood	41	24	28	2	0.4
Hardwood	39	35	20	3	0.3
Pine bark	34	16	34	14	2
Straw (wheat)	40	28	17	11	7
Rice husks	30	25	12	18	16
Peat	10	32	44	11	6

ethanol as a motor fuel has as long a history as the car itself. It began with the use of ethanol in internal combustion engines.

Fermentation feedstocks require pretreatment by chemical, physical, or biological means to open up the structure of biomass and reduce the complex carbohydrates to simple sugars. This type of pretreatment is often referred to as hydrolysis. The resulting sugars can then be fermented by the yeast and bacteria employed in the process. Feedstocks high in starch and sugar are most easily hydrolyzed. Cellulosic feedstocks, including the major fraction of organics in MSW, are more difficult to hydrolyze, requiring more extensive pretreatment. Fermentation is commonly used industrially to convert substrates such as glucose into ethanol for use in beverage, fuel, and chemical applications and into other chemicals (e.g. lactic acid used for producing renewable plastics) and products (e.g. enzymes for detergents). Strictly speaking, fermentation is an enzymatically controlled anaerobic process although the term is sometimes more loosely applied to include aerobic processing as well.

The *bioconversion platform* is an industrial option (as might be used in a biorefinery) for producing fuels from biomass using biochemical reactions and/or biochemical agents. For example, fermentation or anaerobic digestion to produce fuels and chemicals from organic sources is a bioconversion platform. The bioconversion platform therefore has the ability to serve as the basis for wood-based biorefining operations, generating value-added bioproducts as well as fuel and energy for the forest sector.

In addition to the bioconversion platform and the direct combustion of the feedstock, there are other thermal methods of conversion that are grouped into the *thermochemical platform*, which typically uses a combination of pyrolysis, gasification, and catalysis to transform the feedstock wood first into gas products, one of which is synthesis gas, and then into fuels or chemicals (Luque and Speight, 2015; Speight, 2109). Synthesis gas (also referred to as syngas) production through pyrolysis is accompanied by the generation of char, which can then be gasified to provide heat and energy for the thermochemical platform. A variety of commercial-scale processes are available to transform fossil fuels such as coal or natural gas into liquid fuels, including Fischer–Tropsch fuels. However, the use of biomass instead of fossil fuels changes the composition of synthesis gas, creating a more heterogeneous intermediate product and increasing the difficulty in downstream catalysis. A range of technical problems must be overcome before biomass can become a commercially viable substitute for fossil feedstocks in second-generation biofuel production, but, in general, it is felt that elements of the thermochemical platform are highly suitable for bioenergy production.

This platform combines process elements of pretreatment, pyrolysis, gasification, cleanup and conditioning to generate a mixture of hydrogen, carbon monoxide, carbon dioxide, and other gases. The products of this platform may be viewed as intermediate products, which can then be assembled into chemical building blocks and eventually end products. In this platform, the only pretreatment required involves drying, grinding, and screening the material in order to create a substrate that can easily be fed into the reaction chamber. The technology required for this stage is already available on a commercial basis, and is often associated with primary or secondary wood processing, or agricultural residue collection and distribution. The thermochemical platform provides the opportunity to produce a number of additional coproducts, as well as energy in the form of heat or electricity and biofuels. Each component of synthesis gas (i.e. carbon monoxide, carbon dioxide, hydrogen, and methane) may be recovered, separated (Chapter 9), and utilized.

6.2 Production of Alternate Fuels

Biorenewable feedstocks can be converted into liquid or gaseous forms for the production of electric power, heat, chemicals, or gaseous and liquid fuels. The main biomass conversion processes are as follows (arranged alphabetically rather than by preference): (i) anaerobic digestion, (ii) direct combustion, (iii) fermentation, (iv) gasification, and (v) pyrolysis. Each process has its own particular features and the process application is dependent on the type of feedstock and the desired product(s).

The amount of hemicellulose and cellulose in wood and the chemical products desired determine the general type of process that might be used to hydrolyze wood. Hardwoods yield more five-carbon sugars than softwoods. Since, at this time, only the six-carbon sugars from cellulose are readily fermentable, softwoods are desired for ethanol production, but they are not as widely available as hardwoods. Hardwoods are more widely available now, so considerable effort has been expended to develop processes to utilize their unique constituents.

The main components of wood cells are cellulose (an insoluble substance which is the main constituent of plant cell walls and vegetable fibers such as cotton; it is a polysaccharide consisting of chains of glucose monomers), hemicellulose (a class of substances which occur as constituents of the cell walls of plants and are polysaccharides with a structure simpler than cellulose), and lignin (a complex organic polymer deposited in the cell walls of many plants, making them rigid and woody), forming about 99% w/w of the wood material. Cellulose and hemicellulose are formed by long chains of carbohydrates, whereas lignin is a complicated component of polymeric phenolics. Lignin is rich in carbon and hydrogen, which are the main heat-producing elements. Thus, the calorific value of lignin is higher than that of cellulose and hemicellulose (both are carbohydrate derivatives). Wood and bark also contain the so-called extractives, such as terpenes, fats, and phenols. The amount of extractives from wood is relatively small compared to the amount of extractives from bark and foliage.

The nitrogen (N) content of wood is approximately 0.75%, varying somewhat from one tree species to another. For example, the nitrogen-fixing alder (*alnus* sp.) contains twice as much nitrogen as most coniferous trees. Wood has practically no sulfur (S) and, compared to many other fuels, wood has a relatively low carbon content (about 50% of the dry weight) and high oxygen content (about 40% w/w), which leads to relatively low heating value per dry weight.

The collection of solid wastes is usually organized on a community basis; in developing countries however, it may be organized (to a greater or lesser extent) on an informal basis. The treatment and disposal of solid wastes are definitely interconnected. Treatment is applied to recover useful substances or energy, reduce waste volume, or stabilize waste that remains to be dumped or disposed of in landfills. Wastes may be treated before disposal to reduce the volume or alter the characteristics of the waste, which can be achieved by various physical, chemical, and biological processes, while combustion can be used to destroy some toxic organic chemicals. Where the method of waste disposal is not specified, the choice of the disposal route will typically depend on the following factors: (i) the availability of facilities, (ii) the volume of waste material, and (iii) the hydro-geological characteristics; the influence of industrial and environmental lobby groups must also be taken into account.

Provided that there is no shortage of lands with suitable geological formations, landfilling remains the principal final disposal route for the majority of wastes even in highly industrialized countries. Where there is treatment, it is usually designed to reduce the volume of waste to be landfilled and includes compaction, shredding, baling, and combustion. Most solid wastes therefore will directly be disposed of in sanitary landfills.

The term *sanitary* is mainly to be understood as providing some protection for the population against airborne dust and litter, stench, rodents, and insects. Most of these nuisances can be prevented by the prompt covering of freshly dumped waste with soil. However, to be able to call a waste tip or landfill *sanitary* from an environmental viewpoint requires more measures to be taken. The main environmental problem associated with landfilling is pollution of groundwater. The rainwater percolating through solid waste tends to carry large amounts of pollutants to groundwater aquifers if the underlying strata are pervious or fissured. Thus, wells drawing from the aquifers will be extracting groundwater contaminated with leachate; such a situation is often difficult to remedy. Studies have shown that the leachate from solid wastes may have a pollution load up to 15–20 times higher than that of domestic wastewater.

Landfilling tends to be predominant as a waste disposal mode because it is regarded as an effective but low-cost method of disposal even for hazardous waste. Even if other methods are more suitable for environmental reasons, the higher capital and (short-term) running costs mean that they cannot compete without government intervention. However, such cost calculations take no account of the longer term. In the long run, landfill of hazardous materials may impose a larger financial burden than other methods because of the high cost of ensuring that the site remains secure for the time it takes for the waste to be rendered harmless.

Disposal to landfills of untreated wastes other than inert material is becoming less acceptable, partly in response to a few well-documented instances where poorly designed and operated landfills are giving rise to pollution problems and partly as a result of greater public awareness of the issues involved.

6.2.1 *Anaerobic digestion*

Anaerobic digestion is the decomposition of biological wastes by microorganisms, usually under wet conditions, in the absence of air (oxygen), to produce a gas comprising mostly methane and carbon dioxide. A digester system (the anaerobic digester) is a device that promotes the decomposition of manure or digestion of the organics in manure to simple organics and gaseous biogas products.

During the anaerobic digestion of an organic material such as biomass, a varied mixture of complex compounds is converted into a very narrow range of simple compounds, mainly methane and carbon dioxide. The anaerobic bacteria are responsible for the biochemical transformation of the biodegradable organic fraction (BOF). The anaerobic digestion of organic material basically consists of hydrolysis, acidogenesis, acetogenesis, and methanogenesis. These transformations are involved in the breakdown of complex polymers, such as cellulose, fats, and proteins, to long- and short-chain fatty acids, and finally to methane, carbon dioxide, and water.

Any organic substance can be subjected to anaerobic digestion as long as there are warm, wet, and airless conditions. For example, *marsh gas* is a product of the anaerobic digestion of vegetation at the bottom of ponds; this gas rises to the surface as bubbles, and the gas is also combustible.

With the aid of human intervention, there are two products of this process: biogas and landfill gas. The chemical processes behind the production of these gases are very complex.

The anaerobic digestion process focuses on hastening the natural process of biomass conversion to a gaseous fuel (biogas). Research has been conducted to ascertain optimal conditions for anaerobic digestion; these include the following: (i) the feedstock, (ii) the nutrients, (iii) the temperature, (iv) the moisture content of the feedstock, (v) the pH of the system, and (vi) the atmospheric conditions. Most of the biomass waste feedstocks (MSW, WWTPS, agricultural, farm, crop, and forestry waste) studied have produced a biogas rich in CH_4. This medium-to-high-Btu gas can, in some instances, be upgraded to a substitute natural gas (SNG). Depending on the feedstock, sulfur may also be produced.

Anaerobic digestion is a multi-stage biological waste treatment process whereby bacteria, in the absence of oxygen, decompose organic matter into carbon dioxide, methane, and water. In this way, the waste sludge is stabilized and the obnoxious odor is removed. The process can, however, be described adequately and simply as occurring in two stages, involving two different types of bacteria. The process occurs in the absence of air; the decomposition in this case is caused not by heat but by bacterial action. In the first stage, the organic material present in the feed sludge is converted into organic acids (also called volatile fatty acids) by acid-forming bacteria. In the second stage, these organic acids serve as the substrate (food) for the strictly anaerobic methane-producing bacteria, which convert the acids into methane and carbon dioxide. The end result of the process is a well-established sludge in which 40–60% of the volatile solids are destroyed. Finally, a combustible gas is produced consisting of 60–75% methane, with the remainder being largely carbon dioxide.

Anaerobic digestion is a complex process which requires strict anaerobic conditions [oxidation reduction potential (ORP) < –200 mV] to proceed and depends on the coordinated activity of a complex microbial association to transform organic material into mostly carbon dioxide and methane (Table 6.2). Despite the successive steps, hydrolysis is generally considered to be rate limiting. The hydrolysis step degrades both insoluble organic material and high molecular-weight compounds (lipids, polysaccharides, proteins, and nucleic acids) into soluble organic substances (amino acids and fatty acids). The components formed during hydrolysis are further split during acidogenesis, the second step. Volatile fatty acids (VFAs) are produced by acidogenic bacteria along with ammonia (NH_3), carbon dioxide (CO_2), hydrogen sulfide (H_2S), and other by-products. The third stage in anaerobic digestion is *acetogenesis*, where the higher organic acids and alcohols produced by acidogenesis are further digested by acetogens to produce mainly acetic acid as well as carbon dioxide and hydrogen. The final stage of methanogenesis produces methane through two groups of methanogenic bacteria: (i) the first group splits acetate into methane and carbon dioxide and (ii) the second group uses hydrogen as the electron donor and carbon dioxide as the acceptor to produce methane. Thus,

$$CH_3CO_2H \rightarrow CH_4 + H_2$$

$$CO_2 + H_2 \rightarrow CH_4 + 2H_2O$$

The digestion process is continuous — fresh feedstock must be added continuously or at pre-determined frequent intervals. The gas formed during digestion is removed continuously. In high-rate digestion, stabilized sludge is displaced from the digester during feeding. In low-rate digestion, supernatant sludge is typically removed as the feed sludge is added, and stabilized sludge is removed at less frequent intervals.

In a typical anaerobic digestion process, biomass is simply allowed to degrade in an anaerobic environment. A number of factors affect the entire process. They include the following: (i) the temperature of the substrate, (ii) the loading rate, (iii) the pH, (iv) the residence time, (v) the concentration of nutrients, and (vi) the presence of any toxic substances.

In terms of temperature, the optimal temperature (where digestion and gasification proceed at the highest rate) is 35°C (95°F). Below 15°C (59°F), the rate is so slow that very little gas is produced. Temperature is also dependent on the bacterial populations: mesophilic or thermophilic. Mesophilic bacteria prefer temperatures ranging 30–40°C. Thermophilic bacteria prefer temperatures ranging 50–60°C (140°F).

The loading rate is the amount of fermentable matter that is fed into the digester per cubic meter of digester capacity. Any change in loading rate affects the balance inside the digester, so it should be kept constant. For a given capacity, if the loading rate is increased, the fermentation period is correspondingly increased. The common range of solid concentrations is 7–9% and digression from this range can cause fermentation to be retarded. In terms of the alkalinity/acidity (pH), the optimal gas formation occurs at a pH of 7–8. If the pH becomes too acidic, gas production could stop altogether.

The retention time is the amount of time fermentable material resides inside the digester. It has been observed that maximum gas production takes place within the first four weeks, and then gradually tapers off. Detention time can be significantly reduced if the temperature is raised or the contents of the digester are agitated, or the supply of nutrients in the digester is augmented. Usually, some type of nutrient source is needed to help stimulate gas production. Bacteria use nitrogen, phosphorus, and potassium for their nutrients. Once these are available, fermentation proceeds very quickly. However, caution is advised that, although rare, toxic substances such as copper can inhibit gas production if found in large quantities.

It is essential that the organic acids formed in the first stage of the waste treatment process are converted into methane at the same rate at which they are formed. If not, they accumulate and ultimately lower the pH, leading to inhibition of the second stage of the digestion process and digester failure. The temperature must be maintained within certain ranges: heating increases the activity of the anaerobic bacteria, reducing the required digestion time. A pH of 7.0–7.5 is recommended to encourage the methane-producing stage. A correctly operating digester will have sufficient buffer capacity (alkalinity) introduced from the breakdown of organic matter. Also, the gas leaving the digester is almost saturated with water vapor, which condenses as the gas stream cools there, causing problems — the problem is more severe when digesters are heated. To mitigate the problem, it is essential to remove as much of the moisture as possible before the gas comes into contact with the gas system devices. For this reason, water traps should be located as close to the digester as possible and all piping should be sloped a minimum of 1% toward the water trap, which should be situated at a low point in the gas line.

Several types of digesters have been developed including the floating drum, the fixed dome, the bag, the plastic tube, the plug flow, and the upflow anaerobic sludge blanket digesters. A *digester* is an airtight vessel or enclosure in which bacteria decompose biomass in water to produce biogas. Both batch and continuous digesters are commercially available. While batch systems may be suitable for some applications, they are labor intensive, and in developed countries are probably not appropriate to the steady production of gas on any significant scale.

Mixing may be used in both batch and continuous digesters to enhance contact between the bacteria and the substrate. It has, however, been argued that while mixing may increase conversion rates, it is not energy efficient. In two-stage systems, only the first stage is mixed. Multistage systems provide optimum conditions for the successive digestive steps, and further development may prove them to be economical for some biomass materials.

In a process of manure and straw mixture digestion, for the first 72 h, the yield of methane was minimal (essentially 0%) and the yield of carbon dioxide was virtually quantitative (ca. 100%) — in this period, digestion occurred as aerobic fermentation to carbon dioxide. The yields of methane and carbon dioxide gases were approximately equal after 250 h, and after 500 h, the digestion reached the stationary phase. The methane content of the biogas was in the range of 73–79% for the runs, with the remainder being principally carbon dioxide. During a 720-h digestion period, approximately 80–85% v/v of the biogas was produced in the first 360–430 h, which is an indication that the digester retention time can be designed to a shorter period (360–430 h) instead of the longer period (720 h).

The final stage in the process is the disposal of any waste. Essentially all the water entering the system will be present in the effluent. Direct recycling of water after its separation from the sludge cannot be practiced, as this will result in the buildup of toxic substances and eventual failure of the system. It is possible that some fraction of the water may be safely recycled, although data on this aspect have not been seen. Treatment of this large volume of water to remove the toxic constituents is costly and is likely to be uneconomical other than in large-scale operations.

Typically, it can be assumed that the obtained water, which is considered to have some nutritional value, can be used for the irrigation of growing crops. However, serious consideration must be given to the effect of trace metals and other toxic materials present in the water and the effects of these constituents on the crops. In addition, subject to the presence of non-toxic constituents (or the lack of toxic constituents) in the sludge, it may also be used for land application and will generally have higher nitrogen content than the original material. On an absolute (feedstock-to-sludge) basis there may be less nitrogen retention due to losses of the nitrogen as ammonia (NH_3), and there may be additional loss on land application if the nitrogen in the sludge is in a more volatile form than the nitrogen in the original feedstock. Nevertheless, the sludge resulting from the digestion of animal manure is generally considered (subject to a critical analysis of the constituents of the sludge) to have an improved fertilizer value over the original feedstock. During digestion, the volatile fatty acid concentration is lower and the pH is higher (Demirbaş and Ozturk, 2004). Nearly all digester plants have ancillary processes to treat and manage all of the by-products. Before storage and use, the gas stream is dried and sweetened by removal of sulfur compounds (Chapter 9). The sludge liquor mixture has to be separated by any one of a variety of methods, the most common being filtration. Excess water is also treated sometimes in sequencing batch reactors (SBR) for discharge into irrigation systems.

Essentially, all organic materials can be digested except for stable woody materials since anaerobic microorganisms are unable to degrade lignin. The biogas which is formed has a high calorific value and is considered to be a renewable energy source. The main disadvantages of this process are as follows: (i) the possible presence of volatile siloxanes in the biogas that can cause serious damage to the energy users' engine and boiler due to the formation of microcrystalline silica; (ii) the increased concentration of heavy metals and various industrial organic chemicals in the residual sludge due to a significant reduction of the organic fraction during digestion, leaving the mineral and non-degradable fraction untouched.

Anaerobic digestion produces a clean and environmentally friendly fuel, although it contains carbon dioxide, water vapor, hydrogen, hydrogen sulfide, and siloxane derivatives (i.e. compounds having a molecular structure based on a chain of alternate silicon and oxygen atoms, –O–Si–O–Si–, especially — as in silicone derivatives — with organic groups attached to the silicon atoms).

As a note of caution, a mixture of biogas (predominantly methane) and air can be explosive. Methane in concentrations of 5–15% v/v in air is explosive, and air should not be allowed to enter the digester or gasholder. All piping and equipment must be sealed properly to prevent gas from escaping to the outside. In addition, all electrical installations must be of explosion-proof type, as the smallest spark could ignite escaped gases.

6.2.2 *Biophotolysis*

The photosynthetic production of gas (e.g. hydrogen) employs microorganisms such as cyanobacteria, which have been genetically modified to produce pure hydrogen rather than the metabolically relevant substances (notably NADPH2). The conversion efficiency from sunlight to hydrogen is very small, usually under 0.1%, indicating the need for very large collection areas.

Current thinking favors ocean locations of the bioreactors. They have to float on the surface (due to rapidly decreasing solar radiation as a function of depth), and they have to be closed entities with a transparent surface (e.g. glass) in order that the hydrogen produced is retained and in order for sunlight to reach the bacteria. Because hydrogen buildup hinders further production, there has to be a continuous removal of the hydrogen produced, by pipelines to, e.g. a shore location, where gas treatment and purification can take place. These requirements make it little likely that equipment cost can be kept so low that the very low efficiency can be tolerated.

A further problem is that if bacteria are modified to produce maximum amount of hydrogen, their own growth and reproduction are quenched. Presumably, there has to be made a compromise between the requirements of the organisms and the amount of hydrogen produced for export, so that replacement of organisms (produced at some central biofactory) does not have to be made at frequent intervals. The implication of this is probably an overall efficiency lower than 0.05%.

6.2.3 *Combustion*

Combustion (incineration) involves burning the feedstock and then using the resulting hot combustion gases to raise steam. The steam is, in turn, used to drive a steam turbine, which drives a generator to produce electricity. Combustion facilities can burn many types of fuel, including wood, agricultural residues, wood pulping liquor, municipal solid waste (MSW), and refuse-derived fuel. Combustion technologies convert biomass fuels into several forms of useful energy for commercial or industrial use: hot air, hot water, steam, and electricity.

The concept of combustion (incineration) of domestic waste and industrial waste to produce energy (heat) and/or gaseous fuels is one of the first suggestions toward the disposal of waste for the generation of energy. With this in mind, waste incineration has been and continues to be practice in many parts of the world. In fact, cocombustion of waste in (large and efficient) coal-fired power plants is an especially attractive option because of the high conversion efficiency of these plants. However, there remain issues that need attention and while the following comments are not intended to be a critique of waste incineration, there are lessons to be learned and changes to be made in this manner of waste removal.

Although the main goal of incineration is volume reduction (with the sterilization of the waste as a significant side effect), it can be a source of energy generally. Thus, the incineration process may also be used to produce steam and electricity. However, as the moisture content of the waste increases, a self-sustaining combustion process is not possible. Nevertheless, domestic waste and industrial waste are feedstocks for combustors. Nevertheless, incineration is recognized as a waste treatment technology for disposal of the organic constituents of the waste. Incineration and other high-temperature waste treatment systems convert the waste into flue gas, particulate matter, ash, and heat, which can in turn be used to generate electricity. The flue gases are cleaned of pollutants before they are dispersed in the atmosphere. Furthermore, combustion with energy recovery is one of the several waste-to-energy technologies such as gasification and anaerobic digestion that can be applied for the production of alternate fuels.

Modern combustors reduce the volume of the original waste by 95–96%, depending on the composition and degree of recovery of materials such as metals from the ash for recycling. While combustion does not replace landfilling, it does reduce the necessary volume significantly. In fact, combustion has particularly strong benefits for the treatment of certain types of waste in niche areas such as clinical waste and hazardous waste where pathogens and toxins can be destroyed by high temperatures.

The combustion process is a controlled process that is used for reducing solid wastes to carbon dioxide, water, flue gases, and ash. The ash residue may require disposal at a mono-fill or sanitary landfill given its toxicity, leaching potential, and contaminants. In addition, atmospheric emissions must be controlled to maintain good air quality. Not all solid waste can be incinerated; wastes such as white goods, industrial process waste, street and drain cleanings and scrap steel cannot be fed to a combustor and require alternative disposal methods. One of the most difficult aspects of waste as feedstock, whether for combustion or gasification, is its heterogeneous nature. Waste composition is influenced by numerous local conditions, such as population structure and local waste separation and recycling regulations. In addition, the composition of local waste is subject to major seasonal or even daily variations. The variability in chemical composition, water and ash content, heating value, and the presence of a number of substances, such as sulfur, chlorides, or metals, affect the performance of thermal conversion processes.

Modern combustion plants reflect significant advancements in addressing the technical and practical difficulties of material handling, combustion control, flue-gas cleanup and conformance with increasingly stringent air pollution control regulations. Energy recovery and production offer a direct economic benefit and lessen demands on processing and disposal requirements. The types, quantities, and distribution of solid waste with fuel potential, particularly urban solid wastes, have received increased attention over the past decade due to rapidly escalating energy costs.

A furnace is the simplest combustion technology — in a furnace, the biomass fuel is combusted in the combustion chamber in which the biomass feedstock is converted into heat energy. As the biomass burns, hot gases are released and these hot gases contain approximately 85% of the potential energy of the fuel. Thus, a biomass-fired boiler is a more adaptable direct combustion technology because a boiler transfers the heat of combustion into steam. Steam can be used for electricity, mechanical energy, or heat. Biomass boilers supply energy at low cost for many industrial and commercial uses.

However, conventional combustion equipment is not designed for burning agricultural residues. Straw and grass contain alkali (potassium and sodium) compounds, which are also present in all annual crops and crop residues and in the annual growth of trees and plants. During combustion, alkali combines with silica, which is also present in agricultural residues. This reaction causes slagging and fouling problems in conventional combustion equipment designed for burning wood at higher temperatures. The volatile alkali constituents lower the fusion temperature of ash, and in conventional combustion equipment having furnace gas exit temperatures above 790°C (1,450°F), combustion of agricultural residue causes the formation of a troublesome slag that can deposit on heat transfer surfaces. Specially designed boilers with lower furnace exit temperatures could reduce slagging and fouling from the combustion of these fuels. Low-temperature gasification is an option for using these fuels for efficient energy production while avoiding the slagging and fouling problems encountered in direct combustion.

The direct-fired gas turbine is another combustion technology for converting biomass into electricity. In this technology, fuel pretreatment reduces biomass to a particle size of less than 2 mm and a moisture content of less than 25%. Then the fuel is burned with compressed air. Cleanup of the combustion gas reduces particulate matter before the gas expands through the turbine stage. The turbine drives a generator to produce electricity.

Cofiring a biomass-type feedstock (as defined in this text) as a secondary fuel in a coal-burning power plant using high-sulfur coal could help reduce the overall sulfur content of the feedstocks, thereby offering a partial mitigation of the emissions of sulfur dioxide and nitrogen oxide. Also, cofiring decreases net carbon dioxide emissions from the power plant if the biomass fuel comes from a sustainable source (remember that this is a theoretical exercise and does not occur in real time). Also, cofiring of such feedstocks may require recombustion of the wood fuel or boiler modifications to maintain boiler efficiency.

6.2.3.1 *Types of combustors*

The combustor (incinerator) is, simply, a furnace for burning refuse, and modern combustors include pollution mitigation equipment such as flue-gas cleaning. There are various types of combustor plant designs: (i) the simple combustor, (ii) the fixed or moving grate combustor, (iii) the rotary-kiln combustor, and (iv) the fluidized-bed combustor.

The *simple combustor* is an older and simpler kind of combustor that is, essentially, a brick-lined cell with a metal *grate* over a lower ash pit, with one opening in the top or side for loading and another opening in the side for removing incombustible solids (clinker). Many small combustors formerly found in apartments have now been replaced by waste compactors.

The *fixed* or *moving grate combustor* is a large fixed hearth combustor with a moving grate. The moving grate enables the movement of waste through the combustion chamber to be optimized to allow a more efficient and complete combustion. These combustors are typically used for the combustion of municipal waste and are thus referred to as MSWCs (municipal solid waste combustors).

In the *fixed* or *moving grate combustor*, the waste is introduced by a *waste crane* through the "throat" at one end of the grate, from where it moves down over the descending grate to the ash pit in the other end. Here the ash is removed through a water lock. Part of the combustion air (primary combustion air) is supplied through the grate from below. This air flow also has the purpose of cooling the grate itself. Cooling is important for the mechanical strength of the grate, and many moving grates are also water-cooled internally. Secondary combustion air is supplied into the boiler at high speed through nozzles over the grate. It facilitates complete combustion of the flue gases by introducing *turbulence* for better mixing and by ensuring a surplus of oxygen.

The combustor must be designed to ensure that the flue gases reach a temperature of at least 850°C (1,560°F) in order to ensure proper breakdown of organic toxins. This includes backup auxiliary burners (often fueled by oil), which are fired into the boiler in case the heating value of the waste becomes too low to reach this temperature alone. The flue gas is then cooled by heat transfer and by heating the steam to typically 400°C (750°F) at a pressure of 550–600 psi for electricity generation in the turbine. At this point, the flue gas has a temperature of around 200°C (390°F) and is passed to the flue-gas cleaning system.

In the fixed or traveling grate combustor, an automatic feeder distributes the fuel onto a grate, where the fuel burns. Combustion air enters from below the grate. In the stationary grate design, ashes fall into a pit for collection. In contrast, a traveling grate system has a moving grate that drops the ash into a hopper. In another configuration, a *pile burner* consists of separate cells and each cell consists of an upper and a lower combustion chamber for use in a step-wise burner configuration. In this type of equipment, the biomass feedstock is combusted and burns on a grate in the lower chamber, thereby releasing volatile gases, which are then combusted in the upper (secondary) combustion chamber. However, pile burners must be shut down periodically to remove the mineral ash.

The rotary kiln *combustor* has a primary chamber and a secondary chamber. The primary chamber consists of an inclined refractory lined cylindrical tube. The movement of the cylinder on its axis facilitates the movement of waste. In the primary chamber, there is conversion of solid fractions to gases through volatilization, destructive distillation and partial combustion reactions. The secondary chamber is necessary to complete gas phase combustion reactions.

The clinker spills out at the end of the cylinder. A tall flue-gas stack, fan, or steam jet supplies the needed draft. Ash drops through the grate, but many particles are carried along with the hot gases. The particles and any combustible gases may be combusted in an *afterburner*. To control air pollution, the combustion product gases are further treated with acid gas scrubbers to remove sulfuric acid and emissions of nitric acid, and then routed through baghouses to remove particulate matter before the gas is released into the atmosphere.

The *fluidized-bed combustor* uses a strong airflow through a sand bed until a point is reached where the sand particles separate to let the air through and mixing and churning occurs, thus a fluidized bed is created and fuel and waste can now be introduced. The sand with the pre-treated waste and/or fuel is kept suspended on pumped air currents and takes on a fluid-like character. The bed is thereby thoroughly mixed and agitated keeping small inert particles and air in a fluid-like state. This allows all of the mass of waste, fuel, and sand to be fully circulated through the furnace.

Fluidized-bed combustors burn biomass fuel in a hot bed of granular materials, such as sand. Injection of air into the bed creates turbulence resembling a boiling liquid. The turbulence distributes and suspends the fuel. This design increases heat transfer and allows for operating temperatures up to 970°C (1700°F), reducing the nitrogen oxide (NO_x) emissions. Fluidized-bed combustors can handle high-ash fuels and agricultural biomass residue.

The heat produced by a combustor can be used to generate steam which may then be used to drive a turbine in order to produce electricity. However, the amount of energy produced is very much dependent on the type and composition of the waste used as the feedstock. Much of the current thinking involves cocombustion of a specified amount of the waste (calculated on the basis of the carbon content of the waste) with, for example, coal to produce a consistent amount of energy. However, the combustion process has a number of outputs such as the following: (i) the mineral ash and (ii) the emission to the atmosphere of flue gas. Before flue-gas cleaning, the flue gases may contain significant amounts of particulate matter, heavy metals, dioxin derivatives, furan derivatives, sulfur dioxide, hydrochloric acid, and polynuclear aromatic hydrocarbon derivatives (carcinogens).

The most publicized concerns related to the combustion of municipal solid wastes involve the fear that it produces significant amounts of emissions of dioxin and furan derivatives, which are considered by many to be serious health hazards. Older generation combustors that were not equipped with modern gas cleaning technologies were indeed significant sources of dioxin emissions. However, modern combustors (due to advancements in emission control designs and stringent regulations) must limit and even mitigate such emissions.

The quantity of pollutants in the flue gas from combustion plants is reduced by several processes. Taking an example from the gas processing industry (Chapter 9), particulate matter is collected by particle filtration, most often electrostatic precipitators and/or baghouse filters (Mokhatab *et al.*, 2006; Speight, 2019). Hydrogen chloride and sulfur dioxide are removed in scrubbers or in a dry desulfurization process by injection of limestone ($CaCO_3$) as a slurry into the flue-gas stream before the particle filtration step. Waste water from scrubbers must subsequently pass through a waste water treatment plant. Nitrogen oxide (NO_x) emissions are either reduced by catalytic reduction with ammonia in a catalytic converter (a selective catalytic reduction process) or by a high-temperature reaction with ammonia in the furnace (a selective non-catalytic reduction process). Heavy metals are often adsorbed on injected active carbon powder (such as activated charcoal), which is collected by particle filtration. As is the case when coal is combusted, combustion of biomass or organic waste also produces fly ash and bottom ash.

By way of explanation, heavy metals are generally defined as metals with relatively high density, atomic weight, or atomic number. Fly ash is a combustion product that is composed of the particulates (fine particles of burned fuel) that are driven out of coal-fired boilers together with flue gases. Fly ash is, thus, the portion of the ash that escapes the chimney or stack. On the other hand, bottom ash is part of the non-combustible residue of combustion in a power plant, boiler, furnace or combustor. In an industrial context, it typically comprises traces of combustibles embedded in forming clinkers and sticking to the hot side walls of a furnace during its operation. The clinker materials fall by themselves into the bottom hopper of a coal-burning furnace and are cooled. The above portion of the ash is also referred to as bottom ash.

The total amount of ash produced by municipal solid waste combustion ranges from 15% to 20% w/w of the waste and fly ash amounts to 0–10% w/w of the total ash. Fly ash, by far, constitutes more of a potential health hazard than the bottom ash because it often contains high concentrations of heavy metals such as lead, cadmium, copper, and zinc as well as small amounts of dioxin derivatives and furan derivatives. Bottom ash seldom contains significant levels of heavy metals.

While fly ash is always regarded as hazardous waste, bottom ash is generally considered to be safe for regular landfills after a certain level of testing defined by the local legislation. Ash, often considered to be hazardous, may generally only be disposed of in landfills which are carefully designed to prevent the pollutants in the ash from leaching into underground aquifers, or after a chemical treatment to reduce the leaching characteristics of the ash.

In spite of the promise shown by the use of combustors for waste disposal and, in some cases, energy production (in the form of fuel gases), the use of combustors for waste combustion is often controversial. On the one hand, the concerns over the health effects of the emissions of furan derivatives and dioxin derivatives have been significantly minimized by the advancements in emission control designs and very stringent new governmental regulations that have resulted in a large reduction in the amount of dioxin derivatives and furan derivatives emissions into the environment.

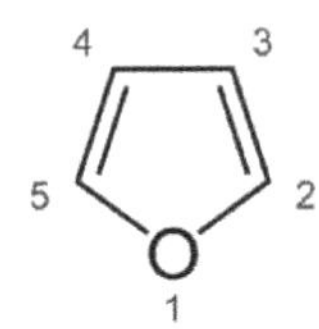

Furan

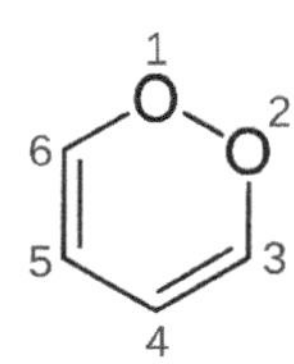

1.2-Dioxin

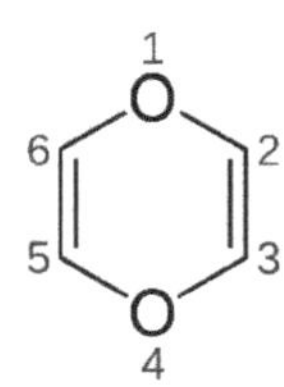

1,4-Dioxin

However, there may be concerns related to the health effects of furan and dioxin emissions into the atmosphere from old combustors. Also, combustors emit varying levels of heavy metals such as (alphabetically) arsenic, cadmium, chromium, lead manganese, mercury, nickel, and vanadium, which can be toxic even at very minute levels. Other advanced alternative technologies are available such as biological treatment combined with anaerobic digestion and gasification.

6.2.3.2 *Combustion plants*

Modern combustion plants generate electricity and heat that can be sold to the regional electric grid and can sell steam to district heating systems or industrial customers. The bottom ash residue remaining after combustion has been shown to be a non-hazardous solid waste that can be safely landfilled or possibly reused. In densely populated areas, finding space for additional landfills is becoming increasingly difficult.

Combustion of municipal solid waste avoids the release of carbon dioxide and methane. Every ton of municipal solid waste incinerated prevents approximately one ton of carbon dioxide equivalents from being released to the atmosphere. Combustion of medical waste and sewage sludge produces an end product ash that is sterile and non-hazardous. On the other hand, the end product ash must still be safely sent for disposal. Many countries exclude the use of combustion as an option for the treatment of their municipal solid wastes because of the potential risk of environmental pollution. In addition, combustors produce soot particles ($PM_{2.5}$, or nanoparticles, i.e. particles with a size of less than 2.5 microns, $< 1 \times 10^{-6}$ meter) in the furnace. Even with modern particle filtering of the flue gases, such as baghouse filters, some of these small particles are emitted into the atmosphere.

In summary, combustion is not a complete method of disposal; its main advantage is that it produces a residue that is substantially reduced in volume and that may be relatively inert. In addition, the efficiency of a combustor is, on occasion, due to excessive throughput or choice of operating parameters that are less than desired. For example, the operating temperature inside the plant should be at the specified temperature (usually of the order of 850°C, 1,560°F) to reduce the possible effects of the most dangerous air pollutants and the production of dioxins. A lower than specified combustion temperature implies incomplete combustion, leaving behind a part of the waste in the ashes and other combustion residues. Furthermore, the exhaust gases may be unnecessarily polluting because of the low combustion temperature and the by-products of such combustion may be sent to an unprotected landfill for disposal.

Combustion plants are usually classified as mass burn systems or refuse-derived fuel (RDF) systems. In the mass burn system, all waste is incinerated; they are usually large facilities, over 200 tons (4,000 lbs) per day. Most mass burn systems burn waste in a single combustion chamber with excess air. The waste is burned on a sloping, moving grate, which helps agitate the municipal solid waste and mixes the waste with combustion air; many different proprietary grate systems exist. Grate design is critical because it must agitate the waste sufficiently to provide complete combustion and minimize ash carryover in the flue gas which contributes to emissions.

In the boiler/furnace portion of the combustor, the waste characteristics of interest are the calorific value, moisture content, proportion of non-combustibles, and other components such as heavy metals, chlorine and sulfur whose presence during combustion will result in the need for flue-gas cleanup. The capacity of the furnace/boiler of the combustor is approximately proportional to the calorific value of the waste. Mass burn produces by-products which have damaging consequences on the environment. Pollution control systems can be used to mitigate such risks. The technologies used to recover energy from municipal solid waste include the following: (i) water-wall combustion and (ii) modular combustion.

Water-wall combustion involves the combustion of unprocessed municipal solid waste (mass burning) or processed municipal solid waste in a furnace with integral boiler tubes. Water-wall units have been built as small as 75–100 tons per day capacity. However, the cost per ton of rated capacity of such units is relatively high. A more common unit size is 250–300 tons per day while water-wall mass fired units have been built as large as 750–1200 tons per day capacity.

The advantages of water-wall combustion are as follows: (i) pre-processing of waste is not required, except for large bulky items; (ii) the ash produced could be used for block manufacture; (iii) auxiliary fuel is not required during normal steady-state operation. On the other hand, the challenges of water-wall combustion are as follows: (i) capital and operating costs are high; (ii) complexity of operation, control and maintenance; (iii) reserve furnace capacity is not available during downtime. Modular combustion is used in smaller plants and is an assembly of prefabricated major components assembled in the field to form a total operation. Modular combustors have been built in individual unit sizes of up to 100 tons per day, combined into plants of just over 400 tons per day.

Modular systems are similar to mass burn systems in that these systems not only combust unprocessed municipal solid waste but also feature two combustion chambers and the municipal solid waste is charged into the system by a hydraulic ram and combustion takes place on a series of stationary hearths. The municipal solid waste is pushed from one hearth to the next by hydraulic rams. Two types of modular systems have been built and operated: starved air (slightly oxygen deficient) and excess air (greater than stoichiometric).

The advantages of modular combustion are low-cost and shorter field construction time given their modular construction. The disadvantages of modular combustion are as follows: waste burnout is not always complete (increased ash quantities and reduced energy recovery); combustion control is less effective (possibility of discharge of trace organics); lower quality of equipment; less redundancy than larger mass-fired water-wall and refuse-derived fuel waste-to-energy plants.

Feedstock preparation plays an important role in any thermal waste conversion approach. Refuse-derived fuel (RDF) is produced when municipal solid waste is processed by mechanical means (picking, separation of inert materials, such as glass, minerals, ferrous and non-ferrous metals, flailing, screening, shredding, and conditioning) to produce a more homogeneous material prior to combustion. Refuse-derived fuel systems utilize pre-treated municipal solid waste. Several types of refuse-derived fuels are generated: coarse, fluff, powder, and densified. They differ in complexity, power requirements, particle size and whether the material is compacted into pellets or briquettes. Refuse-derived fuel can be used as the primary fuel source in a dedicated boiler or cofired with a conventional fossil fuel as natural gas or landfill gas. The heating value of nearly all solid waste fuels is a function of carbon content. Ash content is generally low, but the amount of moisture is highly variable and depends on moisture generation plus the effects of processing, handling, and storage.

Scrap tires can be processed into a waste-derived fuel (WDF) and mixed with other fuels without major alterations to the existing combustion facility. Tire-derived fuels (TDFs) are commonly used as supplemental fuel in cement kilns, which are technologically and environmentally superior for selective waste disposal due to favorable internal kiln conditions, such as flame temperature (1,850°C, 3,360°F), residence time, alkaline environment, high gas turbulence and mixing which ensure complete combustion, and the added benefit of having no bottom ash disposal problem since the residual materials (e.g. steel from steel belted tires) are incorporated into the clinker. The rationale behind the use of tire-derived fuel is as follows: (i) excellent combustion characteristics and high heating value (HHV), (ii) lower cost and good availability, (iii) relatively low sulfur content, especially on a Btu basis, and (iv) a reduction in the environmental burden and health effects of tire stockpiles. Typically, the sulfur content of scrap tires and tire-derived fuels is of the order of 1.5–2.0% w/w. Considering the emission potential of sulfur dioxide per generated energy value of the fuel, the result is closer to that for low sulfur-containing coals.

6.2.4 *Fermentation*

Fermentation is the conversion of organic carbon-containing compounds (and not mineral carbonates) by microorganisms for the production of fuels and chemicals such as alcohols, acids, or energy-rich gases. Traditional fermentation plants producing biogas are in routine use, ranging from farms to large municipal plants. As feedstock they use manure, agricultural residues, urban sewage, and waste from households, and the output gas is typically 64% v/v methane. The biomass conversion process is accomplished by a large number of different agents, from the microbes decomposing and hydrolyzing plant material, over the acidophilic bacteria dissolving the biomass in aquatic solution, to the strictly anaerobic methane bacteria responsible for the gas formation. Operating a biogas plant for a period of some months usually makes the bacterial composition stabilize in a way suitable for obtaining a high conversion efficiency (typically above 60%, the theoretical limit being nearly 100%), and it is important not to vary the feedstock compositions abruptly if optimal operation needs to be maintained. Operating temperatures for the bacterial processes are only slightly above ambient temperatures, e.g. in the mesophilic region around 30°C (86°F).

One method is to select bacteria that produce hydrogen directly. Candidates would include *Clostridium* and *Rhodobacter* species. The best reactor operating temperatures are often in the thermophilic interval or slightly above (50–80°C, 120–175°F). Typical yield is 2 mol of hydrogen per mol of glucose, corresponding to 17% conversion efficiency. The theoretical maximum efficiency is around 35%, but there are also acetic or butanoic acids formed, which could be used to produce methane and thus additional energy, although not necessarily additional hydrogen. Operation of this type of gas-producing plant would require pure feedstock biomass (here sugar) because of the specific bacteria needed for hydrogen production and because contamination can cause decreased yields. Even the hydrogen produced has this negative effect and must therefore be removed continually.

In the *separate hydrolysis and fermentation* (SHF) process, the hydrolysis is carried out in one vessel and the hydrolysate is then fermented in a second reactor. Approximately 73% of the cellulose is converted into ethanol in 48 h, while the remainder of the cellulose, hemicelluloses, and lignin are burned. The most important parameters are the hydrolysis section yield, the product quality, and the required enzyme loading, which are all interrelated. Yields are typically higher in more dilute systems where the inhibition of enzymes by glucose and cellobiose is minimized. Increasing the amount of enzyme loading can help overcome inhibition and increase the yield and concentration. Increased reaction times also lead to higher yields and concentrations. Cellulase enzymes from different organisms can result in markedly different performances.

In the *simultaneous saccharification and fermentation* (SSF) process, both hydrolysis and fermentation are carried out in the same vessel. In this process, yeast ferments the glucose to ethanol as soon as the glucose is produced, preventing the sugars from accumulating and causing the end product inhibition.

6.2.5 *Gasification*

Gasification is a high-temperature process that produces gas, which can then be used in an internal combustion engine or a fuel cell. Briefly, gasification was used as long ago as the early 1800s. The process was rather crude and the fuel was most often coal. The gas product (Speight, 2008) was used for heating and lighting. In the current context, the gasification process is amenable to a variety of biomass feedstocks such as waste rice hulls, wood waste, grass, and the dedicated energy crops. Gasification is a clean process with few air emissions and, when crops are used as the feedstock, little or no ash. The process produces a fuel gas from crops by heating them under carefully controlled temperature, pressure, and atmospheric conditions. A key to gasification is using less air or oxygen than is usually found in a combustor. The product (biogas or fuel gas), similar to natural gas, can be burned in high-efficiency gas turbines.

The gasifier is the heart of the gasification process, no matter what the feedstock is (Speight, 2013, 2014a, 2014b; Luque and Speight, 2015). Gasifiers are designed to process the feedstock in a variety of ways consistent with the type of fuel, the end use of the gas, the size of the process, and the source of oxygen. The oxygen may be introduced as a pure gas or may come from air or steam. Some gasifiers operate under pressure; others do not. In principle, numerous process configurations exist for gasification products to fuels, depending on the feedstock, gasifier type, the gas cleaning process, the product fuel, and whether electricity is cogenerated using part of the synthesis gas output from the gasifier.

In *single-stage* gasifiers, conversion takes place in a single reactor using steam, air, or oxygen. The common single-stage technologies are fixed-bed, fluidized-bed and entrained flow reactors. In fixed-bed gasifiers, the fuel is fed from the top and the gasification medium is injected at the bottom. The fuel moves slowly downward through the reactor with the gas moving upward in a countercurrent direction, whereas in cocurrent designs, the fuel and the gasification agent move in the same direction and the fuel must pass successively through the drying, pyrolysis, oxidation and reduction zones. The advantages of countercurrent designs are fewer restrictions on fuel moisture and particle size, no special fuel preparation required, and a wide range of fuels can be used. By comparison, cocurrent gasifiers produce a better quality (higher heat content) gas but place strict requirements on fuel properties. Fluid-bed gasifiers allow for more efficient gasification due to the elimination of hotspots in the reactor. They are suitable for a wide range of feedstocks and can be scaled up to relatively large plants. They are more expensive to build and need better gas cleaning due to high particulate content in the product gas. Fluidized beds have no defined reaction zones and the conversion of fuel and secondary pyrolytic reactions take place in the same volume. Tar conversion can be supported by introducing catalytically active-bed materials.

The bed temperature is determined by the ash softening point and is usually in the range 815–1,040°C (1,500–2,100°F). Entrained flow gasifiers require pulverized fuels and the fuels are introduced with the steam/oxygen feed and converted into gaseous products in a few seconds. Such gasifiers can be operated at lower temperatures (maintain ash as a dry solid) or at temperatures above the ash fusion temperature in a slagging mode (ash removed as a liquid). At higher temperatures, it is possible to produce a product gas, which is virtually free from oil and tar. Although entrained flow gasifiers seem to be attractive for the production of tar-free synthesis gas for petrochemical applications, challenges arise in achieving satisfactory size reduction (milling and grinding) of the waste. Single-stage gasifiers can be used for biomass and waste gasification if the requirements placed on the product gas quality are low: direct cocombustion of hot raw gas or used as a fuel gas.

Various *multi-stage* processes that are under development or in operation use the fuel conversion steps of drying, devolatilization, gasification and combustion zones to enhance process efficiency and product gas quality by influencing and optimizing the operating parameters. These concepts are categorized as single line (one mainstream of mass, which passes through reactors arranged in series) or "double line" (the mass mainstream is divided into at least two partial streams, which pass through parallel-arranged reactors). In the single-line process, fuel is dried and pyrolyzed in the first stage in an indirectly heated pyrolysis unit.

The pyrolysis products undergo partial oxidation in the narrow zone between the pyrolysis zone and the char gasification zone. The product gas passes over the hot char bed and undergoes substantial tar cracking, resulting in a higher quality product gas. In the "double-line" process, combustion and gasification reactors are separate and are only connected by heat transfer; a pyrolysis stage is used to split the fuel into char and gas. To provide the heat necessary for operation, the char or a part of the pyrolysis gas must be oxidized outside the pyrolysis reactor. In the *molten salt batch reactor*, the feedstock is fed along with steam or oxygen in the molten bath of salt or metal at 1,000–1400°C (1830–2,550°F). Residual products such as ash and sulfur are removed as slag.

The simplest type of gasifier is the *fixed-bed countercurrent gasifier*. The biomass is fed at the top of the reactor and moves downward as a result of the conversion of the biomass and the removal of ashes. The air intake is at the bottom and the gas leaves at the top. The biomass moves in a direction countercurrent to the gas flow and passes through the drying zone, the distillation zone, the reduction zone and the oxidation zone. The major advantages of this type of gasifier are as follows: (i) the simplicity of the unit, (ii) the high charcoal burnout, and (iii) the internal heat exchange leading to low gas exit temperatures and high gasification efficiency. In this way, feedstocks with a high moisture content (up to 50% w/w) can be used. However, major disadvantages of the process are as follows: (i) high yield of tar and (ii) high yields of pyrolysis products because the pyrolysis gas is not taken into the oxidation zone. This may be of minor importance if the gas is used for direct heat applications, in which the tar is burned. In case the gas is used for engines, gas cleaning is required, resulting in problems with tar-containing condensates.

In the conventional *downdraft gasifier* (sometimes called the *coflow* gasifier), biomass is fed at the top of the reactor and air intake is at the top or from the sides. The gas leaves at the bottom of the reactor, so the fuel and the gas move in the same direction. The pyrolysis gases are led through the oxidation zone (with high temperatures) and are more or less burned or cracked. Therefore, the producer gas has low tar content and is suitable for engine applications. In practice, however, a tar-free gas is seldom if ever achieved over the whole operating range of the equipment.

Because of the lower level of organic components in the condensate, downdraft gasifiers suffer less from environmental objections than updraft gasifiers.

Successful operation of a downdraft gasifier requires drying the biomass fuel to a moisture content of less than 20% w/w. The key advantage of the downdraft design is the very low tar content of the producer gas; however, there are also disadvantages such as the following: (i) the high amounts of ash and dust particles in the gas, (ii) the inability to operate on a number of unprocessed fuels, often pelletization or briquetting of the biomass is necessary, (iii) the outlet gas has a high temperature leading to a lower gasification efficiency, and (iv) the moisture content of the biomass must be less than 25% w/w.

A more recent development is the *open core* gasifier design for the gasification of small-sized biomass with high ash content. The producer gas is not tar-free; it contains approximately 0.05 kg tar per kg gas. In the open core gasifier the air is sucked over the whole cross-section from the top of the bed. This facilitates better oxygen distribution since the oxygen will be consumed over the whole cross-section, so that the solid-bed temperature will not reach the local extremes (hotspots) observed in the oxidation zone of conventional gasifiers due to poor heat transfer. Moreover, the air nozzles in conventional gasifiers generate caves and create obstacles that may obstruct solid flow, especially for solids of low bulk like rice husk. On the other hand, the entry of air at the top of the bed creates a downward flow of the pyrolysis gases, and transports the tar products to the combustion zone. Thus, flow problems due to the cracking of rice husk caused by backmixing of tar are avoided.

When applied to biomass, a key advantage of gasification is that it can convert all of the organic matter into gases and then liquids. In particular, lignin (a component of lignocellulose resistant to enzyme attack) is readily converted into gaseous products and made available as a fuel feedstock.

Synthesis gas can be combusted to produce process heat and steam or can be used in gas turbines to produce electricity. It can also be converted into a variety of fuels, such as hydrogen, methanol or dimethyl ether (DME), as well as synthetic diesel and gasoline (via the Fischer–Tropsch process). Gasification combined with the Fischer–Tropsch process has the potential to produce more fuel per ton of biomass.

Agricultural, industrial and urban waste holds considerable promise as a feedstock for gasification because it contains relatively more lignin, which biological processes cannot convert. Unusable or low-value wastes such as furniture-making wood waste, beechwood, nutshells, olive husks, grape residues, bajra stalks, sweet sorghum stalks, bagasse, sugarcane leaves, sawdust, wood chips, corn cobs, nut shells, and rice hulls can be used as feedstock to produce high calorific-value gas. Such waste is abundant in most developing countries and can be harnessed for the production of fuels and petrochemical intermediates. The knowledge of the potential of the waste for gasification and subsequent fuel production is essential for reducing pressure on traditional energy sources.

Gasification technology permits the production of synthesis gas for fuel and chemical manufacture and its application is mainly limited by gas quality (cleanliness). The research focus has been on reactor design, optimization of operating parameters and addition of catalytically active materials to improve the gas quality. Downstream clean-up of hot gas and catalytic tar conversion measures have been incorporated into modern technological designs. Gasification processes can be classified as follows: (i) the Btu content of the product gas such as high-Btu gas, medium-Btu gas, and low-Btu gas, (ii) the type of reactor hardware configuration, such as single-stage design, multi stage design, and molten salt batch reactor, as well as (iii) whether or not the reactor is pressurized.

The heating value of the product gas is a function of the chemical composition, volatile constituents, and moisture content of the waste feedstock, and is carefully controlled given the application of the gas. Low-Btu gas comprises a mixture of carbon monoxide, hydrogen, and some other gases, which limits the applicability of the gas to many chemical syntheses. Medium-Btu gas consists of a mixture of methane (CH_4), carbon monoxide (CO), hydrogen (H_2) and various other gases and is primarily used in the synthesis of methanol, higher molecular-weight hydrocarbon derivatives by means of the Fischer–Tropsch synthesis, and a variety of other chemicals. It can also be used as a fuel to generate steam or drive a gas turbine. The hydrogen-to-carbon monoxide ratio in medium-Btu gas varies from 2:3 (CO-rich) to more than 3:1 (H_2-rich). The increased heating value is attributed to the higher concentrations of methane and hydrogen and the low concentration of carbon dioxide and the absence of nitrogen. High-Btu gas consists predominantly of methane (>95% v/v) and it is compatible with natural gas and can be used as a synthetic or substitute natural gas (SNG) (Lee *et al.*, 2007; Speight, 2014a).

However, before biomass can be gasified it must be pre-treated to meet the processing constraints of the gasifier. This typically involves size reduction and drying to keep moisture contents below specific levels. Thereafter, biomass gasification involves heating biomass in the presence of low levels of oxygen (i.e. less than required for complete combustion into carbon dioxide and water). Above certain temperatures the biomass will break down into a gas stream and a solid residue. The composition of the gas stream is influenced by the operating parameters of the gasifier, with some gasification processes more suited than others to producing a gas for methanol production. In particular, simple gasification with air creates a synthesis gas stream that is diluted with large quantities of nitrogen. This nitrogen is detrimental to subsequent processing to methanol and so techniques using indirect gasification or an oxygen feed are preferred. For large-scale gasification, pressurized systems are considered to be more economical than atmospheric systems. Once the economically optimum synthesis gas is available the methanol synthesis takes place. This typically uses a copper–zinc catalyst at temperatures of 200–280°C and pressures of 750–1,500 psi.

However, most crop or agricultural residues are not found throughout the year but are available only at the time of harvest. This makes collection easy, but creates storage problems if the residues have to be saved for use during other months of the year, especially due to its low bulk density. The amount available depends on the harvesting time, storage-related characteristics, and the storage facility.

In the gasifier, biomass is converted into a gaseous mixture of hydrogen, carbon monoxide, carbon dioxide, and other compounds by applying heat under pressure in the presence of steam and a controlled amount of oxygen. The biomass produces synthesis gas (Chapter 8).

$$C_6H_{12}O_6 + O_2 + H_2O \rightarrow CO + CO_2 + H_2 + \text{other products}$$

The above reaction uses glucose as a surrogate for cellulose. Biomass has highly variable composition and complexity, with cellulose as one major component.

Willow biomass crops have been shown to be a good fuel for farm-based power production using advanced gasification technology (Pian *et al.*, 2006). The fuel gas can be used for generating electricity, using microturbines modified to operate on low-BTU gas, and for other farm energy needs. Willow biomass was found to make a fuel for ash-rejection gasifiers with a predicted net gasification efficiency of the order of 85%. Analysis showed that developing a method to cogasify willow with various amounts of low-cost wastes, such as dairy farm animal waste, can be an excellent way to reduce the fuel cost, to increase the overall fuel availability, and to help work around problems resulting from seasonal availability of bioenergy crops. Cogasification of dairy farm wastes along with willow offers an economical way to dispose of the wastes and manage nutrient flows on a dairy farm.

Agricultural residues can be divided into two groups: crop residues and agro-industrial residues. Crop residues are plant materials left behind in the farm after removal of the main crop produce. The remaining materials could be of different sizes, shapes, forms, and densities like straw, stalks, sticks, leaves, haulms, fibrous materials, roots, branches, and twigs. The agro-industrial residues are by-products of the post-harvest processes of crops such as cleaning, threshing, sieving, and crushing. These could be in the form of husk, dust, and straw. Furthermore, the quantity of agricultural residues produced differs from crop to crop and is affected by soil type and irrigation conditions. Production of agricultural residues is directly related to the corresponding crop production and the ratio of the main crop produce to the residues, which varies from crop to crop and, at times, with the variety of the seeds in one crop itself. Thus, for known amounts of crop production, it may be possible to estimate the amounts of agricultural residues produced using the residue-to-crop ratio.

The gasification process could play a significant role in meeting the goal of greenhouse gas mitigation insofar as it is possible that both in the transition phase to a hydrogen economy and in the steady state, a significant fraction of hydrogen might be derived from domestically abundant crops. In addition, the cofiring applications of crops (and other biomass) with coal can provide up to 15% of the total energy input of the fuel mixture. Such concepts address greenhouse gas mitigation by cofiring biomass with coal to offset the losses of carbon dioxide to the atmosphere that are inherent in coal combustion processes (even with the best-engineered capture and storage of carbon). Since growth of biomass fixes atmospheric carbon, its combustion leads to no net addition of atmospheric carbon dioxide even if vented. Thus, cofiring of crops or crop residues (or other biomass) with coal is an efficient coal gasification process, affording the opportunity for the capture and storage of carbon dioxide, and could lead to a net reduction of atmospheric carbon dioxide. Being cheaper, although less plentiful, biomass residues could supplant crops as a gasifier feedstock leading to less significant impact on the environment than farming bioenergy crops.

Finally, environmental concerns include the disposal of associated tars and ashes, particularly for the fluidized-bed reactors, where these substances must be separated from the flue-gas stream (in contrast to pyrolysis plants, where most tar and ash deposits at the bottom of the reactor). Concerns over biomass transportation are similar to those mentioned above for fermentation, and a positive fertilizer effect can also in many cases be derived from the gasification residues.

Biomass ash has also the potential to be used as a clarifying agent in water treatment, as a wastewater adsorbent, as a liquid waste adsorbent, as a hazardous waste solidification agent, as a lightweight fill for roadways, parking areas, and structures, as asphalt mineral filler, or as a mine spoil amendment.

6.2.6 *Pyrolysis*

Pyrolysis is the thermal decomposition of (in the current context) organic materials at elevated temperatures under an inert atmosphere. The process involves a change in the chemical composition of the material to be pyrolyzed and is irreversible.

Pyrolysis is a thermochemical decomposition process involving the controlled burning or heating of a material in the absence of oxygen, the product being synthesis gas, liquid and char. In addition to liquid fuel or oil, pyrolysis also generates a medium-BTU gas that can be used as a fuel gas after purification. All the pyrolysis products have the potential of being fuels or intermediaries for producing other valuable products for the petrochemical industry.

Generally, pyrolysis is a medium-temperature method which produces gas, oil, and char from crops, which can then be further processed into useful fuels or feedstock (Boateng *et al.*, 2007). Pyrolysis is often considered to be the gasification of biomass in the absence of oxygen but the chemistry of each process may differ significantly. In general, biomass does not gasify as easily as coal, and it produces other hydrocarbon compounds in the gas mixture exiting the gasifier; this is especially true when no oxygen is used. The products of biomass pyrolysis include gases such as methane, hydrogen, carbon monoxide, carbon dioxide, bio-oil, and char. As a result, typically an extra step must be taken to reform these products with a catalyst to yield a clean synthesis gas mixture of hydrogen, carbon monoxide, and carbon dioxide. Thus, when organic matter is heated at increasing temperatures in open containers,

Table 6.4: Examples of the stages that occur during pyrolysis.

Temperature	Event
<100°C	Volatiles, including some water, evaporate. Heat-sensitive substances may partially decompose.
100°C	Any remaining water that is absorbed in the material is evolved. Water trapped in the crystal structure of hydrates may be evolved at somewhat higher temperatures. Some solid substances, like fats, sugars, and waxes, may melt.
100–500°C	Organic molecules break down. Most sugar derivatives start decomposing at 160–180°C. Cellulose decomposes at approximately 350°C. Lignin starts decomposing at approximately 350°C and continues releasing volatile products up to 500°C. The decomposition products usually include water, carbon monoxide and/or carbon dioxide as well as a large number of organic compounds. The non-volatile residues typically become richer in carbon and form large disordered molecules, with colors ranging between brown and black.
200–300°C	If oxygen has not been excluded, the carbonaceous residue may start to burn in a highly exothermic reaction releasing carbon dioxide and/or monoxide. At this stage, some of the nitrogen still remaining in the residue may be oxidized into nitrogen oxides (such as NO_2 and N_2O). Sulfur and other elements (such as chlorine and arsenic) may be oxidized and volatilized.

a variety of processes generally occur in successive or overlapping stages (Table 6.4).

Depending on the operating conditions (temperature, heating rate, particle size and solid residence time), pyrolysis can be divided into three subclasses: conventional, thermal or catalytic, and fast or flash. For example, to maximize bio-oil production, fast/flash pyrolysis is used, heating the biomass to approximately 500°C (930°F) for less than ten seconds. In the pyrolysis of plastics, the macromolecular structures are broken down into smaller molecules (oligomers and monomers); further degradation depends on temperature, residence time, presence of catalysts and other process conditions. Pyrolysis aims at converting waste into other valuable chemicals, which can then be used as feedstocks for downstream industrial processes or as transportation fuels. The approaches, molten salts, cracking (catalytic, hydrocracking, and thermal), and liquefaction are also discussed.

Molten salts such as lithium chloride–potassium chloride with 10% w/w cuprous chloride (LiCl–KCl–10% CuCl) at 520°C (970°F) are used to pyrolyze polymeric waste to yield a fuel oil. At a process temperature of 420°C (790°F) the gaseous fraction is minimized; this enables a larger liquid and solid fraction to be recovered. The recovered fractions consist of low-boiling oils, aromatic derivatives, paraffin waxes, and monomers (Lee *et al.*, 2007).

In general, pyrolysis of organic substances produces volatile products and leaves a solid residue enriched in carbon (char). As an example, pyrolysis is one of the processes involved in the reduction of char from wood. The pyrolysis process is used frequently in the chemical industry, for example, to produce ethylene, many forms of carbon, and other chemicals from crude oil, coal, and wood, and to produce coke from coal. In the current context, other applications of pyrolysis would convert biomass into synthesis gas (Chapter 8) and waste plastics into usable oil and char.

Fast pyrolysis is a thermal decomposition process that occurs at moderate temperatures with a high heat transfer rate to the biomass particles and a short hot vapor residence time in the reaction zone. Several reactor configurations have been shown to assure this condition and to achieve yields of liquid product as high as 75% based on the starting dry biomass weight. Examples include bubbling fluid beds, circulating and transported beds, cyclonic reactors, and ablative reactors.

Briefly, ablative pyrolysis is conceptually very different from other fast pyrolysis methods. Most pyrolysis processes depend on having a high heat-transfer rate, which determines the rate of the reaction, and therefore, these systems employ small biomass particles. Ablative pyrolysis depends on heat transfer from the hot reactor wall to melt the wood particles contacting the wall under pressure. Unlike other pyrolysis processes, because the reaction rates are not limited by heat transfer through the biomass particle, large biomass particles up to lumber size may also be processed by using an ablative pyrolysis reactor.

Fast pyrolysis of biomass produces a liquid product (referred to as pyrolysis oil or bio-oil) that can be readily stored and transported. Pyrolysis oil is a renewable liquid fuel and can also be used for the production of chemicals. Fast pyrolysis has now achieved commercial success in the production of chemicals and is being actively developed for producing liquid fuels. Pyrolysis oil has been successfully tested in engines, turbines, and boilers, and has been upgraded to high-quality hydrocarbon fuels. In the 1990s, several fast pyrolysis technologies reached near-commercial status and the yields and properties of the generated liquid products depend on the feedstock, the process type and conditions, and the product collection efficiency.

The direct hydrothermal liquefaction process involves converting biomass into an oily liquid by contacting the biomass with water at elevated temperatures (300–350°C, 570–660°F) with sufficient pressure to maintain the water primarily in the liquid phase for residence times up to 30 minutes. Alkali may be added to promote organic conversion. The primary product is an organic liquid with reduced oxygen content (approximately 10%) and the primary by-product is water-containing soluble organic compounds.

The importance of the provisions for the supply of feedstocks such as crops and other biomass is often underestimated since it is assumed that the supplies are inexhaustible. While this may hold true over the long term, short-term supply of feedstocks can be as much risky as any venture.

Hydrocracking of polymer waste involves reaction with hydrogen over a catalyst in a stirred batch autoclave at moderate temperatures and pressures. Typical plastic waste feedstocks such as polyethylene (PE), polyethylene terephthalate (PET), polystyrene, polyvinyl chloride (PVC), and mixed polymers are reacted in the presence of a catalyst (platinum, nickel, molybdenum, or iron on acid solids, such as alumina, zeolites, and sulfonated zirconia) at 150–400°C (300–750°F)and hydrogen at 3–10 MPa to produce a high-quality gasoline.

Catalytic cracking involves the use of a suitable catalyst to carry out the cracking reaction. The presence of a catalyst lowers the reaction temperature and time. Catalytic degradation yields a narrow product distribution with a peak in lower molecular-weight hydrocarbon derivatives and occurs at lower temperatures. Using solid acid catalysts, HZSM-5, HY, rare-earth metal-exchanged Y-type (REY) zeolite and silica–alumina for polymeric waste yields 30% gasoline and 20% high-boiling oil. The differences among catalysts arise in the production of coke and gas fractions. Formation of coke is not desirable as it results in catalytic deactivation by fouling and plugging of the pore paths. Catalytic cracking, owing to its lower operating temperature, seems to be more technologically sound than thermal pyrolysis; however, it is costlier. The addition of a catalyst enhances the conversion and fuel quality compared to pure thermal pyrolysis.

There are numerous mechanisms proposed for catalytic degradation. These include the carbonium ion mechanism and the free radical mechanism. In the carbonium ion (R^+) mechanism, there are four steps: initiation, depropagation, isomerization, and aromatization. In the initiation step, the carbonium ion is generated through proton addition to an olefinic linkage, beta scission, or through the abstraction of a hydride ion by low molecular-weight carbonium ions.

In the depropagation step, the molecular weight of the main polymer chain can be successively reduced through beta scission. In the isomerization step, the carbonium ion intermediates can undergo a hydrogen atom shift or a carbon atom shift, leading to rearrangement of the molecules. Some of the carbonium ion intermediates can also undergo cyclization reactions, which lead to the formation of aromatics (aromatization step).

In the free radical mechanism, there are three known radical mechanism steps: initiation, propagation, and termination. Normally, random bond breaking generates a radical in the initiation step after which the radical is propagated via a number of reactions such as beta scission, radical abstraction, and radical decomposition. Finally, recombination of two radicals leads to the termination step. The sequence can be represented more fully as follows:

Initiation:

$R_1R_2 \rightarrow R_1\cdot + R_2\cdot$

Example: $RCH_2CH{=}CHCH_2R + HX \rightarrow RCH_2CH_2{}^+CHCH_2R + X$

Propagation:

$R_1 + R_3 \rightarrow R_3$

$R_1 \rightarrow C_x + R_4$

Example: $RCH_2CH_2{}^+CHCH_2CH_2CH_2R \rightarrow RCH_2CH_2CH{=}CH_2 + {}^+CH_2CH_2R$

$RCH_2CH_2CH_2R + R^+ \rightarrow RCH_2{}^+CHCH_2R + RH$

Termination:

$R_4\cdot + R_5\cdot \rightarrow R_4R_5$

Thermal cracking (*pyrolysis*) involves the degradation of polymeric materials by heating in the absence of oxygen and is usually conducted between 350°C and 900°C (650°F and 1,650°F). The products formed are a carbonized char and a volatile fraction that may be separated into a condensable hydrocarbon oil consisting of paraffins, iso-paraffins, olefins, naphthenes, aromatics, and a non-condensable high calorific-value gas.

The thermal pyrolysis of polymeric waste at temperatures higher than 480°C (895°F) yields a gas–liquid hydrocarbon. At 650–760°C (1,200–1,400°F), more gaseous product is generated; conversely, at lower temperatures, up to 85% of the product is liquid hydrocarbon derivatives. The proportion of each fraction and their precise composition depend primarily on the properties of waste plastic material(s) and the process parameters (temperature, pressure, residence time of the feedstock in the thermal zone). Both the gaseous and the liquid forms of the polymeric waste can be utilized as a feedstock for petrochemical facilities (Speight, 2019b). Generally, thermal cracking results in liquids with low octane value and high residue content at moderate temperatures; this is an inefficient process for producing gasoline ranges. The gaseous products obtained by thermal pyrolysis are not suitable for use as fuel products and require further refining to be upgraded to usable fuel products. A few researchers have sought to improve thermal pyrolysis of waste plastics without the use of catalysts. However, these changes either yielded insignificant improvements or added another level of complexity and costs to the system.

Typically, discarded tires can be reduced in size by grinding, chipping, and pelletizing and passed through a classifier to remove the steel belting after which the chips are pyrolyzed for 1 h at a temperature range of 300–500°C (570–930°F) and then heated for 2 h in a closed retort. The products are then distilled (purification and analysis); this yields oil, gas and char. However, discarded tires can also be debeaded, shredded to 25 mm, and ground to 24 mesh as a feed preparation step for occidental flash pyrolysis, which involves flash pyrolysis and product collection. The pyrolytic reaction occurs without the introduction of hydrogen or the use of a catalyst. This yields a gaseous stream that is passed to a quench tower from which fuel oil and gas (recycled to char fluidized and pyrolysis reactor as a supplemental fuel) and carbon black (35% w/w) are produced.

The terms *hydrothermal liquefaction* and *pyrolysis* are often confused since both are thermochemical processes where organic compounds are converted into liquid products. However, in hydrothermal liquefaction, macromolecules are decomposed into unstable, reactive low molecular-weight products in the presence of a catalyst. These fragments repolymerize into high molecular-weight oily compounds.

The hydrothermal liquefaction process involves solvolysis, decarboxylation, hydrogenolysis, and hydrogenation, leading to challenging mechanistic transformation studies. During solvolysis, the solvent reacts with the molecule, helping to begin the degradation of the biomass. Depolymerization of hemicelluloses, cellulose, and lignin further degrades the biomass to smaller molecules. In decarboxylation, loss of carbon dioxide can lead to rearrangement of the molecule. Molecules can also be further transformed via hydrogenolysis (water molecules split into a hydrogen ion, H^+, and a hydroxyl ion, OH^-, and react with feedstock molecules) and hydrogenation (hydrogen is added to the molecule).

6.2.7 *Other processes*

There are several other processes by which organic waste material can be converted into potentially useful products. The most common composting techniques in the current context (arranged alphabetically rather than by preference) are as follows: (i) acid hydrolysis, (ii) briquetting, (iii) enzymatic hydrolysis, and (iv) transesterification.

6.2.7.1 *Acid hydrolysis*

Hydrolysis is a chemical reaction in which a molecule of water ruptures one or more chemical bonds. The term is used broadly for substitution, elimination, and fragmentation reactions in which water is the nucleophile:

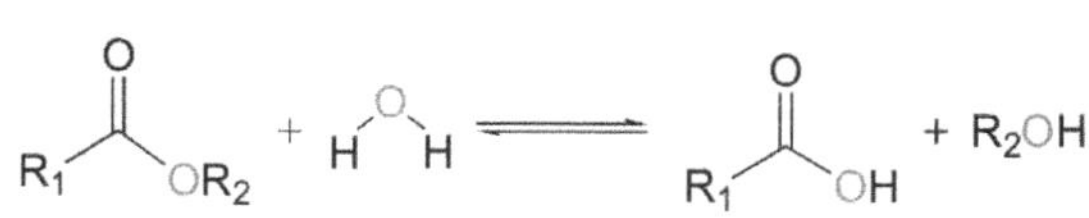

Acid hydrolysis is the means by which cellulosic material can be converted into lower molecular-weight products and thence to fuels. Enormous amounts of inexpensive and renewable lignocellulose wastes are generated each year from forestry and agricultural production. These include fruit processing residues, dairy industry wastes, corn and sugar by-products, and paper industry wastes. Lignocellulosic biomass can be converted into biofuels such as ethanol and hydrogen via simple sugars. Lignocellulose is a composite of cellulose fibers embedded in a cross-linked lignin–hemicellulose matrix. It gives structure and strength to plants. Although abundant, lignocellulosic waste is difficult to convert into fermentable sugars because of its complex chemical structure: its three major components (cellulose (crystalline and amorphous), hemicellulose and lignin) must be processed separately (Lee *et al.*, 2007).

Cellulose consists of linear, highly ordered chains of glucose and is one of the most abundant biopolymers on the planet. When produced by plants, it has both highly amorphous regions containing large voids and other irregularities as well as tightly packed crystalline regions. It accumulates in the environment because it is resistant to most forms of degradation. Hemicellulose is a complex polymer of a variety of sugars. This mix of sugars mainly consists of six-carbon and five-carbon sugars. Lignin is a polyphenolic polymer that fills the spaces in the cell wall between cellulose, hemicellulose, and pectin components. It is covalently linked to hemicellulose and crosslinks different plant polysaccharides, conferring mechanical strength to the cell wall, and by extension, the plant as a whole. The content of cellulose, hemicellulose and lignin in common agricultural residues depends on the source and origin of the feedstocks (Table 6.1). However, because lignocellulosic biomass is very complex, it is difficult to use it as a feedstock for biofuel. A variety of physical, chemical, and enzymatic processes have been developed to fractionate lignocellulose into the major plant components, such as hemicellulose, cellulose and lignin.

Ethanol is produced from lignocellulose via pre-treatment, hydrolysis, fermentation, and distillation. The goal of pre-treatment is to increase the surface area of lignocellulosic material, making the polysaccharides more susceptible to hydrolysis, by separating the xylose and lignin from the crystalline cellulose. Pre-treatment must also avoid the degradation or loss of precious carbohydrates and avoid the formation of by-products that can inhibit subsequent steps. However, pre-treatment often produces biological inhibitors, which affect fermentation. A large variety of pre-treatment processes have been developed. Common pre-treatments are steam-explosion, acid treatment, biological methods, and comminution. These methods can be used singly or in combination.

Steam explosion involves saturation of the pores of plant materials with steam followed by rapid decompression. The explosive expansion of steam reduces the plant material to separate fibers, increasing accessibility of polysaccharides to subsequent hydrolysis. Ammonia fiber explosion (AFEX) is similar to steam explosion except that liquid ammonia is used. It is very effective on agricultural residues but has not been successful in pre-treating woody biomass. *Biological pre-treatments* employ microorganisms that produce lignin-degrading enzymes (ligninase). *Comminution* is an integral part of pre-treatment and uses a hammer mill to produce particle sizes that can pass through 3 mm screen openings.

The mechanisms by which pre-treatments improve the digestibility of lignocellulose are not well understood. Pre-treatment effectiveness has been correlated with removal of hemicellulose and lignin (lignin solubilization is beneficial for subsequent hydrolysis but may also produce derivatives that inhibit enzyme activity). Some pre-treatments reduce the crystallinity of cellulose, which improves reactivity, but this does not appear to be the key for many successful pre-treatments.

Hemicellulose is readily hydrolyzed to pentoses (five-carbon sugars) but pentoses are difficult to ferment. The cellulose hydrolyzes to hexoses (six-carbon sugars). Crystalline cellulose is difficult to hydrolyze, but the resulting hexose derivatives are readily fermented. The three basic methods for hydrolyzing structural polysaccharides to fermentable sugars (glucose, xylose, arabinose) are concentrated acid hydrolysis, dilute acid hydrolysis and enzymatic hydrolysis.

Acid treatment is the use of an acid to hydrolyze cellulosic materials. Two acid processes hydrolyze both hemicellulose and cellulose with very little pre-treatment beyond comminution of the lignocellulosic material to particles of approximately 1 mm size. These are concentrated and dilute acid hydrolysis. *Concentrated acid hydrolysis* dissolves the carbohydrates in woody biomass to form a homogeneous gelatine with acid where cellulose is extremely susceptible to hydrolysis. Typically, this is achieved with 70–90% v/v sulfuric acid (H_2SO_4) at room temperature, leaving lignin. Before fermentation, the solution of oligosaccharides is diluted to 4% v/v sulfuric acid and (i) heated (at the boiling point) for 4 h or (ii) processed using an autoclave at 120°C (248°F) for 1 h to yield monosaccharides. Following neutralization with limestone, the sugar solution is fermented. The procedure takes 10–12 h. Concentrated acid hydrolysis is attractive because it is relatively simple, and high sugar yields (approximately 100% of theoretical hexose yields in pure samples and 90% in mixed samples which replicate municipal waste mixtures) are achieved. However, corrosion-resistant equipment is necessary and acid recovery is expensive.

Dilute acid hydrolysis (1% acid w/w) greatly reduces the amount of acid required to hydrolyze lignocellulose. The process is accelerated at elevated temperatures: 100–160°C (212–320°F) for hemicellulose and 180–220°C (356–428°F) for cellulose. The high temperatures cause the oligosaccharides released from the lignocellulose to decompose, greatly reducing the yields of simple sugars to only 55–60% w/w of the theoretical yield. The decomposition products include toxins (acetic acid and furfural), which inhibit fermentation.

6.2.7.2 *Briquetting*

A briquette (also spelled briquet) is a compressed block of coal dust (Speight, 2013) or other combustible biomass materials (such as charcoal, sawdust, wood chips, peat, or paper) used for fuel as well as for kindling a fire.

Historically, briquettes (especially coal and coke briquettes) have been used as fuel for approximately 100 years. Traditionally, briquetting technology was established in developing countries to produce briquettes of local residues for use in household cooking stoves and restaurants. Later, as the capacities of the machines increased, briquettes were used in industrial boilers to create heat, steam and power for industry and power plants. Within the past three decades, briquetting has also found its way into households in industrialized countries as consumer logs for wood-burning stoves and fireplaces. However, with the advent of modern fuel systems, the use of briquettes has declined and the use of these products is most often found in the operation of barbecue units. While still a marketable product for such uses, briquettes (such as *smokeless fuel*) are not used often as fuels for domestic and industrial heating. In more recent years, as the focus on renewable energy has grown, the applications of briquettes have grown concurrently, as have different technologies and new applications.

Briquette manufacture (*briquetting*) involves the collection of combustible materials that are not usable as such because of their low density, and compressing them into a solid fuel product of any convenient shape that can be burned. The use of waste materials as a means to produce briquettes is investigated in this section. Briquettes from waste materials are commonly made using a combustible material and binder. Combustible materials include char, low-grade biomass (LGB), and bagasse. Low-grade biomass includes grasses, weeds, thinning branches (forest waste resulting from logging), agricultural waste, sawdust, wood shavings, and leaves.

A binding agent is usually necessary to increase the cohesion of the combustible materials. If the combustible material is not well bound, the briquette will crumble when removed from the mould. The binding agents can be sourced based on cost, local sources, and combustion properties. They can include animal manure, treated and dewatered sewage sludge, starch, wax, clay, molasses, cement, wood pitch glue, and local plant resin or synthetic resin. A binder must not cause smoke or gummy deposits, and the creation of excess dust must also be avoided. For this reason, the use of non-combustible binding agents such as clay, cement, and other adhesive minerals is kept to a minimum. Typically, starch is used because it is relatively cheap and easily available. Since bagasse briquetting is often done on site or close to a sugar factory, molasses is usually the binding agent used for bagasse briquettes.

In the briquetting step, the mixture is tightly compacted through a manually operated or automatic press or extruder. The press or extruder for briquetting must be well designed, strongly built, and capable of agglomerating the mixture sufficiently for it to be handled through the drying process. The extruder forms a roll of charcoal while the press favors the production of large pieces (chunks) of charcoal. As explained earlier, sawdust briquettes are formed under sufficiently high pressure to produce cohesion between wood particles. Briquettes often need to be further dried after the briquetting step. Briquettes are dried in the sunlight approximately three days before use. Rolls of charcoal formed from extruders will break into chunks during the drying process.

6.2.7.3 *Enzymatic hydrolysis*

Enzymatic hydrolysis is a process in which enzymes facilitate cleavage of bonds in molecules with the addition of the elements of water and plays an important role in the human system for the digestion of food. In the current context, the process involves the addition of cellulases (a group of enzymes which, by acting together, hydrolyze cellulose) to hydrolyze pretreated lignocellulosic biomass into fermentable sugars. The process involves several key steps: (i) transfer of enzymes from the bulk aqueous phase to the surface of the cellulose, (ii) adsorption of the enzymes and formation of enzyme–substrate complexes, (iii) hydrolysis of the cellulose, (iv) transfer of the hydrolysis products from the surface of the cellulosic particles to the bulk aqueous phase, and (v) hydrolysis of the sugar-type products to glucose in the aqueous phase. The overall rate of the process is influenced by the structural features of lignocellulosic biomass and the composition and source of the cellulases.

Thus, the enzymatic hydrolysis process was developed to better utilize both cellulose and hemicellulose from lignocellulosic materials. The enzymatic process must be preceded by extensive pre-treatment to separate cellulose, hemicellulose, and lignin. Because milder conditions are used compared to acid hydrolysis of cellulose, subsequent enzymatic hydrolysis of the cellulose does not degrade pentoses during pre-hydrolysis. Cellulose is a homopolysaccharide of glucose linked by 1,4-glycoside-type bonds (bonds in which a sugar is bound to another functional group, also known as glycosidic bonds). Enzymatic hydrolysis of cellulose proceeds in several steps to break glycoside-type bonds by the action of a system of enzymes known as cellulase. The system of enzymes also contains hemicellulase to hydrolyze any hemicellulose not solubilized by prehydrolysis. Unfortunately, enzymatic technologies are not likely to be suitable for unsorted municipal solid waste due to its high heterogeneity, and further research is necessary to convert wastes that are very high in hemicellulose.

In a fermentation process, glucose is fermented to produce a dilute ethanol-water solution (dilute beer). Xylose is difficult to convert into ethanol, but it does occur. Lignin, which is not susceptible to biological transformation, can be chemically upgraded or, more frequently, simply burned as a boiler fuel.

Simultaneous saccharification and fermentation (SSF) has been developed for fermenting sugars released from lignocellulose. The process combines hydrolysis and fermentation to overcome end product inhibition. By combining hydrolysis and fermentation, glucose is rapidly removed before it can inhibit further hydrolysis. In the process, biomass feedstock is milled and then pre-hydrolyzed to yield a mixture of pentoses (xylose and arabinose) and fiber. The mixture is neutralized with limestone and mixed with cellulose and hemicellulose enzymes. The cellulose and any remaining hemicellulose are solubilized to hexose (glucose) and pentoses (xylose and arabinose), which are immediately fermented to ethanol. The optimum temperature for the hydrolysis/fermentation reaction is between the optimum temperature for cellulase activity and the best temperature for the yeast. Lignin, as usual, is separated from the mixture and used as boiler fuel or converted into high-value octane enhancers that can be blended with gasoline.

6.2.7.4 *Transesterification*

In the 1970s, it was found that the viscosity of fresh vegetable oils (bio-oils) could be reduced through a simple process, referred to as *transesterification*. Transesterification is the process by which triglycerides (vegetable oils, waste cooking oils, and fats) are reacted with methanol or ethanol to produce methyl or ethyl esters, respectively, along with the coproduct glycerol. This process allowed bio-oils from natural sources to be converted into biodiesel (Chapter 3).

Biodiesel is diesel fuel produced from biomass or some alternative source as opposed to petroleum-derived biodiesel. Biodiesel is advantageous compared to conventional biodiesel in that it is renewable, biodegradable, and non-toxic, and has low emissions. Recent research has suggested that biodiesel can be produced from waste oils such as used cooking vegetable oil and from inedible animal tallow collected from slaughterhouses and other meat processing units.

Vegetable oil-based biodiesel has been shown to be comparable to conventional diesel. It has low viscosity, low flash points, and high vapor pressure, and is easy to process. Animal fats are preferred to vegetable oils because they are much cheaper and there is a much larger supply. Animal fats cannot be used directly because their physical properties cause poor atomization, poor vapor-air mixing, low pressure, incomplete combustion, and engine deposits. Transesterification is a means to improve all of these disadvantages. Biodiesels from animal fats have higher calorific values and cetane numbers but are less stable to oxidation and have a higher cold filter plugging point as compared to biodiesel from vegetable oils.

In the conventional synthesis of biodiesel, the transesterification occurs in the presence of methanol and base (usually sodium hydroxide). In this reaction, the bio-oil (a tri-ester or triglyceride) reacts with methanol and a catalyst to produce glycerol and three new methyl-substituted esters. Mechanistically, the triglycerides are converted into diglycerides, which are in turn converted into monoglycerides. Finally, the monoglyceride is converted into glycerol. A high ratio of methanol to oil is required for transesterification. The high ratio of methanol also helps the glycerol that is formed to phase separate. Once the glycerol phase separates from the biodiesel, the biodiesel is purified via water washing, vacuum drying, and filtration. The energy intensive purification of biodiesel is one of the many reasons its processing is not energy efficient.

Production of biodiesel is dependent on a number of variables. These include the following: (i) the feedstock, (ii) the catalyst, (iii) the solvent, (iv) the temperature and the pressure, and (v) the ratio of alcohol to the oil.

The feedstock usually includes animal tallow (animal fat) and waste cooking vegetable oil. Most studies on animal tallow focus on freshly collected animal tallow, which is low in free fatty acid value. If the animal tallow is not immediately converted, the acid value increases due to hydrolysis caused by the presence of water and degradation. Free fatty acids can react with basic catalysts, thus using up the catalyst, and the so-called soap products are difficult to separate from the desired biodiesel. Waste cooking oils also have higher amounts of free fatty acids and higher levels of moisture.

The high levels of moisture and free fatty acids are usually dealt with through some sort of acidic pre-treatment that causes acid catalyzed transesterification of the free fatty acids, followed by alkaline transesterification of the bio-oils. Numerous studies, in fact, have introduced an acidic pre-treatment or simply used an acidic catalyst when the feedstock had a large amount of free fatty acids.

Conventionally, alkaline bases are used as the *catalyst* for transesterification; however, acidic catalysts are also beginning to be used. Common basic catalysts used include sodium hydroxide, potassium hydroxide, or their alkoxides or carbonates. The commonly used acidic catalysts are sulfuric acid and hydrochloric acid. Novel enzymatic catalysts are now being tested and the scientific literature indicates that they show excellent activity and reduction in undesired coproducts; however, they have significantly higher production costs.

Typically, homogeneous strong bases are used as the catalyst for biodiesel production; however, there have been numerous studies on heterogeneous catalysts. Heterogeneous catalysts are preferred in large-scale productions because of their ease of separation. Obviously, any new catalyst must be inexpensive since the final product (biodiesel) must be inexpensive. Thus, egg shells have been proposed as potential catalysts for biodiesel production. When 3% w/w calcined egg shells (1,000°C, 1,830°F, 2 h) were added to a 9:1 molar ratio mixture of methanol and oil, the biodiesel yield exceeded 95% after 3 h. The reaction conditions and yield are similar to those of the homogeneous catalytic reaction and there is, in this case, the benefit of using a low-cost, heterogeneous catalyst.

The most commonly used alcohols are methanol and ethanol; however, methanol is used more frequently because it is lower in cost. When methanol is used as the solvent, the products are methyl esters. Ethanol is being further investigated because the viscosities of ethanol-ethyl ester mixtures are lower, and ethanol can be produced from renewable or waste resources. Furthermore, ethanol separates from glycerol more effectively.

Transesterification can occur at room temperature if the reaction is left long enough. However, usually, the temperature is kept close to the boiling point of the alcohol used at atmospheric pressure. It has been noted, however, that optimal reaction temperature and pressure depend on the type and quality of the waste cooking oil being processed. Alcohol in excess of the stoichiometric proportion necessary is commonly used. For acid catalyzed conversion of oils with high amounts of free fatty acids, much higher ratios are used compared to base-catalyzed transesterification. Specifically, ratios can range from 6:1 alcohol to oil for base-catalyzed esterification all the way to 250:1 for acid catalyzed reactions.

Using wastes as the sources for biodiesel has several major advantages: (i) avoid the fuel vs. food debate since the feedstock is non-edible, (ii) reduce the final cost of the fuel since a non-edible feedstock is typically cheaper, (iii) recycle waste, and (iv) the produced fuel is biodegradable. However, the production of biodiesel also has its disadvantages. Transesterification is normally achieved using alkaline catalysts. These catalysts are corrosive and form unwanted by-products that are expensive to separate. Homogeneous acid catalysts can also be used, but these are difficult to recycle, operate at very high temperatures, and have significant corrosion and environmental issues.

References

Demirbaş, A. and Ozturk, T. 2004. Anaerobic digestion of agricultural solid residues. *Int. J. Green Energy* 1: 483–494.

Luque, R. and Speight, J.G. (Editors). 2015. *Gasification for Synthetic Fuel Production: Fundamentals, Processes, and Applications*. Woodhead Publishing, Elsevier, Cambridge, United Kingdom.

Mokhatab, S., Poe, W.A., and Speight, J.G. 2006. *Handbook of Natural Gas Transmission and Processing*. Elsevier, Amsterdam, Netherlands.

Prakash, N. and Karunanithi, T. 2008. Kinetic modeling in biomass pyrolysis — a review. *J. Appl. Sci. Res.* 4: 1627–1636.

Speight, J.G. 2013. *The Chemistry and Technology of Coal*, 3rd Edition. CRC Press, Taylor & Francis Group, Boca Raton, Florida.

Speight, J.G. 2014a. *The Chemistry and Technology of Petroleum*, 5th Edition. CRC Press, Taylor & Francis Group, Boca Raton, Florida.

Speight, J.G. 2014b. *Gasification of Unconventional Feedstocks*. Gulf Professional Publishing Company, Elsevier, Oxford, United Kingdom.

Speight, J.G. 2019. *Natural Gas: A Basic Handbook*, 2nd Edition. Gulf Publishing Company, Elsevier, Cambridge, Massachusetts.

Chapter 7

Landfill Gas

7.1 Introduction

One aspect of the use of waste as a feedstock for alternate fuels (Chapter 6) that deserves special mention is the waste that is sent to a landfill site particularly, in the present context, the organic constituents of the waste. A landfill site is an area of land that has been specifically engineered to allow for the deposition of waste on to and into the site. Municipal solid waste landfill sites are a large source of human-related methane emissions and, in some countries such as the United States, account for up to 25% of these emissions. At the same time, methane emissions from landfills represent a lost opportunity to capture and use a significant energy resource (Johnson *et al.*, 2000).

Historically, a landfill has been called a *dump* or a *midden* and has been considered as the most convenient method of waste disposal for millennia. Landfills include internal waste disposal sites (where producers of waste carries out their own waste disposal at the place of production) as well as sites used by many producers. Many landfills are also used for other waste management purposes, such as temporary storage, consolidation and transfer, or processing of waste material (sorting, treatment, or recycling). A landfill also may refer to a ground that has been filled in with soil and rocks instead of waste materials, so that it can be used for a specific purpose, such as for building houses.

Landfill gas is the gas produced at a landfill site through the conversion of the organic constituents of the waste and varies in composition, which also affects the properties of the gas (Table 7.1) with variation in

Table 7.1: An example of the composition range of a landfill gas.

Component	% v/v
Methane, CH_4	50–75
Carbon dioxide, CO_2	25–50
Hydrogen sulfide, H_2S	0–3
Nitrogen, N_2	0–10*
Hydrogen, H_2	0–1
Oxygen, O_2	0–2

the relative amounts of the pollutants in the gas (Tchobanoglous *et al.*, 1993; US EPA, 1995).

7.2 Landfill Classification

A well-developed approach to solid waste management usually incorporates landfilling as an essential element of the final disposal. The need for new landfills cannot be overstated as waste quantities are increasing rapidly and large amounts of waste require disposal in landfills. Poor management of solid waste management practices in the past has made permitting and siting a daunting task due to limited availability of suitable sites and challenges mounted by residents who are strongly opposed to constructing a landfill in their neighborhood.

Landfilling represents disposal of all types of waste in engineered facilities. In addition to the general classification presented above, there are other classifications, including six types of engineered waste disposal facilities presented below alphabetically rather than by preference or by the level of importance: (i) bioreactor landfill sites, (ii) construction and demolition debris landfill sites, (iii) conventional municipal solid waste landfill sites, (iv) hazardous waste landfill sites, (v) manual landfill sites, (vi) sanitary landfill sites, and (vii) surface impoundment sites.

7.2.1 *Bioreactor landfill sites*

An innovative approach to municipal solid waste disposal which actually promotes rapid waste decomposition and speeds up stabilization is the bioreactor landfill. The enhancement of the waste degradation and stabilization by the indigenous microbial populations within the waste is accomplished through the control of moisture, temperature, pH, nutrients, etc. The enhanced microbiological processes within a bioreactor can transform and stabilize decomposable organic waste within 5–10 years of implementation, compared with many decades for conventional landfills where wastes are essentially sealed off from air and moisture.

A bioreactor landfill was initially defined by the Solid Waste Association of North America (SWANA) as "a sanitary landfill operated for the purpose of transforming and stabilizing readily any moderately decomposable organic waste constituents within 5–10 years following closure by purposeful control to enhance microbiological processes". A bioreactor landfill significantly increases the extent of waste decomposition, conversion rates, and process effectiveness compared to what would otherwise occur within the landfill. SWANA now defines a bioreactor landfill as "any permitted Subtitle D landfill or landfill cell, subject to New Source Performance Standards/Emissions Guidelines (NSPS/EG), where liquid or air, in addition to leachate and landfill gas (landfill gas) condensate, is injected in a controlled fashion into the waste mass in order to accelerate or enhance biostabilization of the waste." However, there is still disagreement among scientists and engineers as to the precise definition of a bioreactor landfill.

A bioreactor landfill is typically a municipal solid waste landfill site that operates to rapidly transform and degrade organic waste. The increase in waste degradation and stabilization is accomplished through the addition of liquid and air to enhance microbial processes. There are three types of bioreactor landfills: (i) an aerobic landfill, (ii) an anaerobic landfill, and (iii) a hybrid landfill, a combination of the aerobic and anaerobic processes.

In an aerobic bioreactor landfill, the leachate is removed from the bottom layer, piped to liquid storage tanks, and recirculated into the landfill in a controlled manner. Air is injected into the waste mass using vertical or horizontal wells to promote aerobic activity and accelerate waste stabilization. In an anaerobic bioreactor landfill, moisture is added to the waste mass in the form of recirculated leachate and other sources to obtain optimal moisture levels. Biodegradation occurs in the absence of oxygen (anaerobically) and produces landfill gas. Landfill gas (primarily methane) can be captured and used for energy projects to minimize greenhouse gas emissions. A hybrid bioreactor landfill accelerates waste degradation by employing a sequential aerobic–anaerobic treatment to rapidly degrade the organics in the upper sections of the landfill and collect the gas from the lower sections. The operation of a hybrid results in earlier onset of methanogenesis, compared to aerobic landfills (Speight, 2019a). In all cases, the leachate is formed when rain water percolates through wastes in a landfill. When this liquid comes in contact with buried wastes, it leaches, or draws out, chemicals or other constituents of those wastes. Thus, in contrast to conventional municipal solid waste landfills, bioreactors are designed and operated to significantly enhance microbial processes to degrade and stabilize organic constituents of the waste.

In a bioreactor landfill, the leachate, an additional liquid, is returned to the landfill until the liquid-holding capacity of the waste is achieved. The leachate drained from the waste is recovered and recirculated to maintain the moisture content of the waste and distribute nutrients for microbial degradation. A distinction should be made between the bioreactor landfill operation and the conventional leachate recirculation performed at conventional municipal solid waste landfill sites. The bioreactor process requires significant liquid addition to maintain optimal conditions.

The leachate alone is usually insufficient to sustain the bioreactor needs. Water or other non-toxic, non-hazardous liquids and semi-liquids are suitable candidates to supplement leachate. At the closure of a bioreactor landfill site, the following should be observed: (i) the leachate quantity will be amenable to on-site treatment with limited necessity for off-site transfer, treatment, and/or disposal, (ii) the generation of the landfill gas will be on a decline, and (iii) the long-term environmental risk will be minimized.

7.2.2 *Construction and demolition debris landfill sites*

A construction and demolition debris landfill site (sometimes designated as an *inert landfill site*) is devoted exclusively to construction demolition and debris wastes which are chemically and physically stable and do not undergo decomposition. Construction and demolition debris wastes include sand, bricks, concrete, or gravel. They are so called because they are generated from construction, renovation, repair, and demolition structures, such as residential and commercial buildings, roads and bridges. The composition of construction and demolition debris waste varies across different activities and structures.

Overall, construction and demolition debris waste is composed mainly of wood products, asphalt, drywall and masonry; the other components often present in significant quantities are metals, plastics, earth, shingles, insulation, paper and cardboard. Construction and demolitions debris waste also contains wastes that may be hazardous, potentially toxic, and problematic, for examples, adhesives, leftover paint, excess roofing cement, waste oils, grease and fluids, batteries, fluorescent bulbs, appliances, carpet, and treated or untreated wood. It is not prudent to dispose of such wastes in municipal solid waste landfill sites. Construction and demolition debris landfills are easier to site because demolition materials are mostly inert unless they are contaminated by an external source introduced at the construction/demolition site; therefore, their environmental impact is considered minimal.

7.2.3 *Conventional municipal solid waste landfill sites*

Conventional municipal solid waste landfill (MSWLF) sites generally contain less toxic wastes from sources such as private homes, institutions, schools, and businesses without hazardous wastes. Modern sanitary landfills are engineered to protect public health and the environment and the measures usually include detection and prevention of hazardous waste from entering the landfill, appropriate cover material for active cells and closed landfill, disease vector and explosive gas control, air monitoring, facility access, run-on control, run-off control and surface water requirements, restrictions on liquids entering the cells, and record keeping requirements. An engineered municipal solid waste landfill site comprises a liner, a leachate collection and removal system, a final cover, a gas management and a groundwater monitoring system.

7.2.4 *Hazardous waste landfill sites*

A hazardous waste *landfill site* is a waste disposal site that has units constructed according to specific design criteria and that receives wastes meeting the local definition of hazardous waste. These landfills are generally constructed to be secure repositories for materials that pose a serious hazard to human health, such as high-level radioactive waste. They are restricted, by permit or law, to the types of waste they may handle (chemical vs. radioactive, liquid vs. dry). Double liner systems are the norm for hazardous waste landfills.

Thus, these types of facilities are disposal facilities for the more toxic chemicals and dangerous by-products. These landfills must be extremely well engineered to reduce the chances of escape of hazardous compounds into the environment.

7.2.5 *Manual landfill sites*

Manual landfills have been developed in South and Central American cities, where population sizes are less than 50,000 inhabitants. Manual landfills are similar in design to mechanized ones except for the size and the equipment required. These landfills have the capacity to receive 10–50 tons per day. The use of heavy equipment may be required, but only for the periodic preparation of the terrain and the accumulation of cover material. Otherwise, all the key steps of landfill preparation are carried out manually including cell preparation, compaction, and daily cover.

In some cases, a manual landfill site may also be designated as a *dump site*, which is a landfill site that is not engineered with the special protective measures required for a sanitary landfill. Dump sites are most common in rural, remote, and developing areas. Many jurisdictions prohibit the use of a non-sanitary landfill for the disposal of municipal solid waste. Other jurisdictions that do allow dumps may require that they are constructed according to some engineering standards to mitigate the risk for environmental contamination, such as by (i) limiting the angle of the slope, (ii) requiring compaction of the waste in the place, or (iii) ensuring that the cell is high enough above the groundwater table to mitigate the potential for contamination of the water.

7.2.6 *Sanitary landfill sites*

A *sanitary landfill site*, also called a modern, engineered, or secure landfill site, typically has physical barriers, such as liners and leachate collection systems, and procedures to protect the public from exposure to the disposed wastes. The term *sanitary landfill* normally refers to those sites where municipal solid waste is sent for disposal as well as other wastes that are high in organic material. In some countries, all landfills are sanitary landfills.

7.2.7 *Surface impoundment sites*

Surface impoundments (lagoons or ponds) deal with liquid waste disposal. These impoundments must be properly lined to prevent infiltration of chemical constituents into the soil and protect the groundwater quality. The liner is considered as a part of the impoundment, and migration of waste into the liner is allowed. The liner must therefore have sufficient strength and thickness to prevent failure due to pressure gradients, be chemically compatible with the migrating waste constituents, be protected from climatic conditions, installation and operation stresses, and prevent gradients caused by settlements, compression, or uplift.

Whatever the classification, a landfill site should be designed in such a fashion that there is minimal harm to the environment and that the land may be restored in some way after the period of waste disposal. Despite its connotations as an *out of sight, out of mind* solution (Hester and Harrison, 2002, p. 28), a landfill is a widely used method for waste management that is applied to the majority of solid wastes in many countries.

7.3 Landfill Gas

Landfill gas (LFG; sometimes categorized as biogas) is produced by the microbial decomposition of landfill waste in an oxygen-free (anaerobic) atmosphere. This generally occurs under semi-controlled conditions in a landfill. The constituents of landfill gas depend on the composition and age of the waste.

Landfill gas is, in fact, a fuel gas that is a by-product of current landfilling practices and hence occurs only after municipal solid waste has been sent for disposal to a landfill site. The anaerobic digestion of the buried solid organic waste produces landfill gas naturally, as the bacterial decomposition of organic matter continues over time. In the long term, as the use of landfills will necessarily dwindle, landfill gas will disappear as a resource. It is thus of an inherently transient nature, but will be covered here as there are many examples of landfill gas plants.

In a more specific sense, three processes, (i) bacterial decomposition, (ii) volatilization, and (iii) chemical reactions, participate in the formation of the gas (Table 7.2). The gas is approximately 65% v/v methane and 35% v/v carbon dioxide as well as traces of other organic vapors, but the composition of the gas does vary with age and the type of waste. The odor

Table 7.2: Processes involved in the formation of landfill gas.

Process	Description
Bacterial decomposition	Most landfill gas is produced by bacterial decomposition, which occurs when organic waste is broken down by bacteria naturally present in the waste and in the soil used to cover the landfill. Organic wastes include food, garden waste, street sweepings, textiles, and wood and paper products. Bacteria decompose organic waste in four phases, and the composition of the gas changes during each phase.
Volatilization	Landfill gases can be created when certain wastes, particularly organic compounds, change from a liquid or a solid into a vapor. Non-methane organic compounds (NMOCs) in landfill gas may be the result of volatilization of certain chemicals disposed of in the landfill.
Chemical reactions	Landfill gas, including non-methane organic compounds, can be created by the reaction of certain chemicals present in waste. As an example, if chlorine bleach (NaOCl) and ammonia (NH_3) come in contact with each other within the landfill, a harmful gas (HCl) is produced.

of landfill gas is associated with trace compounds such as hydrogen sulphide (H_2S or H–SH)), mercaptan derivatives (RSH), and ethylene (CH_2=CH_2). In addition, there are several physical conditions that affect the production (the rate of production and the volume of the gas) of the gas: (i) the characteristics of the waste, i.e. the composition and age of the refuse, as well as (ii) environmental factors, such as the presence of oxygen in the landfill, moisture content, and temperature (Table 7.3). Other factors can also affect the migration of the gas (Table 7.4).

The rate and duration of gas production depend on the nature of the waste and oxygen availability. Some sites may continue to produce gas for over 50 years. The gas can move (migrate) in any direction within the site, depending on the permeability of the waste layers. It can also migrate off-site if measures are not taken to prevent it. One indication that gas is escaping from the site and venting out through the ground is the presence of yellow and dying vegetation. Landfill gas, being predominantly methane, is flammable. Methane has a flammability range of 4.4–15% v/v in air. As the gas migrates, it can accumulate in voids,

Table 7.3: Factors that affect landfill gas production.

Factor	Result
Waste composition	The more the organic waste present in a landfill, the more the landfill gas (e.g. carbon dioxide, methane, nitrogen, and hydrogen sulfide) produced by bacteria during decomposition. The more the chemicals disposed of in the landfill, the more likely the non-methane organic compounds and other gases will be produced either through volatilization or chemical reactions.
Age of refuse	Typically, more recently buried waste (buried less than 10 years ago) produces more landfill gas through bacterial decomposition, volatilization, and chemical reactions than older waste (buried more than 10 years ago). Peak gas production usually occurs 5–7 years after the waste is buried.
Presence of oxygen	Methane will be produced only when oxygen is no longer present in the landfill.
Moisture content	The presence of moisture (unsaturated conditions) in a landfill increases gas production because it promotes bacterial decomposition. Moisture may also promote chemical reactions that produce gases.
Temperature	As the temperature of the landfill rises, bacterial activity increases, resulting in increased gas production. Increased temperature may also increase the rates of volatilization and chemical reactions.

underground services, buildings, etc., resulting in a risk of fire and explosion. One of the main concerns about landfill gas is the possibility of gas accumulation inside buildings constructed on or close to a landfill site. Protection of new buildings is usually achieved using a dual barrier approach: a low-permeability gas membrane to resist the passage of gas, and some form of ventilation/extraction system to disperse the gas safely to the atmosphere.

The gas is extracted from the landfills using a series of wells and a blower/flare (or vacuum) system. This system directs the collected gas to a central point where it can be processed and treated, depending on the ultimate use of the gas. From this point onward, the gas can be simply flared or used to generate electricity, replace fossil fuels in industrial and manufacturing operations, or fuel greenhouse operations, or can be upgraded to pipeline-quality gas.

Table 7.4: Factors affecting the migration of landfill gas.

Factor	Effect
Landfill cover type	If the landfill cover consists of relatively permeable material, such as gravel or sand, then gas will likely migrate upward through the landfill cover. If the landfill cover consists of silts and clay minerals, it is not very permeable; gas will then tend to migrate horizontally underground. If one area of the landfill is more permeable than the rest, gas will migrate through that area.
Natural and man-made paths	Drains, trenches, and buried utility corridors (such as tunnels and pipelines) can act as conduits for gas movement. The natural geology often provides underground pathways, such as fractured rock, porous soil, and buried stream channels, where the gas can migrate.
Wind speed and direction	Landfill gas naturally vented into the air at the landfill surface is carried by the wind. The wind dilutes the gas with fresh air as it moves it to areas beyond the landfill. Wind speed and direction determine the concentration of the gas in the air, which can vary greatly from day to day, even hour by hour.
Moisture	Wet surface soil conditions may prevent landfill gas from migrating through the top of the landfill into the air above. Rain and moisture may also seep into the pore spaces in the landfill and "push out" gases in these spaces.
Groundwater levels	Gas movement is influenced by variations in the groundwater table; if water is rising into an area, it will force the landfill gas to move upward.
Temperature	Increases in temperature stimulate gas particle movement, tending to also increase gas diffusion, so that the landfill gas might spread more quickly under warmer conditions. Freezing and thawing cycles can cause cracks in the surface of the soil, thereby causing the landfill gas to migrate upward or horizontally. Frozen soil over the landfill may provide a physical barrier to upward landfill gas migration, causing the gas to migrate further from the landfill horizontally through soil.
Pressure	The difference between the soil gas pressure and barometric pressure allows gas to move either vertically or laterally, depending on whether the barometric pressure is higher or lower than the soil gas pressure. When barometric pressure is falling, landfill gas will tend to migrate out of the landfill into surrounding areas. As barometric pressure rises, gas may be retained in the landfill temporarily as new pressure balances are established.

Landfill gas is created as solid waste decomposes in a landfill. This gas consists of approximately 50% v/v methane (CH_4) (the primary component of natural gas), approximately 50% v/v carbon dioxide (CO_2), and a small amount of non-methane organic compounds. Instead of allowing landfill gas to escape into the air, it can be captured, converted, and used as an energy source. The use of landfill gas helps reduce odors and other hazards associated with landfill gas emissions and also helps prevent methane from migrating to the atmosphere and contributing to local smog and global climate change.

LFG is produced from organic waste disposed of in landfills. The waste is covered and compressed mechanically by applying pressure of higher levels. As conditions become anaerobic, the organic waste is broken down and landfill gas is produced. This gas builds up and is slowly released into the atmosphere. This is hazardous for three key reasons: (i) the risk of explosion, (ii) methane as a greenhouse gas can contribute to climate change, and (iii) volatile organic compounds (VOCs) as precursors to photochemical smog.

Landfill gas is produced by the microbial decomposition of landfill waste in an oxygen-free (anaerobic) atmosphere.

The rate and duration of gas production depend on the nature of the waste and oxygen availability.

Landfill gas, being predominantly methane, is flammable; methane has a flammability range of 5–15% v/v in air (Speight, 2017).

Gas monitoring and control measures are the responsibility of the site operator and the aim of gas control is to prevent migration off-site. This can be achieved by using a gas barrier (lining of the excavation) combined with stone-filled vent trenches, which can vent gas to the atmosphere. Venting can also be carried out using borehole pipes spread throughout the site or at selected points where gas can be pumped to them through a gas drainage system inside the tip. If the quantity or quality (odor) of the gas is such that it requires destroying, a flare system may be adopted. Landfill gas can also be exploited for industrial use as a fuel for heating or to produce electricity. The end result of a landfill is generally the restoration of the land to agriculture or similar purposes. This can be achieved by capping the area with an impermeable layer of clay followed by layers of subsoil and soil. Aftercare of the site may be necessary and will involve monitoring the land for gas.

Landfill gas is not the same as natural gas or methane; each has a separate identity and the names should not be used interchangeably.

The term *landfill methane* is deceptive as it seems to imply (incorrectly) that landfill gas is simply methane.

Landfill gas is approximately 40–60% v/v methane and the remainder is predominantly carbon dioxide with varying amounts of nitrogen, oxygen, water vapor, sulfur and other contaminants, most of which are known as non-methane organic compounds (NMOCs). Inorganic contaminants like mercury are also known to be present in landfill gas. Occasionally, even radioactive contaminants such as tritium (radioactive hydrogen) have been found in landfill gas.

Landfill gas is produced from the organic waste in a landfill. As the waste is covered and compressed mechanically by applying pressure of higher levels (perhaps analogous to the formation of crude oil from organic detritus (Chapter 3)) and as the conditions become anaerobic, the organic waste is broken down by microorganisms and gas is produced. Landfill gas often displays a potential for explosion. In addition, its major constituent (methane) is a greenhouse gas and there is also a potential for the evolution of VOCs as photochemical smog.

As already noted, the composition of landfill gas is variable and the evolution rate of the gas from the landfill and the quantity of gas are dependent on a number of factors: (i) the composition of the waste, (ii) age of the waste, (iii) the presence of oxygen, (iv) the moisture content, (v) the temperature, (vi) the waste input rate, (vii) the ambient acidity–alkalinity, the pH, (viii) the waste density — closely packed or loosely packed, and (ix) the site management strategy/strategies. Perhaps, the most important factors are the first four: waste composition, age of refuse, presence of oxygen in the landfill, and moisture content (Crawford and Smith, 1995).

- **Waste composition:** The more the organic waste present in a landfill, the more the landfill gas produced by bacterial decomposition. Some types of organic wastes contain nutrients such as sodium, potassium, calcium, and magnesium that help bacteria to thrive. When these nutrients are present, landfill gas production increases. Alternatively, some wastes contain compounds that harm bacteria, causing less gas to be produced. For example, methane-producing bacteria can be inhibited when waste has high salt concentrations.
- **Oxygen in the landfill:** Only when oxygen is used up will bacteria begin to produce methane. The more oxygen present in a landfill, the longer aerobic bacteria can decompose waste during the first phase of

the decomposition. If waste is loosely buried or frequently disturbed, more oxygen is available, so that oxygen-dependent bacteria live longer and produce carbon dioxide and water for longer periods. If the waste is highly compacted, however, methane production will begin earlier as the aerobic bacteria are replaced by methane-producing anaerobic bacteria in the third phase of the decomposition. Methane gas starts to be produced by the anaerobic bacteria only when the oxygen in the landfill is used up by the aerobic bacteria; therefore, any oxygen remaining in the landfill will slow methane production. Barometric highs will tend to introduce atmospheric oxygen into surface soils in shallow portions of a landfill, possibly altering bacterial activity. In this scenario, waste in the fourth phase of the decomposition, for example, might briefly revert to Phase I until all the oxygen is used up again.

- **Moisture content:** The presence of a certain amount of water in a landfill increases gas production because moisture promotes bacterial growth and transports nutrients and bacteria to all areas within a landfill. A moisture content of 40% or higher, based on wet weight of waste, promotes maximum gas production (e.g. in a capped landfill). Waste compaction slows gas production because it increases the density of the landfill contents, decreasing the rate at which water can infiltrate the waste. The rate of gas production is higher if heavy rainfall and/or permeable landfill covers introduce additional water into a landfill.
- **Temperature:** Warm temperatures increase bacterial activity, which in turn increases the rate of landfill gas production. Colder temperatures inhibit bacterial activity. Typically, bacterial activity drops off dramatically below 50°F. Weather changes have a far greater effect on gas production in shallow landfills because the bacteria are not as insulated against temperature changes as compared to deep landfills where a thick layer of soil covers the waste. A capped landfill usually maintains a stable temperature, maximizing gas production. Bacterial activity releases heat, stabilizing the temperature of a landfill between 77°F and 113°F, although temperatures up to 158°F have been noted. Temperature increases also promote volatilization and chemical reactions. As a general rule, emissions of organic compounds other than methane double with every 10°C (18°F) increase in temperature.
- **Age of refuse:** More recently buried waste will produce more gas than older waste. Landfills usually produce appreciable amounts of

gas within 1–3 years. Peak gas production usually occurs 5–7 years after wastes are dumped. The majority of gas is produced within 20 years after the wastes are dumped; however, small quantities of gas may continue to be emitted from the landfill for 50 or more years. A low-methane-yield scenario, however, estimates that slowly decomposing waste will produce methane after 5 years and will continue to emit gas over a 40-year period. Different parts of the landfill might be in different phases of the decomposition process at the same time, depending on when the waste was originally placed in each area. The amount of organic material in the waste is an important factor in determining how long the gas production will last.

In addition, the pressure of the gas is also affected by the following: (i) the rate of landfill gas evolution since a higher rate of gas evolution will result in an increase in the pressure of the landfill, (ii) the permeability of the surrounding waste types and/or strata (rock type) since more permeable waste and/or rock types will result in lower pressures as they will allow the landfill gas to migrate farther than less permeable types, and (iii) variations in leachate levels since an increase in leachate levels leads to a decrease in volume and thus an increase in the pressure within the landfill.

Landfill gas can pose a potential hazard, thus making accurate landfill monitoring procedures a necessity. Its potential effects, on a local scale, range from fire and/or explosion to toxicity, asphyxiation, and vegetation death. One of the problems is that landfill gas is hard to control as it is not a static medium. Depending on the surrounding pressure levels, landfill gas can move laterally (through lines of weakness in the waste), vertically (around gas/leachate wells) and also through areas of weakness where the landfill waste borders the surrounding strata.

Landfill gas is composed of approximately 50% methane and 50% carbon dioxide and is produced by the decomposition of organic waste under anaerobic conditions. Aside from environmental concerns about methane emissions and unpleasant odors associated with landfill gas, uncontrolled landfill gas can present a serious explosion hazard. It has been known to migrate underground to adjoining property and be set off by such things as furnace pilot lights. For this reason, regulations require that the gas from large landfills be controlled, usually by the drilling of wells into the landfill and establishing a system of collection pipes to draw the gas out of the landfill. As a result of the slight vacuum used in these collection systems, air may be drawn in, adding another contaminant to the methane, particularly in wells near the perimeter of the landfill. Most of this gas is flared, but over the last 20 years an increasing number of landfill operators have found ways to convert a liability into an asset by using landfill gas to produce energy for sale.

In addition, regulations implemented under the Clean Air Act require that many large landfills install gas collection and control systems, which means that at the very least, they will have to collect and flare the gas. The EPA estimates that another 600 landfills have the potential to support landfill gas energy projects. This represents a good opportunity for the use of technology to economically upgrade landfill gas to pipeline-quality gas without relying on an off-and-on tax credit.

A landfill gas facility will typically consist of gas well fields, a gas processing plant, and a gas delivery pipeline to the customer. This will allow (i) recovery of the crude gas being generated within the landfill, (ii) the ability to process the gas to produce medium-Btu gas — approximately 500 Btu/ft^3, and (iii) the delivery to a pipeline of specification gas.

At the facility, the gas is drawn from the wells and sent to a main collection header and thence to the processing plant. Vapor–liquid separators are used to separate water and any liquid condensate from the gas. The condensate is usually a mixture of hydrocarbon derivatives that resembles contaminated gasoline or kerosene. Further vapor contaminant removal is accomplished by solvent scrubbing. The crude gas is compressed and then cooled and scrubbed by countercurrent solvent flow.

The production of alternate fuels from landfill gas is an emerging area. Landfill gas has been successfully delivered to the natural gas pipeline system as both high-Btu and medium-Btu fuels. Landfill gas has also been converted to vehicle fuels in the form of compressed natural gas (CNG) and liquefied natural gas (LNG) and can also be used for methanol production (Chapter 12). Most modern landfill sites are engineered to contain liquid leachate or landfill gas produced by decomposing organic waste, even to the point of requiring a minimum of one landfill liner. This consists of a layer of compacted clay with a minimum required thickness and a maximum allowable hydraulic conductivity. A deep geological repository of high-level radioactive waste is not generally classified as a landfill site.

Municipal solid waste contains significant portions of organic materials that produce a variety of gaseous products when dumped, compacted, and covered in landfills. Anaerobic bacteria thrive in the oxygen-free environment, resulting in the decomposition of the organic materials and the production of primarily carbon dioxide and methane. Carbon dioxide is likely to leach out of the landfill because it is soluble in water. Methane, on the other hand, which is less soluble in water and lighter (less dense) than air, is likely to migrate out of the landfill. Landfill gas energy facilities capture the methane (the principal component of natural gas) and combust it for energy. However, not all landfill sites are capable of producing landfill gas that is usable as an energy gas. Gas production is dependent on the composition of the waste in the landfill and this, in turn, dictates the classification of the landfill. Large variations may exist in the composition of landfill gas due to differences in the sources of municipal solid waste and operating conditions at the landfill. The three main gas constituents [methane (methane), carbon dioxide (CO_2), and hydrogen sulfide (H_2S)] are used to characterize landfill gas (Table 7.5).

Typically, landfill gas is composed of 45–60% v/v methane, 40–60% v/v carbon dioxide, 0–1.0% v/v hydrogen sulfide, 0–0.2% v/v hydrogen (H_2), trace amounts of nitrogen (N_2), low molecular-weight hydrocarbon derivatives (dry volume basis), and water vapor (saturated).

The specific gravity of landfill gas is approximately 1.02–1.06. The complete combustion of 1 m^3 of methane releases approximately 38 MJ; in comparison, combustion of 1 m^3 of landfill gas releases 20–26 MJ/m^3 (dependent on the methane content) — this represents a low- to medium-value fuel in comparison to methane. In addition, landfill gas requires a pressure of 34,450 kPa for liquefaction whereas methane must be successively cooled to below –160°C for liquefaction.

The production rates of landfill gas vary spatially within the waste mass in the landfill and its movement is by diffusion (concentration) and convection (pressure). Landfill gas moves through the path of least resistance and several concerns are noted: (i) landfill gas contains methane, which is highly combustible, making it a potential fire and explosion hazard in the landfill environment or in adjacent structures, (ii) landfill gas is capable of migrating significant distances through the soil, increasing the risk of explosion and exposure: serious accidents (injury, loss of life, and property damage) may occur when landfill conditions favor gas migration, (iii) as landfill gas is produced, the pressure gradient upward may create cracks and disrupt the geo-membrane in the landfill cover, (iv) methane, a component of landfill gas, is an asphyxiant in high concentrations, (v) the migrating gas may cause adverse effects such on stress

Table 7.5: Physical properties of the major constituents of landfill gas.

Characteristics	Unit	CO_2	CH_4	H_2S
Molecular Weight	g	44.1	16.04	34.08
Vapor Pressure @ 21°C	kPa	5719.0		1736.3
Specific volume at 21°C, 101 kPa	m³/kg	0.456	1.746	0.701
Boiling Point at 101 kPa	°C	–164.0	–161.61	–59.6
Freezing Point at 101 kPa	°C	–78.0	–182.5	–82.9
Specific Gravity at 15°C (air = 1.0)		1.53	0.555	1.189
Density at 0°C	kg/m³	1.85	0.719	1.539
Critical temperature	°C	31.0	82.1	100.4
Critical pressure	kPa	7386.0	4640.68	9007.0
Critical density	kg/m³	0.468	0.162	0.349
Latent heat of vaporization at boiling point	kJ/kg	982.72	520.24	548.29
Latent heat of fusion at melting point	kJ/kg	189.0	58.74	69.78
Specific heat, C_p at 21°C, 101 kPa	kJ/kg°C	0.83	2.206	1.06
Specific heat, C_v at 21°C, 101 kPa	kJ/kg°C	0.64	1.688	0.803
Specific heat ratio, C_p/C_v		1.303	1.307	1.32
Thermal conductivity	M/m K	0.8323		0.0131
Flammable limits in air	% by vol	—	5.3–14	4.3–45
Solubility in water	kg/m³	4.0	24.0	3.4
Viscosity	mPa s	0.0148	0.012	0.0116
Net heat of combustion at 25°C	MJ/m³	—	36.71	—
Gross heat of combustion at 25°C	MJ/m³	—	37.97	—
Ignition temperature	°C	—	650	—
Octane rating		—	130	—
Combustion equation		—	$CH_4 + 2O_2 \rightarrow CO_2 + 2H_2O$	$H_2S + 2O_2 \rightarrow SO_3 + H_2O$

on vegetation by lowering the oxygen content of soil gas available to the root zone — plant roots require sufficient oxygen to carry out normal respiration processes, and methane acts as an asphyxiant to roots, (vi) gas generated at landfills and vented to the atmosphere emits nuisance odors (such as the rotten egg smell of hydrogen sulfide) and septic soil causing annoyance to people residing nearby, (vii) emissions of the non-methane organic compounds (NMOCs) present in landfill gas contribute to the degradation of local air quality and the formation of smog, (viii) vinyl chloride (CH_2=CHCl), was found to be present in substantial concentrations in landfill gas, resulting in health and safety concerns, (ix) methane, a *greenhouse gas*, is 23 times more potent than carbon dioxide as a contributor to global warming, and (x) uncontrolled landfill gas represents a loss of potential resources; instead, it can be used as a satisfactory fuel for boilers, gas turbines, and internal combustion engines (ICE) or upgraded to pipeline quality and then distributed to consumers. Harnessing the power of landfill gas provides environmental and economic benefits to landfills, energy users, and the community.

In most countries, government regulations call for control of landfill gas emissions and require gas management systems as a component of the landfill cover. For example, the regulation might include a reference to the lower and upper explosion limits of methane. The owner/operators of municipal solid waste landfill sites can be required to ensure that the concentration of methane generated by the facility does not exceed 25% v/v of the lower explosion limit (LEL) of methane in facility structures (excluding gas control or recovery system components) or the lower explosion limit at the facility's property boundary.

The lower explosion limit is defined as the lowest percent by volume of a mixture of explosive gases in air that will propagate a flame at 25°C (77°F) and at atmospheric pressure. Methane is explosive when present in the range 5–15% v/v in air. At methane concentrations in excess of 15% v/v, the gas mixture will not explode as the gaseous mixture is considered "rich" from a methane perspective. This 15% threshold is the upper explosive limit (UEL), which is the maximum concentration of a gas, above which the substance will not explode when exposed to an ignition source. The explosive hazard range occurs between the lower explosive limit and the upper explosive limit. At methane concentrations below the upper explosive limit, fire may still be possible and asphyxiation will occur. A sudden dilution of methane can bring the mixture back within an explosive range.

Other regulations might also require that landfill facilities install gas collection and control systems to reduce the emission of non-methane organic compounds (NMOCs). In addition, such regulations may require stringent monitoring and control of landfill gas emissions.

7.4 Formation of Landfill Gas

Most landfill gas is produced by bacterial decomposition, which occurs when organic waste is broken down by bacteria naturally present in the waste and in the soil used to cover the landfill. Organic wastes include food, garden waste, street sweepings, textiles, and wood and paper products. Bacteria decompose organic waste in four phases, and the composition of the gas changes during each phase.

During the *first phase* of decomposition, aerobic bacteria (bacteria that live only in the presence of oxygen) consume oxygen while breaking down the long molecular chains of complex carbohydrates, proteins, and lipids that comprise organic waste. The primary by-product of this process is carbon dioxide. Nitrogen content is high at the beginning of this phase, but declines as the landfill moves through the four phases. The chemistry of this first phase continues until the available oxygen is depleted. This phase can last for days or months, depending on how much oxygen is present when the waste is disposed in the landfill. Oxygen levels will vary according to factors such as how loose or compressed the waste was when it was buried.

The *second phase* commences after the oxygen in the landfill has been consumed and is followed by the anaerobic phase (a process that does not require oxygen) in which the bacteria convert compounds created by aerobic bacteria into acetic acid, lactic acid, formic acid, and alcohol derivatives such as methanol and ethanol. The landfill becomes highly acidic and, as the acid derivatives mix with the moisture present in the landfill, the acids cause certain nutrients to dissolve, making nitrogen and phosphorus available to the increasingly diverse species of bacteria in the landfill. The gaseous by-products of these processes are carbon dioxide and hydrogen. If the landfill is disturbed or if oxygen is introduced into the landfill, microbial processes return those chemical processes typical of the *first phase* of decomposition.

The *third phase* commences when certain kinds of anaerobic bacteria consume the organic acids produced in the *second phase* and form acetate, an organic acid. This process causes the landfill to become a more neutral environment in which methane-producing bacteria begin to establish themselves. The methane-producing and acid-producing bacteria have a symbiotic relationship insofar as the acid-producing bacteria create compounds for the methanogenic bacteria to consume. Also, the methanogenic bacteria consume the carbon dioxide and acetate, too much of which would be toxic to the acid-producing bacteria.

The *fourth phase* of decomposition begins when both the composition and production rates of landfill gas remain relatively constant. The gas produced in this phase gas usually contains approximately 45–60% methane by volume, 40–60% carbon dioxide, and 2–9% other gases, such as sulfides. Gas is produced at a stable rate, typically for approximately 20 years. However, gas will continue to be emitted for 50 or more years after the waste is placed in the landfill (Crawford and Smith, 1985). Gas production might last longer, for example, if greater amounts of organics are present in the waste, such as at a landfill receiving higher than average amounts of domestic animal waste.

As might be anticipated from the above paragraphs, there are many reactions that occur amongst the array of waste that is sent to a landfill. Of the wastes that are disposed to a landfill, municipal wastes are likely to play the greatest role in chemical reactions, with perhaps some arising as a result of hazardous waste. However, as the name suggests, inert wastes are chemically un-reactive. The majority of reactions occur as a result of the biodegradation, or breakdown, of organic material via microbial activity, but chemical interactions can also be seen to exist in the case of inorganic materials.

Many of the biochemical reactions in landfill sites occur as a result of the presence of reactive organic wastes, which account for more than half of the typical household waste (Hester and Harrison, 2002, p. 29). There are five major reaction processes involved, with different micro-organisms involved at each stage.

Although much more complex, the chemistry occurring at landfill sites is generally represented in the simplest form as the decomposition of cellulose and hemicellulose.

Cellulose:

$$(C_6H_{10}O_5)_n + nH_2O \rightarrow 3nCO_2 + 3nCH_4$$

Hemicellulose:

$$2(C_5H_8O_4)_n + 2nH_2O \rightarrow 5nCO_2 + 5nCH_4$$

The reactions actually occur in stages, and cellulose (Section 7.2.1) and hemicellulose are not the only starting materials.

7.4.1 *Organic reactions*

Many of the biochemical reactions in landfill sites occur as a result of the presence of reactive organic wastes (Hester and Harrison, 2002, p. 29).

7.4.1.1 *Hydrolysis and aerobic degradation*

The initial stage of organic decomposition occurs during the placement of the waste in the landfill and for the period of time after, when oxygen is available within the waste. There chemical processes are initiated and facilitated by the presence of aerobic micro-biota, which metabolize a fraction of the organic waste to produce simpler hydrocarbon derivatives, water, carbon dioxide, and, since this is an exothermic reaction, heat (the heat generated can raise the temperature of the waste to levels of up to 70–90°C; however, compacted waste achieves lower temperatures due to the reduced availability of oxygen).

In these reactions, water and carbon dioxide are produced in the greatest concentrations. The carbon dioxide can dissolve in water, forming a leachate rich in carbonic acid, which, in turn, lowers the pH of the surroundings. This stage generally lasts for only a few days or weeks, depending on the amount of oxygen that is available in the waste.

7.4.1.2 *Hydrolysis and fermentation*

The removal of oxygen during *hydrolysis and aerobic degradation* facilitates a change in the conditions from aerobic (oxygen present) to anaerobic (oxygen absent). Thus, the majority of micro-biota found in the waste change into anaerobic species. Carbohydrates are hydrolyzed (a chemical process in which a molecule is split into two parts through the addition of a water molecule) into sugars, which further decompose into carbon dioxide, hydrogen, ammonia, and organic acids. Proteins decompose via deaminization [the removal of an amino (NH_2) group] into ammonia, carboxylic acids, and carbon dioxide. The leachate produced at this stage contains ammoniacal nitrogen in high concentrations. Acetic acid is the main organic acid formed, but propionic, butyric, lactic, and formic acids and acid derivative products are also produced, and their formation is influenced by the composition of the initial waste material. The temperatures in the landfill drop to 30–50°C. Gas composition may rise to levels of up to 80% carbon dioxide and 20% hydrogen.

7.4.1.3 *Acetogenesis*

In this stage, anaerobic conditions are still present and the organic acids that were formed in the hydrolysis and fermentation stages are now converted, via specific microorganisms, to acetic acid, acetic acid derivatives, carbon dioxide, and hydrogen. Other microorganisms convert carbohydrates directly to acetic acid in the presence of carbon dioxide and nitrogen. Hydrogen and carbon dioxide levels begin to diminish towards the end of this stage, with the lower hydrogen concentrations promoting methane-generating microorganisms (methanogens), which subsequently generate methane and carbon dioxide from the organic acids and their derivatives generated in the earlier stages.

7.4.1.4 *Methanogenesis*

This stage consists of the main processes leading to the production of landfill gas. At this point, the chemical processes involved are comparatively slow and can take many years to complete. Oxygen-depleted, anaerobic conditions still remain the same as in the previous two stages. Low levels of hydrogen are required to promote methanogenic organisms, which generate carbon dioxide and methane from the organic acids and their derivatives such as acetates and formates formed in the earlier stages. Methane generation may also result from the direct conversion of hydrogen and carbon dioxide (via microorganisms) into methane and water. Hydrogen concentrations produced during stages 2 and 3 therefore fall to low levels during stage 4.

7.4.1.5 *Oxidation*

Oxidation processes mark the final stage of the reactions involved in the biodegradation of waste. As the acids are used up in the production of landfill gas (as seen in Stage 4), new aerobic microorganisms slowly replace the anaerobic forms and reintroduce oxygen into the region.

Microorganisms that convert methane into carbon dioxide and water may also become established.

7.4.2 *Inorganic reactions*

There are two prominent examples of reactions involving inorganic waste: (i) sulfate-containing materials and (ii) heavy metals (Hester and Harrison, 1995). In the former case, plaster board (here, the sulfate is present in the form of gypsum) will be reduced under anaerobic, oxygen-deficient, conditions to produce a metal sulfide (via association with free metal ions). Under acidic conditions, the sulfate will be released as hydrogen sulfide (H_2S), which has the potential to pose a serious hazard. This gas is toxic and can cause respiratory arrest if large concentrations are inhaled. Also, at concentrations above 100 ppm, the human ability to detect the gas is affected by rapid temporary paralysis of the olfactory nerves in the nose, leading to a loss of the sense of smell. This makes it extremely dangerous to rely totally on the sense of smell to warn of the presence of hydrogen sulfide.

Reduced sulfate, in the form of sulfide, will react with heavy metals (see above) to produce a metal sulfide, which is comparatively immobile. Other insoluble salts such as metal carbonates may also be formed. Although these are potential pollutants, their immobility reduces their potential to migrate out of the landfill site boundary.

7.4.3 *Landfill leachate*

In addition to landfill gas, another product of landfill chemistry is landfill leachate, which is liquid formed within a landfill site that comprise the liquids that enter the site (including rain water) and the materials leached from the waste as the infiltrating liquids percolate downward through the waste (Hester and Harrison, 1995, p. 45).

One of the most important problems with designing and maintaining a landfill is managing the leachate that is generated when water passes through the waste. The leachate consists of many different organic and inorganic compounds that may be either dissolved or suspended. Regardless of the nature of the compounds, they pose a potential pollution problem for local ground and surface waters (Barton *et al.*, 2003).

Many factors influence the production and composition of leachate. One major factor is the climate of the landfill. For example, where the climate is prone to higher levels of precipitation there will be more water entering the landfill and therefore more leachate generated. Another factor is the site topography of the landfill that influences the runoff patterns and again the water balance within the site.

Current leachate treatment options include recycling and reinjection, on-site treatment, discharge to a municipal water treatment facility, or a combination of these options. However, with stricter regulations regarding ground and surface water contamination, landfills are forced to find new treatment alternatives.

The vast majority of wastes will produce a leachate of sorts if water is allowed to percolate through it, but the proportions of the leachate constituents are chiefly determined by the composition and solubility of the waste components. For example, if the waste is changing in composition (for example, due to weathering or biodegradation), then leachate quality will change with time. This is particularly the case in landfills containing municipal waste (SEPA, 2003).

At the start, the leachate produced is typically generated under the aerobic, oxygen-rich conditions, and this produces a complex solution with a near-neutral pH of 7.5. This process generally lasts for only a few days or weeks and is relatively unimportant in terms of the quality of the leachate. However, the heat produced during aerobic decomposition can sometimes rise to as high as 80–90°C, and if this heat is retained it can enhance the later stages of leachate production.

As waste decomposition progresses, the oxygen within the waste gradually becomes diminished, and the conditions become anaerobic. This initially creates high concentrations of soluble degradable organic compounds and an acidic pH. Ammonium and metal concentrations also rise during this phase. After several months or years, the onset of methanogenic conditions changes the pH of the leachate, making it more neutral or even slightly alkaline. As biodegradation nears completion, aerobic conditions may return, and the leachate will eventually cease to be hazardous to the environment.

The character of the leachate can be determined by means of the *toxicity characteristic leaching procedure* (TCLP). This procedure (TCLP) is a soil sample extraction method for chemical analysis. It is analytical method to simulate leaching through a landfill. The leachate is analyzed for appropriate substances. The TCLP comprises four fundamental procedures: (i) sample preparation for leaching, (ii) sample leaching, (iii) preparation of leachates for analysis, and (iv) leachate analysis and is generally useful for classifying waste material for disposal options. Extremely contaminated material is expensive to dispose and grading is required to ensure safe disposal and to avoid paying for disposal of *clean fill*. The main problem is that the TCLP test is based on the assumption that the waste material will be buried in the landfill along with organic material. Organic matter is not really buried with other waste anymore (composting usually applies) and other leachate techniques may be more appropriate.

As an example of the method, the pH of the sample material is first established, and then leached with an acetic acid/sodium hydroxide solution at a 1:20 ratio of sample to solvent. The leachate solution is sealed in an extraction vessel for general analytes, or possibly pressure sealed as in zero-headspace extractions (ZHE) for volatile organic compounds, cyanides, or sulfites, and tumbled for 18 hours to simulate an extended leaching time in the ground.

7.4.4 *Landfill gas monitoring*

The one foreseeable disadvantage is that, while the global and local environmental effects of uncontrolled releases would be reduced, emissions such as CO_2, CO, NO_x, SO_x and other components would increase, leading to other detrimental effects. In a sense, the problem of landfill gas is simply being shifted rather than solved and monitoring is essential. Landfill gas monitoring is the process by which gases that are released from a landfill are electronically monitored.

Landfill gas production results from chemical reactions — many of which are initiated by microbial species — which break down the putrescible components of the waste. Due to the constant production of landfill gas, pressure increases within the landfill provoke its release into the atmosphere. Such emissions lead to important environmental issues in the landfill and, due to the risk presented by the landfill gas, there is a clear need to monitor gas produced by landfills. Techniques for the monitoring of landfill gas include the following: (i) surface monitoring: used to check the integrity of caps on waste, check on borehole monitoring which may give preliminary indications of off-site gas migration, (ii) subsurface monitoring using gas probes that can assist in the point source monitoring of gas concentrations in the local environment around the probe — the probes are typically formed in a ring around a landfill and the distance between probes varies (based on several factors) and rarely exceeds 1000 feet; these probes are also known as migration or perimeter probes,

(iii) excavated trenches and pits backfilled around standpipes provide a means of measuring gas concentrations in shallow sites, (iv) gas monitoring at boreholes or wells is a preferred method for measuring gas concentrations and sometimes flow rates both on- and off-site, (v) ambient air samplers are used to sample the air up-wind and down-wind of a landfill over a period of several hours, and (vi) the use of leachate wells may be used for gas monitoring but they are not comparable with or substitutes for specially designed gas monitoring points.

7.4.5 *Landfill mining*

Landfill mining and reclamation (LFMR) is a process whereby solid waste which has previously been landfilled is excavated and processed. The function of landfill mining is to reduce the amount of landfill mass encapsulated within the closed landfill and/or temporarily remove hazardous material to allow protective measures to be taken before the landfill mass is replaced. In the process, mining recovers valuable recyclable materials, a combustible fraction (which may serve as a precursor to fuel), soil, and landfill space. The aeration of the landfill soil is a secondary benefit regrading the landfill's future use. The overall appearance of the landfill mining procedure is a sequence of processing machines laid out in a functional conveyor system. The operating principle is to excavate, sieve, and sort the landfill material.

The concept of landfill mining was introduced when it was recognized that waste contains many resources with high value — the most notable of which are non-ferrous metals such as aluminum cans and scrap metal. The concentration of aluminum in many landfills is higher than the concentration of aluminum in bauxite from which the metal is derived. However, the potential to mine a fuel precursor such as wood or compressed organic material adds extra impetus to the concept of landfill mining.

Mining of landfill sites that were created for specific purposes is often the easiest form of landfill mining. However, mining municipal landfills has to be based on the expected content of the landfill. Older landfills in the United States were often capped and closed, essentially entombing the waste. This can be beneficial for waste recovery. It can also create a higher risk for toxic waste and leachate exposure as the landfill has not fully processed the stewing wastes. Mining of bioreactor landfill sites and properly stabilized modern sanitary landfills has its own benefits.

The biodegradable wastes are more easily sieved out, leaving the non-biodegradable materials readily accessible. The quality of these materials for recycling and reprocessing purposes is not as high as initially recycled materials; however, materials such as aluminum and steel are usually excluded from this.

In the mining process, an excavator or front-end loader uncovers the landfill materials and places them on a moving floor conveyor belt to be taken to the sorting machinery. First, a large trommel (a mechanical screening machine) is used to separate materials such as appliances and fabrics after which a smaller trommel then allows the biodegraded soil fraction to pass through, leaving non-biodegradable, recyclable materials on the screen to be collected. An electromagnet is used to remove the ferrous material from the waste mass as it passes along the conveyor belt. A front-end loader is used to move sorted materials to trucks for further processing. Odor control sprayers are wheeled tractors with a cab and movable spray arm mounted on a rotating platform. A large reservoir tank mounted behind the cab holds neutralizing agents, usually in liquid form, to reduce the smell of exposed wastes.

Excavators dig up waste mass and transport it with the help of front-end loaders onto elevators and moving floor conveyor belts. The conveyor belts empty into a coarse, rotating trommel. The large holes in the screen allow most wastes to pass through, leaving behind the over-sized, non-processable materials. The over-sized wastes are removed from inside the screen. The coarse trommel empties into the fine rotating trommel. The fine rotating trommel allows the soil fraction to pass through, leaving mid-sized, non-biodegradable, mostly recyclable materials. The materials are removed from the screen. These materials are put on a second conveyor belt where an electromagnet removes any metal debris. Depending on the level of resource recovery, material can be put through an air classifier, which separates low-density organic material from the high-density organic material. The separate streams are then loaded, by front-end loaders, onto trucks either for further processing or for sale. Further manual processing can be done on-site if processing facilities are too far away to justify the transportation costs.

7.5 Gas Migration

Once gases are produced under the landfill surface, they generally move away from the landfill. Gases tend to expand and fill the available space, so that they move or *migrate* through the limited pore spaces within the refuse and soils covering the landfill. The tendency of some landfill gases, such as methane that is lighter (less dense) than air, is to move upward, usually through the landfill surface. However, among the low-boiling hydrocarbon derivatives (the hydrocarbon gases), methane is the only one that is lighter than air. Higher molecular-weight hydrocarbon derivatives such as ethane, propane, and butane are denser than air (Table 7.6) and may collect in pockets rather than migrate with the methane. This is an error often made by many of the arm-waving specialists who do not check gas properties, and such deficiencies of knowledge can cause problems when pockets are suddenly opened to the air (for example, during landfill mining) and the flammability limit or flash point of the collected gas is reached.

In addition to the density phenomenon, upward movement of landfill gas can be inhibited by densely compacted waste or landfill cover material (e.g. by daily soil cover and caps). When upward movement is inhibited, the gas tends to migrate horizontally to other areas within the landfill or to areas outside the landfill, where it can resume its upward path. Basically, the gases follow the path of least resistance. Some gases, such as carbon dioxide, are denser than air and will collect in subsurface areas, such as utility corridors. Three main factors influence the migration of landfill gases: diffusion (concentration), pressure, and permeability.

7.5.1 *Diffusion*

Diffusion (uniform concentration) is the natural tendency of a gas to reach a uniform concentration in a given space, whether it is a room or the

Table 7.6: Density of hydrocarbon gases relative to air.

Hydrocarbon gas	Relative density*
Methane	0.554
Ethane	1.038
Propane	1.522
Butane	2.066

Note: *Air = 1.0.

atmosphere of the Earth. Gases in a landfill move from areas of high gas concentrations to areas with lower gas concentrations. Because gas concentrations are generally higher in the landfill than in the surrounding areas, landfill gases diffuse out of the landfill to the surrounding areas with lower gas concentrations.

7.5.2 *Pressure*

Gases accumulating in a landfill create areas of high pressure in which gas movement is restricted by compacted refuse or soil covers and areas of low pressure in which gas movement is unrestricted. The variation in pressure throughout the landfill results in gases moving from areas of high pressure to areas of low pressure. Movement of gases from areas of high pressure to areas of lower pressure is known as *convection*. As more gases are generated, the pressure in the landfill increases, usually causing subsurface pressures in the landfill to be higher than either the atmospheric pressure or indoor air pressure. When pressure in the landfill is higher, gases tend to move to ambient or indoor air.

7.5.3 *Permeability*

Gases will also migrate according to where the pathways of least resistance occur. Permeability is a measure of how well gases and liquids flow through connected spaces or pores in refuse and soils. Dry, sandy soils are highly permeable (many connected pore spaces), while moist clay tends to be much less permeable (fewer connected pore spaces). Gases tend to move through areas of high permeability (e.g. areas of sand or gravel) rather than through areas of low permeability (e.g. areas of clay or silt). Landfill covers are often made of low-permeability soils, such as clay. Gases in a covered landfill, therefore, are more likely to move horizontally than vertically.

7.5.4 *Factors affecting gas migration*

The direction, speed, and distance of landfill gas migration depend on a number of factors including (i) landfill cover, (ii) natural and anthropogenic pathways, (iii) wind speed and direction, (iv) moisture, (v) groundwater levels, (vi) temperature, (vii) barometric and soil gas pressure.

7.5.4.1 *Landfill cover*

If the landfill cover consists of relatively permeable materials, such as gravel or sand, then gas will likely migrate up through the landfill cover. If the landfill cover consists of silts and clays, it is not very permeable; gas will then tend to migrate horizontally underground. If one area of the landfill is more permeable than the rest, gas will migrate through that area.

7.5.4.2 *Natural and anthropogenic pathways*

Natural or man-made (anthropogenic) pathways are structures such as drains, trenches, and buried utility corridors (such as tunnels and pipelines) that can act as conduits for gas movement. The natural geology often provides underground pathways, such as fractured rock, porous soil, and buried stream channels, where the gas can migrate.

7.5.4.3 *Wind speed and direction*

Landfill gas naturally vented into the air at the landfill surface is carried by the wind. The wind dilutes the gas with fresh air as it moves it to areas beyond the landfill. Wind speed and direction determine the concentration of the gas in the air, which can vary greatly from day to day, even hour by hour. For example, at the time of the day or night when the winds tend to be gentle, there is least dilution and dispersion of the gas to other areas.

7.5.4.4 *Moisture*

Wet surface soil conditions may prevent landfill gas from migrating through the top of the landfill into the air above. Rain and moisture may also seep into the pore spaces in the landfill and "push out" gases in these spaces.

7.5.4.5 *Groundwater levels*

Gas movement is influenced by variations in the groundwater table. If the water table is rising into an area, it will force the landfill gas to move upward.

7.5.4.6 *Temperature*

Increases in temperature stimulate gas particle movement, tending also to increase gas diffusion, so that landfill gas might spread more quickly under warmer conditions. Although the landfill generally maintains a stable temperature, freezing and thawing cycles can cause the soil surface to crack, causing landfill gas to migrate in an upward direction and/or horizontal direction. Frozen soil over the landfill may provide a physical barrier to upward landfill gas migration, causing the gas to migrate further from the landfill horizontally through the soil.

7.5.4.7 *Barometric and soil gas pressure*

The difference between the soil gas pressure and barometric pressure allows gas to move either vertically or laterally, depending on whether the barometric pressure is higher or lower than the soil gas pressure. When barometric pressure is falling, landfill gas will tend to migrate out of the landfill into surrounding areas. As barometric pressure rises, gas may be retained in the landfill temporarily as new pressure balances are established.

7.6 Uses

The utilization of landfill gas involves gathering, processing, and treating the methane or another gas emitted from decomposing garbage to produce electricity, heat, fuels, and various chemical compounds (Table 7.7) After fossil fuel and agriculture, landfill gas is the third largest human-generated source of methane. Compared to carbon dioxide, methane is 25 times more effective as a greenhouse gas. It is important to not only control its emission but also, where conditions allow, use it to generate energy. Energy generating projects are popular because they control energy costs and reduce the emissions of greenhouse gas emissions. These projects collect the methane gas and treat it, so it can be used for electricity or upgraded to pipeline-grade gas. These projects power homes, buildings, and vehicles.

7.6.1 *Gas collection*

In order to use the landfill gas, it must be first collected from the landfill. A collection system usually consisting of vertical extraction wells, gas

Table 7.7: Uses of landfill gas.

Application	Advantages	Disadvantages
Power generation	Efficiencies range from good to high. Relatively low cost per kWh produced Waste heat can be recovered and reused in the process thus increasing the efficiency Low nitrogen oxides emissions	Gas must be of a high quality
Boilers, cement kilns, process heaters	Complete removal of condensate is not required.	Changes may be needed to boiler design
High quality alternative fuels	Gas can be mixed with other fuels resulting in an optimized fuel mix that can exceed the performance of existing hydrocarbon fuels	Gas processing can be very expensive
Natural gas quality	Can be piped directly into the existing natural gas pipe network	Processing requirements to bring gas to such high quality increases the cost
Chemical feedstock	Can be used in the production process of diesel and methanol	Pre-treatment process is expensive
Heat recovery from landfill flares	Cheaper alternative to current expensive leachate removal processes	Must be consumed close to landfill site

pipelines, blower units, and possibly a flare to burn-off gases safely is used. These collection wells are placed strategically throughout the landfill site. If the gas production is significant, each extraction well is connected to the main gas line, which carries the gas to a station; landfill operators can either sell the gas or use it for energy.

The gas is extracted from the landfill by means of a gas pump or a compressor that leads the gas to a utilization plant by means of pressure in the transmission pipe. Connection of the single wells to the pump and utilization system can be done in different ways. The most common method is to connect the wells to a main collection pipe which runs throughout the landfill. The main problem with this system is the difficulty involved in the regulation of both the quality and quantity of the gas. Another problem is to find the location of leaks when all the wells are connected in one big system. To save operation costs and to provide good conditions for the workers, the best solution is to have a single pipe running from each well to a pump and regulation house.

7.6.2 *Gas pretreatment*

After collecting the gas from the wells, the methane-rich gas is flared or pre-treated before being used in an energy recovery system. Treatment requirements depend on the end use. If the landfill gas is required for energy or other uses a rigorous pre-treatment of the gas is necessary.

Treatment systems can be divided into primary and secondary treatment processing systems. Most primary processing systems include dewatering and filtration to remove moisture and particulates. Dewatering can be as simple as physical removal of free water or condensate in the landfill gas. However, it is common practice to remove water vapor using gas cooling and compression.

Secondary treatment systems are designed to provide much greater gas cleaning than primary gas cleaning systems. A large portion of the landfill gas consists of carbon dioxide that is a non-combustible gas and this therefore reduces the calorific value of the landfill gas. Furthermore, in the presence of ambient moisture, acid gases produce acid rain, which is a well-known global environmental menace. The various gas cleaning systems may employ multiple clean-up processes, depending on the specifications of the end use. Such processes include both physical and chemical treatments. The most common technologies used for secondary treatment are absorption and adsorption (Kohl and Nielsen, 1997; Mokhatab *et al.*, 2006; Speight, 2013, 2014, 2019b; Miltner *et al.*, 2017).

The removal of these toxic compounds can be done through dry or wet scrubbing processes. The scrubbing process is an operation in which one or more components of a gas stream are selectively absorbed into an absorbent. The term "scrubbing" is used interchangeably with "absorption" when describing this process. In wet scrubbing a relatively non-volatile liquid is used as the absorbent. In dry scrubbing a dry powder or semi-dry slurry are possible absorbents. Adsorption involves the binding of a particle to the surface. In this case the surface is usually carbon or silica gel. There also exist advanced treatment technologies that remove carbon dioxide, non-methane organic compounds (NMOCs), and a variety of other contaminants in LFG to produce a high-Btu gas.

Landfill gas can be upgraded to natural gas quality so that it can be used in the existing natural gas distribution network. One of the indices used to evaluate the quality of gas for comparative use is the Wobbe Index. The Wobbe Index allows for the comparison of the volumetric energy content of different gas fuels at different temperatures. Gases can possess different heating values and relative densities but can have the same Wobbe Index, i.e. they will deliver the same heat and yield the same performance for a given process.

The upgrading process involves a series of complex processes that will result in the removal of compounds such as hydrogen sulphide, carbon dioxide, heavy hydrocarbon, siloxanes, nitrogen, and oxygen. Though the process used to remove such gases is not new when applied to landfill gas, it may not be a cost-effective option. Upgraded landfill gas can also be used as chemical feedstock for the process industry such as the production of diesel fuel and methanol. The process involves the upgrading of the landfill gas to chemical feedstock quality, hence rendering this option economically unattractive.

7.6.3 *Environmental impact*

Harnessing the energy of landfill gas provides environmental benefits to landfills, energy providers, and end users by (i) offsetting the use of non-renewable resources, namely coal, oil, and natural gas, and (ii) reducing the greenhouse gas emissions caused by the escape of methane-rich landfill gas into the atmosphere. However, a process utilizing waste produces substances of concern from two major products of the process: (i) solids, such as mineral ash, slag, heavy metals, and dust, and (ii) gases, such as flue gas and other gaseous chemical constituents.

7.6.3.1 *Ash, slag, heavy metals, and dust*

Ash or slag originates from the incineration or gasification of the waste. It consists of solid incombustible material combined with traces of metals that can be either toxic or benign in nature. Apart from the traces of metals, ash also contains toxic organic compounds (TOCs), e.g. dioxins. These metals can be made harmless through the sintering of ash. Extremely toxic ash is buried in landfills.

Dioxins are formed during waste combustion. The formation of these chemicals must be controlled as a lack of control can lead to serious environmental issues. Heavy metals such as mercury (from coal-fired power plants), cadmium, and lead must be monitored. However, new technologies currently exist that can be applied to mitigate the negative effects of the metals. Another solid element of concern is fly ash, which consists of fine solid particles that often contain high levels of toxic metals. The fly ash escapes along with the flue gas or the exhaust gas from the waste-to-energy plants. Fabric or bag filters and electrostatic precipitators can be used to capture the ash, which can then be buried in the landfill.

7.6.3.2 *Flue gas*

Flue gas and other gaseous constituents are primarily gases leaving the electrical energy generating side of the waste-to-energy system. These gases have a high temperature, consist of toxic chemicals, and can be both cooled and cleaned before being expelled into the atmosphere.

7.6.4 *Gasification of landfill waste*

The advantage of gasification of waste is that using synthesis gas (syngas or H_2/CO) is potentially more efficient than direct combustion of the original fuel because it can be combusted at higher temperatures or even in fuel cells. Also, synthesis gas may be burned directly to produce methanol and hydrogen, or converted by the Fischer–Tropsch process into a variety of products, including synthetic fuel (Chapter 8). In addition, the high-temperature process refines out corrosive ash elements, thereby allowing the production of a clean and usable gas from a variety of feedstocks.

Thus, there is also the potential for the gasification of organic waste material mined from the landfill as a solid or semi-solid. A waste gasification facility converts the waste into a pressurized gas mixture, which is then passed through a cleanup system to remove particulate matter and tar vapor and finally used to turn turbines in a power plant. The remaining gas is a mixture of carbon monoxide, hydrogen, water vapor, and carbon dioxide. Finally, the particulates and tar are thermally treated (pyrolyzed) to produce carbon monoxide.

Any hydrogen produced that is not used as a constituent of synthesis gas (Chapter 8) can be removed and stored for later use in hydrogen fuel cells. Waste gasification does produce some carbon dioxide that needs attention. It is thought that although carbon dioxide is produced from a renewable organic source, it has a short atmospheric residence time and therefore does not contribute to global warming. On this basis, carbon dioxide from biomass is acceptable but carbon dioxide from a fossil fuel is unacceptable.

The goal of using fuels from alternative sources is to not only supplement the dwindling supplies of fossil fuels but also reduce the emissions of greenhouse gases. Carbon dioxide is a greenhouse gas — no matter what the source is. To consider carbon dioxide allowable just because it is from a biomass-based process is, in any sense, playing with numbers and using a statistical means to justify the end. Such thinking is not only irrational but also more than suspect — it is twisted logic!

References

Barton, C.D., Paddock, L., Romanek, C.S., and Seaman, J.C. 2003. Geochemistry of an abandoned landfill containing coal combustion waste: implications for remediation. In *Chemistry of Trace Elements in Fly Ash*, K.S. Sajwan, A.K. Alva, and R.F. Keefer (Editors). Kluwer Academic/Plenum Publishers, pp. 105–141.

Crawford, J.F. and Smith, P.G. 1985. *Landfill Technology*. Butterworths Scientific Press, London, England.

Hester, R.E. and Harrison, R.M. (Editors) 2002. *Environmental and Health Impact of Solid Waste Management Activities*. The Royal Society of Chemistry, London, England.

Johnston, A.G., Edwards, J.S., Owen, J., Marshall, S.J., and Young, S.D. 2000. Research into Methane Emissions from Landfills. Proceedings. Waste 2000: Waste Management at the Dawn of the Third Millennium. Stratford-upon-Avon, United Kingdom, pp. 13–20.

Kohl, A. and Nielsen, R. 1997. *Gas Purification*, 5th Edition. Gulf Publishing Company, Houston, Texas.

Miltner, M., Makaruk, A., and Karasek, M. 2017. Review on available biogas upgrading technologies and innovations towards advanced solutions. *J. Clean. Prod.* 161(10): 1329–1337.

Mokhatab, S., Poe, W.A., and Speight, J.G. 2006. *Handbook of Natural Gas Transmission and Processing*. Elsevier, Amsterdam, Netherlands.

Speight, J.G. 2013. *The Chemistry and Technology of Coal*, 3rd Edition. CRC Press, Taylor & Francis Group, Boca Raton, Florida.

Speight, J.G. 2014. *The Chemistry and Technology of Petroleum*, 5th Edition. CRC Press, Taylor & Francis Group, Boca Raton, Florida.

Speight, J.G. (Editor) 2017. *Lange's Handbook of Chemistry*, 17th Edition. McGraw-Hill Education, New York.

Speight, J.G. 2019a. *Biogas: Production and Properties*. Nova Science Publishers, New York.

Speight, J.G. 2019b. *Natural Gas: A Basic Handbook*, 2nd Edition. Gulf Publishing Company, Elsevier, Cambridge, Massachusetts.

Tchobanoglous, G., Theisen, H., and Vigil, S. 1993. *Integrated Solid Waste Management, Engineering Principles and Management Issues*. McGraw-Hill, Inc. New York, pp. 381–417.

US EPA. 1995. Compilation of Air Pollutant Emissions Factors, AP-42, Fifth Addition, Volume 1: Stationary Point and Area Sources. January 1995. U.S. Environmental Protection Agency, Washington, DC. Section 2.4 — Municipal Solid Waste Landfills, http://www.epa.gov/ttn/chief/ap42/ch02/. http://www.atsdr.cdc.gov/HAC/landfill/PDFs/Landfill_2001_ch2mod.pdf.

Chapter 8

Synthesis Gas

8.1 Introduction

Synthesis gas (also commonly referred to as syngas) is the name given to a gas mixture that contains varying amounts of carbon monoxide and hydrogen, which is generated by the gasification of a carbonaceous (carbon-containing) feedstock. The name comes from the use of the gas as an intermediate in creating synthetic natural gas (SNG) and in producing ammonia or methanol, as well as a host of other chemicals via the Fischer–Tropsch process (Table 8.1) (Speight, 2013, 2014a, 2014b, 2017, 2019a; Luque and Speight, 2015). Examples of such processes include the following: (i) steam reforming of natural gas or higher molecular-weight hydrocarbon derivatives, (ii) the gasification of crude oil resids, (iii) the gasification of extra heavy oil and tar sand bitumen that may or may not have been topped to remove constituents boiling below, say, 345°C (650°F), (iv) the gasification of coal, and (v) the gasification of organic waste in some types of waste-to-energy facilities or waste formerly destined for a landfill site. Synthesis gas is also used as an intermediate in producing synthetic crude oil for use as a fuel or a lubricant by way of the Fischer–Tropsch process. In fact, the route for coal conversion to synthetic hydrocarbon fuels, as practiced by Sasol, is technically proven and products with favorable environmental characteristics are produced, with the potential for expansion of this technology in the future (Couvaras, 1997). As is the case in essentially all coal conversion processes where air or oxygen is used for the utilization or partial conversion of the energy in coal, the carbon dioxide burden is a drawback as compared to crude oil.

Table 8.1: Illustration of the role of synthesis gas in the production of various products.

	Stage 1	Stage 2	Stage 3	Stage 4	Products
Feedstock					
	Gasification				
		Synthesis gas			
			Gas cleaning		
				Fischer–Tropsch process	
					Naphtha*
					Gasoline
					Kerosene
					Diesel fuel
					Jet fuel
					Wax

*Gasoline blend stock.

Currently, in many refineries, hydrogen is produced from a variety of carbonaceous feedstocks by gasification and subsequent processing of the resulting gaseous product to produce synthesis gas (Chapter 9). In its simplest form, coal gasification works by first reacting coal with oxygen and steam under high pressures and temperatures to form a synthesis gas consisting primarily of carbon monoxide and hydrogen. This synthesis gas is cleaned of virtually all of its impurities and shifted to produce additional hydrogen. The clean gas is sent to a separation system to recover hydrogen.

Gasification to produce synthesis gas can proceed from a wide range of carbonaceous organic feedstocks, including biomass and waste plastic (Gay *et al.*, 1980; Bhattacharaya *et al.*, 1999; Baxter, 2005; Brigwater, 2003; EPCI, 2004; Vélez *et al.*, Arena, 2012; Brar *et al.*, 2012; Basu, 2013; Luque and Speight, 2015). The resulting synthesis gas burns cleanly into carbon dioxide and water vapor:

$$CO + H_2 \rightarrow CO_2 + H_2O$$

Alternatively, synthesis gas may be converted efficiently into methane via the Sabatier reaction at 400°C (750°F) under pressure or to a diesel-like synthetic fuel via the Fischer–Tropsch process:

$$CO_2 + 4H_2 \rightarrow CH_4 + 2H_2O$$

The inorganic components of the feedstock, such as metals and minerals, are trapped in an inert char (the carbonaceous residue of the reaction), which may find use as a fertilizer.

When used as an intermediate in the large-scale, industrial synthesis of hydrogen and ammonia, synthesis gas is also produced from natural gas (via the steam reforming reaction) as follows:

$$CH_4 + H_2O \rightarrow CO + 3H_2$$

The manufacture of gas mixtures of carbon monoxide and hydrogen has been an important part of chemical technology for more than 100 years. Originally, such mixtures were obtained by the reaction of steam with incandescent coke and were known as *water gas*. Used first as a fuel, water gas soon attracted attention as a source of hydrogen and carbon monoxide for the production of chemicals, at which time it gradually became known as *synthesis gas*. Eventually, steam reforming processes, in which steam is reacted with natural gas (methane) or crude oil (naphtha) over a nickel catalyst, found wide application in the production of synthesis gas.

A modified version of steam reforming known as autothermal reforming, which is a combination of partial oxidation near the reactor inlet with conventional steam reforming further along the reactor, improves the overall reactor efficiency and increases the flexibility of the process. Partial oxidation processes using oxygen instead of steam also found wide application in synthesis gas manufacture, with the special feature that they could utilize low-value feedstocks such as high-boiling crude oil residua. In recent years, catalytic partial oxidation employing very short reaction times (ms) at high temperatures (850–1000°C, 1,560–1,830°F) is providing yet another approach to synthesis gas manufacture. Nearly complete conversion of methane, with close to 100% selectivity to hydrogen and carbon monoxide, can be obtained with a rhenium monolith under well-controlled conditions. Experiments on the catalytic partial oxidation of *n*-hexane conducted with added steam give much higher yields of H_2 than can be obtained in experiments without steam, a result of much interest in obtaining hydrogen-rich streams for fuel cell applications.

The uses of synthesis gas include use as a chemical feedstock and in gas-to-liquid processes (Mangone, 2002), which use Fischer–Tropsch chemistry to produce liquid fuels as the feedstock for chemical synthesis, as well as being used in the production of fuel additives, including diethyl ether and methyl t-butyl ether (MTBE), acetic acid and its anhydride; synthesis gas could also make an important contribution to chemical synthesis through conversion to methanol. There is also an option in which stranded natural gas is converted into synthesis gas followed by conversion to liquid fuels.

8.2 Coal Gasification

The gasification process evolved from immaturity to maturity using coal as the predominant feedstock, and it is pertinent to present this aspect of the gasification process at this point. Gasification is a process that converts carbonaceous materials, such as coal, crude oil, or biomass, into carbon monoxide and hydrogen by reacting the raw material at high temperatures with a controlled amount of oxygen (Luque and Speight, 2015). In fact, the gasification of coal has been, and continues to be, widely used on industrial scales to generate electricity. However, almost any type of organic carbonaceous material, such as wood, biomass, or even plastic waste, can be used as the raw material for gasification. Thus, gasification may be an important technology for renewable energy. In particular, biomass gasification is carbon-neutral. Furthermore, because of the historical use of coal for gasification purposes (Ishi, 1982; Ricketts *et al.*, 2002; Hotchkiss, 2003; Speight, 2013, 2014; Luque and Speight, 2015), it is the feedstock against which the suitability of all other feedstocks is measured. Therefore, inclusion of coal among the gasification feedstocks in this section is warranted.

In the process, coal or coal char (coke) is converted into gaseous products by reaction with steam, oxygen, air, hydrogen, carbon dioxide, or a mixture of these. The amount of oxidation in the reactor is carefully controlled. The resulting gas mixture (synthesis gas) is itself a fuel gas. The process is an efficient method for extracting energy from many different types of organic materials, and also has application as a clean waste disposal technique.

The advantage of the process is that using the synthesis gas is more efficient than direct combustion of the original fuel; more of the energy contained in the fuel is extracted. Synthesis gas may be burned directly in internal combustion engines, to produce methanol and hydrogen, or converted via the Fischer–Tropsch process into synthetic fuel. Gasification can also begin with materials that are not otherwise useful fuels, such as biomass or organic waste. In addition, the high-temperature combustion refines out corrosive ash elements such as chloride and potassium, allowing clean gas production from otherwise problematic fuels.

For many years, such processes were used to manufacture illuminating gas (coal gas) for gas lighting, before electric lighting became widely available. The simplest method, and the first used, was to heat coal in a retort in the absence of air, partially converting coal into gas with a residue of coke; the Scottish engineer William Murdock used this technique in pioneering the commercial gasification of coal in 1792. Murdock licensed his process to the Gas Light and Coke Company in 1813, and in 1816 the Baltimore Gas Company, the first coal gasification company in the United States, was established. The process of heating coal to produce coke and gas is still used in the metallurgical industry. Another advantage of gasification-based energy systems is that when oxygen is used in the gasifier in place of air, the carbon dioxide produced by the process is in a concentrated gas stream, making it easier and less expensive to separate and capture. Once the carbon dioxide is captured, it can be sequestered and prevented from escaping to the atmosphere, where it could otherwise potentially contribute to the greenhouse effect.

8.2.1 *Chemistry*

Gasification relies on chemical processes at elevated temperatures >700°C (>1,290°F), which distinguishes it from biological processes such as anaerobic digestion that produce biogas. Gasification occurs in stages and the chemistry of each stage can be represented by simple, straightforward chemical equations, and hence, coal gasification processes are reasonably efficient. In a gasifier, the carbonaceous material undergoes several different processes: (i) pyrolysis or devolatilization of the carbonaceous feedstock, (ii) combustion of the pyrolysis products, and (iii) gasification of the remaining char.

In the pyrolysis stage (the devolatilization stage), the thermal degradation of the organic substance in the absence of air commences with the production of gas, pyrolysis oil, and char. The pyrolysis process occurs as the carbonaceous feedstock heats up. Volatile products are released and a char is produced, resulting in a weight loss of about 70% w/w for coal. However, as might be anticipated, the process is highly dependent on the properties of the carbonaceous feedstock, which determines the following: (i) the yield of char, (ii) the composition of the char, and (iii) the structure of the char, which, in turn, determines the behavior of the char in the combustion section and in the gasification section of the reactor.

The combustion process occurs as the volatile products and some of the char react with oxygen to form carbon dioxide and carbon monoxide, which provide heat for the subsequent gasification reactions:

$$2C + O_2 \rightarrow CO_2$$

The gasification process occurs as the char reacts with carbon dioxide and steam to produce carbon monoxide and hydrogen:

$$C + H_2O \rightarrow H_2 + CO_2$$

In addition, the gas-phase water–gas shift reaction reaches equilibrium quickly at the temperatures in a gasifier. This removes carbon dioxide from the reactor and provides water for the gasification reaction:

$$CO + H_2O \leftrightarrow CO_2 + H_2$$

In essence, a limited amount of oxygen or air is introduced into the reactor to allow some of the organic material to be burned to produce carbon monoxide and energy, which drives a second reaction that converts further organic material into hydrogen and additional carbon monoxide.

The most complete conversion of coal or coke to gas that is feasible was achieved by reacting coal continuously in a vertical retort with air and steam. The gas obtained in this manner, called producer gas, has a relatively low heat content per unit volume of gas (100–150 Btu/ft^3). The development of a cyclic steam–air process in 1873 made possible the production of a gas of higher thermal content (300–350 Btu/ft^3), composed chiefly of carbon monoxide and hydrogen, and known as water gas. By adding oil to the reactor, the thermal content of gas was increased to 500–550 Btu/ft^3; this became the standard for gas distributed to residences and industry. Since 1940, processes have been developed to produce continuously a gas equivalent to water gas; this involves the use of steam and essentially pure oxygen as a reactant. A more recently developed process reacts coal with pure oxygen and steam at an elevated pressure (450 psi) to produce a gas that may be converted into synthetic natural gas.

8.2.2 *Processes*

The most common modern coal gasification process uses lump coal in a vertical retort. In the processes, coal is fed at the top with air, and steam is introduced at the bottom and the gas (air and steam) rising up the retort heat the coal in its downward flow and react with the coal to convert it into gas. Ash is removed at the bottom of the retort. Using air and steam as reacting gases results in a producer gas; using oxygen and steam results in a water gas. Increasing operating pressure increases the productivity.

Two other processes currently in commercial use react finely powdered coal with steam and oxygen. One of these, the Winkler process, uses a fluidized bed in which the powdered coal is agitated with the reactant gases. The Koppers–Totzek process operates at a much higher temperature, and the powdered coal is reacted while it is entrained in the gases passing through the reactor. The ash is removed as a molten slag at the bottom of the reactor.

As crude oil supplies decrease, the desirability of producing gas from coal may increase, especially in those areas where natural gas is in short supply. It is also anticipated that costs of natural gas will increase, allowing coal gasification to compete as an economically viable process. Research in progress on a laboratory and pilot-plant scale should lead to the invention of new process technologies by the end of the century, thus accelerating the industrial use of coal gasification.

Thus, the products of coal gasification consist of carbon monoxide, carbon dioxide, hydrogen, methane, and some other gases in proportions dependent on the specific reactants and conditions (temperatures and pressures) employed within the reactors, and the treatment steps which the gases undergo subsequent to leaving the gasifier. A similar chemistry can also be applied to the gasification of coke derived from crude oil and other sources. The reaction of coal or coal char with air or oxygen to produce heat and carbon dioxide could be called gasification, but it is more properly classified under combustion. The principal purposes of such a conversion are the production of synthetic natural gas as a substitute gaseous fuel and synthesis gases for the production of chemicals and plastics.

In all cases of commercial interest, gasification with steam, which is endothermic, is an important chemical reaction. The necessary heat input is typically supplied to the gasifier by combusting a portion of the coal with oxygen added along with the steam. From the industrial viewpoint, the final product is either synthesis gas, medium-Btu gas, or substitute natural gas. Each of the gas types has its potential industrial applications.

In the chemical industry, synthesis gas from coal is a potential alternative source of hydrogen and carbon monoxide. This mixture is obtained primarily from steam reforming of natural gas, natural gas liquids, or other crude oil liquids. Fuel users in the industrial sector have studied the feasibility of using medium-Btu gas instead of natural gas or oil for fuel applications. Finally, the natural gas industry is interested in substitute natural gas, which can be distributed in existing pipeline networks.

The conversion of the gaseous products of coal gasification processes to synthesis gas, a mixture of hydrogen (H_2) and carbon monoxide (CO), in a ratio appropriate to the application, needs additional steps, after purification. The product gases — carbon monoxide, carbon dioxide, hydrogen, methane, and nitrogen — can be used as fuels or as raw materials for chemical or fertilizer manufacture.

8.2.3 *Gasifiers*

The focal point of any gasification-based system is the gasifier. A gasifier converts hydrocarbon feedstock into gaseous components by applying heat under pressure in the presence of steam.

A gasifier differs from a combustor in that the amount of air or oxygen available inside the gasifier is carefully controlled so that only a relatively small portion of the fuel burns completely. The *partial oxidation* process provides the heat and rather than by combustion, most of the carbon-containing feedstock is chemically broken apart by the heat and pressure applied in the gasifier, resulting in the chemical reactions that produce synthesis gas. However, the composition of the synthesis gas will vary depending on the conditions in the gasifier and the type of feedstock.

Minerals in the fuel (i.e. rocks, dirt, and other impurities which do not gasify) separate and leave the bottom of the gasifier either as an inert glass-like slag or as other marketable solid products.

Sulfur impurities in the feedstock are converted into hydrogen sulfide (H_2S) and carbonyl sulfide (COS), from which sulfur can be extracted, typically as elemental sulfur. Nitrogen oxides (NO_x), other potential pollutants, are not formed in the oxygen-deficient (reducing) environment of the gasifier. Instead, ammonia (NH_3) is created by nitrogen–hydrogen reactions and can be washed out of the gas stream.

Four types of gasifiers are currently available for commercial use: (i) the countercurrent fixed-bed gasifier, (ii) the cocurrent fixed-bed gasifier, (iii) the fluidized-bed gasifier, and (iv) the entrained flow, sometimes referred to as the entrained-bed gasifier (Speight, 2013).

The countercurrent fixed-bed (updraft) gasifier consists of a fixed bed of carbonaceous fuel (e.g. coal or biomass) through which the "gasification agent" (steam, oxygen, and/or air) flows in a countercurrent configuration. The ash is either removed dry or as a slag. The slagging gasifiers require a higher ratio of steam and oxygen to carbon in order to reach temperatures higher than the ash fusion temperature. The nature of the gasifier means that the fuel must have high mechanical strength and must be non-caking so that it will form a permeable bed, although recent developments have reduced these restrictions to some extent. The throughput for this type of gasifier is relatively low. The thermal efficiency is high as the gas exit temperatures are relatively low. However, this means that tar and methane production is significant at typical operation temperatures, so product gas must be extensively cleaned before use or recycled to the reactor.

The cocurrent fixed-bed (downdraft) gasifier is similar to the countercurrent type, but the gasification agent gas flows in a cocurrent configuration with the fuel (downward, hence the name downdraft gasifier). Heat needs to be added to the upper part of the bed either by combusting small amounts of the fuel or using external heat sources. The produced gas leaves the gasifier at a high temperature, and most of this heat is often transferred to the gasification agent added at the top of the bed, resulting in an energy efficiency on level with the countercurrent type. Since all tars must pass through a hot bed of char in this configuration, tar levels are much lower than those of the countercurrent type.

In the fluidized-bed gasifier, the fuel is fluidized by oxygen (or air) and steam. The ash is removed dry or as high-density agglomerates. The temperatures are relatively low in dry ash gasifiers, so the fuel must be highly reactive; low-grade coals are particularly suitable. The agglomerating gasifiers have slightly higher temperatures and are suitable for higher rank coals. The fuel throughput is higher than that for the fixed-bed gasifier, but not as high as for the entrained-flow gasifier. The conversion efficiency is rather low, so recycling or subsequent combustion of solids is necessary to increase conversion. Fluidized-bed gasifiers are most useful for fuels that form highly corrosive ash that would damage the walls of slagging gasifiers. Also, a disadvantage of biomass feedstocks is that they generally contain high levels of corrosive ash.

In the entrained-flow gasifier, a dry pulverized solid, an atomized liquid fuel, or a fuel slurry is gasified with oxygen (much less frequently: air) in cocurrent flow. The gasification reactions take place in a dense cloud of very fine particles. Most coal is suitable for this type of gasifier because of the high operating temperatures and because the coal particles are well separated from one another. The high temperatures and pressures also mean that a higher throughput can be achieved, but the thermal efficiency is somewhat lower as the gas must be cooled before it can be cleaned with existing technology. The high temperatures also mean that tar and methane are not present in the product gas; however, the oxygen requirement is higher than that for the other types of gasifiers. All entrained-flow gasifiers remove a major part of the ash as a slag since the operating temperature is well above the ash fusion temperature. A smaller fraction of the ash is produced either as a very fine dry fly ash or as black fly ash slurry. Some fuels, in particular certain types of biomasses, can form a slag that is corrosive for ceramic inner walls that serve to protect the gasifier outer wall. However, some entrained-bed type of gasifiers do not possess a ceramic inner wall but have an inner water- or steam-cooled wall covered with a partially solidified slag. These types of gasifiers do not suffer from corrosive slag. Some fuels have ashes with very high ash fusion temperatures. In this case, mostly limestone is mixed with the fuel prior to gasification. Addition of limestone will usually suffice for lowering the fusion temperatures. The fuel particles must be much smaller than those for other types of gasifiers. This means that the fuel must be pulverized, which requires somewhat more energy than that for the other types of gasifiers. By far, the most energy consumption related to entrained-bed gasification is not the milling of the fuel but the production of oxygen used for the gasification.

In Integrated Gasification Combined-Cycle (IGCC) systems, the synthesis gas is cleaned of its hydrogen sulfide, ammonia, and particulate matter and is burned as fuel in a combustion turbine (much like natural gas is burned in a turbine). The combustion turbine drives an electric generator. And hot air from the combustion turbine can be channeled back to the gasifier or the air separation unit, while exhaust heat from the combustion turbine is recovered and used to boil water, creating steam for a steam turbine-generator.

The use of these two types of turbines — a combustion turbine and a steam turbine — in combination, known as a combined cycle, is one reason why gasification-based power systems can achieve unprecedented power generation efficiencies. Currently, commercially available gasification-based systems can operate at around 42% efficiencies; in the future, these systems may be able to achieve efficiencies approaching 60%. A conventional coal-based boiler plant, by contrast, employs only a steam turbine-generator and is typically limited to 33–40% efficiency.

Higher efficiency means that less fuel is required to generate the rated power, resulting in better fuel economy (which can mean lower costs to the consumer) and the formation of fewer greenhouse gases — a 60%-efficient gasification power plant can cut the formation of carbon dioxide by 40% compared to a typical coal combustion plant.

8.3 Gasification of Crude Oil Fractions

Because of the ever-increasing need for crude oil products and the increasing scarcity of conventional crude oil, high-conversion refineries are moving to gasification of feedstocks for the development of alternative fuels and to enhanced equipment usage (Speight, 2011, 2014a). A major trend in the refining industry market demand for refined products will be in synthesizing fuels from simple basic reactants (e.g. synthesis gas) when it becomes uneconomical to produce superclean transportation fuels through conventional refining processes. Fischer–Tropsch plants together with IGCC systems will be integrated with or even into refineries, which will offer the advantage of high-quality products (Stanislaus *et al.*, 200).

In addition, the concept of using other feedstocks to blend with crude oil feedstocks brings to the fore the concept of a *gasification refinery*, which would use gasification technology as is the case of the Sasol refinery in South Africa (Chadeesingh, 2011). The refinery would produce synthesis gas (from the carbonaceous feedstock) from which liquid fuels would be manufactured using the Fischer–Tropsch synthesis technology. Synthesis gas is used as a source of hydrogen or as an intermediate in producing hydrocarbon derivatives via the Fischer–Tropsch synthesis.

8.3.1 *Feedstocks*

The most common, and perhaps the best, feedstocks for steam reforming are low-boiling saturated hydrocarbon derivatives that have a low sulfur content, including natural gas, refinery gas, liquefied petroleum gas (LPG), and low-boiling naphtha.

Natural gas is the most common feedstock for hydrogen production since it meets all the requirements for reformer feedstock. Natural gas typically contains more than 90 percent methane and ethane with only a few percent of propane and higher boiling hydrocarbon derivatives (Speight, 1993). Natural gas may (or most likely will) contain traces of carbon dioxide with some nitrogen and other impurities. Purification of natural gas, before reforming, is usually relatively straightforward (Speight, 1993). Traces of sulfur must be removed to avoid poisoning the reformer catalyst; zinc oxide treatment in combination with hydrogenation is usually adequate.

Low molecular-weight *refinery gas*, containing a substantial amount of hydrogen, can be an attractive steam reformer feedstock since it is produced as a by-product. Processing of refinery gas will depend on its composition, particularly the levels of olefins and of propane and heavier hydrocarbon derivatives. Olefins, which can cause problems by forming coke in the reformer, are converted into saturated compounds in the hydrogenation unit. Higher boiling hydrocarbon derivatives in refinery gas can also form coke, either on the primary reformer catalyst or in the preheater. If there is more than a few percent of C_3 and higher compounds, a promoted reformer catalyst should be considered in order to avoid carbon deposits.

Refinery gas from different sources varies in suitability as a hydrogen plant feed. Catalytic reformer off-gas (Chapter 24), for example, is saturated, very low in sulfur, and often has high hydrogen content. The process gases from a coking unit (Chapter 17) or from a fluid catalytic cracking unit (Chapter 18) are much less desirable because of the content of unsaturated constituents. In addition to olefins, these gases contain substantial amounts of sulfur that must be removed before the gas is used as a feedstock. These gases are also generally unsuitable for direct hydrogen recovery, since the hydrogen content is usually too low. Hydrotreated off-gas lies in the middle of the range. It is saturated, so it is readily used as hydrogen plant feed. The content of hydrogen and heavier hydrocarbon derivatives depends to a large extent on the upstream pressure. Sulfur removal will generally be required.

As hydrogen use has become more widespread in refineries, hydrogen production has moved from the status of a high-tech specialty operation to an integral feature of most refineries. This has been made necessary by the increase in hydrotreating and hydrocracking, including the treatment of progressively heavier feedstocks (Speight, 2007). The continued increase in hydrogen demand over the last several decades is a result of the conversion of crude oil to match changes in product slate and the supply of viscous, high-sulfur oil, and in order to make lower-boiling, cleaner, and more salable products. There are also many reasons other than product quality for using hydrogen at the relevant stages of the refining process, and most important is the availability of hydrogen.

Hydrogen has historically been produced during catalytic reforming processes as a by-product of the production of the aromatic compounds used in gasoline and in solvents. As reforming processes changed from fixed-bed to cyclic to continuous regeneration, process pressures have dropped and hydrogen production per barrel of reformate has tended to increase. However, hydrogen production as a by-product is not always adequate to meet the needs of the refinery, and other processes are necessary. Thus, hydrogen production by steam reforming or by partial oxidation of residua has also been used, particularly where heavy crude oil, extra heavy crude oil, and tar sand bitumen are available. Steam reforming is the dominant method for hydrogen production and is usually combined with pressure-swing adsorption (PSA) to purify the hydrogen to greater than 99% by volume.

The gasification of crude oil residua and crude oil coke to produce hydrogen and/or power may become an attractive option for refiners (Speight, 2014a, 2016, 2017). In the past decades, the delayed coking process has been considered to be the *garbage can* of the refinery insofar as the process is used to get rid of the less valuable resids by converting these materials into more valuable products (Parkash, 2003; Gary *et al.*, 2007; Speight, 2014a, 2017, 2019a; Hsu and Robinson, 2017). The premise that the gasification section of a refinery will be the *garbage can* for deasphalted residues, high-sulfur coke, as well as other refinery wastes is worthy of consideration.

Among the processes that are available for the production of the hydrogen, many can be considered as dual processes insofar as they also produce carbon monoxide and, therefore, are considered as producers of synthesis gas. For example, most of the external hydrogen is manufactured by steam–methane reforming or by oxidation processes. Other processes such as ammonia dissociation, steam-methanol interaction, or electrolysis are also available for hydrogen production, but economic factors and feedstock availability dictate the choice of processing alternatives.

The processes described in this section are those gasification processes by which hydrogen is produced for use in other parts of the refinery.

8.3.2 *Chemistry*

In *steam reforming*, low-boiling hydrocarbon derivatives, such as methane, are reacted with steam to form hydrogen:

$$CH_4 + H_2O \rightarrow 3H_2 + CO \qquad \Delta H_{298K} = +97{,}400 \text{ Btu/lb}$$

where H is the heat of reaction. A more general form of this equation that shows the chemical balance of higher-boiling hydrocarbon derivatives is as follows:

$$C_nH_m + nH_2O \rightarrow (n + m/2)H_2 + nCO$$

The reaction is typically carried out at approximately 815°C (1500°F) over a nickel catalyst packed into the tubes of a reforming furnace. The high temperature also causes the hydrocarbon feedstock to undergo a series of cracking reactions plus the reaction of carbon with steam:

$$CH_4 \rightarrow 2H_2 + C$$
$$C + H_2O \rightarrow CO + H_2$$

Carbon is produced on the catalyst at the same time that hydrocarbon is reformed to hydrogen and carbon monoxide. With natural gas or similar feedstock, reforming predominates and the carbon can be removed by the reaction with steam as fast as it is formed. When higher boiling feedstocks are used, the carbon is not removed fast enough and builds up, thereby requiring catalyst regeneration or replacement. Carbon buildup on the catalyst (when high-boiling feedstocks are employed) can be avoided by the addition of alkali compounds, such as potash, to the catalyst, thereby encouraging or promoting the carbon-steam reaction.

However, even with an alkali-promoted catalyst, feedstock cracking limits the process to hydrocarbon derivatives with a boiling point of less than 180°C (350°F). Natural gas, propane, butane, and low-boiling naphtha are most suitable. Pre-reforming, a process that uses an adiabatic catalyst bed operating at a lower temperature, can be used as a pretreatment to allow heavier feedstocks to be used with lower potential for carbon deposition (coke formation) on the catalyst.

After reforming, the carbon monoxide in the gas is reacted with steam to form additional hydrogen (the *water–gas shift* reaction):

$$CO + H_2O \rightarrow CO_2 + H_2 \qquad \Delta H_{298K} = -16{,}500 \text{ Btu/lb}$$

This leaves a mixture consisting primarily of hydrogen and carbon monoxide, which is removed by conversion to methane:

$$CO + 3H_2O \rightarrow CH_4 + H_2O$$

$$CO_2 + 4H_2 \rightarrow CH_4 + 2H_2O$$

The critical variables for steam reforming processes are as follows: (i) temperature, (ii) pressure, and (iii) the steam/hydrocarbon ratio. Steam reforming is an equilibrium reaction, and conversion of the hydrocarbon feedstock is favored by high temperature, which in turn requires higher fuel use. Because of the volume increase in the reaction, conversion is also favored by low pressure, which conflicts with the need to supply the hydrogen at high pressure. In practice, materials of construction limit temperature and pressure.

On the other hand, and in contrast to reforming, shift conversion is favored by low temperature. The gas from the reformer is reacted over an iron oxide catalyst at 315–370°C (600–700°F) with the lower limit being dictated by the activity of the catalyst at low temperature.

Hydrogen can also be produced by *partial oxidation* (POX) of hydrocarbon derivatives in which the hydrocarbon is oxidized in a limited or controlled supply of oxygen:

$$2CH_4 + O_2 \rightarrow CO + 4H_2 \qquad \Delta H_{298K} = -10{,}195 \text{ Btu/lb}$$

The shift reaction also occurs and a mixture of carbon monoxide and carbon dioxide is produced in addition to hydrogen. The catalyst tube materials do not limit the reaction temperatures in partial oxidation processes and higher temperatures may be used that enhance the conversion of methane to hydrogen. Indeed, much of the design and operation of hydrogen plants involves protecting the reforming catalyst and the catalyst tubes because of the extreme temperatures and the sensitivity of the catalyst. In fact, minor variations in feedstock composition or operating conditions can have significant effects on the life of the catalyst or the reformer itself. This is particularly true of the changes in molecular weight of the feed gas or poor distribution of heat to the catalyst tubes.

Since the high temperature takes the place of a catalyst, partial oxidation is not limited to the lower-boiling feedstocks that are required for steam reforming. Partial oxidation processes were first considered for hydrogen production because of expected shortages of lower boiling feedstocks and the need to have available a disposal method for higher-boiling, high-sulfur streams, such as asphalt or crude oil coke.

Catalytic partial oxidation, also known as auto-thermal reforming, reacts oxygen with a low-boiling feedstock and passes the resulting hot mixture over a reforming catalyst. The use of a catalyst allows the use of lower temperatures than in non-catalytic partial oxidation and which causes a reduction in oxygen demand.

The feedstock requirements for catalytic partial oxidation processes are similar to the feedstock requirements for steam reforming and low-boiling hydrocarbon derivatives from refinery gas to naphtha are preferred. The oxygen substitutes for much of the steam in preventing coking and a lower steam/carbon ratio is required. In addition, because a large excess of steam is not required, catalytic partial oxidation produces more carbon monoxide and less hydrogen than steam reforming. Thus, the process is better suited to situations where carbon monoxide is the more desirable product, for example, as synthesis gas for chemical feedstocks.

8.3.3 *Commercial processes*

In spite of the use of low-quality hydrogen (that contains up to 40% by volume hydrocarbon gases), a high-purity hydrogen stream (95–99% by volume hydrogen) is required for the hydrodesulfurization, hydrogenation, hydrocracking, and petrochemical processes. Hydrogen, produced as a by-product of refinery processes (principally hydrogen recovery from catalytic reformer product gases), is often not enough to meet the total refinery requirements, necessitating the manufacturing of additional hydrogen or obtaining supply from external sources. The various commercial processes are presented in the following sections and, as is usually the case, the processes are listed alphabetically rather than by preference.

8.3.3.1 *Heavy residue gasification and combined cycle power generation*

High-boiling residua are gasified and the produced gas is purified to a clean fuel gas (Gross and Wolff, 2000). For example, solvent deasphalted residuum is gasified by the partial oxidation method at a pressure of approximately 570 psi and at a temperature of the order of 1,300–1,500°C (2370–2730°F). The high temperature generated gas flows into the specially designed waste heat boiler, in which the hot gas is cooled and high pressure saturated steam is generated. The gas from the waste heat boiler is then heat exchanged with the fuel gas and flows to the carbon scrubber, where unreacted carbon particles are removed from the generated gas by water scrubbing.

The gas from the carbon scrubber is further cooled by the fuel gas and boiler feed water and led into the sulfur compound removal section, where hydrogen sulfide (H_2S) and carbonyl sulfide (COS) are removed from the gas to obtain a clean fuel gas. This clean fuel gas is heated with the hot gas generated in the gasifier and finally supplied to the gas turbine at a temperature of 250–300°C (480–570°F).

The exhaust gas from the gas turbine having a temperature of approximately 550–600°C (1020–1110°F) flows into the heat recovery steam generator consisting of five heat exchange elements. The first element is a superheater, in which the combined stream of the high-pressure saturated steam generated in the waste heat boiler and in the second element (high-pressure steam evaporator) is superheated. The third element is an economizer, the fourth element is a low-pressure steam evaporator, and the final or the fifth element is a deaerator heater. The off-gas from heat recovery steam generator having a temperature of approximately 130°C (265°F) is emitted into the air via stack.

In order to decrease the nitrogen oxide (NO_x) content in the flue gas, two methods can be applied. The first method is the injection of water into the gas turbine combustor. The second method is the selective reduction of the nitrogen oxide content by injecting ammonia gas in the presence of de-NO_x catalyst that is packed in a proper position of the heat recovery steam generator. The latter method is more effective than the former method to lower the nitrogen oxide emissions to the air.

8.3.3.2 *Hybrid gasification process*

In the hybrid gasification process, a slurry of coal and residual oil is injected into the gasifier, where it is pyrolyzed in the upper part of the reactor to produce gas and chars. The chars produced are then partially oxidized to ash, which is removed continuously from the bottom of the reactor.

In this process, coal and vacuum residue are mixed together into a slurry to produce a clean fuel gas. The slurry fed into the pressurized gasifier is thermally cracked at a temperature of 850–950°C (1560–1740°F) and converted into gas, tar, and char. The mixture of oxygen and steam in the lower zone of the gasifier gasifies the char. The gas leaving the gasifier is quenched to a temperature of 450°C (840°F) in the fluidized-bed heat exchanger and is then scrubbed to remove tar, dust, and steam at around 200°C (390°F).

The coal and residual oil slurry is gasified in the fluidized-bed gasifier. The charged slurry is converted into gas and char by thermal cracking reactions in the upper zone of the fluidized bed. The produced char is further gasified with the steam and oxygen that enter the gasifier just below the fluidizing gas distributor. The ash is discharged from the gasifier and is indirectly cooled with steam and then discharged into the ash hopper. It is burned in an incinerator to produce process steam. Coke deposited on the silica sand is removed in the incinerator.

8.3.3.3 *Hydrocarbon gasification*

The gasification of hydrocarbon derivatives to produce hydrogen is a continuous, non-catalytic process that involves partial oxidation of the hydrocarbon. Air or oxygen (with steam or carbon dioxide) is used as the oxidant at 1095–1480°C (2000–2700°F). Any carbon produced (2–3% by weight of the feedstock) during the process is removed as a slurry in a carbon separator and pelletized for use either as a fuel or as raw material for carbon-based products.

8.3.3.4 *Hypro process*

The hypro process is a continuous catalytic method for hydrogen manufacture from natural gas or from refinery effluent gases. The process is designed to convert natural gas:

$$CH_4 \rightarrow C + 2H_2$$

Hydrogen is recovered by phase separation to yield hydrogen of approximately 93% purity; the principal contaminant is methane.

8.3.3.5 *Pyrolysis processes*

Recently, there has been increased interest in the use of pyrolysis processes to produce hydrogen. Specifically, there has been increased focused on the pyrolysis of methane (natural gas) and hydrogen sulfide.

Natural gas is readily available and offers relatively rich stream of methane with smaller amounts of ethane, propane, and butane. The thermocatalytic decomposition of natural gas hydrocarbon derivatives offers an alternate method for the production of hydrogen (Uemura *et al.*, 1999; Weimer *et al.*, 2000; Dahl and Weimer, 2001):

$$C_nH_m \rightarrow nC + (m/2)H_2$$

If a hydrocarbon fuel such as natural gas (methane) is to be used for hydrogen production through direct decomposition, then the process that is optimized to yield hydrogen production may not be suitable for achieving high-quality carbon black by-product intended for the industrial rubber market. Moreover, it appears that the carbon produced from high-temperature (850–950°C; 1560–1740°F) direct thermal decomposition of methane is a soot-like material with a high tendency for catalyst deactivation (Murata, 1997). Thus, if the object of methane decomposition is hydrogen production, then the carbon by-product may not be marketable as high-quality carbon black for rubber and tire applications.

The production of hydrogen by direct decomposition of hydrogen sulfide has been studied extensively and a process proposed (Clark and Wassink, 1990; Zaman and Chakma, 1995; Donini, 1996; Luinstra 1996). Hydrogen sulfide decomposition is a highly endothermic process and equilibrium yields are poor (Clark *et al.*, 1995). At temperatures less than 1500°C (2730°F), the thermodynamic equilibrium is unfavorable toward hydrogen formation. However, in the presence of catalysts such as platinum–cobalt (at 1000°C; 1830°F), disulfides of molybdenum or tungsten (Mo or W) at 800°C (1470°F) (Kotera *et al.*, 1976), or other transition metal sulfides supported on alumina (at 500–800°C, 930–1470°F), decomposition of hydrogen sulfide proceeds rapidly (Kiuchi, 1982, Bishara *et al.*, 1987, Al-Shamma and Naman, 1989, Clark and Wassink, 1990, Megalofonos and Papayannakos, 1997; Arild, 2000; Raissi, 2001). In the temperature range of approximately

800–1500°C (1470–2730°F), thermolysis of hydrogen sulfide can be treated simply:

$$H_2S \rightarrow H_2 + 1/xS_x \qquad \Delta H_{298K} = +34{,}300 \text{ Btu/lb}$$

where $x = 2$. Outside this temperature range, multiple equilibria may be present, depending on temperature, pressure, and relative abundance of hydrogen and sulfur (Clark, 1990).

Above approximately 1000°C (1830°F), there is a limited advantage of using catalysts since the thermal reaction proceeds to equilibrium very rapidly (Clark and Wassink, 1990). The hydrogen yield can be doubled by preferential removal of either H_2 or sulfur from the reaction environment, thereby shifting the equilibrium. The reaction products must be quenched quickly after leaving the reactor to prevent reversible reactions.

8.3.3.6 *Shell gasification (partial oxidation) process*

The Shell Gasification Process is a flexible process for generating synthesis gas, principally hydrogen and carbon monoxide, for the ultimate production of high-purity high-pressure hydrogen, ammonia, methanol, fuel gas, town gas, or reducing gas by reaction of gaseous or liquid hydrocarbon derivatives with oxygen, air, or oxygen-enriched air.

The most important step in converting high-boiling residue into industrial gas is the partial oxidation of the oil using oxygen with the addition of steam. The gasification process takes place in an empty, refractory-lined reactor at temperatures of approximately 1400°C (2550°F) and pressures ranging 29–1140 psi (196–7845 kPa). The chemical reactions in the gasification reactor proceed without catalyst to produce gas containing carbon amounting to about 0.5–2% by weight, based on the feedstock. The carbon is removed from the gas with water, extracted in most cases with feed oil from the water, and returned to the feed oil. The high reformed gas temperature is utilized in a waste heat boiler for generating steam. The steam is generated at 850–1565 psi (5884–10787 kPa). Some of this steam is used as process steam and for oxygen and oil preheating. The surplus steam is used for energy production and heating purposes.

8.3.3.7 *Steam–methane reforming*

Steam–methane reforming is a catalytic process that involves a reaction between natural gas or other low-boiling hydrocarbon derivatives and steam. Steam–methane reforming is the benchmark process that has been employed over a period of several decades for hydrogen production. The process involves reforming natural gas in a continuous catalytic process in which the major reaction is the formation of carbon monoxide and hydrogen from methane and steam.

Thus, the first reforming step catalytically reacts methane (the chief chemical constituent of natural gas) to form hydrogen and carbon monoxide in an endothermic (heat-absorbing) reaction:

$$CH_4 + H_2O = CO + 3H_2 \qquad \Delta H_{298K} = +97{,}400 \text{ Btu/lb}$$

Higher molecular-weight feedstocks can also be reformed to hydrogen:

$$C_3H_8 + 3H_2O \rightarrow 3CO + 7H_2$$

That is,

$$CnHm + nH2O \rightarrow nCO + (0.5m + n)H_2$$

In the actual process, the feedstock is first desulfurized by passage through activated carbon, which may be preceded by caustic and water washes. The desulfurized material is then mixed with steam and passed over a nickel-based catalyst (730–845°C, 1350–1550°F and 400 psi, 2758 kPa). Effluent gases are cooled by the addition of steam or condensate to approximately 370°C (700°F), at which point carbon monoxide reacts with steam in the presence of iron oxide in a shift converter to produce carbon dioxide and hydrogen in which the carbon monoxide is then "shifted" with steam to form additional hydrogen and carbon dioxide in an exothermic (heat-releasing) reaction:

$$CO + H_2O = CO_2 + H_2 \qquad \Delta H_{298K} = -41.16 \text{ kJ/mol} \qquad (3)$$

The carbon dioxide is usually removed by amine washing, leaving behind hydrogen separately for its commercial use; the hydrogen is usually a high-purity (>99%) material.

Since the presence of carbon monoxide or carbon dioxide in the hydrogen stream can interfere with the chemistry of the catalytic application, a third stage is used to convert these gases into methane:

$$CO + 3H_2 \rightarrow CH_4 + H_2O$$

$$CO_2 + 4H_2 \rightarrow CH_4 + 2H_2O$$

For many refiners, sulfur-free natural gas (CH_4) is not always available to produce hydrogen by this process. In that case, higher-boiling hydrocarbon derivatives (such as propane, butane, or naphtha) may be used as the feedstock to generate hydrogen (*q.v.*).

The net chemical process for steam methane reforming is then given by the following:

$$CH_4 + 2H_2O \rightarrow CO_2 + 4H_2 \qquad \Delta H_{298K} = +165.2 \text{ kJ/mol}$$

Indirect heating provides the required overall endothermic heat of reaction for the steam–methane reforming.

One way of overcoming the thermodynamic limitations of steam reforming is to remove either hydrogen or carbon dioxide as it is produced, hence shifting the thermodynamic equilibrium toward the product side. The concept of sorption-enhanced methane steam reforming is based on *in situ* removal of carbon dioxide by using a sorbent such as calcium oxide (CaO):

$$CaO + CO_2 \rightarrow CaCO_3$$

Sorption enhancement results in lower reaction temperatures, which may reduce catalyst coking and sintering, while enabling the use of less expensive reactor wall materials. In addition, heat release by the exothermic carbonation reaction supplies most of the heat required by the endothermic reforming reactions. However, energy is required to regenerate the sorbent to its oxide form by the energy-intensive calcination reaction:

$$CaCO_3 \rightarrow CaO + CO_2$$

Use of a sorbent requires either that there be parallel reactors operated alternatively and out of phase in reforming and sorbent regeneration modes, or that a sorbent be continuously transferred between the reformer/carbonator and the regenerator/calciner (Balasubramanian *et al.*, 1999; Hufton *et al.*, 1999).

In autothermal (or secondary) reformers, the oxidation of methane supplies the necessary energy and is carried out either simultaneously or prior to the reforming reaction (Brandmair *et al.*, 2003; Ehwald *et al.*,

2003; Nagaoka *et al.*, 2003). The equilibrium of the methane steam reaction and the water–gas shift reaction determines the conditions for optimum hydrogen yields. The optimum conditions for hydrogen production require high temperature at the exit of the reforming reactor (800–900°C, 1470–1650°F), high excess of steam (molar steam/carbon ratio of 2.5–3) and relatively low pressures (below 450 psi). Most commercial plants employ supported nickel catalysts for the process.

The steam–methane reforming process described briefly above would be an ideal hydrogen production process if it were not for the fact that large quantities of natural gas, a valuable resource, are required both as feed gas and as combustion fuel. For each mole of methane reformed, more than 1 mole of carbon dioxide is coproduced and must be disposed. This can be a major issue as it results in the same amount of greenhouse gas emission as would be expected from direct combustion of natural gas or methane. In fact, the production of hydrogen as a clean burning fuel by way of steam reforming of methane and other fossil-based hydrocarbon fuels is not in environmental balance if in the process, carbon dioxide and carbon monoxide are generated and released into the atmosphere, although alternate scenarios are available (Gaudernack, 1996). Moreover, as the reforming process is not totally efficient, some of the energy value of the hydrocarbon fuel is lost by conversion to hydrogen but with no tangible environmental benefit, such as a reduction in emission of greenhouse gases. Despite these apparent shortcomings, the process has the following advantages: (i) it produces 4 moles of hydrogen for each mole of methane consumed, (ii) feedstocks for the process (methane and water) are readily available, (iii) the process is adaptable to a wide range of hydrocarbon feedstocks, (iv) it operates at low pressures, less than 450 psi, (v) it requires a low steam/carbon ratio of the order of 2.5–3, (vi) good utilization of input energy of up to 93%, (vii) it can use catalysts that are stable and resist poisoning, and (viii) good process kinetics.

Liquid feedstocks, either liquefied petroleum gas or naphtha, can also provide backup feed if there is a risk of natural gas curtailments. The feed handling system needs to include a surge drum, feed pump, and vaporizer (usually steam-heated) followed by further heating before desulfurization. The sulfur in liquid feedstocks occurs as mercaptans, thiophene derivatives, or higher boiling compounds. These compounds are stable and will not be removed by zinc oxide; therefore, a hydrogenation unit will be required. In addition, as is the case with refinery gas, olefins must also be hydrogenated if they are present.

The reformer will generally use a potash-promoted catalyst to prevent coke buildup due to cracking of the heavier feedstock. If liquefied petroleum gas is to be used only occasionally, it is often possible to use a methane-type catalyst at a higher steam/carbon ratio to avoid coking. Naphtha will require a promoted catalyst unless a pre-former is used.

8.3.3.8 *Steam–naphtha reforming*

Steam–naphtha reforming is a continuous process for the production of hydrogen from liquid hydrocarbon derivatives and is, in fact, similar to steam–methane reforming, one of the several possible processes for the production of hydrogen from low-boiling hydrocarbon derivatives other than ethane (Murata, 1998; Muradov, 1997, 2000; Brandmair *et al.*, 2003; Find *et al.*, 2003). A variety of naphtha types in the gasoline boiling range may be employed, including feeds containing up to 35% aromatics. Thus, following pretreatment to remove sulfur compounds, the feedstock is mixed with steam and routed to the reforming furnace (675–815°C, 1250–1500°F, 300 psi, 2068 kPa), where hydrogen is produced.

8.3.3.9 *Synthesis gas generation*

The synthesis gas generation process is a non-catalytic process for producing synthesis gas (principally hydrogen and carbon monoxide) for the ultimate production of high-purity hydrogen from gaseous or liquid hydrocarbon derivatives.

In this process, a controlled mixture of preheated feedstock and oxygen is fed to the top of the generator where carbon monoxide and hydrogen emerge as the products. Soot, produced in this part of the operation, is removed in a water scrubber from the product gas stream and then extracted from the resulting carbon–water slurry with naphtha and transferred to a fuel oil fraction. The oil–soot mixture is burned in a boiler or recycled to the generator to extinction to eliminate carbon production as part of the process.

The soot-free synthesis gas is then charged to a shift converter where the carbon monoxide reacts with steam to form additional hydrogen and carbon dioxide at the stoichiometric rate of 1 mole of hydrogen for every mole of carbon monoxide charged to the converter.

The reactor temperatures vary from 1095°C to 1490°C (2000°F to 2700°F), while pressures can vary from approximately atmospheric pressure to approximately 2000 psi (13,790 kPa). The process has the capability of producing high-purity hydrogen although the extent of the purification procedure depends on the use to which the hydrogen is to be put. For example, carbon dioxide can be removed by scrubbing with various alkaline reagents, while carbon monoxide can be removed by washing with liquid nitrogen or, if nitrogen is undesirable in the product, the carbon monoxide should be removed by washing with copper–amine solutions.

This particular partial oxidation technique has also been applied to a whole range of liquid feedstocks for hydrogen production. There is now serious consideration being given to hydrogen production by the partial oxidation of solid feedstocks such as crude oil coke (from both delayed and fluid-bed reactors), lignite, and coal, as well as crude oil residua.

The chemistry of the process, using naphthalene as an example, may be simply represented as the selective removal of carbon from the hydrocarbon feedstock and further conversion of a portion of this carbon into hydrogen:

$$C_{10}H_8 + 5O_2 \rightarrow 10CO + 4H_2$$

$$10CO + 10H_2O \rightarrow 10CO_2 + 10H_2$$

Although these reactions may be represented very simply using equations of this type, the reactions can be complex and result in carbon deposition on parts of the equipment, thereby requiring careful inspection of the reactor.

8.3.3.10 *Texaco gasification process*

The Texaco gasification process (the Texaco partial oxidation process) is a gasification process for generating synthetic gas, principally hydrogen and carbon monoxide. The characteristic of the Texaco gasification process is to inject feedstock together with carbon dioxide, steam, or water into the gasifier. Therefore, solvent deasphalted residua or crude oil coke rejected from any coking method can be used as a feedstock for this gasification process. The produced gas from this gasification process can be used for the production of high-purity high-pressurized hydrogen, ammonia, and methanol. The heat recovered from the high-temperature gas is used for the generation of steam in the waste heat boiler. Alternatively, the less expensive quench-type configuration is preferred when high-pressure steam is not needed or when a high degree of shift is needed in the downstream CO converter.

In the process, the feedstock, together with the feedstock carbon slurry recovered in the carbon recovery section, is pressurized to a given pressure, mixed with high-pressure steam and then blown into the gas generator through the burner together with oxygen.

The gasification reaction is a partial oxidation of hydrocarbon derivatives to carbon monoxide and hydrogen:

$$C_xH_{2y} + x/2O_2 \rightarrow xCO + yH_2$$

$$C_xH_{2y} + xH_2O \rightarrow xCO + (x+y)H_2$$

The gasification reaction is instantly completed, thus producing gas mainly consisting of H_2 and CO ($H_2 + CO = >90\%$). The high-temperature gas leaving the reaction chamber of the gas generator enters the quenching chamber linked to the bottom of the gas generator and is quenched to 200–260°C (390–500°F) with water.

8.4 Gasification of Other Feedstocks

Gasification offers more scope to recover products from waste than incineration. When waste is burnt in a modern incinerator, the only practical product obtained is energy, whereas the gases, oils, and solid char from pyrolysis and gasification can not only be used as a fuel but also purified and used as a feedstock for petrochemicals and other applications. Many processes also produce a stable granulate instead of an ash, which can be more easily and safely utilized. In addition, some processes are targeted at producing specific recyclables such as metal alloys and carbon black. From waste gasification, in particular, it is feasible to produce hydrogen, which many see as an increasingly valuable resource.

The process can be used in conjunction with gas engines (and potentially gas turbines) to obtain higher conversion efficiency than conventional fossil-fuel energy generation. By displacing fossil fuels, waste pyrolysis and gasification can help meet renewable energy targets, address concerns related to global climate change, and contribute to achieving Kyoto Protocol commitments. Conventional incineration, when used in conjunction with steam-cycle boilers and turbine generators, achieves lower efficiency.

In addition to the feedstock derived above, the gasification process can accept a variety of feedstocks, but the reactor must be selected on the basis of the feedstock properties and behavior in the process, especially when coal and biomass are considered as gasification feedstocks (Demirbaş, 2011; Speight, 2014b). Further, due to the historical use of coal for gasification purposes (Speight, 2013), it remains to be the main feedstock against which the suitability of all other feedstocks is measured. Therefore, inclusion of coal among the gasification feedstocks is warranted.

Many of the processes fit well into a modern integrated approach to waste management. They can be designed to handle the waste residues and are fully compatible with an active program of composting for the waste fraction that is subject to decay and putrefaction. Thus, by analogy with coal, the high-temperature conversion of waste (Chapter 10) is a downdraft gasification process, which gasifies the feed material within a controlled and limited oxygen supply. Combustion of the feed material is prevented by the limited oxygen supply. The temperature within the reactor reaches 2,700°C (4,890°F), at which point molecular dissociation takes place. The pollutants that were contained within the feedstock such as dioxin derivatives and furan derivatives as well as pathogens are completely cracked into harmless compounds.

All metal components in the waste stream are converted into a castable iron alloy/pig iron for use in steel foundries. The mineral fraction is reduced to a non-leaching vitrified glass, which can be used in road construction and/or further processed into a mineral wool for insulation. All of the organic material is fully converted into a fuel quality synthesis gas, which can be used to produce electrical energy, heat, methanol, or various other chemical compounds. The resultant synthesis gas, with a hydrogen/carbon monoxide ratio approximately equal to one, is also capable of being used for the production of Fischer–Tropsch fuels. Under certain conditions, heat from the reactor could be used for district heating, industrial steam production, or water desalination plants.

A wide range of materials can be handled by gasification technologies and specific processes have been optimized to handle particular feedstock (e.g. tire pyrolysis and sewage sludge gasification), while others have been designed to process mixed wastes. For example, recovering energy from agricultural and forestry residues, household and commercial waste, materials recycling (auto-shredder residue, electrical and electronic scrap, tires, mixed plastic waste, and packaging residues) are feasible processes.

8.5 Fuels and Petrochemicals

The fuels and petrochemical industries, as the names imply, is based on the production of chemicals from, initially, crude oil. However, there is more to the industry than just crude oil products. The petrochemical industry also deals with chemicals manufactured from the by-products of crude oil refining, such as natural gas, natural gas liquids, and (in the context of this book) other feedstocks such as coal, oil shale, and biomass. The structure of the industry is extremely complex, involving thousands of chemicals and processes, and there are many inter-relationships within the industry with products of one process being the feedstocks of many others. For most chemicals, the production route from feedstock to final products is not unique, but includes many possible alternatives. As complicated as it may seem, however, this structure is comprehensible, at least in the general form.

At the beginning of the production chain are the raw feedstocks: crude oil, natural gas, and alternate carbonaceous feedstocks tar. From these are produced a relatively small number of important building blocks, which include primarily, but not exclusively, the lower-boiling olefins and aromatic derivatives, such as ethylene, propylene, butylene isomers, butadiene, benzene, toluene, and the xylene isomers. These building blocks are then converted into a complex array of thousands of intermediate chemicals. Some of these intermediates have commercial value in and of themselves, and others are purely intermediate compounds in the production chains. The final products of the petrochemical industry are generally not consumed directly by the public, but are used by other industries to manufacture consumer goods.

Thus, on a scientific basis, as might be expected, the petrochemical industry is concerned with the production and trade of petrochemicals that have a wide influence on lifestyles through the production of commodity chemicals and specialty chemicals that have a marked influence on lifestyles:

Crude oil/natural gas → bulk chemicals (commodity chemicals) → specialty chemicals

The basis of the petrochemical industry and, therefore, petrochemical production consists of two steps: (i) feedstock production from primary energy sources to feedstocks and (ii) petrochemical production from feedstocks.

Crude oil/natural gas → feedstock production → petrochemical products

This simplified equation encompasses the multitude of production routes available for most chemicals. In the actual industry, many chemicals are products of more than one method, depending on the local conditions, corporate policies, and desired by-products. There are also additional methods available, which have either become obsolete and are no longer used, or which have never been used commercially but could become important as technology, supplies, and other factors change. Thus, versatility, adaptability, and dynamic nature are three of the important features of the modern petrochemical industry.

Thus, the fuel and petrochemical industry began as suitable by-products became available through improvements in the refining processes. As the decades of 1920s and 1930s ended, the industry had developed in parallel with the crude oil industry and has continued to expand rapidly since the 1940s as the crude oil refining industry was able to provide relatively cheap and plentiful raw materials (Speight, 2002; Hsu and Robinson, 2006; Gary *et al.*, 2007; Lee *et al.*, 2007; Speight, 2011). The supply–demand scenario as well as the introduction of many innovations have resulted in basic chemicals and plastics becoming the key building blocks for the manufacture of a wide variety of durable and non-durable consumer goods. Chemicals and plastic materials provide the fundamental building blocks that enable the manufacture of the vast majority of consumer goods. Moreover, the demand for chemicals and plastics is driven by global economic conditions, which are directly linked to the demand for consumer goods.

At the start of the production chain is the selection and preparation of the feedstock from which the petrochemicals will be produced. Typically, primary energy sources, such as crude oil, natural gas, coal, and biomass are extracted and then converted into feedstocks, such as naphtha, gas oil, and/or methanol. In the production of petrochemicals, the feedstocks are converted into basic petrochemicals, such as ethylene ($CH_2{=}CH_2$) and aromatic derivatives, which are then separated from each other. Thus, petrochemicals or products derived from these feedstocks, along with other raw materials, are converted into a wide range of products.

Therefore, the history of the industry has always been strongly influenced by the supply of primary energy sources and feedstocks. Thus, the petrochemical industry directly interfaces with the crude oil industry and the natural gas industry, which provides the feedstocks (Speight, 2014a, 2017, 2019a), and especially the downstream sector, as well as the potential for the introduction and use of non-conventional feedstocks (Chapter 2). A major part of the petrochemical industry is the polymer (plastics) industry (Speight, 2014a, 2019a).

The petrochemical industry is currently the biggest of the industrial chemicals sectors and petrochemicals represent the majority of all chemicals shipped between the continents of the world (EPCA, 2016). Petrochemicals have a history that began in the 19th century and has experienced many changes since then. However, from the beginning, there have been underlying trends which shaped the evolution of the industry to modern times. From the start, it was an industry that was destined to become a global sector because of the contribution of the products to raise the standards of living of much of the population of the world. These same influences have also shaped the rate and nature of the expansion and the structure of the industry as it exists in the 21st century.

In the petrochemical industry, the organic chemicals produced in largest volumes are methanol (methyl alcohol, CH_3OH), ethylene ($CH_2=CH_2$), propylene ($CH_3CH=CH_2$), butadiene $CH_2=CHCH=CH_2$), benzene (C_6H_6), toluene ($C_6H_5CH_3$), and the xylene isomers ($H_3CC_6H_4CH_3$). Ethylene, propylene, and butadiene, along with butylene isomers, are collectively called olefins, which belong to a class of unsaturated aliphatic hydrocarbon derivatives having the general formula C_nH_{2n}. Olefin derivatives contain one or more double bonds ($>C=C<$), which make them chemically reactive and, hence, starting materials for many products. Benzene, toluene, and xylenes, commonly referred to as aromatics, are unsaturated cyclic hydrocarbon derivatives containing one or more rings.

As stated above, some of the chemicals and compounds produced in a refinery are destined for further processing and as raw material feedstocks for the fast growing petrochemical industry. Such non-fuel uses of crude oil products are sometimes referred to as their non-energy uses. Crude oil products and natural gas provide two of the basic starting points for this industry: methane, naphtha, including benzene, toluene, and the xylene isomers, and refinery gases, which contain olefin derivatives, such as ethylene, propylene, and, potentially, all of the butylene isomers (Parkash, 2003; Gary *et al.*, 2007; Speight, 2014a, 2017, 2019a; Hsu and Robinson, 2017).

Petrochemical intermediates are generally produced by chemical conversion of primary petrochemicals into more complicated derivative products. Petrochemical derivative products can be produced in a variety of ways: directly from primary petrochemicals; through intermediate products that still contain only carbon and hydrogen; through intermediates that incorporate chlorine, nitrogen or oxygen in the finished derivative. In some cases, they are finished products; in others, more steps are needed to arrive at the desired composition.

Of all the processes used, one of the most important is polymerization (Speight, 2014a, 2019a). It is used in the production of products, such as plastics, fibers, and synthetic rubber, the main finished petrochemical derivatives. Some typical petrochemical intermediates are as follows: (i) vinyl acetate, $CH_2=CHOCOCH_3$, which is used in the manufacture of paint, paper, and textile coatings, (ii) vinyl chloride, $CH_2=CHCl$, which is used in the manufacture of polyvinyl chloride (PVC), resin, (iii) ethylene glycol, CH_2OHCH_2OH, which is used for polyester textile fibers, and (iv) styrene, which is important in rubber, and the manufacture of plastics. The end products number in thousands, and some are used as inputs into the chemical industry for further processing. The more common products made from petrochemicals include adhesives, plastics, soaps, detergents, solvents, paints, drugs, fertilizers, pesticides, insecticides, explosives, synthetic fibers, synthetic rubber, and flooring and insulating materials.

Petrochemical products include such items as plastics, soaps, and detergents, solvents, drugs, fertilizers, pesticides, explosives, synthetic fibers and rubbers, paints, epoxy resins, and flooring and insulating materials. Petrochemicals are found in products as diverse as aspirin, luggage, boats, automobiles, aircraft, polyester clothes, and recording discs and tapes.

The petrochemical industry has grown with the crude oil industry (Goldstein, 1949; Steiner, 1961; Hahn, 1970) and is considered by some to be a mature industry. However, as is the case with the latest trends in the changing crude oil types, it must also evolve to meet the changing technological needs (Speight, 2011). The manufacture of chemicals or chemical intermediates from a variety of raw materials is well established (Wittcoff and Reuben, 1996). The use of natural gas and crude oil is an excellent example of the conversion of such raw materials to more valuable products. The individual chemicals made from natural gas and crude oil are numerous and include industrial chemicals, household chemicals, fertilizers, and paints, as well as intermediates for the manufacture of products, such as synthetic rubber and plastics.

Petrochemicals are generally considered as chemical compounds derived from crude oil either by direct manufacture or indirect manufacture as by-products from the variety of processes that are used during the refining of crude oil. Gasoline, kerosene, fuel oil, lubricating oil, wax, asphalt, and the like are excluded from the definition of petrochemicals, since they are not, in the true sense, chemical compounds but are in fact intimate mixtures of hydrocarbon derivatives.

The classification of materials as petrochemicals is used to indicate the source of the chemical compounds, but it should be remembered that many common petrochemicals can be made from other sources, and the terminology is therefore a matter of source identification.

The starting materials for the petrochemical industry are obtained from crude oil in one of two general ways. They may be present in the raw crude oil and, as such, are isolated by physical methods, such as distillation or solvent extraction. On the other hand, they may be present, if at all, in trace amounts and are synthesized during the refining operations. In fact, unsaturated (olefin) hydrocarbon derivatives, which are not usually present in crude oil, are nearly always manufactured as intermediates during the various refining sequences (Parkash, 2003; Gary *et al.*, 2007; Speight, 2014a, 2017, 2019a; Hsu and Robinson, 2017).

The manufacture of chemicals from crude oil is based on the ready response of the various compound types to basic chemical reactions, such as oxidation, halogenation, nitration, dehydrogenation addition, polymerization, and alkylation. The low molecular-weight paraffins and olefins, as found in natural gas and refinery gases, and the simple aromatic hydrocarbon derivatives have so far been of most interest because it is individual species that can be readily be isolated and dealt with. The identification of a wide range of compounds is possible, many are being manufactured and a sizable group of products is being prepared from the heavier fractions of crude oil (Speight, 2019a). For example, the various reactions of asphaltene constituents (Speight, 1994, 2014a, 2019b) indicate that these materials may be regarded as constituents that contain chemical functions and are therefore different and are able to participate in numerous chemical or physical conversions to, perhaps, more useful materials. The overall effect of these modifications is the production of materials that afford either good-grade aromatic cokes comparatively easily or the formation of products bearing functional groups that may be employed as a non-fuel material.

For example, sulfonated and sulfomethylated materials and their derivatives have satisfactorily undergone tests as drilling mud thinners, and the results are comparable to those obtained with commercial mud thinners. In addition, these compounds may also find use as emulsifiers for the *in situ* recovery of heavy oils. There are also indications that these materials and other similar derivatives of the asphaltene constituents, especially those containing functions such as carboxylic or hydroxyl, readily exchange cations and could compete well with synthetic zeolites. Other uses of the hydroxyl derivatives and/or the chloro-asphaltenes include high-temperature packing or heat transfer media.

Reactions incorporating nitrogen and phosphorus into the asphaltene constituents are particularly significant at a time when the effects of many materials containing these elements on the environment are receiving considerable attention. An example is the potential to produce slow-release soil conditioners that only release the nitrogen or phosphorus only after considerable weathering or bacteriological action (Moschopedis and Speight, 1977). One may proceed a step further and suggest that the carbonaceous residue remaining after the release of the hetero-elements may be beneficial to humus-depleted soils, such as the gray-wooded and solonetzic soils. It is also feasible that coating a conventional quick-release inorganic fertilizer with a water-soluble or water-dispersible derivative will provide a slower release fertilizer and an organic humus-like residue. In fact, variations on this theme are multiple.

Nevertheless, the main objective in producing chemicals from crude oil is the formation of a variety of well-defined chemical compounds that are the basis of the petrochemical industry. It must be remembered, however, that ease of separation of a particular compound from crude oil does not guarantee its use as a petrochemical building block. Other parameters, particularly the economical aspects of the reaction sequences, including the costs of the reactant equipment, must be taken into consideration.

Petrochemicals are made, or recovered from, the entire range of crude oil fractions, but the bulk of petrochemical products are formed from the lighter (C_1–C_4) hydrocarbon gases used as raw materials. These materials generally occur in natural gas, but they are also recovered from the gas streams produced during refinery, especially cracking, operations. Refinery gases are also particularly valuable because they contain substantial amounts of olefins that, owing to the double bonds, are much more reactive than saturated (paraffin) hydrocarbon derivatives. Also important as raw materials are aromatic hydrocarbon derivatives (benzene, toluene, and xylene), which are obtained in rare cases from crude oil and, more likely, from the various product streams. By means of the catalytic reforming process, non-aromatic hydrocarbon derivatives can be converted into aromatics by dehydrogenation and cyclization (Parkash, 2003; Gary *et al.*, 2007; Speight, 2014a, 2017, 2019a; Hsu and Robinson, 2017).

A highly significant proportion of these basic petrochemicals is converted into plastics, synthetic rubbers, and synthetic fibers. Together these materials are known as polymers because their molecules are high-molecular-weight compounds made up of repeated structural units that have combined chemically. The major products are polyethylene, polyvinyl chloride, and polystyrene, all derived from ethylene, and polypropylene, derived from monomer propylene. Major raw materials for synthetic rubbers include butadiene, ethylene, benzene, and propylene. Among synthetic fibers, the polyesters, which are a combination of ethylene glycol and terephthalic acid (made from xylene), are the most widely used. They account for about one-half of all synthetic fibers. The second major synthetic fiber is nylon, its most important raw material being benzene. Acrylic fibers, in which the major raw material is the propylene derivative acrylonitrile, make up most of the remainder of the synthetic fibers.

8.5.1 *Gaseous fuels and chemicals*

Synthesis gas is a very important chemical intermediate for many relevant processes including the production of methanol and Fischer–Tropsch synthesis of synthetic fuels (Chapter 10) and, thus, is an essential industrial feedstock in several catalytic processes.

8.5.1.1 *Ammonia*

In the process, natural gas (methane) or liquefied petroleum gases (such as propane) or crude oil naphtha is first converted into hydrogen by steam reforming (Chapter 8). The hydrogen is then combined with nitrogen to produce ammonia by the Haber–Bosch process:

Steam reforming process:

$$CH_4 + H_2O \rightarrow CO + 3H_2$$

$$C_3H_8 + 3H_2O \rightarrow 3CO + 7H_2$$

$$C_nH_m + nH_2O \rightarrow nCO + (0.5m + n)H_2$$

The Haber–Bosch process, also called the Haber process, which was developed in the first decade of the 20th century, is a nitrogen fixation process and is the main industrial procedure for the production of ammonia. The process converts atmospheric nitrogen (N_2) into ammonia (NH_3) by a reaction with hydrogen (H_2) using a metal catalyst under high temperatures and pressures:

$$N_2 + 3H_2 \rightarrow 2NH_3$$

Starting with the feedstock (natural gas), the first process leading to hydrogen production is the removal of sulfur compounds from the feedstock because sulfur-containing compounds deactivate the catalysts used in subsequent steps. Sulfur removal requires catalytic hydrogenation to convert sulfur compounds in the feedstock into hydrogen sulfide:

$$H_2 + RSH \rightarrow RH + H_2S \text{ (gas)}$$

The gaseous hydrogen sulfide is then adsorbed (Chapter 9) and removed by passing it through beds of zinc oxide where it is converted into solid zinc sulfide:

$$H_2S + ZnO \rightarrow ZnS + H_2O$$

Catalytic steam reforming of the sulfur-free feedstock is then used to form hydrogen plus carbon monoxide (synthesis gas):

$$CH_4 + H_2O \rightarrow CO + 3H_2$$

The next step is the shift conversion to convert the carbon monoxide into carbon dioxide and more hydrogen:

$$CO + H_2O \rightarrow CO_2 + H_2$$

The carbon dioxide is then removed either by absorption in aqueous ethanolamine solution or by adsorption in a pressure swing adsorption unit using a solid adsorbent (Chapter 9). The final step in producing the hydrogen is to use catalytic methanation to remove any small residual amounts of carbon monoxide or carbon dioxide from the hydrogen:

$$CO + 3H_2 \rightarrow CH_4 + H_2O$$
$$CO_2 + 4H_2 \rightarrow CH_4 + 2H_2O$$

To produce the desired end-product ammonia, the hydrogen is then catalytically reacted with nitrogen (derived from process air) to form anhydrous liquid ammonia — the Haber–Bosch process (also known as the ammonia synthesis loop):

$$3H_2 + N_2 \rightarrow 2NH_3$$

Due to the nature of the (typically multi-promoted magnetite) catalyst used in the ammonia synthesis reaction, only very low levels of oxygen-containing compounds (especially carbon monoxide, carbon dioxide, and water) can be tolerated in the synthesis gas. Relatively pure nitrogen can be obtained by an air separation step, but additional oxygen removal may be required.

The steam reforming, shift conversion, carbon dioxide removal and methanation steps each operate at absolute pressures of approximately 375–525 psi, and the ammonia synthesis loop operates at absolute pressures ranging 900–2,700 psi, depending on the process design.

8.5.1.2 *Hydrogen*

Hydrogen can be produced from the gasification product gas through the steam–methane reforming and water–gas shift reactions.

Steam–methane reforming reaction:

$$CH_4 + H_2O\ CO + 3H_2$$

Water–gas shift reaction:

$$CO + H_2O \rightleftharpoons CO_2 + H_2$$

Using a dual fluidized-bed (DFB) gasification system with carbon dioxide adsorption along with suitable catalysts, it is possible to achieve a hydrogen yield up to 70 vol% directly in the gasifier (Soukup *et al.*, 2009) Another source of hydrogen is glycerol, a by-product derived from the production of biodiesel. One approach to alleviate this problem is to transform glycerol into valuable chemicals, such as hydrogen and syngas (Lin, 2013).

8.5.1.3 *Synthetic natural gas*

Natural gas produced from coal or biomass is known as *synthetic natural gas* or *substitute natural gas* (SNG). The typical catalyst for methanation is nickel and the main reaction is represented by the following equation:

$$CO + 3H_2 \rightarrow CH_4 + H_2O$$

Nickel-based catalysts are also active in water–gas shift and hydration of higher hydrocarbon derivatives, such as olefin derivatives.

Typically a hydrogen/carbon monoxide ratio of 3 is necessary, and hence this is obtained using a water–gas shift reactor before methanation. In some types of reactors, such as fluidized-bed reactors, the water–gas shift can also be carried out in parallel with the methanation, so no external adjustment of the hydrogen/carbon monoxide ratio is necessary. Methanation can be done at atmospheric pressure, although according to the thermodynamics higher pressure is preferred.

Since nickel-based catalysts are sensitive to poisoning by sulfur-containing compounds, gas treatment (Chapter 9) is quite important before the methanation, and sulfur compounds have to be removed to below 0.1 ppm.

8.5.2 *Liquid fuels and chemicals*

Liquid hydrocarbon transportation fuels and various other chemical products can be produced from syngas via the well-known and established catalytic Fischer–Tropsch process (Chapter 10). Depending on the source of the synthesis gas, this technology is often referred to as coal-to-liquids (CTLs) technology or gas-to-liquids (GTL) technology.

8.5.2.1 *Fischer–Tropsch liquids*

Fischer–Tropsch liquids are well known for being virtually sulfur-, nitrogen-, and heteroatom-free, and they are said to carry a product premium in excess of conventional crude oil-derived liquid products, such as diesels and gas oils.

Conventional refinery processes can be used for upgrading Fischer–Tropsch liquids and Fischer–Tropsch waxes. The low-temperature Fischer–Tropsch (LTFT) process is employed in the production of waxes, which are converted into naphtha or kerosene (diesel oil) after a hydroprocessing step. The high-temperature Fischer–Tropsch (HTFT) process is employed in the production of alpha-olefins (i.e. $RCH{=}CH_2$) (Chapter 10). Fuels produced with Fischer–Tropsch synthesis are of a high quality due to a very low content of aromatic constituents and zero content of sulfur constituents. The product stream consists of various fuel types: liquefied petroleum gas (LPG), gasoline, diesel fuel, and jet fuel.

8.5.2.2 *Methanol*

Methanol (CH_3OH, MeOH, methyl alcohol, wood alcohol, or wood spirits) can be produced from fossil or renewable resources and can either be used directly as a transportation fuel or be converted further into hydrocarbon derivatives (Olsbye *et al.*, 2012; Tian *et al.*, 2015).

Methanol is produced from synthesis gas by the hydrogenation of carbon oxides over a suitable (copper oxide-, zinc oxide-, or chromium oxide-based) catalyst:

$$CO + 2H_2 \leftrightarrow CH_3OH$$

$$CO_2 + 3H_2 \leftrightarrow CH_3OH + H_2O$$

The first reaction is the primary methanol synthesis reaction and a small amount of carbon dioxide (2–10% v/v) in the feedstock acts as a promoter of this primary reaction and helps maintain catalyst activity. The reactions are exothermic and give a net decrease in molar volume. Therefore, the equilibrium is favored by high pressure and low temperature. During production, heat is released and has to be removed to keep optimum catalyst life and reaction rate; 0.3% of the produced methanol reacts further to form side products such as dimethyl ether, formaldehyde, or higher molecular-weight alcohols.

The catalyst deactivates primarily because of loss of active copper due to physical blockage of the active sites by large by-product molecules; poisoning by halogens or sulfur in the synthesis gas, which irreversibly form inactive copper salts; and sintering of the copper crystallites into larger crystals, which then have a lower surface-to-volume ratio. Conventionally, methanol is produced in two-phase systems, in which the reactants and products form the gas phase and the catalyst form the solid phase.

8.5.2.3 *Dimethyl ether*

Dimethyl ether (DME, CH_3OCH_3) is generally produced by dehydration of methanol. The methanol production and dehydration processes are combined in one reactor, such that dimethyl ether is produced directly from synthesis gas slightly more efficiently than methanol. In addition, the direct synthesis of dimethyl ether allows a hydrogen/carbon monoxide ratio of approximately 1, which is an advantage for oxygen-blown gasifiers.

Like methanol, dimethyl ether has promising features as a fuel candidate with both the Otto and the diesel engine. With adaptations to the engine and the fuel system, dimethyl ether can be used in diesel engines, leading to higher fuel efficiency and lower emissions. In Otto engines, dimethyl ether can be used with liquefied petroleum gas (LPG). Also, since dimethyl ether is as easily reformed as methanol, it has a great potential as fuel for fuel cell vehicles. Dimethyl ether has physical properties similar to liquefied petroleum gas (LPG) and can be handled as a liquid using the same infrastructure as liquefied petroleum gas.

8.5.2.4 *Methanol-to-gasoline and olefins*

In the 1970s, at the time of the oil embargos, Mobil developed and commercialized a methanol-to-gasoline (MtG) process. The methanol-to-olefins (MtO) synthesis is a commercially attractive process because of the high demand for ethylene ($CH_2{=}CH_2$) and propylene ($CH_3CH{=}CH_2$).

Currently, these compounds are produced mainly through non-catalytic steam cracking of fossil fuels (naphtha). The methanol-to-olefins process uses zeolite-based catalysts that efficiently convert methanol into ethylene and propylene.

8.5.2.5 *Other processes*

The *methanol-to-diesel* (MtD) process first converts methanol into propylene, which is followed by olefin oligomerization (conversion to distillates), and then product separation-plus-hydrogenation. The process would yield mostly kerosene and diesel, along with a small yield of naphtha (or gasoline) and low-boiling hydrocarbon derivatives. The near-zero sulfur/polyaromatics diesel fuel resulting from this process would differ from the more conventional Fischer–Tropsch diesel product only in cetane number (>52 via methanol-to-synfuel compared to >70 cetane for Fischer–Tropsch diesel). The incidental gasoline stream not only would be near-zero sulfur but also have commercially acceptable octane ratings (92 RON, 80 MON) and maximally 11% v/v aromatic derivatives.

The catalytic conversion of synthesis gas to *mixed alcohols* is of great importance because mixed alcohols are valuable additives to gasoline to increase the octane number and reduce the environmental pollution. Furthermore, a benefit of the mixed alcohol synthesis is the high resistance of the catalysts against sulfur poisoning and the fact that the gas cleaning facilities can be simpler than in other syntheses. Mixed alcohols can also be converted into high-quality fuels via dehydration and oligomerization. Typically, the alkali-doped oxides (zinc and chromium oxides) and alkali-doped sulfides (molybdenum sulfides) are used as catalysts for mixed alcohol synthesis.

Depending on the process conditions and catalysts, the main primary products are generally methanol and water. The reaction mechanism of mixed alcohol synthesis is represented simply as follows:

$$n\mathrm{CO} + 2n\mathrm{H_2} \rightarrow \mathrm{C}_n\mathrm{H}_{2n} + {}_1\mathrm{OH} + (n\text{–}1)\mathrm{H_2O}$$

Owing to reaction stoichiometry the proposed carbon monoxide ratio is 2, but the optimal ratio is in practice closer to 1 because of the water–gas shift reaction that occurs in parallel with the alcohol formation.

8.6 The Future

The petrochemical industry is concerned with the production and trade of petrochemicals and has a direct relationship with the crude oil industry, especially the downstream sector of the industry. The petrochemical industries are specialized in the production of petrochemicals that have various industrial applications. The petrochemical industry can be considered to be a sub-sector of the crude oil industry since without the crude oil industry the petrochemical industry cannot exist. Thus, crude oil is the major prerequisite raw material for the production of petrochemicals in either quality or quantity. In addition, the petrochemical industry is subject to the geopolitics of the crude oil industry, with each industry being reliant on the other for sustained survival.

In the 1970s, as a result of various oil embargos, coal liquefaction processes seemed on the point of commercialization and would have provided new sources of coal liquids for chemical use, as well as fulfilling the principal intended function of producing alternate fuels. Because of the varying price of crude oil, this prospect is unlikely to come to fruition in the immediate future, due to the question of economic viability rather than not technical feasibility. The combination of these and other factors has contributed to sharpening the focus on the use of coal for the production of heat and power, and lessening or eclipsing its possible use as a starting point for other processes.

The growth and development of petrochemical industries depends on a number of factors and also varies from one country to another based on technical know-how, marketability and applicability of these petrochemicals for the manufacture of petrochemical products through petrochemical processes which are made feasible by knowledge and application of petrochemistry. Moreover, petrochemistry is a branch of chemistry (chemistry being a branch of natural science concerned with the study of the composition and constitution of substances and the changes such substances undergo because of changes in the molecules that make up such substances), which deals with crude oil, natural gas, and their derivatives.

However, not all of the petrochemical or commodity chemical materials produced by the chemical industry are made in a single location but groups of related materials are often made in adjacent manufacturing plants to induce industrial symbiosis as well as material and utility efficiency and other economies of scale (integrated manufacturing). Specialty and fine chemical companies are sometimes found in manufacturing locations similar to petrochemicals, but, in most cases, they do not need the same level of large-scale infrastructure (e.g. pipelines, storage, ports, and power) and therefore can be found in multi-sector business parks. This will continue as long as the refining industry continues to exist in its present form (Favennec, 2001; Speight, 2011).

The petrochemical industry continues to be impacted by the globalization and integration of the world economy. For example, world-scale petrochemical plants built during the past several years are substantially larger than those built over two decades ago. As a result, smaller, older, and less efficient units are being shut down, expanded, or, in some cases, retrofitted to produce different chemical products. In addition, crude oil prices had been on the rise during the past decade and petrochemical markets are impacted during sharp price fluctuations, creating a cloud of uncertainty in upstream and downstream investments. Also, increasing concerns over fossil fuel supply and consumption, with respect to their impact on health and the environment, have led to the passage of legislation globally that will affect chemical and energy production and processing in the foreseeable future.

The recent shift from local markets to a large global market has led to an increase in the competitive pressures on petrochemical industries. Furthermore, because of fluctuations in product prices and the high price of feedstocks, the economical attractiveness of petrochemical plants can be considered as the main challenge. The ever-increasing cost of energy and more stringent environmental regulations has impacted the operational costs. When cheap feedstocks are not available, the best method for profitability is to apply integration and optimization in petrochemical complexes with the adjacent refineries. This is valid for installed plants and plants under construction. Petrochemical–refinery integration is an important factor in reducing costs and increasing efficiencies because integration guarantees the supply of feedstocks for petrochemical industries. Also, integrated schemes take the advantage of the economy of scale as well, and an integrated complex can produce more diverse products. Petrochemical–refinery integration avoids selling crude oil, optimizes products, economizes costs, and increases benefits.

On an innovation and technological basis (Hassani *et al.*, 2017), manufacturing processes introduced in recent years have resulted in raw material replacement, shifts in the ratio of coproduct(s) produced, and cost. This has led to a supply/demand imbalance, particularly for smaller downstream petrochemical derivatives. In addition, growing environmental concerns and higher crude oil prices have expedited the development and commercialization of renewably derived chemical products and technologies previously considered to be economically impractical. Among the various technological advances, the combination of vertical hydraulic fracturing ("fracking") and horizontal drilling in multistage hydraulic fracturing resulted in a considerable rise in natural gas production in the United States. This new potential has caused many countries to reexamine their natural gas reserves and pursue development of their own gas plays.

Currently, natural gas and crude oil are the main sources of the raw materials for the production of petrochemicals because they (natural gas and crude oil) are the least expensive, the most readily available, and could be processed most easily into the primary petrochemicals. However, as the current century progresses and given the changes to crude oil supply that might be anticipated during the next five decades (Speight, 2011), there is a continuing need to assess the potential of other sources of petrochemicals.

For example, coal could well undergo a revitalization of use, given that there is a need to adhere to the various environmental regulations that apply to the use of any fossil fuel. Coal carbonization is the earliest and most important method to produce chemicals. For many years, chemicals that have been used for the manufacture of diverse materials, such as nylon, styrene, fertilizers, activated carbon, drugs, and medicine, as well as many others, have been made from coal. These products will expand in the future as natural gas and crude oil resources become strained to supply petrochemical feedstocks and coal will become a predominant chemical feedstock once more. The ways in which coal may be converted into chemicals include carbonization, hydrogenation, oxidation, solvent extraction, hydrolysis, halogenation, and gasification (followed by conversion of the synthesis gas to chemical products) (Speight, 2013, 2014a). In some cases such processing does not produce chemicals in the sense that the products are relatively pure and can be marketed as even industrial-grade chemicals. Thus, although many traditional markets for coal tar chemicals have been taken over by the petrochemical industry, the position can change suddenly as oil prices fluctuate upward. Therefore, the concept of using coal as a major source of chemicals can be very real indeed.

Compared to crude oil, shale oils obtained by retorting of oil shale from deposits are characterized by a wide boiling range, high concentrations of heteroelements, and a high content of oxygen-, nitrogen-, or sulfur-containing compounds. The chemical potential of oil shales as a retort fuel to produce shale oil and from that liquid fuel and specialty chemicals has so far been used to a relatively small extent. While the majority of countries are discovering the real practical value of shale oil, in Estonia, retorting of its national resource, kukersite oil, obtained for the production of a variety of products is in practice for 75 years. Using stepwise cracking, motor fuels have been produced and even exported before World War II. At the same time, shale oils possess molecular structures of interest to the specialty chemicals industry and also a number of non-fuel specialty products have been marketed based on the functional group, broad-range concentrate, or even pure-compound values.

Based on the large quantity of oxygen-containing compounds in heavy fraction, asphalt blending material, road asphalt and road oils, construction mastics, anticorrosion oils, and rubber softeners are produced, Benzene and toluene for the production of benzoic acid as well as solvent mixtures on pyrolysis of lighter fractions of shale oil are produced. Middle shale oil fractions with antiseptic properties are used to produce an effective oil for the impregnation of wood as a major shale oil-derived specialty product. Water-soluble phenols are selectively extracted from shale oil, fractionated and crystallized for the production of pure 5-methyl resorcinol and other alkyl resorcinol derivatives and high-value intermediates to produce tanning agents, epoxy resins, and adhesives, diphenyl ketone and phenol-formaldehyde adhesive resins, rubber modifiers, chemicals, and pesticides. Some conventional products such as coke and distillate boiler fuels are produced from shale oil as by-products. New market opportunities for shale oil and its fractions may be found by improving the oil conversion and separation techniques. In the petrochemical industry, the organic chemicals produced in the largest volumes are methanol, ethylene, propylene, butadiene, benzene, toluene, and xylene isomers.

Basic chemicals and plastics are the key building blocks for the manufacture of a wide variety of durable and non-durable consumer goods. The petrochemical industry continues to be impacted by the globalization and integration of the world economy. In the future, manufacturing processes introduced in recent years will continue to result in the adaptation of the industry to new feedstocks, which will cause shifts in the ratio of products produced. This, in turn, will lead to the potential for a supply/demand imbalance, particularly for smaller downstream petrochemical derivatives. In addition, the growing environmental concerns and the variability of crude oil prices (usually upward) will expedite the development and commercialization of chemical products from sources other than natural gas and crude oil. As a result, feedstocks and technologies previously considered to be economically impractical will rise to meet the increasing demand.

There is, however, the ever-present political uncertainty that arises from the occurrence of natural gas and crude oil resources in countries (provider countries) other than the user countries. This has serious global implications for the supply and demand of petrochemicals and raw materials. In addition, the overall expansion of the population and the increase in individual purchasing power have resulted in an increase in the demand for finished goods and greater consumption of energy in China, India, and Latin America.

However, the continued development of shale gas (tight gas) resources as well as crude oil from tight formations as well as the various technological advances to recover these resources (such as the combination of vertical hydraulic fracturing and horizontal drilling) will lead to a considerable rise in natural gas production and crude oil production. This new potential will cause many countries to reexamine their natural gas reserves and crude oil reserves to pursue development of their own nationally occurring gas plays and crude oil plays.

The production of chemicals from biomass is becoming an attractive area of investment for industries in the framework of a more sustainable economy. From a technical point of view, a large fraction of industrial chemicals and materials from fossil resources can be replaced by their bio-based counterparts. Nevertheless, fossil-based chemistry is still dominant because of optimized production processes and lower costs. The best approach to maximize the valorization of biomass is the processing of biological feedstocks in integrated biorefineries where both bio-based chemicals and energy carriers can be produced, similar to a traditional crude oil refinery. The challenge is to prove, together with the technical and economic feasibility, an environmental feasibility, in terms of lower impact over the entire production chain.

Biomass is essentially a rich mixture of chemicals and materials and, as such, has tremendous potential as a feedstock for making a wide range of chemicals and materials with applications in industries from pharmaceuticals to furniture. Various types of available biomass feedstocks, including waste, and the different pre-treatment and processing technologies are being developed to convert these feedstocks into platform chemicals, polymers, materials and energy.

There are several viable biological and chemical transformation pathways to convert sugars into building blocks. A large number of sugar-to-building block transformations can be carried out by aerobic fermentation employing fungi, yeast or bacteria. Chemical and enzymatic transformations are also important process options. It should be noted, however, that pathways with more challenges and barriers are less likely to be considered as viable industrial processes. In addition to gasification followed by Fischer–Tropsch chemistry of the gaseous product (synthesis gas), chemical reduction, oxidation, dehydration, bond cleavage, and direct polymerization predominate. Enzymatic biotransformations comprise the largest group of biological conversions and some biological conversions can be accomplished without the need for an intermediate building block. 1,3-Propanediol ($HOCH_2CH_2CH_2OH$) is an example where a set of successive biological processes convert sugar directly into an end product. Each pathway has its own set of advantages and disadvantages. Biological conversions, of course, can be tailored to result in a specific molecular structure but the operating conditions must be relatively mild. Chemical transformations can operate at high throughput but, unfortunately, less conversion specificity is achieved.

Bio-based feedstocks may present a sustainable alternative to petrochemical sources to satisfy the ever-increasing demand for chemicals. However, the conversion processes needed for these future biorefineries will likely differ from those currently used in the petrochemical industry. Biotechnology and chemo-catalysis offer routes for converting biomass into a variety of chemicals that can serve as starting-point chemicals. While a host of technologies can be leveraged for biomass upgrading, the outcome can be significant because there is the potential to upgrade the bio-derived feedstocks while minimizing the loss of carbon and the generation of by-products.

In fact, biomass offers a source of carbon from the biosphere as an alternative to fossilized carbon laid down tens of millions of years ago. Anything that grows and is available in non-fossilized form can be classified as biomass, including arable crops, trees, bushes, animal by-products, human and animal waste, waste food, and any other waste stream that rots quickly and which can be replenished on a rolling time-frame of years or decades. One of the attractions of biomass is its versatility: under the right circumstances, it can be used to provide a sustainable supply of electricity, heat, transport fuels, or chemical feedstocks in addition to its many other uses. One of the drawbacks of biomass, especially in the face of so many potential end uses, is its limited availability, even though the precise limitation is the subject of debate. Compared with the level of attention given to biomass as a source of electricity or heat, relatively little attention has been paid to biomass as a chemical feedstock. However, in a world in which conventional feedstocks are becoming constrained and countries are endeavoring to meet targets for reducing carbon dioxide emissions, there is a question as to whether biomass is too good to burn.

Developments in homogeneous and heterogeneous catalysis have led the way to effective approaches to utilizing renewable sources; however, further advances are needed to realize technologies that are competitive with established petrochemical processes. Catalysis will play a key role, with new reactions, processes, and concepts that leverage both traditional and emerging chemo- and biocatalytic technologies.

Thus, new knowledge and better technologies are needed to deal with chemical transformations that involve milder oxidation conditions, selective reduction, and dehydration, better control of bond cleavage, and improvements in direct polymerization of multifunctional monomers. For biological transformations, a better understanding of metabolic pathways and cell biology, lower downstream recovery costs, increased utility of mixed sugar streams, and improved molecular thermal stability is necessary. While it is possible to prepare a very large number of molecular structures from the top building blocks, there is a scarcity of information about these behaviors of the molecular products and industrial processing properties. A comprehensive database on biomolecular performance characteristics would prove extremely useful to both the public sector and the private sector. Nevertheless, here is a significant market opportunity for the development of bio-based products from four-carbon building blocks. In order to be competitive with petrochemical-derived products, there is a significant technical challenge, which should be addressed from a long-term perspective.

The petrochemical industry, which is based on natural gas and crude oil, competes with the energy-providing industry for the same fossil raw material. The dwindling oil and gas reserves, concern regarding the greenhouse effect (carbon dioxide emissions) and rising energy demand worldwide raise the question of the future availability of fossil raw materials. Biotechnological, chemical and engineering solutions are needed for the utilization of this second-generation biorenewable-based supply chain. One approach consists of the concept of a biorefinery. Also, gasification followed by Fischer–Tropsch chemistry is a promising pathway. In the short and medium term, a feedstock mix with natural gas and crude oil dominating can most likely be expected. In the long term, due to the final limited availability of oil and gas, biomass will prevail. Prior to this change, a great deal of research and development efforts must be carried out to have the necessary technology available when needed.

In summary, the petrochemical industry gives a series of value-added products to the natural gas and crude oil industry, but, like any other business, it suffers from issues relating to maturity of the industry such as the following: (i) expired patents, (ii) varying demand, (iii) matching of demand with capacity and intense competition. Actions needed to combat the aches and pains of maturity include restructuring the capacity for achieving megasizes, downstream, and restructuring business practices. Strategies followed by some companies to combat maturity include exit, focus on core business and exploit a competitive advantage.

Nevertheless, the petrochemical industry is and will remain a necessary industry for the support of modern and emerging lifestyles. In order to maintain an established petrochemical industry, strategic planning is the dominating practice to maintain the industry (replace imports, export new products, use alternate feedstocks such as the return to the chemicals-from-coal concept and accept feedstocks such as oil shale and biomass), including developing criteria for selecting products/projects. After the oil crises of the 1970s (even though it has now been four decades since these crises), it is necessary to cope with the new environment of product demand through the response to new growth markets and security of feedstock supply. Mergers, alliances, and acquisitions could well be the dominating practice to combat industry maturity and increased market demand as one of the major activities. Other strategies include the focus on core business (the production of chemicals) and, last but not least, the emergence (or, in the case of coal, the re-emergence) of alternate feedstocks to ensure industry survival.

In summary, synthesis gas offers many routes to industrial chemicals. They can be classified in a direct and an indirect path. The direct path involves methanation, Fischer–Tropsch chemistry, and synthesis of oxygenates. The direct conversion deals with the straight hydrogenation of carbon monoxide to paraffins, olefins, and oxygen-containing products. Best known in the direct hydrogenation of carbon monoxide is the Fischer–Tropsch synthesis, yielding mixtures of mainly linear alkanes and/or alkenes. Mechanistically, it can be described as a reductive oligomerization of carbon monoxide following a geometric progression (Schulz–Flory distribution). With a-values close to one, broad product distributions are obtained, whereas small a-values predominantly yield methane. The indirect path embraces carbonylation, methanol, and methyl formate chemistry (Keim, 1986).

The great desire for the synthesis of ethylene glycol based on carbon monoxide chemistry is also underlined by the research efforts being reported by various companies. Besides the direct hydrogenation, routes via methanol or formaldehyde are also possible. Also, the reductive carbonylation of formaldehyde yields glycol dialdehyde, which can be hydrogenated to ethylene glycol (Keim, 1986).

References

Arena, U. 2012. Process and technological aspects of municipal solid waste gasification. A review. *Waste Management* 32: 625–639.

Baxter, L. 2005. Biomass-coal co-combustion: Opportunity for affordable renewable energy. *Fuel* 84(10): 1295–1302.

Basu, P. 2013. *Biomass Gasification, Pyrolysis and Torrefaction*, 2nd Edition: Practical Design and Theory. Academic Press, Inc., New York.

Bhattacharaya, S., Md. Mizanur Rahman Siddique, A.H., and Pham, H-L. 1999. A study in wood gasification on low tar production. *Energy* 24: 285–296.

Brar, J.S., Singh, K., Wang, J., and Kumar, S. 2012. Cogasification of coal and biomass: A review. *International Journal of Forestry Research* 2012: 1–10.

Brigwater, A.V. (Editor) 2003. *Pyrolysis and Gasification of Biomass and Waste*. CPL Press, Newbury, Berkshire, United Kingdom.

Chadeesingh, R. 2011. The Fischer–Tropsch Process. In *The Biofuels Handbook*. J.G. Speight (Editor). The Royal Society of Chemistry, London, United Kingdom. Part 3, Chapter 5.

Clark, J.H. and Deswarte, F. (Editors) 2015. *Introduction to Chemicals from Biomass*, 2nd Edition. John Wiley & Sons Inc., Hoboken, New Jersey.

Couvaras, G. 1997. Sasol's slurry phase distillate process and future applications. In *Proceedings of the Monetizing Stranded Gas Reserves Conference*, Houston, December 1997.

EPCA. 2016. 50 Years of Chemistry for You. European Petrochemical Association. Brussels, Belgium. https://epca.eu/.

EPCI. 2004. The Gasification of Residual Plastics Derived from Municipal Recycling Facilities. The Environment and Industry Council, Canadian Plastics Industry Association, Ottawa, Ontario, Canada.

Favennec, J.-P. (Editor) (2001). *Petroleum Refining: Refinery Operation and Management*. Editions Technip, Paris, France.

Gary, J.H., Handwerk, G.E., and Kaiser, M.J. 2007. *Petroleum Refining: Technology and Economics*, 5th Edition. CRC Press, Taylor & Francis Group, Boca Raton, Florida.

Gay, R.L., Barclay, K.M., Grantham, L.F., and Yosim, S.J. 1980. Fuel production from solid waste. In *Symposium on Thermal Conversion of Solid Waste and Biomass*. Symposium Series No. 130, American Chemical Society, Washington, DC. Chapter 17, pp. 227–236.

Goldstein, R.F. 1949. *The Petrochemical Industry*. E. & F. N. Spon, London.

Hahn, A.V. 1970. *The Petrochemical Industry: Market and Economics*. McGraw-Hill, New York.

Hassani, H., Silva, E.S., and Al Kaabi, A.M. 2107. The role of innovation and technology in sustaining the petroleum and petrochemical industry. *Technological Forecasting and Social Change*, 119 (June): 1–17.

Hotchkiss, R. 2003. Coal gasification technologies. *Proceedings. Institute of Mechanical Engineers Part A* 217(1): 27–33.

Hsu, C.S. and Robinson, P.R. (Editors). 2017. *Handbook of Petroleum Technology*. Springer International Publishing AG, Cham, Switzerland.

Ishi, S. 1982. Coal gasification technology. *Energy* 15(7): 40–48.

Keim, W. 1986. C1 Chemistry: Potential and developments. *Pure & Appl. Chem.* 58(6): 825–832.

Lee, S., Speight, J.G., and Loyalka, S. 2007. *Handbook of Alternative Fuel Technologies*. CRC Press, Taylor & Francis Group, Boca Raton, Florida.

Lin, Y.-C. 2013. Catalytic Valorization of glycerol to hydrogen and syngas. *Int. J. Hydrog. Energy* 38(6): 2678–2700.

Luque, R. and Speight, J.G. (Editors) 2015. *Gasification for Synthetic Fuel Production: Fundamentals, Processes, and Applications*. Woodhead Publishing, Elsevier, Cambridge, United Kingdom.

Mangone, C. 2002. Gas to Liquids — Conversions Produce Extremely Pure Base Oils. *Machinery Lubrication Magazine*, Independent Lubricant Manufacturers Association (ILMA). November.

Moschopedis, S.E. and Speight, J.G. 1977. A Contribution to the chemistry of petroleum asphaltenes — Non-fuel uses of asphaltenes. *Chimika Chronika* 42(7): 38.

Olsbye, U., Svelle, S., Bjorgen, M., Beato, P., Janssens, T.V.W., Joensen, F., Bordiga, S., and Lillerud, K.P. 2012. Conversion of methanol to hydrocarbons: how zeolite cavity and pore size controls product selectivity. *Angew. Chem. Int. Ed.* 51(24): 5810–5831.

Parkash, S. 2003. *Refining Processes Handbook*. Gulf Professional Publishing, Elsevier, Amsterdam, Netherlands.

Ricketts, B., Hotchkiss, R., Livingston, W., and Hall, M. 2002. Technology status review of waste/biomass co-gasification with coal. In *Proceedings. Inst. Chem. Eng. Fifth European Gasification Conference*. Noordwijk, Netherlands, April 8–10.

Soukup, G., Pfeifer, C., Kreuzeder, A., and Hofbauer, H. 2009. In situ carbon dioxide capture in a dual fluidized bed biomass steam gasifier bed material and fuel variation. *Chem. Eng. Technol.* 32: 348–354.

Speight, J.G. 1994. Chemical and physical studies of petroleum asphaltene constituents. In *Asphaltene constituents and Asphalts. I. Developments in Petroleum Science*, 40. T.F. Yen and G.V. Chilingarian (Editors). Elsevier, Amsterdam, Netherlands. Chapter 2.

Speight, J.G. 2002. *Chemical Process and Design Handbook*. McGraw-Hill Publishers, New York.

Speight, J.G. 2011. *The Refinery of the Future*. Gulf Professional Publishing Company, Elsevier, Oxford, United Kingdom.

Speight. J.G. 2013. *The Chemistry and Technology of Coal*, 3rd Edition. CRC Press, Taylor & Francis Group, Boca Raton, Florida.

Speight, J.G. 2014a. *The Chemistry and Technology of Petroleum*, 5th Edition. CRC Press, Taylor & Francis Group, Boca Raton, Florida.

Speight, J.G. 2014b. *Gasification of Unconventional Feedstocks*. Gulf Professional Publishing Company, Elsevier, Oxford, United Kingdom.

Speight, J.G. 2016. Hydrogen in Refineries. In *Hydrogen Science and Engineering: Materials, Processes, Systems, and Technology*, D. Stolten and B. Emonts (Editors). Wiley-VCH Verlag GmbH & Co., Weinheim, Germany. Chapter 1. pp. 3–18.

Speight, J.G. 2017. *Handbook of Petroleum Refining*. CRC Press, Taylor & Francis Group, Boca Raton, Florida.

Speight, J.G. 2019a. *Handbook of Petrochemical Processes*. CRC Press, Taylor & Francis Group, Boca Raton, Florida.

Speight, J.G. 2019b. The asphaltene fraction — demystified. *Petroleum and Chemical Industry International*, 2019, 2(4). https://opastonline.com/wp-content/uploads/2019/10/the-asphaltene-fraction-demystified-pcii-19.pdf.

Steiner, H. 1961. *Introduction to Petroleum Chemicals*. Pergamon Press, New York.

Tian, P., Wei, Y., Ye, M., and Liu, Z. 2015. Methanol to olefins (MTO): From fundamentals to commercialization. *ACS Catal.* 5(3): 1922–1938.

Vélez, F.F., Chejne, F., Valdés, C.F., Emery, E.J., and Londoño, C.A. 2009. Cogasification of colombian coal and biomass in a fluidized bed: An experimental study. *Fuel* 88(3): 424–430.

Wittcoff, H.A. and Reuben, B.G. 1996. *Industrial Organic Chemicals*. John Wiley & Sons Inc., New York.

Chapter 9

Gas Cleaning

9.1 Introduction

Gas cleaning is a necessary, important step in many process industries to control particulate matter emissions and maintain a high level of environmental protection. Furthermore, in terms of their use as fuels or as feedstocks for fuel production, gaseous products are experiencing a period of rapid development and, as a result, gas upgrading technology is attracting increasing attention (Mokhatab *et al.*, 2006; Ryckebosch *et al.*, 2011; Niesner *et al.*, 2013; Andriani *et al.*, 2014; Speight, 2014, 2019). Currently, many of the gaseous products are used as on-site (refinery) fuel in power generators and boilers and, for these uses, it is required that the hydrogen sulfide content in the gas be less than 200 parts per million (ppm v/v) to ensure a long life for the power and heat generators. Consequently, the market for clean (i.e. free of pollutants) gas is facing significant challenges in terms of energy consumption and operating costs.

Gas cleaning systems typically consist of different processes and methods, depending on the specific requirements (Mokhatab *et al.*, 2006; Speight, 2019). The most commonly used technologies are venturi-type scrubbers, disintegrators, bag filters, and electrostatic precipitators. However, the selection of a cleaning technology (or technologies) is site-specific, case-sensitive and dependent on the gas utilization requirements and local circumstances (Ramírez *et al.*, 2015). But it is important to first determine the composition of the gas so that appropriate technologies can be selected for removing specific contaminants. Thus, when any biomass-derived gas (either synthesis gas or biosynthesis gas) is used for biofuel production, cleaning of the raw gas is strictly necessary in order to remove contaminants and potential catalyst poisons as well as to achieve the qualitative composition required before the gas is sold for use.

A gas stream gas, as produced by any of the methods described in earlier chapters (i.e. by anaerobic digestion, pyrolysis, or gasification), is typically saturated with water vapor and contains, in addition to methane (CH_4) and carbon dioxide (CO_2), various amounts of hydrogen sulfide (H_2S) as well as other contaminants in varying amounts depending on the composition of the feedstock and the process parameters employed for the production of the gas. The properties of the contaminants are such that failure to remove the contaminants will make the gas unusable and even poisonous to the user. For example, hydrogen sulfide is a toxic gas with a specific, unpleasant odor similar to rotten eggs and forms acidic products by combining with water vapor in the gas stream, which could also result in corrosion of equipment. To prevent this and similar examples of the adverse effects of other contaminants, the gas stream must be dried and any contaminants removed before the gas stream is sent to one or more conversion units or for sales.

Typically, in the gas processing industry (also called the *gas cleaning industry* or the *gas refining industry*), the feedstock is not used directly as fuel because of the complex nature of the feedstock and the presence of one or more of the aforementioned impurities that are corrosive or poisonous. It is, therefore, essential that any gas is contaminant free when it enters one of the various reactors (Parkash, 2003; Gary *et al.*, 2007; Speight, 2014, 2017; Hsu and Robinson, 2017). Typically, the primary raw gas stream has been subjected to chemical and/or physical changes (refining) after being produced. Thus, for many applications, the quality of the gas stream has to be improved. In addition, landfill gas (Chapter 7) often contains significant amounts of halogenated compounds that need to be removed prior to use. Occasionally, the oxygen content is high if air is allowed to invade the gas producer during collection of the landfill gas or if the biomass from which the gas is produced contains an abundance of oxygen-containing organic constituents (Table 9.1).

The main (but not necessarily the only) contaminants that may require removal using gas cleaning systems are hydrogen sulfide, water, carbon dioxide, and halogenated compounds. Desulfurization aims to prevent corrosion and reduce the concentration of the toxic hydrogen sulfide for safety in use and in the workplace. Also, if hydrogen sulfide and other sulfur-containing species such as thiol derivatives (RSH; also called

Table 9.1: General properties of the potential organic constituents of natural gas up to and including *n*-octane (n-C_8H_{18}) as well as toluene, ethylbenzene, and the xylene isomers.[a]

Constituent	Molecular weight	Specific gravity	Vapor density (air = 1)	Boiling point (°C)	Ignition temperature (°C)	Flash point (°C)
Methane	16	0.553	0.56	–160	537	–221
Ethane	30	0.572	1.04	–89	515	–135
Propane	44	0.504	1.50	–42	468	–104
Butane	58	0.601	2.11	–1	405	–60
Pentane	72	0.626	2.48	36	260	–40
Hexane	86	0.659	3.00	69	225	–23
Benzene	78	0.879	2.80	80	560	–11
Heptane	100	0.668	3.50	98	215	–4
Octane	114	0.707	3.90	126	220	13
Toluene	92	0.867	3.20	161	533	4
Ethyl benzene	106	0.867	3.70	136	432	15
o-Xylene	106	0.88	3.7	144	463	17
m-Xylene	106	0.86	3.7	139	527	25
p-Xylene	106	0.861	3.7	138	528	25

Notes: [a]Other constituents such as carbon dioxide and sulfur compounds are not included in this table.

mercaptan derivatives) and carbonyl sulfide (COS) are not removed, then combustion of the gas will produce sulfur dioxide and sulfur trioxide, which are even more poisonous than hydrogen sulfide. At the same time, the presence of sulfur dioxide in the gas stream lowers the dew point (i.e. the temperature to which the gas must be cooled to become saturated with water vapor) of the stack gas. Sulfurous acid (H_2SO_3) is formed by the reaction of sulfur dioxide with water vapor whereas the reaction of carbon dioxide with water vapor forms highly corrosive acidic species:

$$SO_2 + H_2O \rightarrow H_2SO_3$$
$$CO_2 + H_2O \rightarrow H_2CO_3$$

Thus, the removal of water from the gas stream is essential not only because of the potential for a reaction of water with contaminants in the gas stream but also because of a potential accumulation of water vapor condensing in the gas line, the formation of a corrosive acidic solution when hydrogen sulfide is dissolved, or to achieve low dew points when the gas stream is stored under elevated pressures in order to avoid condensation and freezing. Also, if the non-combustible carbon dioxide is not removed, the energy content of the gas stream is diluted and there may also be an environmental impact due to the presence and emission of this contaminant.

Generally, the contaminants — not in order of importance — are classified into (i) condensable hydrocarbon derivatives, including tar products, (ii) alkali metals, such as sodium and potassium, (iii) nitrogen-containing derivatives, including ammonia and hydrogen cyanide, (iv) sulfur-containing derivatives, such as hydrogen sulfide, carbonyl sulfide, and carbon disulfide, (v) halogen-containing derivatives, such as hydrogen chloride, hydrogen bromide, and hydrogen fluoride, and (vi) particulate matter.

The tar products, which are formed during gasification when biomass decomposes in pyrolysis and gasification reactions, are mostly composed of oxygenated compounds, which are decomposition products of biomass. These compounds react further to form secondary and tertiary tar compounds, which consist of compounds that do not exist in the source biomass. Secondary tar typically consists of alkylated one-ring and two-ring aromatic compounds (including heterocyclic compounds) whereas tertiary tar consists of aromatic hydrocarbon derivatives such as benzene, naphthalene, and various polycyclic aromatic hydrocarbon derivatives [PAHs; also called polynuclear aromatic compounds (PNAs)]. Tar constituents are typically classified on the basis of the number of rings (or the size of the condensed-ring systems) in the constituents or by the boiling point distribution or physical properties. More generally, tar is defined as an aromatic compound that has a higher boiling point than benzene; in addition, an operational definition of tar depending on the end-use application has been used.

Process selectivity is the preference of the process for removal of one acid gas component relative to (or in preference to) another. For example, some processes remove both hydrogen sulfide and carbon dioxide, while other processes are designed to remove hydrogen sulfide only. It is very important to consider that the process selectivity for, say, hydrogen sulfide removal compared to carbon dioxide removal ensures minimal concentrations of these components in the product; thus there is a need for consideration of the ratio of carbon dioxide to hydrogen sulfide in the gas stream.

Furthermore, it is the purpose of this chapter to present the methods by which various gaseous and liquid feedstocks can be processed and prepared for petrochemical production. This requires removal of impurities that would otherwise be deleterious to petrochemical production and analytical assurance that feedstocks are, indeed, free of deleterious contaminants (Speight, 2015, 2018).

To include a description of all possible process variations for gas cleaning is beyond the scope of this book. Therefore, this chapter focuses on a selection of processes forming an integral part within the concept of production of a specification-grade product (methane) for sale to the consumer.

9.2 Gas Streams

The processes that have so far been developed to accomplish gas purification range from a simple once-through wash operation to more complex multi-step recycling systems (Speight, 1993). In many cases, the process complexities arise because of the need for recovery of the materials used to remove the contaminants or even recovery of the contaminants in the original form or in the altered form (Katz, 1959; Kohl and Riesenfeld, 1985; Newman. 1985).

There are many variables involved in treating refinery gas or natural gas. The precise area of application of a given process is difficult to define and so it is necessary to consider several factors: (i) the types of contaminants in the gas stream, (ii) the concentration of each contaminant in the gas stream, (iii) the degree of contaminant removal desired, (iv) the selectivity of acid gas removal required, (v) the temperature of the gas stream to be processed, (vi) the pressure of the gas stream to be processed, (vii) the volume of the gas stream to be processed, (viii) the composition of the gas to be processed, (ix) the carbon dioxide/hydrogen sulfide ratio in the gas, and (x) the desirability of sulfur recovery due to economic or environmental issues.

In addition to hydrogen sulfide (H_2S) and carbon dioxide (CO_2), the gas stream may contain other contaminants, such as mercaptans (RSH) and carbonyl sulfide (COS). The presence of these impurities may eliminate some of the sweetening processes from use, since some of these processes remove large amounts of acid gas but do so not to a low enough concentration. On the other hand, there are certain processes not designed to remove (or incapable of removing) large amounts of acid gases but capable of removing the acid gas impurities to very low levels when the acid gases are present only in low-to-medium concentrations in the gas.

Gas processing involves the use of several different types of processes but there is always overlap between the various processing concepts. In addition, the terminology used for gas processing can often be confusing and/or misleading because of the overlap (Nonhebel, 1964; Curry, 1981; Maddox, 1982). There are four general processes used for emission control (often referred to in another, more specific context as flue gas desulfurization): (i) adsorption, (ii) absorption, (iii) catalytic oxidation, and (iv) thermal oxidation (Soud and Takeshita, 1994).

Adsorption is a physical–chemical phenomenon in which the gas is concentrated on the surface of a solid or liquid to remove impurities. Usually, carbon is the adsorbing medium, which can be regenerated upon *desorption* (Fulker, 1972; Mokhatab *et al.*, 2006; Speight, 2007). The quantity of material adsorbed is proportional to the surface area of the solid and, consequently, adsorbents are usually granular solids with a large surface area per unit mass. Subsequently, the captured gas can be desorbed with hot air or steam for either recovery or thermal destruction.

Adsorbing units (adsorbers) are widely used to increase a low gas concentration prior to incineration unless the gas concentration is very high in the inlet air stream. Adsorption is also used for reducing problematic odors from gases. There are several disadvantages of the use of adsorption systems, but it is generally felt that the major disadvantage is the need to minimize particulate matter emission and/or condensation of liquids (e.g. water vapor), which could otherwise mask the adsorption surface and reduce its efficiency drastically (Mokhatab *et al.*, 2006; Speight, 2007).

Absorption differs from *adsorption* in that it is not a physical–chemical surface phenomenon but an approach by which the absorbed gas is ultimately distributed throughout the absorbent (liquid). The process depends only on physical solubility and may include chemical reactions in the liquid phase (*chemisorption*). The commonly used absorbing media are water, aqueous amine solutions, caustic, sodium carbonate, and nonvolatile hydrocarbon oils, depending on the type of gas to be absorbed. Usually, the gas–liquid contactor designs employed are plate columns or packed beds (Mokhatab *et al.*, 2006; Speight, 2019).

Absorption is achieved by dissolution (a physical phenomenon) or by reaction (a chemical phenomenon) (Barbouteau and Galaud, 1972; Ward, 1972; Mokhatab *et al.*, 2006; Speight, 2007). Chemical adsorption processes adsorb sulfur dioxide onto a carbon surface, where it is oxidized (by oxygen in flue gas) and absorbs moisture to give sulfuric acid impregnated into and on the adsorbent.

Liquid absorption processes (which usually employ temperatures below 50°C (120°F)) are classified as *physical solvent processes* and *chemical solvent processes*. The former processes employ an organic solvent, and absorption is enhanced by low temperatures, or high pressures, or both. Regeneration of the solvent is often accomplished readily (Staton *et al.*, 1985). In chemical solvent processes, absorption of the acid gases is achieved mainly by the use of alkaline solutions such as amines or carbonates (Kohl and Riesenfeld, 1985). Regeneration (desorption) can be brought about by the use of reduced pressures and/or high temperatures, whereby the acid gases are stripped from the solvent.

The solvents used for emission control processes should have the following (Mokhatab *et al.*, 2006; Speight, 2007):

1. A high capacity for removal of acid gas.
2. A low tendency to dissolve hydrogen.
3. A low tendency to dissolve low-molecular-weight hydrocarbon derivatives.
4. Low vapor pressure at operating temperatures to minimize solvent losses.
5. Low viscosity.
6. Low thermal stability.
7. Absence of reactivity toward gas components.
8. Low tendency for fouling.
9. Low tendency for corrosion.
10. Economically acceptable.

Amine washing of gas emissions involves chemical reaction of the amine with any acid gases with the liberation of an appreciable amount of heat and it is necessary to compensate for the absorption of heat. Amine derivatives such as ethanolamine [monoethanolamine (MEA)], diethanolamine (DEA), triethanolamine (TEA), methyldiethanolamine (MDEA), diisopropanolamine (DIPA), and diglycolamine (DGA) have been used in commercial applications (Katz, 1959; Kohl and Riesenfeld, 1985;

Maddox *et al.*, 1985; Polasek and Bullin, 1985; Jou *et al.*, 1985; Pitsinigos and Lygeros, 1989; Speight, 1993, 2018; Mokhatab *et al.*, 2006).

The chemistry can be represented by simple equations for low partial pressures of the acid gases:

$$2RNH_2 + H_2S \rightarrow (RNH_3)_2S$$

$$2RHN_2 + CO_2 + H_2O \rightarrow (RNH_3)_2CO_3$$

At a high acid gas partial pressure, the reactions will lead to the formation of other products:

$$(RNH_3)_2S + H_2S \rightarrow 2RNH_3HS$$

$$(RNH_3)_2CO_3 + H_2O \rightarrow 2RNH_3HCO_3$$

The reaction is extremely fast, the absorption of hydrogen sulfide being limited only by mass transfer; this is not the case with carbon dioxide.

Regeneration of the solution leads to near-complete desorption of carbon dioxide and hydrogen sulfide. A comparison of monoethanolamine, diethanolamine, and diisopropanolamine reveals that monoethanolamine is the cheapest of the three but has the highest heat of reaction and corrosion; the reverse is true for diisopropanolamine.

Carbonate washing is a mild alkali process for emission control by the removal of acid gases (such as carbon dioxide and hydrogen sulfide) from gas streams (Speight, 1993) and uses the principle that the rate of absorption of carbon dioxide by potassium carbonate increases with temperature. It has been demonstrated that the process works best near the temperature of reversibility of the reactions:

$$K_2CO_3 + CO_2 + H_2O \rightarrow 2KHCO_3$$

$$K_2CO_3 + H_2S \rightarrow KHS + KHCO_3$$

Water washing, in terms of the outcome, is analogous to washing with potassium carbonate (Kohl and Riesenfeld, 1985), and it is also possible to carry out the desorption step by pressure reduction. The absorption is purely physical and there is also a relatively high absorption of hydrocarbon derivatives, which are liberated at the same time as the acid gases.

In *chemical conversion processes*, contaminants in gas emissions are converted to compounds that are not objectionable or that can be removed from the stream with greater ease than the original constituents. For example, a number of processes have been developed that remove hydrogen sulfide and sulfur dioxide from gas streams by absorption in an alkaline solution.

Catalytic oxidation is a chemical conversion process that is used predominantly for destruction of volatile organic compounds and carbon monoxide. These systems operate in the temperature range of 205–595°C (400–1100°F) in the presence of a catalyst. Without the catalyst, the system would require higher temperatures. Typically, the catalysts used are a combination of noble metals deposited on a ceramic base in a variety of configurations (e.g. honeycomb-shaped) to ensure good surface contact.

Catalytic systems are usually classified on the basis of bed types such as *fixed bed* (or *packed bed*) and *fluid bed* (*fluidized bed*). They generally show very high efficiencies for destruction of most volatile organic compounds, resulting in the formation of carbon dioxide, water, and varying amounts of hydrogen chloride (from halogenated hydrocarbon derivatives). The presence in emissions of chemicals such as heavy metals, phosphorus, sulfur, chlorine, and most halogens in the incoming air stream acts as a poison to the system and can foul up the catalyst.

Thermal oxidation systems, without the use of catalysts, also involve chemical conversion (more correctly, chemical destruction) and operate at temperatures in excess of 815°C (1500°F), or 220–610°C (395–1100°F) higher than catalytic systems.

Historically, *particulate matter control* (*dust control*) (Mody and Jakhete, 1988) has been one of the primary concerns of industries, since the emission of particulate matter is readily observed through the deposition of fly ash and soot as well as the impairment of visibility. Different ranges of control can be achieved by the use of various types of equipment. Upon a proper characterization of the particulate matter emitted by a specific process, the appropriate piece of equipment can be selected, sized, and installed, and performance tested. The general types of control devices for particulate matter are listed as follows.

Cyclone collectors are the most common type of inertial collectors. Cyclones are effective in removing coarser fractions of particulate matter. The particle-laden gas stream enters an upper cylindrical section tangentially and proceeds downward through a conical section. Particles migrate by centrifugal force generated by providing a path for the carrier gas to be subjected to a vortex-like spin. The particles are forced to the wall and are removed through a seal at the apex of the inverted cone.

A reverse-direction vortex moves upward through the cyclone and discharges through a top center opening. Cyclones are often used as primary collectors because of their relatively low efficiency (50–90% is usual). Some small-diameter high-efficiency cyclones are utilized. The equipment can be arranged either in parallel or in series to both increase efficiency and decrease the pressure drop. However, there are disadvantages that must be recognized (Mokhatab *et al.*, 2006; Speight, 2019). These units for particulate matter operate by contacting the particles in the gas stream with a liquid. In principle, the particles are incorporated in a liquid bath or in liquid particles, which are much larger and therefore more easily collected.

Fabric filters are typically designed with non-disposable filter bags. As the dusty emissions flow through the filter media (typically cotton, polypropylene, Teflon, or fiberglass), particulate matter is collected on the bag surface as a dust cake. Fabric filters are generally classified on the basis of the filter bag cleaning mechanism employed. Fabric filters operate with collection efficiencies up to 99.9% although other advantages are evident. There are several issues that arise during the use of such equipment (Mokhatab *et al.*, 2006; Speight, 2019).

Wet scrubbers are devices in which a counter-current spray liquid is used to remove particles from an air stream. Device configurations include plate scrubbers, packed beds, orifice scrubbers, venturi scrubbers, and spray towers, individually or in various combinations. Wet scrubbers can achieve high collection efficiencies at the expense of prohibitive pressure drops (Mokhatab *et al.*, 2006; Speight, 2019).

Other methods include the use of high-energy-input *venturi scrubbers* or electrostatic scrubbers where particles or water droplets are charged and flux force/condensation scrubbers where a hot humid gas is contacted with cooled liquid or where steam is injected into saturated gas. In the latter scrubber the movement of water vapor toward the cold water surface carries the particles with it (*diffusiophoresis*), while the condensation of water vapor on the particles causes the particle size to increase, thus facilitating collection of fine particles.

The *foam scrubber* is a modification of a wet scrubber in which the particle-laden gas is passed through a foam generator, where the gas and particles are enclosed by small bubbles of foam.

Electrostatic precipitators (Mokhatab *et al.*, 2006; Speight, 2019) operate on the principle of imparting an electric charge to particles in the incoming air stream, which are then collected on an oppositely charged plate across a high-voltage field. Particles of high resistivity create the most difficulty in collection. Conditioning agents such as sulfur trioxide (SO_3) have been used to lower resistivity.

Important parameters include the design of electrodes, spacing of collection plates, minimization of air channeling, and collection electrode rapping techniques (used to dislodge particles). Techniques under study include the use of high-voltage pulse energy to enhance particle charging, electron beam ionization, and wide plate spacing. Electrical precipitators are capable of efficiencies >99% under optimum conditions, but performance is still difficult to predict in new situations.

9.3 Gas Cleaning

Whatever the source of the gas stream, methane (in varying amounts) is the predominant hydrocarbon derivative in the gas stream. However, there are other hydrocarbon derivatives in addition to water vapor, hydrogen sulfide (H_2S), carbon dioxide, nitrogen, and other compounds.

A treated gas stream may consist mainly of methane (often referred to as biomethane). However, the gas stream is not typically pure methane, and its properties are modified by the presence of impurities, such as nitrogen, carbon dioxide, and small amounts of unrecovered higher boiling (non-gaseous at STP) hydrocarbon derivatives. An important property of any gas stream is its heating value — relatively large amounts of nitrogen and/or carbon dioxide reduce the heating value of the gas. The gas stream, if it was pure methane, would have a heating value of 1,010 Btu/ft^3. This value is reduced to approximately 900 Btu/ft^3 if the gas contains approximately 10% v/v nitrogen and carbon dioxide — the heating value of carbon dioxide is zero.

In order to remove the unwanted by-products and contaminants in the product gas, primary treatments, such as optimization of the properties of biomass feedstock, design of the gasifier, and the operational parameters of the gasifier, are first implemented. Secondary treatments, such as a downstream cleaning system based on physical (scrubbers, filters) or catalytic strategies, have to be incorporated in the hot gas cleaning to achieve a more satisfactory reduction for end- use applications and also to meet sales specification, which include meeting environmental regulations.

Thus, to reach the conditions of acceptability, the gas stream must be the end-product after being treated by a series of processes that have successfully removed the contaminants present in the raw (untreated) gas. Gas processing consists of separating all of the non-methane derivatives (both the combustible and non-combustible environmentally-unfriendly and performance-unfriendly constituents) and fluids from the gas (Kidnay and Parrish, 2006; Mokhatab *et al.*, 2006; Speight, 2014, 2019). Although often assumed to be hydrocarbon derivatives in nature, there are also components of the gaseous products that must be removed before release of the gases to the atmosphere or before use of the gas, i.e. as a fuel gas or as a process feedstock.

The actual practice of processing gas streams to pipeline quality levels can be quite complex, but usually involves several processes to remove the various impurities. Gas streams produced during natural gas and crude oil refining, while ostensibly being hydrocarbon in nature, may contain substantial proportions of acid gases such as hydrogen sulfide and carbon dioxide. Most commercial plants employ hydrogenation to convert organic sulfur compounds into hydrogen sulfide. Hydrogenation is effected by means of recycled hydrogen-containing gases or external hydrogen over a nickel molybdate or cobalt molybdate catalyst. Also, refinery process gas, in addition to hydrocarbon derivatives, may contain other contaminants, such as carbon oxides (CO_x, where $x = 1$ and/or 2), sulfur oxides (So_x, where $x = 2$ and/or 3), as well as ammonia (NH_3), mercaptans (R-SH), and carbonyl sulfide (COS).

The presence of these impurities may eliminate some of the sweetening processes from use, since some of these processes remove large amounts of acid gas but do so not to a low enough concentration. On the other hand, there are certain processes not designed to remove (or incapable of removing) large amounts of acid gases by capable of removing the acid gas impurities to very low levels when the acid gases are present only in low-to-medium concentration in the gas (Katz, 1959).

The typical processes used for gas cleaning operations include (i) water removal, (ii) acid gas removal, (iii) removal of condensable hydrocarbons, (iv) tar removal, and (v) various other processes such as nitrogen removal, ammonia removal, particulate matter removal, siloxane removal, alkali metal salt removal, and biological methods (such as biofiltration, bioscrubbing, and biooxidation). But first, the selection of any particular cleaning process or series of processes is based on the composition of the gas stream.

9.3.1 *Effect of gas composition*

Trace quantities of contaminants in hydrocarbon products can be harmful to many catalytic chemical processes in which these products are used. In general, the raw product gas of biomass gasification contains a range of minor species and contaminants including particles, tar, alkali metals, chlorine compounds such as hydrogen chloride (HCl), nitrogen compounds such as ammonia (NH_3) and hydrogen cyanide (HCN), sulfur compounds such as hydrogen sulfide (H_2S) and carbonyl sulfide (COS), as well as other species.

The maximum permissible levels of total contaminants are normally included in the specifications for such hydrocarbon derivatives. It is recommended that this test method be used to provide a basis for agreement between two laboratories when the determination of sulfur in hydrocarbon gases is important. In the case of liquefied petroleum gas (LPG), total volatile sulfur is measured on an injected gas sample. One test method (ASTM D3246) describes a procedure for the determination of sulfur in the range of 1.5–100 mg/kg (ppm w/w) in hydrocarbon products that are gaseous at normal room temperature and pressure. There are also a variety of other standard test methods that can be applied to the determination of the properties of the gas stream and, hence, the suitability of the gas for use (ASTM, 2018; Speight, 2018).

Acidic constituents such as carbon dioxide and hydrogen sulfide as well as thiol derivatives (RSH; also called mercaptan derivatives) can contribute to corrosion of refining equipment, harm catalysts, pollute the atmosphere, and prevent the use of hydrocarbon components in petrochemical manufacture (Mokhatab *et al.*, 2006; Speight, 2014, 2019). When the amount of hydrogen sulfide is large, it may be removed from a gas stream and converted to sulfur or sulfuric acid; a recent option for hydrogen sulfide removal is the use of chemical scavengers (Kenreck, 2014). Some gases contain sufficient carbon dioxide to warrant recovery as dry ice (Bartoo, 1985).

Many gas streams are not always truly hydrocarbon in nature and may contain contaminants, such as carbon oxides (CO_x, where $x = 1$ and/or 2), sulfur oxides (SO_x, where $x = 2$ and/or 3), as well as ammonia (NH_3), mercaptan derivatives (RSH), carbonyl sulfide (COS), and mercaptan derivatives (RSH). The presence of these impurities may eliminate some of the sweetening processes from use, since some of these processes remove considerable amounts of acid gas but do so not to a low enough concentration. On the other hand, there are certain processes not designed to remove (or incapable of removing) large amounts of acid gases but capable of removing the acid gas impurities to very low levels when the acid gases are present only in low-to-medium concentration in the gas (Katz, 1959; Mokhatab *et al.*, 2006; Speight, 2014, 2019).

The sources of the various gas streams are varied but, in terms of gas cleaning (i.e. removal of the contaminants before petrochemical production), the processes are largely the same but it is a question of degree. For example, gas streams from some sources may produce gases that may contain larger amounts of carbon dioxide and/or hydrogen sulfide and the processes will have to be selected accordingly.

In addition to its primary importance as a fuel, a gas stream — such as synthesis gas — is also a source of hydrocarbon derivatives for petrochemical feedstocks. While the major hydrocarbon constituent of a gas stream is methane, there are components such as carbon dioxide (CO_2), hydrogen sulfide (H_2S), and mercaptan derivatives (thiols, also called mercaptans, RSH), as well as trace amounts of sundry other emissions such as carbonyl sulfide (COS). The fact that methane has a foreseen and valuable end use makes it a desirable product, but in several other situations it is considered a pollutant, having been identified as a greenhouse gas.

In practice, heaters and scrubbers are usually installed at or near the source. The scrubbers serve primarily to remove sand and other large-particle impurities and the heaters ensure that the temperature of the gas does not drop too low. With gas that contains even lower quantities of water, gas hydrates ($C_nH_{2n+2}.xH_2O$) tend to form when temperatures drop. These hydrates are solid or semi-solid compounds, resembling ice-like crystals. If the hydrates accumulate, they can impede the passage of gas through valves and gathering systems (Zhang *et al.*, 2007). To reduce the occurrence of hydrates, small gas-fired heating units are typically installed along the gathering pipe wherever it is likely that hydrates may form.

9.3.2 *Process types*

A gas stream gas can also be upgraded to pipeline-quality gas and this upgraded gas may be used for residential heating and as vehicle fuel. When distributing the gas stream using pipelines, gas pipeline standards will, more than likely, become applicable. Removing water vapor is easier than removing carbon dioxide and hydrogen sulfide from the gas stream (Table 9.2). A condensate trap at a proper location on the gas pipeline can remove water vapor as warm gas stream cools by itself after leaving the digester or the gas-generating equipment.

Gas processing involves the use of several different types of processes to remove contaminants from gas streams but there is always overlap between the various processing concepts. In addition, the terminology used for gas processing can often be confusing and/or misleading because of the overlap (Curry, 1981; Maddox, 1982). Gas processing is necessary to ensure that the gas prepared for transportation (usually by pipeline) and for sales must be as clean and pure as the specifications dictate. Thus, a gas stream, as it is used by consumers, is much different from the gas that is brought from, for example, an anaerobic digester (Chapter 3). Moreover, although the gas produced in the digester may be composed primarily of methane, it is by no means as pure.

In addition to the corrosion of equipment by acid gases (Speight, 2014) the escape into the atmosphere of sulfur-containing gases can eventually lead to the formation of the constituents of acid rain, i.e. the oxides of sulfur [sulfur dioxide, (SO_2) and sulfur trioxide, (SO_3)]. Similarly, the nitrogen-containing gases can also lead to nitrous and nitric acids (through the formation of the oxides NO_x, where $x = 1$ or 2), which are the other major contributors to acid rain. The release of carbon dioxide and hydrocarbon derivatives as constituents of refinery effluents can also influence the behavior and integrity of the ozone layer.

However, the precise area of application of a given process is difficult to predict or define and a careful consideration of several process factors is required before process selection is done: (i) the types of contaminants in the gas, (ii) the concentrations of contaminants in the gas, (iii) the degree of contaminant removal desired, (iv) the selectivity of acid gas removal required, (v) the temperature of the gas to be processed, (vi) the pressure of the gas to be processed, (vii) the volume of the gas to be processed, (viii) the composition of the gas to be processed, (ix) the ratio of carbon dioxide to hydrogen sulfide in the gas feedstock, and (x) the desirability of sulfur recovery due to environmental or economic issues.

Finally, during the various cleaning processes, it is important to minimize the loss of methane in order to achieve an economically viable cleaning operation. However, it is also important to minimize the methane slip since methane is a strong greenhouse gas. Thus, the release of methane to the atmosphere should be minimized by treating the off-gas, air, or water streams leaving the plant even though the methane cannot be utilized.

Table 9.2: Common methods for the removal of carbon dioxide and hydrogen sulfide from gas streams.

Method	Description
Carbon dioxide removal	
Water scrubbing	Uses the principle of the higher solubility of carbon dioxide in water to separate the carbon dioxide from the gas. The process uses high pressure and removes hydrogen sulfide as well as carbon dioxide. The main disadvantage of this process is that it requires a large volume of water that must be purified and recycled.
Polyethylene glycol scrubbing	This process is more efficient than water scrubbing. It also requires the regeneration of a large volume of polyethylene glycol.
Chemical absorption	Chemical reaction between carbon dioxide and amine-based solvents (olamines).
Carbon molecular sieves	Uses differential adsorption characteristics to separate methane and carbon dioxide, which is carried out at high pressure (pressure swing adsorption). Hydrogen sulfide should be removed before the adsorption process.
Membrane separation	Selectively separates hydrogen sulfide and carbon dioxide from methane. The carbon dioxide and the hydrogen sulfide dissolve while the methane is collected for use.
Cryogenic separation	Cooling until condensation or sublimation of the carbon dioxide.
Hydrogen sulfide removal	
Biological desulfurization	Natural bacteria can be used to convert hydrogen sulfide into elemental sulfur in the presence of oxygen and iron. This can be done by introducing a small amount (2–5%) of air into the head space of the digester. As a result, deposits of elemental sulfur will be formed in the digester. This process may be optimized using a more sophisticated design where air is bubbled through the digester feed material. It is critical that the introduction of the air is carefully controlled so as to avoid reducing the amount of gas produced.
Iron/iron oxide reaction	Hydrogen sulfide reacts readily with either iron oxide or iron chloride to form insoluble iron sulfide. The reaction can be exploited by adding the iron chloride to the digester feed material or passing the gas through a bed of iron oxide-containing material. The iron oxide medium needs to be replaced periodically. The regeneration process is highly exothermic and must be controlled to avoid problems.
Activated carbon	Activated carbon impregnated with potassium iodide can catalytically react with oxygen and hydrogen sulfide to form water and sulfur. The activated carbon beds need regeneration or replacement when saturated.
Scrubbing/membrane separation	The carbon dioxide and hydrogen sulfide can be removed by washing with water, or glycol solutions, or separated using the membrane technique.

Methane can be present in the off-gas leaving a pressure swing adsorption (PSA) column, in air from a water scrubber with water recirculation, or in water in a water scrubber without water recirculation. The off-gas from an upgrading plant seldom contains a methane concentration high enough to maintain a flame without the addition of natural gas (methane). One way of limiting the methane slip is to mix the off-gas with air, which is used for combustion. Alternatively, the methane can be oxidized by thermal or catalytic oxidation. Thus,

$$CH_4 + O_2 \rightarrow CO_2 + H_2O$$

In the catalytic process, the oxidation takes place at the surface of the catalyst, which lowers the energy needed to oxidize the methane, thus enabling oxidation at a lower temperature. The active component of the catalyst is platinum, palladium, or cobalt.

9.4 Water Removal

The relative humidity of the gas stream inside the digester is 100%, so the gas is saturated with water vapor. To protect the energy conversion equipment from wear and eventual damage, water must be removed from the produced gas stream. The quantity of water contained by the gas stream depends on temperature. A part of the water vapor can be condensed by the cooling of the gas. This is frequently done in the gas pipelines transporting the gas stream from the digester to a unit in which the water condensates on the walls of the sloping pipes and can be collected in a condensation separator at the lowest point of the pipeline.

Water in the liquid phase causes corrosion or erosion problems in the pipelines and equipment, particularly when carbon dioxide and hydrogen sulfide are present in the gas. The simplest method for water removal (refrigeration or cryogenic separation) is to cool the gas to a temperature at least equal to or (preferentially) below the dew point (Mokhatab *et al.*, 2006; Speight, 2014, 2019).

In addition to separating any condensate from the wet gas stream, it is necessary to remove most of the associated water. Most of the liquid, free water associated with the extracted gas stream is removed by using simple separation methods; however, the removal of the water vapor present in solution in a gas stream requires a more complex treatment. This treatment consists of dehydrating the gas stream, which usually involves either absorption or adsorption.

Moisture may be removed from hydrocarbon gases at the same time that hydrogen sulfide is removed. Moisture removal is necessary to prevent harm to anhydrous catalysts and to prevent the formation of hydrocarbon hydrates (e.g. $C_3H_8.18H_2O$) at low temperatures. A widely used dehydration and desulfurization process is the glycol amine process, in which the treatment solution is a mixture of ethanolamine and a large amount of glycol. The mixture is circulated through an absorber and a reactivator in the same way as ethanolamine is circulated in the Girbotol process. The glycol absorbs moisture from the hydrocarbon gas passing through the absorber; the ethanolamine absorbs hydrogen sulfide and carbon dioxide. The treated gas leaves the top of the absorber; the spent ethanolamine–glycol mixture enters the reactivator tower, where heat drives off the absorbed acid gases and water.

Absorption occurs when water vapor is taken out by using a dehydrating agent, whereas adsorption occurs when water vapor is condensed and collected on the surface. In a majority of cases, cooling alone would be insufficient and, for the most part, impractical to use in field operations. Other, more convenient water removal options use (i) *hygroscopic* liquids (e.g. diethylene glycol or triethylene glycol) and (ii) solid adsorbents or desiccants (e.g. alumina, silica gel, and molecular sieves). Ethylene glycol can be injected directly into the gas stream in refrigeration plants.

9.4.1 *Absorption*

One example of absorption dehydration is *glycol dehydration*, the principal agent of this process being diethylene glycol, which has a chemical affinity for water (Mokhatab *et al.*, 2006; Abdel-Aal *et al.*, 2016; Speight, 2019). Glycol dehydration uses a solution of a glycol, such as diethylene glycol (DEG) or triethylene glycol (TEG), which is brought into contact with the wet gas stream in a *contactor*. In practice, absorption systems recover 90–99% by volume of methane that would otherwise be flared into the atmosphere.

In this process, a liquid desiccant dehydrator serves to absorb water vapor from the gas stream. The glycol solution absorbs water from the wet gas and, once absorbed, the glycol particles become heavier and sink to the bottom of the contactor, where they are removed. The dry gas stream is then transported out of the dehydrator. The glycol solution, bearing all of the water stripped from the gas stream, is recycled through a specialized boiler designed to vaporize only the water out of the solution. The boiling point differential between water (100°C, 212°F) and glycol (204°C, 400°F) makes it relatively easy to remove water from the glycol solution.

In addition to absorbing water, the glycol solution occasionally contains small amounts of methane and other compounds absorbed from the wet gas. In order to present methane and other compounds from being lost, flash tank separator-condensers are employed to remove these compounds before the glycol solution reaches the boiler. A flash tank separator consists of a device to reduce the pressure of the glycol solution stream, thus allowing the methane and other hydrocarbon derivatives to vaporize (*flash*). The glycol solution then reaches the boiler, which may also be fitted with air- or water-cooled condensers that serve to capture any remaining organic compounds in the glycol solution. The regeneration (stripping) of the glycol is limited by the temperature: diethylene glycol and triethylene glycol decompose at or even below their respective boiling points. Such techniques as stripping of hot triethylene glycol with dry gas (e.g. heavy hydrocarbon vapors, the *Drizo process*) or vacuum distillation are recommended for use.

Another absorption process, namely the Rectisol process, is a physical acid gas removal process using an organic solvent (typically methanol) at subzero temperatures, and as is characteristic of physical acid gas removal processes, it can purify a gas stream down to 0.1 ppm v/v of total sulfur, including hydrogen sulfide (H_2S) and carbonyl sulfide (COS) and carbon dioxide (CO_2) in the ppm v/v range (Mokhatab *et al.*, 2006; Liu *et al.*, 2010; Abdel-Aal *et al.*, 2016). The process uses methanol as a wash solvent and the wash unit operates under favorable temperatures below 0°C (32°F). To lower the temperature of the feed gas, it is cooled against the cold-product streams before entering the absorber tower. At the absorber tower, carbon dioxide and hydrogen sulfide (with carbonyl sulfide) are removed. By the use of an intermediate flash, co-absorbed products such as hydrogen and carbon monoxide are recovered, thus increasing the product recovery rate. To reduce the required energy demand for the carbon dioxide compressor, the carbon dioxide product is recovered in two different pressure steps (medium pressure and lower pressure). The carbon dioxide product is essentially sulfur-free (hydrogen sulfide-free, carbonyl sulfide-free) and water-free. The carbon dioxide products can be used for enhanced oil recovery (EOR) and/or sequestration or as pure carbon dioxide for other processes.

9.4.2 *Adsorption*

It must be emphasized that *adsorption* differs from *absorption* in that absorption is not a physical–chemical surface phenomenon but a process by which the absorbed gas is ultimately distributed throughout the absorbent (liquid). Dehydration using a solid adsorbent or solid desiccant is the primary form of dehydrating the gas stream using adsorption, and usually consists of two or more adsorption towers, which are filled with a solid desiccant (Mokhatab *et al.*, 2006; Abdel-Aal *et al.*, 2016; Speight, 2019). Typical desiccants include activated alumina or a granular silica gel material. A wet gas stream is passed through these towers, from top to bottom. As the wet gas passes around the particles of desiccant material, water is retained on the surface of these desiccant particles. Passing through the entire desiccant bed, almost all of the water is adsorbed onto the desiccant material, allowing the dry gas to exit the bottom of the tower. There are several solid desiccants that possess the physical characteristic to adsorb water from the gas stream. These desiccants are generally used in dehydration systems consisting of two or more towers and associated regeneration equipment.

Molecular sieves — a class of aluminosilicates that produce the lowest water dew points and which can be used to simultaneously sweeten dry gases and liquids (Mokhatab *et al.*, 2006; Maple and Williams, 2008; Abdel-Aal *et al.*, 2016; Speight, 2019) — are commonly used in dehydrators ahead of plants designed to recover ethane and other higher-boiling hydrocarbon derivatives. These plants operate at very cold temperatures and require very dry feed gas to prevent the formation of hydrates. Dehydration to –100°C (–148°F) dew point is possible with molecular sieves. Water dew points less than –100°C (–148°F) can be accomplished with a special design and definitive operating parameters (Mokhatab *et al.*, 2006).

Molecular sieves are commonly used to selectively adsorb water and sulfur compounds from light hydrocarbon streams such as liquefied petroleum gas, propane, butane, pentane, light olefin derivatives, and alkylation feed. Sulfur compounds that can be removed are hydrogen sulfide, mercaptan derivatives, sulfide derivatives, and disulfide derivatives. In the process, the sulfur-containing feedstock is passed through a bed of sieves at ambient temperature. The operating pressure must be high enough to keep the feed in the liquid phase. The operation is cyclic in that the adsorption step is stopped at a predetermined time before sulfur breakthrough occurs. Sulfur and water are removed from the sieves by purging with fuel gas at 205–315°C (400–600°F).

Solid-adsorbent dehydrators are typically more effective than liquid absorption dehydrators (e.g. glycol dehydrators) and are usually installed as a type of straddle system along gas pipelines. These types of dehydration systems are best suited for large volumes of gas under very high pressure, and are thus usually located on a pipeline downstream of a compressor station. Two or more towers are required due to the fact that after a certain period of use, the desiccant in a particular tower becomes saturated with water. To regenerate and recycle the desiccant, a high-temperature heater is used to heat the gas to a very high temperature and the passage of the heated gas stream through a saturated desiccant bed vaporizes the water in the desiccant tower, allowing it to dry and allowing for further dehydration of the gas stream.

Although two-bed adsorbent treaters have become more common (while one bed is removing water from the gas, the other undergoes alternate heating and cooling), on occasion, a three-bed system is used: one bed adsorbs, one is being heated, and one is being cooled. An additional advantage of the three-bed system is the facile conversion of a two-bed system so that the third bed can be maintained or replaced, thereby ensuring continuity of the operations and reducing the risk of a costly plant shutdown.

Silica gel (SiO_2) and alumina (Al_2O_3) have good capacities for water adsorption (up to 8% by weight). Bauxite (crude alumina, Al_2O_3) adsorbs up to 6% by weight water, and molecular sieves adsorb up to 15% by weight water. Silica is usually selected for dehydration of sour gas because of its high tolerance to hydrogen sulfide and to protect molecular sieve beds from plugging by sulfur. Alumina *guard beds* serve as protectors by the act of attrition and may be referred to as an *attrition reactor* containing an *attrition catalyst* (Speight, 2000, 2014, 2017) that may be placed ahead of the molecular sieves to remove the sulfur compounds. Downflow reactors are commonly used for adsorption processes, with an upward flow regeneration of the adsorbent and cooling using gas flow in the same direction as adsorption flow.

Solid desiccant units generally cost more to buy and operate than glycol units. Therefore, their use is typically limited to applications such as gases having a high hydrogen sulfide content, very low water dew point requirements, and simultaneous control of water and hydrocarbon dew points. In processes where cryogenic temperatures are encountered, solid desiccant dehydration is usually preferred over conventional methanol injection in order to prevent hydrate and ice formation (Kidnay and Parrish, 2006).

9.4.3 *Cryogenics*

Another route for drying a gas stream is by cooling the gas in electrically powered gas coolers at temperatures below 10°C (50°F), which allows a lot of humidity to be removed. In order to minimize the relative humidity but not the absolute humidity, the gas can be warmed up again after cooling in order to prevent condensation along the gas pipelines.

Thus, cryogenic upgrading makes use of the distinct boiling/sublimation points of the different gases, particularly for the separation of carbon dioxide and methane. In the process, the raw (untreated) gas stream is cooled down to the temperatures where the carbon dioxide in the gas condenses or sublimates and can be separated as a liquid or a solid fraction, while methane accumulates in the gas phase. Water and siloxane derivatives are also removed during cooling of the gas. However, the content of methane in the gas stream affects the characteristics of the gas, i.e. higher pressures and/or lower temperatures are needed to condense or sublime carbon dioxide when it is in a mixture with methane. Cooling usually takes place in several steps so as to remove the different gases present in the gas stream individually and to optimize the energy recovery.

As an example, a hot gas stream containing water and carbon dioxide can be upgraded by cooling the gas to 40°C (104°F) when most of the water is condensed. The remaining water is removed in the carbon dioxide removal step — the carbon dioxide has to be removed from the gas stream to meet the specifications. The final concentration of carbon dioxide in the gas stream is determined by the specifications of the Wobbe index. For carbon dioxide removal, considering the high partial pressure of carbon dioxide both membranes and physical solvents can be chosen, where membranes are at their maximum scale and physical solvents are at their minimum scale (Alderton, 1993; Hall and Lokhandwala, 2004). The main advantage of the cryogenic process is that it is possible to obtain a gas stream with a high methane content of up to 99% v/v. The main disadvantage is that for the upgrading process it is necessary to use many of the technological equipment, especially compressors, turbines, and heat exchangers. This significant demand for additional equipment can make the cryogenic separation process extremely expensive.

9.4.4 *Methanol-based processes*

Methanol is probably one of the most versatile solvents in the natural gas processing industry. Historically, methanol was the first commercial organic physical solvent and has been used for dehydration, hydrate inhibition, gas sweetening, and liquid recovery (Kohl and Nielsen, 1997). Most of these applications involve low temperatures, where the physical properties of methanol are advantageous compared with other solvents, which that exhibit high viscosity problems or even solids formation. Operation at low temperatures tends to suppress the most significant disadvantage of methanol — i.e. high solvent loss. Furthermore, methanol is relatively inexpensive and easy to produce, making the solvent a very attractive alternative for gas processing applications.

The use of methanol has been further exploited in the development of the Rectisol process either alone or as toluene–methanol mixtures that are used to more selectively remove hydrogen sulfide and slip carbon dioxide to the overhead product (Ranke and Mohr, 1985). Toluene has an additional advantage insofar as carbonyl sulfide is more soluble in toluene than in methanol. The Rectisol process was primarily developed to remove both carbon dioxide and hydrogen sulfide (along with other sulfur-containing species) from gas streams resulting from the partial oxidation of coal, oil, and crude oil residua. The ability of methanol to absorb these unwanted components makes it the natural choice as solvent. Unfortunately, at cold temperatures, methanol also has a high affinity for hydrocarbon constituents of the gas streams. For example, propane is more soluble in methanol than in carbon dioxide. There are two versions of the Rectisol process: the two-stage and the once-through. In the two-stage process the first step is desulfurization before shift conversion; the concentrations of hydrogen sulfide and carbon dioxide are about 1% and 5% v/v, respectively. The regeneration of methanol following the desulfurization of the feed gas produces high-sulfur feed for sulfur recovery. The once-through process is applicable to high-pressure partial oxidation products and is also applicable when the hydrogen sulfide to carbon dioxide content ratio is unfavorable, in the neighborhood of 1:50 (Esteban *et al.*, 2000).

Recently, a process using methanol has been developed that has the simultaneous capability to dehydrate, to remove acid gas, and to control the hydrocarbon dew point (Rojey and Larue, 1988; Rojey *et al.*, 1990). The IFPEXOL-1 process is used for water removal and for hydrocarbon dew point control, while the IFPEXOL-2 process is used for acid gas removal. The novel idea behind the IFPEXOL-l process is to use a portion of the water-saturated inlet feed to recover methanol from the aqueous portion of the low-temperature separator. This approach has solved a major problem of methanol injection in large facilities, namely methanol recovery via distillation. But for this very simple discovery, the cold section of the process is remarkably similar to the basic methanol injection process. Modifications made to the process include water washing of the hydrocarbon liquid from the low-temperature separator in order to enhance the methanol recovery. The IFPEXOL-2 process for acid gas removal is very similar to an amine-type process except for the operating temperature. The absorber operates below –20°F to minimize methanol losses, while the regenerator operates at about 90 psi. Cooling is required in the regenerator condenser to recover the methanol. This process usually follows the IFPEXOL-1 process, so excessive hydrocarbon absorption is not so much great a problem (Minkkinen and Jonchere, 1997).

9.5 Acid Gas Removal

In addition to water removal, one of the most important parts of gas processing involves the removal of hydrogen sulfide and carbon dioxide, which are generally referred to as contaminants. Gas streams from some sources contain significant amounts of hydrogen sulfide and carbon dioxide and, analogous to natural gas that contains these same impurities, are usually referred to as *sour gas*. Sour gas is undesirable because the sulfur compounds it contains can be extremely harmful, even lethal, to breathe and the gas can also be extremely corrosive. The process for removing hydrogen sulfide from sour gas is commonly referred to as *sweetening* the gas. Thus, sulfur compounds, mainly hydrogen sulfide, must be removed from the synthesis gas to the best possible extent since they can poison catalysts. The methanation catalyst is particularly prone to sulfur poisoning. Regenerative sorbents (the so-called sulfur guards) can be used to reduce sulfur concentrations to the necessary limits, well below 1 ppm v/v. Washing techniques (based on physical or chemical adsorption) can also be implemented, making sulfur recovery via a Claus process possible if economically viable.

There are four general processes used for emission control (often referred to in another, more specific context as flue gas desulfurization: (i) physical adsorption in which a solid adsorbent is used, (ii) physical absorption in which a selective absorption solvent is used, (iii) chemical absorption in which a selective absorption solvent is used, and (iv) catalytic oxidation and thermal oxidation (Soud and Takeshita, 1994; Mokhatab *et al.*, 2006; Speight, 2014, 2019).

9.5.1 *Adsorption*

Activated carbon and molecular sieves are also capable of adsorbing water in addition to acid gases.

In these processes, a solid with a high surface area is used. Molecular sieves (zeolites) are widely used and are capable of adsorbing large amounts of gas. In practice, more than one adsorption bed is used to allow for continuous operation — one bed is in use while the other is being regenerated.

On the other hand, adsorption is usually a gas–solid interaction by which an adsorbent such as activated carbon (the *adsorbent* or *adsorbing medium*) can be regenerated upon *desorption* (Mokhatab *et al.*, 2006; Boulinguiez and Le Cloirec, 2009; Speight, 2014, 2019). Activated carbon impregnated with potassium iodide can catalytically react with oxygen and hydrogen sulfide to form water and sulfur. The reaction is best achieved at 50–70°C (122–158°F) and 100–120 psi. The activated carbon adsorption beds must be regenerated or replaced before they become ineffective due to oversaturation.

Regeneration is accomplished by passing hot dry fuel gas through the bed. Molecular sieves are competitive only when the quantities of hydrogen sulfide and carbon disulfide are low. The captured (adsorbed) gas can be desorbed with hot air or steam for either recovery or thermal destruction. Adsorber units are widely used to increase a low gas concentration prior to incineration unless the gas concentration is very high in the inlet air stream and the process is also used for reducing problematic odors (or obnoxious odors) from gases. There are several disadvantages of the use of adsorption systems, but it is generally the case that the major disadvantage is the need for minimize particulate matter emission and/or condensation of liquids (e.g. water vapor), which could otherwise mask the adsorption surface and reduce its efficiency drastically.

9.5.2 *Absorption*

In addition to economic issues or constraints, the solvents used for gas processing should have the following: (i) a high capacity for removal of acid gas, (ii) a low tendency to dissolve hydrogen, (iii) a low tendency to dissolve low-molecular weight hydrocarbon derivatives, (iv) low vapor pressure at operating temperatures to minimize solvent losses, (v) low viscosity, (vi) low thermal stability, (vii) absence of reactivity toward gas components, (viii) low tendency for fouling, and (ix) a low tendency for corrosion, and (x) economically acceptable (Mokhatab *et al.*, 2006; Speight, 2014, 2019).

Among the commercial processes being used, the Selexol process, the Sulfinol process, and the Rectisol process are noteworthy (Mokhatab *et al.*, 2006; Speight, 2014, 2019). In these processes, no chemical reaction occurs between the acid gas and the solvent. The solvent, or absorbent, is a liquid that selectively absorbs acid gases and leaves out hydrocarbons.

The Selexol process uses dimethyl ether or propylene glycol as a solvent. It is nontoxic and its boiling point is not high enough for amine formulation. The selectivity of the solvent for hydrogen sulfide (H_2S) is much higher than that for carbon dioxide (CO_2), so it can be used to selectively remove these different acid gases, minimizing carbon dioxide content in the hydrogen sulfide stream sent to the sulfur recovery unit (SRU) and enabling regeneration of solvent for carbon dioxide recovery by economical flashing. In the process, a gas stream is injected at the bottom of the absorption tower operated at 1,000 psi. The rich solvent is flashed in a flash drum (flash reactor) at 200 psi where methane is flashed and recycled back to the absorber and joins the sweet (low-sulfur or non-sulfur) gas stream. The solvent is then flashed at atmospheric pressure and acid gases are flashed off. The solvent is then stripped by steam to completely regenerate the solvent, which is recycled back to the absorber. Any hydrocarbon derivatives are condensed and any remaining acid gases are flashed from the condenser drum. This process is used when there is a high acid gas partial pressure and no heavy hydrocarbon derivatives. Diisopropanolamine (DIPA) can be added to this solvent to remove carbon dioxide to a level suitable for pipeline transportation.

The raw gas stream passes countercurrently to the descending solvent. When the solvent becomes saturated with the acid gases, the pressure is reduced, and hydrogen sulfide and carbon dioxide are desorbed. The solvent is then recycled to the absorption tower.

The sulfinol process uses a solvent that is a composite solvent consisting of a mixture of diisopropanolamine (30–45% v/v) or methyl diethanolamine (MDEA), sulfolane (tetrahydrothiophene dioxide) (40–60% v/v), and water (5–15% v/v). The acid gas loading of the sulfinol solvent is higher and the energy required for its regeneration is lower than for purely chemical solvents. At the same time, it has the advantage over purely physical solvents that severe product specifications can be fulfilled more easily and the extent of co-absorption of hydrocarbon derivatives is relatively low. Aromatic compounds, higher molecular weight hydrocarbon derivatives, and carbon dioxide are soluble to a lesser extent. The process is typically used when the hydrogen sulfide–carbon dioxide ratio is greater than 1:1 or when carbon dioxide removal is not required to the same extent as hydrogen sulfide removal. The process uses a conventional solvent absorption and regeneration cycle in which the sour gas components are removed from the feed gas by countercurrent contact with a lean solvent stream under pressure. The absorbed impurities are then removed from the rich solvent by stripping with steam in a heated regenerator column. The hot, lean solvent is then cooled for reuse in the absorber. A part of the cooling may be via heat exchange with the rich solvent for partial recovery of heat energy. The solvent reclaimer is used in a small ancillary facility for recovering solvent components from higher boiling products of alkanolamine degradation or from other high-boiling or solid impurities.

The major difference between water scrubbing and the Selexol process is that carbon dioxide and hydrogen sulfide are more soluble in Selexol, which results in a lower solvent demand and reduced pumping. In addition, water and halogenated hydrocarbon derivatives (contaminants in gas streams from landfill sites) are removed when scrubbing the gas stream with Selexol in which the Selexol solvents is designed for recirculation. Due to the formation of elemental sulfur on stripping, the Selexol solvent with air is not recommended but that with steam or inert gas (an upgraded gas stream or natural gas). Removal of hydrogen sulfide beforehand is one alternative.

9.5.3 *Chemisorption*

Chemisorption (chemical absorption) processes are characterized by a high capacity for absorbing large amounts of acid gases. They employ a solution of a relatively weak base, such as monoethanolamine. The acid gas forms a weak bond with the base, which can be regenerated easily. Mono- and di-ethanolamine derivatives are frequently used for this purpose. The amine concentration normally ranges between 15 and 30%.

The gas stream is passed through the amine solution, where sulfides, carbonates, and bicarbonates are formed. Diethanolamine is a favored absorbent because of its lower corrosion rate, smaller amine loss potential, fewer utility requirements, and minimal reclaiming needs. Diethanolamine reacts reversibly with 75% of carbonyl sulfides (COS), whereas monoethanolamine reacts irreversibly with 95% of carbonyl sulfide to form a degradation product, which must be disposed of in an environmentally acceptable manner.

The ethanolamine process, also known as the *Girbotol* process, removes the acid gases (hydrogen sulfide and carbon dioxide) from gases. The Girbotol process uses an aqueous solution of ethanolamine ($H_2NCH_2CH_2OH$) that reacts with hydrogen sulfide at low temperatures and releases hydrogen sulfide at high temperatures. The ethanolamine solution fills a tower (the absorber) through which the sour gas is bubbled. Purified gas leaves the top of the tower, and the ethanolamine solution leaves the bottom of the tower with the absorbed acid gases. The ethanolamine solution enters a reactivator tower where heat drives the acid gases from the solution. Ethanolamine solution, restored to its original condition, leaves the bottom of the reactivator tower to go to the top of the absorber tower, and acid gases are released from the top of the reactivator.

Alkanolamine scrubbers can be used to remove hydrogen sulfide or hydrogen sulfide and carbon dioxide simultaneously (Katz, 1959; Kohl and Riesenfeld, 1985; Maddox *et al.*, 1985; Polasek and Bullin, 1985; Jou *et al.*, 1985; Pitsinigos and Lygeros, 1989; Kohl and Nielsen, 1997; Mokhatab *et al.*, 2006; Speight, 2014; Abdel-Aal *et al.*, 2016; Speight, 2019). For instance, monoethanolamine (MEA), diethanolamine (DEA), methyldiethanolamine (MDEA), diisopropanolamine (DIPA), and triethanolamine (TEA) can be used. The alkanolamine scrubbing process is also carried out in two steps: the first step involves the amine contactor and the second step involves the regenerator. In a process with diethanolamine and methyldiethanolamine the main reactions in equilibrium in the system are as follows:

$$2H_2O \rightarrow H_3O^+ + OH^-$$

$$CO_2 + 2H_2O \rightarrow HCO_3^- + H_3O^+$$

$$HCO_3^- + H_2O \rightarrow H_3O^+ + CO_3^{2-}$$

$$H_2S + H_2O \rightarrow H_3O^+ + HS^-$$

$$HS^- + H_2O \rightarrow H_3O^+ + S^{2-}$$
$$DEA^+ + H_2O \rightarrow DEA + H_3O^+$$
$$MDEA^+ + H_2O \rightarrow MDEA + H_3O^+$$
$$DEA^+COO^- + H_2O \rightarrow DEA + HCO_3^-$$

The pH is an important parameter because it affects the concentration of the ionic species and therefore the reactions with the amines. Moreover, the temperature and pressure also have a significant effect on the process.

The reactions of carbon dioxide and hydrogen sulfide with amines are as follows:

$$\begin{gathered} DEA + CO_2 + H_2O \rightarrow DEA^+COO^- + H_3O^+ \\ MDEA + CO_2 + H_2O \rightarrow MDEA^+ + HCO_3^- \\ MDEA + H_2S \rightarrow MDEA^- + HS^- \\ DEA + H_2S \rightarrow DEA^+ + HS^- \end{gathered} \tag{9.1}$$

This system is very complex because of the large number of reactions involved and other reactions may also occur in the system. Tertiary amines such as DMEA do not react directly with carbon dioxide but reaction of hydrogen sulfide with amines is almost instantaneous by proton transfer. Thus, tertiary amines are used for selective hydrogen sulfide removal, whereas for selective hydrogen sulfide and carbon dioxide removal, a mixed amine system (tertiary and primary or secondary) is employed. Also, other compounds can be mixed with amine solutions.

Thus, treatment of a gas stream to remove the acid gas constituents (hydrogen sulfide and carbon dioxide) is most often accomplished by contacting the gas stream with an alkaline solution. The most commonly used treating solutions are aqueous solutions of ethanolamine or alkali carbonates, although in recent years a considerable number of other treating agents have been developed (Mokhatab *et al.*, 2006; Speight, 2014, 2019), most of which rely upon physical absorption and chemical reaction. If just carbon dioxide has to be removed in large quantities or if just partial removal is needed, a hot carbonate solution or one of the physical solvents would be the most economical choice.

The primary process for sweetening sour gas uses an amine (*olamine*) solution to remove the hydrogen sulfide (the *amine process*). The sour gas is run through a tower, which contains the olamine solution. There are two principal amine solutions used: monoethanolamine (MEA) and diethanolamine (DEA). Either of these compounds, in liquid form, will absorb sulfur compounds from the gas stream as it passes through. The effluent gas is virtually free of sulfur compounds and thus loses its sour gas status. The amine solution used can be regenerated for reuse and, although most sour gas sweetening involves the amine absorption process, it is also possible to use solid desiccants like iron sponge to remove hydrogen sulfide and carbon dioxide (Mokhatab *et al.*, 2006; Abdel-Aal *et al.*, 2016; Speight, 2019), including biobased iron sponge (Cherosky and Li, 2013).

Diglycolamine (DGA) is another amine solvent used in the Econamine process in which absorption of acid gases occurs in an absorber containing an aqueous solution of diglycolamine, and the heated rich solution (saturated with acid gases) is pumped to the regenerator (Reddy and Gilmartin, 2008). Diglycolamine solutions are characterized by low freezing points, which makes them suitable for use in cold climates.

The most well-known hydrogen sulfide removal process is based on the reaction of hydrogen sulfide with iron oxide (often called the iron sponge process or the dry box method) in which the gas is passed through a bed of wood chips impregnated with iron oxide. The iron oxide is converted to the corresponding sulfur, which is reaerated by oxidation (Mokhatab *et al.*, 2006; Speight, 2014, 2019):

$$Fe_2O_3 + 3H_2S \rightarrow Fe_2S_3 + 3H_2O$$
$$2Fe_2S_3 + 3O_2 \rightarrow 2Fe_2O_3 + 6S$$

The iron oxide process (which was implemented during the 19th century and is also referred to as the iron sponge process) is the oldest and still the most widely used batch process for sweetening the gas stream (Zapffe, 1963; Anerousis and Whitman, 1984; Mokhatab *et al.*, 2006; Speight, 2006, 2014). The reaction can be exploited by adding the iron chloride to the digester feed material or passing the gas stream through a bed of iron oxide-containing material. The iron oxide comes in different forms such as rusty steel wool, iron oxide pellets or wood pellets coated with iron oxide. The iron oxide medium needs to be replaced periodically. The regeneration process is highly exothermic and must be controlled to avoid problems.

In the process, the sour gas is passed down through the bed. In the case where continuous regeneration has to be utilized a small concentration of air is added to the sour gas before it is processed. This air serves to continuously regenerate the iron oxide, which has reacted with hydrogen sulfide, which serves to extend the on-stream life of a given tower but probably serves to decrease the total amount of sulfur that a given weight of bed will remove. Ferric hydroxide ($Fe(OH)_3$) can also be used in this process.

The process is usually best applied to gases containing low-to-medium concentrations (300 ppm v/v) of hydrogen sulfide or mercaptan derivatives. This process tends to be highly selective and does not normally remove significant quantities of carbon dioxide. As a result, the hydrogen sulfide stream from the process is usually of high purity. The use of the iron oxide process to sweeten sour gas is based on adsorption of the acid gases on the surface of the solid sweetening agent, followed by chemical reaction of ferric oxide (Fe_2O_3) with hydrogen sulfide:

$$2Fe_2O_3 + 6H_2S \rightarrow 2Fe_2S_3 + 6H_2O$$

The reaction requires the presence of slightly alkaline water and a temperature below 43°C (110°F), and bed alkalinity (pH 8–10) should be checked regularly, usually on a daily basis. The pH level has be maintained through the injection of caustic soda with water. If the gas does not contain a sufficient amount of water vapor, water may need to be injected into the in-coming gas stream (i.e. the gas stream at the inlet).

The ferric sulfide produced by the reaction of hydrogen sulfide with ferric oxide can be oxidized with air to produce sulfur and regenerate ferric oxide:

$$2Fe_2S_3 + 3O_2 \rightarrow 2Fe_2O_3 + 6S$$
$$2S + 2O_2 \rightarrow 2SO_2$$

The regeneration step is exothermic, and air must be introduced slowly so that the heat of the reaction can be dissipated. If air is introduced quickly the heat of the reaction may ignite the bed. Some of the elemental sulfur produced in the regeneration step remains in the bed. After several cycles this sulfur will form a cake over the ferric oxide, decreasing the reactivity of the bed. Typically, after 10 cycles the bed must be removed and a new bed introduced into the vessel.

The iron oxide process is one of several metal oxide-based processes that scavenge hydrogen sulfide and organic sulfur compounds (mercaptan derivatives) from gas streams through reactions with a solid-based chemical adsorbent (Kohl and Riesenfeld, 1985). They are typically non-regenerable, although some are partially regenerable, losing activity upon each regeneration cycle. Most of the processes are governed by the reaction of a metal oxide with hydrogen sulfide to form the metal sulfide. For regeneration, the metal oxide is reacted with oxygen to produce elemental sulfur and the regenerated metal oxide. In addition to iron oxide, the primary metal oxide used for dry sorption processes is zinc oxide.

In the zinc oxide process, the zinc oxide medium particles are extruded cylinders 3–4 mm in diameter and 4–8 mm in length (Kohl and Nielsen, 1997; Mokhatab *et al.*, 2006; Abdel-Aal *et al.*, 2016; Speight, 2019) and react readily with the hydrogen sulfide:

$$ZnO + H_2S \rightarrow ZnS + H_2O$$

At increased temperatures (205–370°C, 400–700°F), zinc oxide has a high reaction rate, therefore providing a short mass transfer zone, resulting in a short length of unused bed and improved efficiency.

Removal of larger amounts of hydrogen sulfide from gas streams requires a continuous process, such as the *Ferrox* process or the Stretford process. The *Ferrox process* is based on the same chemistry as the iron oxide process except that it is fluid and continuous. The *Stretford* process employs a solution containing vanadium salts and anthraquinone disulfonic acid (Maddox, 1974; Mokhatab *et al.*, 2006; Abdel-Aal *et al.*, 2016).

Most hydrogen sulfide removal processes return the hydrogen sulfide unchanged, but if the quantity involved does not justify installation of a sulfur recovery plant (usually a Claus plant) it is necessary to select a process that directly produces elemental sulfur. In the *Beavon–Stretford* process, a hydrotreating reactor converts sulfur dioxide in the off-gas to hydrogen sulfide, which is contacted with Stretford solution (a mixture of vanadium salt, anthraquinone disulfonic acid, sodium carbonate, and sodium hydroxide) in a liquid–gas absorber. The hydrogen sulfide reacts stepwise with sodium carbonate and anthraquinone disulfonic acid to produce elemental sulfur, with vanadium serving as a catalyst. The solution proceeds to a tank where oxygen is added to regenerate the reactants. One or more froth or slurry tanks are used to skim the product sulfur from the solution, which is recirculated to the absorber.

Even though the removal of hydrogen sulfide can be achieved by chemical or physical absorption, chemical scrubbers are the most commonly used systems for hydrogen sulfide removal (Mokhatab *et al.*, 2006; Speight, 2014, 2019). The most important processes are iron-based chelation methods and absorption by alkanolamines. In iron-based chelation processes there are two main reactions:

$$H_2S + 2Fe^{3+}\ \text{Chelant}^{n-} \rightarrow S + 2H^+ + 2\ Fe^{2+}\ \text{Chelant}^{n-}$$

$$O_2 + 4Fe^{2+}\ \text{Chelant}- + 2H_2O \rightarrow 4Fe^{3+}\ \text{Chelant}- + 4\ OH^-$$

The first step involves the chemical reaction to produce elemental sulfur and in the second step iron(III) is regenerated. Several commercial processes are based on iron chelation, such as the Lo-cat and Sulferox processes. Ethylenediaminetetraacetic acid (EDTA), hydroxyethylenediaminetriacetic acid (HEDTA), diethylenetriaminepentaacetic acid (DTPA), and nitrilotriacetic acid (NTA) are the most conventional ligands used in this process. The operating conditions are typically the following: 4–8°C (39.5–46.5°F), pH 4–8, iron concentration of the order of 1,000–10,000 ppm v/v with a chelant/iron ratio in the range of 1.1–2.0 (Deshmukh and Shete, 2013; Ramírez *et al.*, 2015).

The process using potassium phosphate is known as phosphate desulfurization, and it is used in the same way as the Girbotol process to remove acid gases from liquid hydrocarbon derivatives as well as from gas streams. The treatment solution is a water solution of potassium phosphate (K_3PO_4), which is circulated through an absorber tower and a reactivator tower in much the same way as ethanolamine is circulated in the Girbotol process; the solution is regenerated thermally. Processes using ethanolamine and potassium phosphate are now widely used.

9.5.4 *Other processes*

There are a series of alternative processes: (i) the use of chemical reactions to remove contaminants from gas streams or (ii) the use of specialized equipment to physically remove contaminants from gas streams.

As an example of the first category, i.e. the use of chemical reactions to remove contaminants from gas streams, strong basic solutions are effective solvents for acid gases. However, these solutions are not normally used to treat large volumes of a gas stream because the acid gases form stable salts, which are not easily regenerated. For example, carbon dioxide and hydrogen sulfide react with aqueous sodium hydroxide to yield sodium carbonate and sodium sulfide, respectively:

$$CO_2 + 2NaOH \rightarrow Na_2CO_3 + H_2O$$

$$H_2S + 2NaOH \rightarrow Na_2S + 2H_2O$$

However, a strong caustic solution is used to remove mercaptans from gas and liquid streams. In the Merox process, for example, a caustic solvent containing a catalyst such as cobalt, which is capable of converting mercaptans (RSH) to caustic insoluble disulfides (RSSR), is used for streams rich in mercaptans after removal of hydrogen sulfide. Air is used to oxidize the mercaptan derivatives to disulfide derivatives. The caustic solution is then recycled for regeneration. The Merox process is mainly used for the treatment of refinery gas streams.

As one of the major contaminants in gas streams, carbon dioxide must optimally be removed as it reduces the energy content of the gas and affects the selling price of the gas. Moreover, it becomes acidic and corrosive in the presence of water that has the potential to damage the pipeline and the equipment system. Hence, the presence of carbon dioxide in gas streams remains one of the challenging gas separation problems in process engineering for carbon dioxide/methane systems. Therefore, the removal of carbon dioxide from the gas stream through the purification processes is vital for the improvement of the quality of the product (Mokhatab *et al.*, 2006; Speight, 2014, 2019).

Carbonate washing is a mild alkali process (typically the alkali is potassium carbonate, K_2CO_3) for gas processing for the removal of acid gases (such as carbon dioxide and hydrogen sulfide) from gas streams and uses the principle that the rate of absorption of carbon dioxide by potassium carbonate increases with temperature (Mokhatab *et al.*, 2006; Speight, 2014, 2019).

The Fluor process uses propylene carbonate to remove carbon dioxide, hydrogen sulfide, carbonyl sulfide, water, and higher boiling hydrocarbon derivatives (C_2^+) from the gas stream (Abdel-Aal *et al.*, 2016).

Water washing, in terms of the outcome, is almost analogous to (but often less effective than) washing with potassium carbonate (Kohl and Riesenfeld, 1985; Kohl and Nielsen, 1997), and it is also possible to carry out the desorption step by pressure reduction. The water scrubbing processes use the higher solubility of carbon dioxide in water to separate the carbon dioxide from the gas stream. This process is done under high pressure and removes hydrogen sulfide as well as carbon dioxide. The main disadvantage of this process is that it requires a large volume of water that must be purified and recycled. An analogous process, polyethylene glycol scrubbing, is similar to water scrubbing but is more efficient. However, the process does require the regeneration of a large volume of polyethylene glycol.

These systems operate in a temperature range of 205–595°C (400–1100°F) in the presence of a catalyst — in the absence of the catalyst, the system would require a higher operating temperature.

Other processes include the *Alkazid process* for the removal of hydrogen sulfide and carbon dioxide using concentrated aqueous solutions of amino acids. The hot potassium carbonate process decreases the acid content of natural and refinery gas from as much as 50% to as low as 0.5% and operates in a unit similar to that used for amine treating. The *Giammarco–Vetrocoke* process is used for hydrogen sulfide and/or carbon dioxide removal. In the hydrogen sulfide removal section, the reagent consists of sodium carbonate (Na_2CO_3) or potassium carbonate (K_2CO_3) or a mixture of the carbonates which contains a mixture of arsenite derivatives and arsenate derivatives; the carbon dioxide removal section utilizes hot aqueous alkali carbonate solution activated by arsenic trioxide (As_2O_3) or selenous acid (H_2SeO_3) or tellurous acid (H_2TeO_3). A word of caution might be added about the last three chemicals, which are toxic and can involve stringent environmental-related disposal protocols.

Molecular sieves are highly selective for the removal of hydrogen sulfide (as well as other sulfur compounds) from gas streams and over a continuously high absorption efficiency. They are also an effective means for water removal and thus offer a process for the simultaneous dehydration and desulfurization of gas. Gas that has excessively high water content may require upstream dehydration, however (Mokhatab *et al.*, 2006; Speight, 2014, 2019; Abdel-Aal *et al.*, 2016). The carbon molecular sieve method uses differential adsorption characteristics to separate methane and carbon dioxide.

In addition, the *molecular sieve process* is similar to the iron oxide process. Regeneration of the bed is achieved by passing heated clean gas over the bed. As the temperature of the bed increases, it releases the adsorbed hydrogen sulfide and the sour effluent regeneration gas is sent to a flare stack, and up to 2% v/v of the gas seated can be lost in the regeneration process. A portion of the gas stream may also be lost by the adsorption of hydrocarbon components by the sieve (Mokhatab *et al.*, 2006; Speight, 2014, 2019).

In this process, unsaturated hydrocarbon components, such as olefin derivatives and aromatic derivatives, tend to be strongly adsorbed by the molecular sieve. Molecular sieves are susceptible to poisoning by chemicals such as glycols and require thorough gas cleaning methods before the adsorption step. Alternatively, the sieve can be offered some degree of protection by the use of *guard beds* in which a less expensive catalyst is placed in the gas stream before contact of the gas with the sieve, thereby protecting the catalyst from poisoning. This concept is analogous to the use of guard beds or attrition catalysts in the crude oil industry (Speight, 2000, 2014, 2017).

Carbon molecular sieves are excellent products to separate specifically a number of different gaseous compounds in the gas stream. Thereby the molecules are usually loosely adsorbed in the cavities of the carbon sieve but not irreversibly bound. The selectivity of adsorption is achieved by different mesh sizes and/or the application of different gas pressures. When the pressure is released the compounds extracted from the gas stream are desorbed. This process — pressure swing adsorption (PSA) technology — can be used to enrich methane from the gas stream, in which the molecular sieve is applied which is produced from coke rich in pores in the micrometer range. The pores are then further reduced by cracking of the hydrocarbon derivatives.

In order to reduce the energy consumption for gas compression, a series of vessels are linked together. The gas pressure released from one vessel is subsequently used by the others. Usually four vessels in series are used filled with the molecular sieve, which removes at the same time carbon dioxide and any water vapor. After removal of hydrogen sulfide, i.e. using activated carbon and water condensation in a cooler at 4°C, the gas stream flows at a pressure of 90 psi into the adsorption unit. The first column cleans the raw gas at 90 psi to an upgraded gas stream with a vapor pressure of less than 10 ppm water and a methane content of 96% or more.

In the second column the pressure of 90 psi is first reduced to approximately 45 psi by pressure communication with column 4, which was previously degassed by a slight vacuum. In a second step the pressure is then reduced to atmospheric pressure. The released gas flows back to the digester in order to recover the methane. The third column is evacuated from 15 to 0.15 psi. The desorbed gas consists predominantly of carbon dioxide and also some methane and is therefore normally released to the environment. In order to reduce methane losses the system can be designed with recirculation of the desorbed gases.

A number of other possible impurities, such as organic acids (e.g. formic and acetic acid) and unsaturated and higher boiling hydrocarbon derivatives, are normally not included in purity specifications for synthesis processes because they are already undesirable in upstream process stages of synthesis gas production (e.g. compression steps). However, the organic compounds present must be below their dew point at the pressure of the gas application in order to prevent condensation and fouling in the system. For organic compounds with sulfur or nitrogen heteroatoms (such as thiophene derivatives and pyridine derivatives) the additional specification that they need to be removed below 1 ppm v/v level applies, as they are intrinsically poisonous to the catalyst.

Thiophene

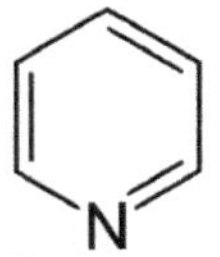

Pyridine

9.6 Removal of Condensable Hydrocarbons

Hydrocarbon derivatives of higher molecular weight than methane present in the gas streams are valuable raw materials and important fuels. They can be recovered by lean oil extraction. The first step in this process is to cool the treated gas by exchange with liquid propane after which the cooled gas is washed with a cold hydrocarbon liquid, which dissolves most of the condensable hydrocarbon derivatives. The uncondensed gas is a dry gas stream composed mainly of methane. The dry gas stream may then be used either as a fuel or as a chemical feedstock. Another way of recovering any higher molecular weight hydrocarbon constituents from a gas stream is by cryogenic cooling (cooling to very low temperatures of the order of –100°C to –115°C (–150°F to –175°F), which are achieved primarily by lowering the temperatures to below the dew point.

To prevent hydrate formation, a gas stream may be treated with glycols, which dissolve water efficiently. Ethylene glycol (EG), diethylene glycol (DEG), and triethylene glycol (TEG) are typical solvents for water removal. Triethylene glycol is preferable in vapor phase processes because of its low vapor pressure, which results in less glycol loss. The triethylene glycol absorber unit typically contains 6–12 bubble-cap trays to accomplish the water absorption. However, more contact stages may be required to reach dew points below –40°C (–40°F). Calculations to determine the number of trays or feet of packing, the required glycol concentration, or the glycol circulation rate require vapor–liquid equilibrium data. In addition, predicting the interaction between triethylene glycol and water vapor in the gas stream over a broad range allows the design of ultra-low-dew point applications.

One alternative to using bubble-cap trays is by the use of the adiabatic expansion of the inlet gas. In this process, the inlet gas is first treated to remove water and acid gases and then cooled via heat exchange and refrigeration. Further cooling of the gas is accomplished through turbo expanders, and the gas is sent to a demethanizer to separate methane from higher-boiling hydrocarbon derivatives. Improved recovery of the higher-boiling hydrocarbon derivatives could be achieved through better control strategies and by the use of online gas chromatography analysis.

Membrane separation processes are very versatile and are designed to process a wide range of feedstocks, and offer a simple method for the removal and recovery of higher boiling hydrocarbon derivatives from gas streams (Alderton, 1993; Foglietta, 2004; May-Britt, 2008; Basu *et al.*, 2010; Rongwong *et al.*, 2012; Ozturk and Demirciyeva, 2013; Abdel-Aal *et al.*, 2016). There are two membrane separation techniques: (i) high-pressure gas separation and (ii) gas–liquid adsorption. The high-pressure separation process selectively separates hydrogen sulfide and carbon dioxide from methane. Usually, this separation is performed in three stages and produces methane with a high degree of purity (of the order of 96% v/v methane).

The separation process is based on high-flux membranes that selectively permeate higher boiling hydrocarbon derivatives (compared to methane) and are recovered as a liquid after recompression and condensation. The residue stream from the membrane is partially depleted of higher boiling hydrocarbon derivatives, and is then sent to a sales gas stream. Gas permeation membranes are usually made of vitreous polymers that exhibit good selectivity but, in order to be effective, the membrane must be very permeable with respect to the separation process (Rojey *et al.*, 1997).

Membrane separation is based on the selectivity properties of the membrane material. The driving force is the difference in chemical potential, which includes the effect of temperature and partial pressure regimes. Membranes can be operated at high pressure (>300 psi) or low pressure (120–150 psi). Several membranes are used for gas stream upgrading and can be classified based on the type of material into non-polymeric membranes (such as alumina, zeolites, and carbon), ceramic membranes, palladium membranes, and carbon molecular sieve membranes. Nevertheless, polymeric membranes are most widely used because they are cheaper than inorganic membranes. Moreover, polymeric materials can be easily fabricated into flat sheet or asymmetric hollow fibers; both these shapes are usually used in industrial applications.

The material properties that define the performance are the molecular structure, glass transition temperature, crystallinity, and degree of crosslinking. However, the commercially available polymeric membranes are usually degraded by hydrogen sulfide, siloxanes, and/or volatile organic compounds (Accettola *et al.*, 2008; Ajhar *et al.*, 2010; Gislon *et al.*, 2013). For instance, hydrogen sulfide can cause plasticization of glassy polymeric membranes and this alters the polymeric structure and weakens the mechanical strength of the membrane.

In the high-pressure separation option, pressurized gas (36 bar) is first cleaned over, for example, an activated carbon bed to remove (halogenated) hydrocarbon derivatives and hydrogen sulfide from the raw gas as well as oil vapor from the compressors. The carbon bed is followed by a particle filter and a heater. The membranes made of acetate–cellulose separate small polar molecules such as carbon dioxide, moisture, and the remaining hydrogen sulfide. These membranes are not effective in separating nitrogen from methane.

The raw gas is upgraded in three stages into a clean gas with 96% or more methane. The waste gas from the first two stages is recycled and the methane is recovered, while the waste gas from stage 3 (and from a part of stage 2) is flared or used in a steam boiler as it still contains 10–20% v/v methane. First experiences have shown that the membranes can last for up to 3 years, which is comparable to the lifetime of the membranes for natural gas purification (a primary market for membrane technology), which last typically 2–5 years.

The clean gas is further compressed to 3,600 psi and stored in steel cylinders with capacities of 276 m^3 divided into high-, medium- and low-pressure banks. The membranes are very selective for the given molecules, i.e. hydrogen sulfide and carbon dioxide are upgraded by different modules. Hollow-fiber membranes are used to construct very compact modules working in crossflow.

The gas–liquid absorption membrane method is a separation technique that was developed for gas stream upgrading just recently. The essential element is a microporous hydrophobic membrane separating the gaseous phase from the liquid phase. The molecules from the gas stream flowing in one direction, which are able to diffuse through the membrane, will be absorbed on the other side by the liquid flowing in countercurrent.

The absorption membranes work at approximately atmospheric pressure (1 bar), which allows low-cost construction. The removal of gaseous components is very efficient. At a temperature of 25–35°C (77–95°F), the hydrogen sulfide concentration in the raw gas of 2% v/v is reduced to less than 0.025% v/v (250 ppm v/v) — the absorbent is sodium hydroxide (NaOH) or another alkaline material. Sodium hydroxide saturated with hydrogen sulfide can be used in water treatment to remove heavy metals. The concentrated hydrogen sulfide solution is fed into a Claus reaction or oxidized into elementary sulfur. The gas stream is upgraded very efficiently from 55% v/v methane (43% v/v carbon dioxide) to more than 96% methane. The amine solution is regenerated by heating, and the carbon dioxide released is pure and can be sold for industrial applications.

The main advantages of membranes are (i) the safety and simplicity of operation, (ii) the relative ease of maintenance and scale-up, (iii) excellent reliability, (iv) small footprint, and (v) operation without hazardous chemicals. Several operation modes can be employed — as an example, gas stream upgrading can be achieved by a two-stage cascade process with recycling, and a single stage provides good flexibility for integration into upgraded gas streams.

9.6.1 *Extraction*

There are two principal techniques for removing higher-boiling hydrocarbon constituents from gas streams: (i) the absorption method and (ii) the cryogenic expander process. In the latter process, a turboexpander is used to produce the necessary refrigeration and very low temperatures, and a high recovery of light components, such as ethane and propane, can be attained. The gas stream is first dehydrated using a molecular sieve, followed by the cooling of the dry stream. The separated liquid containing most of the heavy fractions is then demethanized, and the cold gases are expanded through a turbine that produces the desired cooling for the process. The expander outlet is a two-phase stream that is fed to the top of the demethanizer column. This serves as a separator, in which (i) the liquid is used as the column reflux, and the separator vapors combined with vapors stripped in the demethanizer are exchanged with the feed gas, and (ii) the heated gas, which is partially recompressed by the expander compressor, is further recompressed to the desired distribution pressure in a separate compressor.

The extraction of higher-boiling hydrocarbon constituents from the gas stream produces both a cleaner, purer gas stream and valuable hydrocarbon derivatives for use in, say, a petrochemical plant (Speight, 2019). This process allows for the recovery of approximately 90–95% v/v of the ethane originally present in the gas stream. In addition, the expansion turbine is able to harness some of the energy released when the gas stream is expanded to recompress the gaseous methane effluent, thus reducing energy costs associated with extraction of ethane.

9.6.2 *Absorption*

Using the absorption method for recovery of high-molecular-weight hydrocarbon derivatives is very similar to using absorption for dehydration (Mokhatab *et al.*, 2006; Speight, 2014, 2019). The main difference is that in the absorption of higher-boiling hydrocarbon constituents from the gas stream, absorbing oil is used as opposed to glycol. This absorbing oil has an affinity for higher-boiling hydrocarbon constituents in much the same manner as glycol has an affinity for water. Before the oil has picked up any higher-boiling hydrocarbon constituents, it is termed *lean* absorption oil.

The *oil absorption process* involves the countercurrent contact of the lean (or stripped) oil with the incoming wet gas with the temperature and pressure conditions optimized to maximize the dissolution of the liquefiable components present in the oil. The *rich* absorption oil (sometimes referred to as *fat* oil), containing higher-boiling hydrocarbon constituents, exits the absorption tower through the bottom. It is now a mixture of absorption oil, propane, butanes, pentanes, and other higher boiling hydrocarbon derivatives. The rich oil is fed into lean oil stills, where the mixture is heated to a temperature that is above the boiling point of the higher-boiling hydrocarbon constituents but below that of the oil. This process allows for the recovery of higher boiling constituents from the gas stream.

The basic absorption process has been subjected to modifications to improve the process effectiveness and even target the extraction of specific higher-boiling hydrocarbon constituents. In the refrigerated oil absorption method, where the lean oil is cooled through refrigeration, propane recovery can be of the order of 90%+ v/v, and approximately 40% v/v of the ethane can be extracted from the gas stream. Extraction of the other, higher-boiling hydrocarbon constituents is typically near-quantitative using this process.

9.6.3 *Fractionation*

Fractionation processes are very similar to those processes classed as *liquids removal* processes but often appear to be more specific in terms of the objectives; hence the need to place the fractionation processes into a separate category. Fractionation processes are defined as those processes used to remove (i) the more significant product stream first or (ii) any unwanted light ends from the higher-boiling liquid products.

As a general practice in gas processing, the first unit is a de-ethanizer followed by a depropanizer, a debutanizer, and finally, a butane fractionator. Thus, each column can operate at a successively lower pressure, which allows the different gas streams to flow from column to column by virtue of the pressure gradient, without necessarily the use of pumps. The purification of hydrocarbon gases by any of these processes is an important part of gas processing operations.

Thus, after the higher-boiling hydrocarbon constituents have been removed from the gas stream, they must be separated (fractionated) into their individual constituents prior to sales. The process of fractionation occurs in stages with each stage involving separation of the hydrocarbon derivatives as individual products. The process commences with the removal of the lower-boiling hydrocarbon derivatives from the feedstock. The particular fractionators are used in the following order: (i) the de-ethanizer, which is used to separate the ethane from the gas stream, (ii) the depropanizer, which is used to separate the propane from the de-ethanized gas stream, (iii) the debutanizer, which is used to separate the butane isomers, leaving the pentane isomers and higher boiling hydrocarbon derivatives in the gas stream, and (iv) the butane splitter or deisobutanizer, which is used to separate *n*-butane and isobutane.

After the recovery of the higher-boiling hydrocarbon constituents, sulfur-free dry gas (methane) may be liquefied for its transportation through cryogenic tankers. Further treatment may be required in order to reduce the water vapor below 10 ppm and carbon dioxide and hydrogen sulfide to less than 100 and 50 ppm v/v, respectively. There are two methods generally used to liquefy the gas: (i) the expander cycle and (ii) mechanical refrigeration. In the expander cycle, a part of the gas is expanded from a high transmission pressure to a lower pressure. This lowers the temperature of the gas. Through heat exchange, the cold gas cools the incoming gas, which in turn cools more incoming gas until the liquefaction temperature of methane is reached.

In mechanical refrigeration, a multi-component refrigerant consisting of nitrogen, methane, ethane, and propane is used through a cascade cycle. When these liquids evaporate, the heat required for this is obtained from the gas, which loses energy/temperature till it is liquefied. The refrigerant gases are recompressed and recycled.

9.6.4 *Enrichment*

The gas product must meet specific quality measures in order for the pipeline grid to operate properly. Consequently, a gas stream, which in most cases contains contaminants, must be processed, i.e. cleaned, before it can be safely delivered to the high-pressure, long-distance pipelines, which transport the product to the consuming public. A gas stream not within certain specific gravities, pressures, Btu content range, or water content levels will cause operational problems, or pipeline deterioration, or can even cause pipeline rupture. Thus, the purpose of *enrichment* is to produce a gas stream for sale. Therefore, the process concept is essentially the separation of higher-boiling hydrocarbon constituents from the methane to produce a lean, dry gas.

As an example, carbon dioxide is to some extent soluble in water and therefore some carbon dioxide will be dissolved in the liquid phase of the digester tank. In upgrading with the *in situ* methane enrichment process, sludge from the digester is circulated to a desorption column and then back to the digester. In the desorption column, carbon dioxide is desorbed by pumping air through the sludge. The constant removal of carbon dioxide from the sludge leads to an increased concentration of methane in the gas stream phase leaving the digester.

The gas stream received and transported must (especially in the United States and many other countries) meet the quality standards specified by the pipeline. These quality standards vary from pipeline to pipeline and are usually a function of (i) the design of the pipeline system, (ii) the design of any downstream interconnecting pipelines, and (iii) the requirements of the customer. In general, these standards specify that the gas stream should (i) be within a specific Btu content range, typically 1,035 Btu $ft^3 \pm 50$ Btu ft^3, (ii) be delivered at a specified hydrocarbon dew point temperature level to prevent any vaporized gas liquid in the mix from condensing at pipeline pressure, (iii) contain no more than trace amounts of elements such as hydrogen sulfide, carbon dioxide, nitrogen, water vapor, and oxygen, (iv) be free of particulate solids and liquid water that could be detrimental to the pipeline or its ancillary operating equipment. Gas processing equipment, whether in the field or at processing/treatment plants, assures that these specifications can be met.

In most cases processing facilities extract contaminants and higher-boiling hydrocarbon derivatives from the gas stream but, in some cases, the gas processors blend some higher-boiling hydrocarbon derivatives into the gas stream in order to bring it within acceptable Btu levels. For instance, in some areas the produced gas (including coalbed methane) does not meet (is below) the Btu requirements of the pipeline operator, in which case a blend of higher Btu content gas stream or a propane–air mixture is injected to enrich the heat content (Btu value) prior to delivery to the pipeline.

The number of steps and the type of technique used in the process of creating a pipeline-quality gas stream most often depends on the source and composition of the well-head production stream. Among the several stages of gas processing are the following: (i) gas–oil separation, (ii) water removal, (iii) liquids removal, (iv) nitrogen removal, (v) acid gas removal, and (vi) fractionation.

In many instances, pressure relief at the well-head will cause a natural separation of gas from oil (using a conventional closed tank, where gravity separates the gas hydrocarbon derivatives from the heavier oil). In some cases, however, a multi-stage gas–oil separation process is needed to separate the gas stream from the crude oil. These gas–oil separators are commonly closed cylindrical shells, horizontally mounted with inlets at one end, an outlet at the top for the removal of gas, and an outlet at the bottom for the removal of oil. Separation is accomplished by alternately heating and cooling (by compression) the flow stream through multiple steps. In some cases, several of the steps may be integrated into one unit or operation, performed in a different order or at alternative locations (lease/plant), or not required at all.

9.7 Tar Removal

Tar is a highly viscous liquid product (typically produced in thermal processes) that condenses in the low-temperature zones of the equipment (the pyrolizer or the gasifier) thereby interfering with the gas flow and leading to system disruption. It is, perhaps, the most undesirable product of the process and a high residual tar concentration in the gas can prevent its utilization (Devi *et al.*, 2003; Torres *et al.*, 2007; Anis and Zainal, 2011; Huang *et al.*, 2011). Methods for reduction or elimination of tar can be divided into two broad groups: (i) *in situ* tar reduction, also called primary tar reduction, which focuses on reducing or avoiding tar formation in the reactor, and (ii) post-gasification tar reduction, also called secondary tar reduction, which strips the tar from the product gas. A combination of *in situ* and post-gasification tar reductions can prove to be more effective than either of the single stages alone. The two basic post-gasification methods are physical removal and cracking (catalytic or thermal).

9.7.1 *Physical methods*

In a process for the physical removal of tar, the tar is treated as dust particles or mist and the tar is condensed before separation. Physical tar removal can be accomplished by using cyclones, barrier filters, wet electrostatic precipitators (ESPs), or wet scrubbers. The selection of the equipment for application depends on the load concentrations of particulate matter (PM) and tar, particle size distribution, and particulate tolerance of downstream users.

Cyclones may not be efficient in removing small tar droplets, and barrier filters are the porous material that can capture a certain amount of tar as the product gas passes through the filters. As an aid, catalyst grains can be integrated as a fixed bed inside the filter to promote the simultaneous removal of particulate matter and tar (Rapagna *et al.*, 2010).

Fabric filters are typically designed with non-disposable filter bags. As the gaseous (dust-containing) emissions flow through the filter medium (typically cotton, polypropylene, fiberglass, or Teflon), particulate matter is collected on the bag surface as a dust cake. Fabric filters operate with collection efficiencies up to 99.9% and other advantages are evident, but there are several issues that arise during use of such equipment.

Wet electrostatic precipitator units have a high collection efficiency (of the order of 90%) over the entire range of particle sizes down to 0.5 mm with a very low pressure drop. Wet scrubbers offer a high collection efficiency (also of the order of 90%) as well.

9.7.2 *Thermal methods*

Cracking methods are more advantageous in terms of recovering the energy content in tar by converting the high-molecular-weight constituents of tar into lower molecular weight (gaseous) products such as hydrocarbon derivatives and hydrogen at high temperature (up to 1200°C, 2190°F) or catalytic reactions (up to 800°C, 1470°F). Catalytic cracking is commercially used in many refineries and has been demonstrated to be one of the most effective processes for the conversion of high-molecular-weight products (such as crude oil residua, heavy oi, extra heavy oil, and tar sand bitumen) into gases and lower molecular weight, distillable liquids (Parkash, 2003; Gary *et al.*, 2007; Speight, 2014, 2017; Hsu and Robinson, 2017).

Non-metallic catalysts such as dolomite ($CaCO_3.MgCO_3$), calcite ($CaCO_3$), and zeolite and metallic catalysts such as nickel-based catalysts, nickel/molybdenum-based catalysts, nickel/cobalt-based catalysts, molybdenum-based catalysts, platinum-based catalysts, and ruthenium-based catalysts (based on Mo, NiO, Pt and Rt) have been used for various tar conversion processes.

Tar removal can also be achieved by catalytic reforming, which has the advantage of keeping the carbon contained in the tars available for further conversion to fuel. Another option to reduce the tar content of the synthesis gas is the use of catalytic bed material in the gasification reactor. Olivine sand has been shown to effectively reduce the tar content in a gas stream from steam gasification.

9.8 Other Methods of Gas Cleaning

The types and amounts of the contaminants that occur in gas streams are, like the contaminants presented above, dependent on the composition of the biomass feedstock and the type of process used to produce the gas as well as the individual process parameters. In addition to the gas from anaerobic digesters, landfill gas varies widely from landfill site to landfill site as the items disposed of also vary widely (i.e. there is site specificity). Moreover, there is variation in how frequently gas is extracted and in what volumes, and the stage of decomposition of the waste. Section 9.8 presents some contaminants not dealt with in other works on contaminant presence and removal. For example, siloxanes derivatives generally occur to a greater extent in the gas streams from landfill sites than in gas streams from other sources.

9.8.1 *Nitrogen removal*

Nitrogen may often occur in sufficient quantities in gas streams and, consequently, may lower the heating value of the gas and so nitrogen must be removed (Baker *et al.*, 2002). However, it must be recognized that nitrogen removal requires liquefaction and fractionation of the entire gas stream, which may affect process economics. In some cases, the nitrogen-containing gas stream is blended with a gas having a higher heating value and sold at a reduced price depending on the thermal value (Btu/ft^3).

For high-flow-rate gas streams, a cryogenic process is typical and involves the use of the different volatility of methane (b.p. –161.6°C/–258.9°F) and nitrogen (b.p. –195.7°C/–320.3°F) to achieve separation. In the process, a system of compression and distillation columns drastically reduces the temperature of the gas mixture to a point where methane is liquefied and the nitrogen is not. On the other hand, for smaller volumes of gas, a system utilizing pressure swing adsorption (PSA) is a more typical method of separation.

The PSA process is a fixed-bed adsorption process carried out at constant temperature and variable pressure. In the process, carbon dioxide is separated from the gas stream by adsorption on a surface under elevated pressure. The adsorbing material, usually activated carbon or zeolites, is regenerated by a sequential decrease in pressure before the column is reloaded again. An upgrading plant, using this technique, typically has four, six or nine vessels working in parallel. When the adsorbing material in one vessel becomes saturated, the raw gas flow is switched to another vessel in which the adsorbing material has been regenerated. During regeneration the pressure is decreased in several steps. The gas that is desorbed during the first and eventually the second pressure drop may be returned to the inlet of the raw gas, since it will contain some methane that was adsorbed together with carbon dioxide. The gas desorbed in the following pressure reduction step is either led to the next column or, if it is almost entirely methane free, released to the atmosphere. If hydrogen sulphide is present in the raw gas, it will be irreversibly adsorbed on the adsorbing material. In addition, water present in the raw gas can destroy the structure of the material. Therefore, hydrogen sulphide and water need to be removed before the PSA column.

Thus, the PSA process enables the separation of carbon dioxide from a gas stream. Nevertheless, it can also be used for the separation of other compounds such as nitrogen, oxygen, and carbon monoxide. Several packing materials can be used for the removal of carbon dioxide and these include zeolite, silica gel, and activated carbon. In order to be effective continuously, the PSA process is composed of various fixed-bed columns running in alternative cycles of adsorption, regeneration, and pressure build-up (Nikolić *et al.*, 2009; Alonso-Vicario *et al.*, 2010; Santos *et al.*, 2013). In some cases, the regeneration step is carried out under vacuum and this technology is known as vacuum swing adsorption (VSA), which requires a vacuum pump at the outlet.

Moreover, the regeneration step can be carried out by increasing the temperature to enhance the desorption of compounds such as hydrogen sulfide (which is strongly adsorbed on the packing material). On increasing the temperature at constant pressure, the process is also known as temperature swing adsorption (TSA). The TSA process is used for gas sweetening. The gas stream must be fed dry into PSA, vacuum swing adsorption, and TSA processes and the water content must therefore be reduced prior to these treatments.

The PSA operation can be based on equilibrium or kinetic separation. The more strongly adsorbed and faster diffusing compounds are retained on the packing material in equilibrium and kinetic separation, respectively. Natural zeolites have been used for hydrogen sulfide and carbon dioxide removal by PSA with thermal desorption working at 25°C (77°F) in the pressure range 45–100 psi. However, variations in the concentration of carbon dioxide can affect the performance of the PSA process. For example, an increase of 5% in the flow rate of the carbon dioxide can lead to an increase in the temperature at the top of the packed bed — thus, the use of a thermocouple could enhance the control strategy.

Also, in the PSA method, methane and nitrogen can be separated by using an adsorbent with an aperture size very close to the molecular diameter of the larger species (the methane) which allows nitrogen to diffuse through the adsorbent. This results in a purified gas stream that is suitable for pipeline specifications. The adsorbent can then be regenerated, leaving a highly pure nitrogen stream. The PSA method is a flexible method for nitrogen rejection and is used for both small and large flow rates.

9.8.2 *Ammonia removal*

Biomass with a relatively high nitrogen content will generate a product gas that contains ammonia (NH_3) and hydrogen cyanide (HCN) (Yuan *et al.*, 2010), which, in turn, will generate nitrogen oxides (NO_x) during combustion. In the case of the gasification of biomass (to produce a gaseous product), the concentration of ammonia in the product gas depends on not only the type of biomass feedstock used but also the gasifier design parameters and operating conditions.

Ammonia can be removed by wet scrubbing technology, which has been widely adopted in the existing biomass gasification processes (Dou *et al.*, 2002; Proell *et al.*, 2005) and the ammonia can be removed when the gas is dried. As a result, a separate cleaning step is usually not necessary.

Compared with wet scrubbing technology, hot-gas cleanup technology, preferably employing catalysts, is more advantageous with respect to energy efficiencies as it eliminates the need for cooling the product gas and reheating it for synthesis gas applications (Torres *et al.*, 2007). Catalytic processes effectively remove ammonia by converting it to nitrogen, hydrogen, and water. Nickel (Ni)-based catalysts have higher activity, whereas other catalysts do not have good potential for ammonia removal.

9.8.3 *Particulate matter removal*

The particulate matter present in the product gas can also pose a serious problem for some end-users, and the catalysts used for cleaning product gases have been demonstrated to be negatively affected by particulate matter (Gustafsson *et al.*, 2011).

Particulate matter can be removed using standard technologies such as cyclones, filters, and separators, which also reduce the tar content of the gas stream, with the extent of tar removal depending on the particle separation technology used. The presence of alkali is of major importance, because alkali can form silicates with low melting temperatures that may negatively affect the filter operation.

9.8.4 *Siloxane removal*

Siloxane compounds are a subgroup of silicone derivatives containing silicon–oxygen (Si–O) bonds with organic radical; they are widely used in a variety of industrial processes. Siloxanes are used in products such as deodorants and shampoos and therefore can be found in the gas streams from sewage sludge treatment plants and in landfill gas. When siloxanes are burned, silicon oxide, a white powder, is formed, which can create problems in gas engines. Although most siloxanes disperse into the atmosphere, where they are decomposed, some end up in wastewater; more generally, siloxane derivatives end up as a significant component in the sludge.

As sludge undergoes anaerobic digestion, it may be subjected to temperatures of up to 60°C (140°F). At this point the siloxanes contained in the sludge will volatize and become an unwanted constituent of the resulting gas stream. This problem can get exacerbated by the fact that silicone-based anti-foaming agents are frequently added to the anaerobic digesters; these silicones sometimes biodegrade into siloxanes. Unfortunately, when siloxanes gases are burned, they are usually converted into silicon dioxide particles, which are chemically and physically similar to sand.

Currently, there are six primary technologies for removing siloxanes from gas streams: (i) activated carbon is widely used to remove organic substances from gases and liquids due to its superior adsorbent properties, (ii) activated alumina (Al_2O_3) absorbs siloxanes from the gas stream; when the alumina becomes saturated, the absorption capability can be recovered by passing a regeneration gas through it, (iii) refrigeration with condensation in combination can be used to selectively remove specific compounds by lowering the temperature or pressure of the gas and then allowing the compound to precipitate out to a liquid and then settle out, (iv) synthetic resins remove volatile materials through adsorption; they can be specially formulated to remove specific classes of compounds, (v) liquid absorbents are used by a small number of landfill operators to treat the gas stream prior to use in combustion devices such as gas turbines, and (vi) membrane technology is a relatively recent development for siloxane removal; however, membranes are subject to acid deterioration from the acidic content usually found in the raw gas stream. Of these technologies, activated carbon appears to be among the most dominant in the industry.

9.8.5 *Alkali metal salt removal*

Compared with fossil fuels, biomass is rich in alkali salts that typically vaporize at high gasifier temperatures but condense downstream below 600°C (1110°F). Methods are available to strip the alkali contents since condensation of alkali salts can cause serious corrosion problems. The alkali will condense onto fine solid particles and can be subsequently captured in a cyclone, electrostatic precipitators, or filters when the gas temperature is below 600°C (1110°F). The hot gas can also be passed through a bed of active bauxite to remove alkali when the cooling of gas is not permitted.

9.8.6 *Biological methods*

Biological processes have been used mainly to remove hydrogen sulfide (Syed *et al.*, 2006). Hydrogen sulfide can be oxidized by microorganisms of the species Thiobacillus and Sulfolobus. The degradation requires oxygen and therefore a small amount of air (or pure oxygen if levels of nitrogen should be minimized) is added for biological desulfurization to take place. The degradation can occur inside the digester and can be facilitated by immobilizing the microorganisms occurring naturally in the digestate. An alternative is to use a trickling filter through which the gas stream passes when leaving the digester. In the trickling filter the microorganisms grow on a packing material. A gas stream with added air meets a counterflow of water containing nutrients. The sulfur-containing solution is removed and replaced when the pH drops below a certain level. Both methods are widely applied; however, they are not suitable if the gas stream is used as a vehicle fuel or for grid injection because of the remaining traces of oxygen. An alternative system is available in which the absorption of the hydrogen sulfide is separated from the biological oxidation to sulfur — hence, the gas stream flow remains free of oxygen.

There has also been a focus on the removal of siloxane derivatives and carbon dioxide (Jensen and Webb, 1995; Soreanu *et al.*, 2011).

As another example, natural bacteria can be used to convert hydrogen sulfide into elemental sulfur in the presence of oxygen and iron. This can be done by introducing a small amount (2–5%) of air into the head space of the digester. As a result, deposits of elemental sulfur will be formed in the digester. Even though this situation will reduce the hydrogen level, it will not lower it below the level recommended for pipeline-quality gas. This process may be optimized using a more sophisticated design where air is bubbled through the digester feed material. It is critical that the introduction of the air is carefully controlled so as to avoid reducing the amount of gas stream produced.

The main biological processes for the removal of hydrogen sulfide are presented below.

9.8.6.1 *Biofiltration*

Traditional gas cleaning and air pollution control technologies for pollutant gases, for example, adsorption, absorption and combustion, were developed to treat high-concentration waste gas streams associated with process emissions from stationary point sources (Mokhatab *et al.*, 2006). Although these technologies rely on established physico-chemical principles to achieve effective control of gaseous pollutants, in many cases the control technique yields products that require further treatment before disposal or recycling. For the treatment of dilute waste gas streams, however, such traditional methods are relatively less effective, and more expensive and wasteful in terms of energy consumption, and so the identification of alternative control measures is warranted. A suitable alternative air pollution control technology is biofiltration, which utilizes naturally occurring micro-organisms supported on a stationary bed (filter) to continuously treat contaminants in a flowing waste gas stream (Allen and Uang, 1992; Accettola *et al.*, 2008).

Biofiltration, by definition, is the aerobic degradation of pollutants from (in the current context) a gas stream in the presence of a carrier medium. The early work on the biofiltration technology concentrated on organic media, such as peat, compost, and tree bark. In general terms, organic compounds are degraded to carbon dioxide and water, while inorganic compounds, such as sulfur compounds, are oxidized to form oxygenated derivatives. The formation of these acidic compounds can lead to a lowering of the pH of the filtration media, which in turn impacts on the performance of the system. Removal of the oxidized compound from the media is an important consideration in the design of biofiltration systems. Biofiltration, like many biotreatment-based processes in the crude oil refining and gas processing industries (El-Gendy and Speight, 2016; Speight and El-Gendy, 2018), is successfully emerging as a reliable, low-cost option for a wide range of air treatment applications. It is now becoming apparent that biological treatment will play a far more significant role in achieving environmental control of air emissions.

In the biofiltration process, three types of bioreactor designs are usually considered: the biofilter, (ii) the biotrickling filter, and (iii) the bioscrubber. The main differences between these systems concern their design and mode of operation: micro-organism conditioning, the nature of the fluid phase (gas or liquid), and the presence or absence of stationary solid phases. Nevertheless, gas stream desulfurization has been carried out by the biotrickling filter and the bioscrubber.

The bio-filter is a pollution control technique which involves using a bioreactor that contains living material to capture and biologically degrade pollutants such as hydrogen sulfide. Common uses include processing wastewater, capturing harmful chemicals or silt from surface runoff, and the macrobiotic oxidation of contaminants in gas streams. The technology finds its greatest application in treating malodorous compounds and water-soluble volatile organic compounds (VOCs). Compounds treated are typically mixed VOCs and various sulfur compounds, including hydrogen sulfide. Very large airflows may be treated although a large area (footprint) has typically been required. Engineered biofilters, designed and built since the early 1990s, have provided significant footprint reductions over the conventional flat-bed, organic media type.

In the process, the polluted gas flow is purified with biofiltration by forcing the gas flow upwards through a filter bed, which consists of biological material, e.g. compost, tree bark, or peat. The filter material is a carrier of a thin water film in which micro-organisms live. The pollution in the gas flow is held back by ad- and absorption on the filter material, and then broken by the microorganisms present. The filter material serves as a supplier of the necessary nutrients. The products of the conversion are carbon dioxide, sulfate, and nitrate. The dry weight of the filter varies typically from 40% to 60%. To reduce desiccation of the bed, the gas flow must be saturated with water. For this reason, polluted gas flow has to be moistened before it goes through the biofilter, which is achieved by using a pre-scrubber. The relative humidity of the gas must be 95%. In practice, it is always better to apply a moistener in order to protect the biofilter against dehydration.

A biotrickling filter is a packed bed bioreactor with immobilized biomass (Montebello *et al.*, 2012, 2013; Fernández *et al.*, 2013a, 2013b). The gas flows through a fixed bed co- or counter-currently to a mobile liquid phase. Synthetic carriers are usually used and these include plastics, ceramics, lava rocks, polyurethane foam, etc. The synthetic carrier does not provide any nutrients, so the mobile liquid phase must contain nutrients for the growth and maintenance of the biomass. Programmed or continuous discharge of the recirculation medium helps to remove the oxidation products. The hydrogen sulfide must be transferred from the gas to the liquid phase, and degradation is finally carried out in the biofilm.

In the microbial cycling of sulfur, biological oxidation of hydrogen sulfide to sulfate is one of the major reactions involved. There are numerous sulfur-oxidizing micro-organisms, but hydrogen sulfide is exclusively oxidized by prokaryotes. In any case, the biotrickling filter is typically inoculated with aerobic active sludge and during the process a specific biomass population is developed.

A biotrickling filter can be operated under aerobic or anoxic conditions. The removal of hydrogen sulfide from the gas stream has been mainly studied under aerobic conditions and the overall reaction is

$$HS^- + 2O_2 \rightarrow SO_4^{2-} + 2H^+$$

In summary, two compounds can be produced, sulfate and elemental sulfur, and the production ratio of sulfate (SO_4^{2-}) will be dependent on the oxygen/hydrogen sulfide ratio and the trickling liquid velocity. Air is used to supply oxygen and must, therefore, be supplied in excess. However, control of air supply is an important issue for safety and operational reasons — the lower and upper explosive limits for methane in air are 5% v/v and 15% v/v, respectively.

On the other hand, the gas stream can be diluted (mainly by the nitrogen content in air) and undesired residual oxygen can be found in the outlet stream. Therefore, a high oxygen supply is a drawback because the calorific power is also decreased and, moreover, a lack of oxygen increases the clogging problems caused by the formation of elemental sulfur. The gas stream dilution depends on the efficiency in the mass transfer and the concentration of hydrogen sulfide. An alternative to the direct supply of air mixed with the gas stream involves the use of venture-based devices to increase the oxygen mass transfer. However, the installation of an aerated liquid recirculation system could produce hydrogen sulfide stripping of the dissolved sulfide.

The anoxic biotrickling filter is based on dissimilatory nitrate reduction. Dissimilatory nitrate reduction is carried out by certain bacteria that can use nitrate and/or nitrite as electron acceptors instead of oxygen:

- Complete denitrification vs. partial hydrogen sulfide oxidation

$$5H_2S + NO_3^- \rightarrow 5S + N_2 + 4H_2O + 2OH^-$$

- Complete denitrification vs. complete hydrogen sulfide oxidation

$$5H_2S + 8NO_3^- \rightarrow 5SO_4^{2-} + 4N_2 + 4H_2O + 2H^+$$

- Partial denitrification *vs.* partial hydrogen sulfide oxidation

$$H_2S + 2NO_3^- \rightarrow S + 4NO_2^- + H_2O$$

- Partial denitrification vs. complete hydrogen sulfide oxidation

$$H_2S + 4NO_3^- \rightarrow SO_4^{2-} + 4NO_2^- + 2H^+$$

9.8.6.2 *Bio-scrubbing*

Bio-scrubber systems have been used for hydrogen sulfide removal from gas streams (Nishimura and Yoda, 1997). The bioscrubber involves a two-stage process with an absorption tower and a bioreactor, in which the sulfide is oxidized to sulfur and/or sulfate. For example, the Shell-Paques THIOPAQ® process employs alkaline conditions to produce elemental sulfur. In the first step of this process, the hydrogen sulfide is absorbed into an alkaline solution by reaction with hydroxyl and bicarbonate ions. In the second step, the hydrosulfide is oxidized to elemental sulfur under oxygen-limiting conditions. Thus,

$$H_2S + OH^- \rightarrow HS^- + H_2O$$

$$H_2S + HCO_3^- \rightarrow HS^- + CO_2 + H_2O$$

$$H_2S + CO_3^{2-} \rightarrow HS^- + HCO_3^-$$

9.8.6.3 *Biooxidation*

Biooxidation (biological oxidation) is one of the most used methods for desulfurization, based on the injection of a small amount of air (2–8% v/v) into the raw gas stream. In this way, the hydrogen sulfide is biologically oxidized to either solid free sulfur or liquid sulfurous acid (H_2SO_3) (Mokhatab *et al.*, 2006; Speight, 2018):

$$2H_2S + O_2 \rightarrow 2H_2O + 2S$$

$$2H_2S + 3O_2 \rightarrow 2H_2SO_3$$

In practice, the precipitated sulfur is collected and added to the storage tanks where it is mixed with digestate in order to improve the fertilizer properties of digestate. Biological desulfurization is frequently carried out inside the digester and, for this kind of desulfurization, oxygen and *Sulfobacter oxydans* bacteria must be present to convert hydrogen sulfide into elementary sulfur in the presence of oxygen. Typically, *Sulfobacter oxydans* is present inside the digester (does not have to be added) as the anaerobic digester substrate contains the necessary nutrients for their metabolism. In the process, the air is injected directly into the head space of the digester, and the reactions occur in the reactor head space, on the floating layer (if existing), and on reactor walls. Due to the acidic nature of the products, there is a risk of corrosion. The process is dependent on the existence of a stable floating layer inside the digester and the process often takes place in a separate reactor.

Chemical desulfurization of gas streams can take place outside of the digester, using e.g. a base (usually sodium hydroxide). Another chemical method to reduce the content of hydrogen sulfide is to add commercial ferrous solution (Fe^{2+}) to the feedstock. Ferrous compounds bind sulfur in an insoluble compound in the liquid phase, thereby preventing the production of gaseous hydrogen sulfide.

9.9 Tail Gas Cleaning

Tail gas cleaning (also called tail gas treating) involves the removal of the remaining sulfur compounds from the gases remaining after sulfur recovery. Tail gas from a typical Claus process, whether a conventional Claus or one of the extended versions of the process (Parkash, 2003; Mokhatab *et al.*, 2006; Gary *et al.*, 2007; Speight, 2014, 2017, 2019;

Hsu and Robinson, 2017), usually contains small but different quantities of carbonyl sulfide, carbon disulfide, hydrogen sulfide, and sulfur dioxide as well as sulfur vapor. In addition, there may be hydrogen, carbon monoxide, and carbon dioxide present in the tail gas. In order to remove the rest of the sulfur compounds from the tail gas, all of the sulfur-bearing species must first be converted to hydrogen sulfide, which is then absorbed into a solvent and the clean gas vented or recycled for further processing.

It is necessary to develop and implement reliable and cost-effective technologies for coping with the various sales specifications and environmental requirements. In response to this need, several new technologies are emerging to comply with the most stringent regulations. The typical sulfur recovery efficiencies of a Claus plant are of the order of 90–96% w/w for a two-stage plant and 95–98% w/w for a three-stage plant. Most environmental agencies require the sulfur recovery efficiency to be in the range of 98.5–99.9% w/w and, as a result, the sulfur content of the Claus plant tail gas needs to be reduced further.

Moreover, a sulfur removal process must be very precise, since some gas streams may contain only a small quantity of sulfur-containing compounds that must be reduced by several orders of magnitude. Most consumers of gas require less than 4 ppm v/v in the gas — a characteristic feature of a gas stream that contains hydrogen sulfide is the presence of carbon dioxide (generally in the range of 1–4% v/v). In the cases where the gas stream does not contain hydrogen sulfide, there may also exist a relative lack of carbon dioxide.

9.9.1 *Claus process*

The Claus process is not so much a gas cleaning process but a process for the disposal of hydrogen sulfide, a toxic gas that originates from most gas streams. Burning hydrogen sulfide as a fuel gas component or as a flare gas component is precluded by safety and environmental considerations since one of the combustion products is sulfur dioxide (SO_2), which is also toxic. As described above, hydrogen sulfide is typically removed from gas streams by means of an olamine process after which the application of heat regenerates the olamine and forms an acid gas stream. Following from this, the acid gas stream is treated to convert the hydrogen sulfide to elemental sulfur and water. The conversion process utilized in most modern refineries is the Claus process or a variant thereof.

The Claus process involves combustion of approximately one-third of the hydrogen sulfide to sulfur dioxide and then reaction of the sulfur dioxide with the remaining hydrogen sulfide in the presence of a fixed bed of activated alumina, cobalt molybdenum catalyst, resulting in the formation of elemental sulfur:

$$2H_2S + 3O_2 \rightarrow 2SO_2 + 2H_2O$$
$$2H_2S + SO_2 \rightarrow 3S + 2H_2O$$

Different process flow configurations are in use to achieve the correct hydrogen sulfide/sulfur dioxide ratio in the conversion reactors.

In a split-flow configuration, one-third split of the acid gas stream is completely combusted and the combustion products are then combined with the non-combusted acid gas upstream of the conversion reactors. In a once-through configuration, the acid gas stream is partially combusted by providing sufficient oxygen in the combustion chamber to combust only one-third of the acid gas.

Two or three conversion reactors may be required depending on the level of hydrogen sulfide conversion required. Each additional stage provides incrementally less conversion than the previous stage. Overall, conversion of 96–97% v/v of the hydrogen sulfide to elemental sulfur is achievable in a Claus process. If this is insufficient to meet air quality regulations, a Claus process tail gas treater is utilized to remove essentially the entire remaining hydrogen sulfide in the tail gas from the Claus unit. The tail gas treater may employ a proprietary solution for absorbing the hydrogen sulfide and converting it elemental sulfur.

9.9.2 *SCOT process*

The Shell Claus Off-gas Treating (SCOT) unit is the most common type of tail gas unit and uses a hydrotreating reactor followed by amine scrubbing to recover and recycle sulfur, in the form of hydrogen, to the Claus unit (Nederland, 2004).

Early SCOT units used diisopropanolamine in a Sulfinol (Sulfinol-D) solution. Methyl diethanolamine-based Sulfinol (Sulfinol-M) was used later to enhance hydrogen sulfide removal and to allow for selective rejection of carbon dioxide in the absorber. To achieve the lowest possible hydrogen sulfide content in the treated gas, the Super-SCOT configuration was introduced. In this version, the loaded Sulfinol-M solution is regenerated in two stages. The partially stripped solvent goes to the middle of the absorber, while the fully stripped solvent goes to the top of the absorber. The solvent going to the top of the absorber is cooled below that used in the conventional SCOT process.

In the process, tail gas (containing hydrogen sulfide and sulfur dioxide) is contacted with hydrogen and reduced in a hydrotreating reactor to form hydrogen sulfide and water. The catalyst is typically cobalt/molybdenum on alumina. The gas is then cooled in a water contractor. The hydrogen sulfide-containing gas enters an amine absorber, which is typically in a system segregated from the other refinery amine systems. The purpose of segregation is two-fold: (i) the tail gas treater frequently uses a different amine than the rest of the plant, and (ii) the tail gas is frequently cleaner than the refinery fuel gas (with regard to contaminants) and segregation of the systems reduces maintenance requirements for the SCOT unit. The amines chosen for use in the tail gas system tend to be more selective for hydrogen sulfide and are not affected by the high levels of carbon dioxide in the off-gas.

However, the SCOT process can be configured in various ways. For example, it can be integrated with the upstream acid gas removal unit if the same solvent is used in both units. Another configuration has been used to cascade the upstream gas cleanup with the SCOT unit. A hydrogen sulfide-lean acid gas from the upstream gas treating unit is sent to a SCOT process with two absorbers. In the first absorber, the hydrogen sulfide-lean acid gas is enriched, while the second absorber treats the Claus tail gas. A common stripper is used for both SCOT absorbers and, in this latter configuration, different solvents could be used in both the upstream and the SCOT units.

A hydrotreating reactor converts sulfur dioxide in the off-gas to hydrogen sulfide that is then contacted with a Stretford solution (a mixture of a vanadium salt, anthraquinone disulfonic acid, sodium carbonate, and sodium hydroxide) in a liquid-gas absorber (Abdel-Aal *et al.*, 2016). Other tail gas treating processes include (i) caustic scrubbing, (ii) polyethylene glycol treatment, (iii) the Selectox process, and (iv) sulfite/bisulfite tail gas treatment (Mokhatab *et al.*, 2006; Speight, 2014, 2019).

References

Abdel-Aal, H.K., Aggour, M.A., and Fahim, M.A. 2016. *Petroleum and Gas Field Processing*. CRC Press, Taylor & Francis Publishers, Boca Raton, Florida.

Accettola, F., Guebitz, G.M., and Schoeftner, R. 2008. Siloxane removal from biogas by biofiltration: Biodegradation studies. *Clean Technol. Environ. Policy* 10(2): 211–218.

Ajhar, M., Travesset, M., Yüce, S., and Melin, T. 2010. Siloxane removal from landfill and digester gas — A technology overview. *Bioresour. Technol.* 101(9): 2913–2923.

Alderton, P.D. 1993. Natural Gas Treating Using Membranes. *Proceedings of the 2nd GPA Technical Meeting*, GCC Chapter, Bahrain, October 27.

Allen, R. and Yang, Y. 1992. Biofiltration: An Air Pollution Control Technology for Hydrogen Sulfide Emissions. In *Industrial Environmental Chemistry: Waste minimization in Industrial Processes and Remediation of Hazardous Waste*. D.T. Sawyer and A.E. Martell (Editors). Springer, Boston, Massachusetts, pp. 273–287.

Alonso-Vicario, A., Ochoa-Gómez, J.R., Gil-Río, S., Gómez-Jiménez-Aberasturi, O., Ramírez-López, C.A., and Torrecilla-Soria, J. 2010. Purification and upgrading of biogas by pressure swing adsorption on synthetic and natural zeolites. *Microporous Mesoporous Mater* 134(1–3): 100–107.

Andriani, D., Wresta, A., Atmaja, T.D., and Saepudin, A. 2014. A review on optimization production and upgrading biogas through CO removal using various techniques. *Appl. Biochem. Biotechnol.* 172(4): 1909–1928.

Anis, S. and Zainal, Z.A. 2011. Tar reduction in biomass producer gas via mechanical, catalytic and thermal methods: A Review. *Renew Sustain Energy Rev*. 15: 2355–2377.

ASTM. 2019. *Annual Book of Standards*, ASTM International, West Conshohocken, Pennsylvania.

ASTM D3246. 2019. Standard Test Method for Sulfur in Petroleum Gas by Oxidative Microcoulometry. *Annual Book of Standards*, ASTM International, West Conshohocken, Pennsylvania.

Baker, R.W., Lokhandwala, K.A., Wijmans, J.G., and Da Costa, A.R. 2002. United States Patent 6,425,267. Two-Step Process for Nitrogen Removal from Natural Gas.

Barbouteau, L. and Dalaud, R. 1972. In *Gas Purification Processes for Air Pollution Control*, G. Nonhebel (Editor). Butterworth and Co., London, United Kingdom, Chapter 7.

Basu, S., Khan, A.L., Cano-Odena, A., Liu, C., and Vankelecom, I.F. 2010. Membrane-based technologies for biogas separations. *Chem. Soc. Rev.* 39(2): 750–768.

Boulinguiez, B. and Le Cloirec, P. 2009. Biogas pre-upgrading by adsorption of trace compounds onto granular activated carbons and an activated carbon fiber cloth. *Water Sci. Technol.* 59(5): 935–944.

Cherosky, P. and Li, Y. 2013. Hydrogen sulfide removal from biogas by bio-based iron sponge. *Biosyst. Eng.* 114(1): 55–59.

Curry, R.N. 1981. *Fundamentals of Natural Gas Conditioning*. PennWell Publishing Co., Tulsa, Oklahoma.

Deshmukh, M.G. and Shete, A. 2013. Oxidative absorption of hydrogen sulfide using iron-chelate based process: Chelate degradation. *J. Anal. Bioanal. Tech.* 88(3): 432–36.

Devi, L., Ptasinski, K.J., and Jenssen, F.J.J.G. 2003. A review of the primary measures for tar elimination in biomass gasification processes. *Biomass Bioenergy* 24: 125–140.

Dou, B., Zhang, M., and Gao, J. 2002. High-temperature removal of NH3, organic sulfur, HCl, and tar component from coal-derived gas. *Ind. Eng. Chem. Res.* 41: 4195–4200.

El-Gendy, N. Sh. and Speight, J.G. 2016. *Handbook of Refinery Desulfurization*. CRC Press, Taylor & Francis Group, Boca Raton, Florida.

Esteban, A., Hernandez, V., and Lunsford, K. 2000. Exploit the Benefits of Methanol. *Proceedings of the 79th GPA Annual Convention*, Atlanta, Georgia, March.

Fernández, M., Ramírez, M., Gómez, J.M., and Cantero, D. 2013a. Biogas biodesulfurization in an anoxic biotrickling filter packed with open-pore polyurethane foam. *J. Hazard. Mater.* 264: 529–535.

Fernández, M., Ramírez, M., Pérez, R.M., Gómez, J.M., and Cantero, D. 2013b. Hydrogen sulfide removal from biogas by an anoxic biotrickling filter packed with pall rings. *Chem. Eng. J.* 225: 456–463.

Gary, J.G., Handwerk, G.E., and Kaiser, M.J. 2007. *Petroleum Refining: Technology and Economics*, 5th Edition. CRC Press, Taylor & Francis Group, Boca Raton, Florida.

Gislon, P., Galli, S., and Monteleone, G. 2013. Siloxanes removal from biogas by high surface area adsorbents. *Waste Management* 33(12): 2687–2693.

Gustafsson, E., Lin, L., and Seemann, M. 2011. Characterization of particulate matter in the hot product gas from indirect steam bubbling fluidized bed gasification of wood pellets. *Energy Fuels* 25: 1781–1789.

Hall, P. and Lokhandwala, K.A. 2004. Advances in Membrane Materials Provide New Gas Processing Solutions. *Proceedings of the GPA Annual Convention*, New Orleans, Louisiana, March.

Hsu, C.S. and Robinson, P.R. (Editors). 2017. *Handbook of Petroleum Technology*. Springer International Publishing AG, Cham, Switzerland.

Huang, J., Schmidt, K.G., and Bian, Z. 2011. Removal and conversion of tar in syngas from woody biomass gasification for power utilization using catalytic hydrocracking. *Energies* 4: 1163–1177.

Jensen, A.B. and Webb, C. 1995. Treatment of H_2S-containing Gases — A Review of Microbiological Alternatives. *Enzyme Microb. Tech.* 17(1): 2–10.

Jou, F.Y., Otto, F.D., and Mather, A.E. 1985. In *Acid and Sour Gas Treating Processes*, S.A. Newman (Editor). Gulf Publishing Company, Houston, Texas, Chapter 10.

Katz, D.K. 1959. *Handbook of Natural Gas Engineering*. McGraw-Hill Book Company, New York.

Kidnay, A.J. and Parrish, W.R. 2006. *Fundamentals of Natural Gas Processing*. CRC Press, Taylor & Francis Group, Boca Raton, Florida.

Kohl, A.L. and Nielsen, R.B., 1997. *Gas Purification*. Gulf Publishing Company, Houston, Texas.

Kohl, A.L. and Riesenfeld, F.C. 1985. *Gas Purification*. 4th Edition. Gulf Publishing Company, Houston, Texas.

Maddox, R.N. 1982. *Gas Conditioning and Processing, Volume 4. Gas and Liquid Sweetening*. Campbell Publishing Co., Norman, Oklahoma.

Maddox, R.N., Bhairi, A., Mains, G.J., and Shariat, A. 1985. In *Acid and Sour Gas Treating Processes*. S.A. Newman (Editor). Gulf Publishing Company, Houston, Texas, Chapter 8.

May-Britt, H. 2008. Membranes in Gas Separation. In *Handbook of Membrane Separations*. CRC Press, Taylor & Francis Group, Boca Raton, Florida, pp. 65–106.

Minkkinen, A. and Jonchere, J.P. 1997. Methanol Simplifies Gas Processing. *Proceedings of the 5th GPA-GCC Chapter Technical Conference*, Bahrain, May 6.

Mody, V. and Jakhete, R. 1988. *Dust Control Handbook*. Noyes Data Corp., Park Ridge, New Jersey.

Mokhatab, S., Poe, W.A., and Speight, J.G. 2006. *Handbook of Natural Gas Transmission and Processing*. Elsevier, Amsterdam, Netherlands.

Montebello, A.M., Fernández, M., Almenglo, F., Ramírez, M., Cantero, D., and Baeza, M. 2012. Simultaneous methyl mercaptan and hydrogen sulfide removal in the desulfurization of biogas in aerobic and anoxic biotrickling filters. *Chem. Eng. J.* 200/202(0): 237–246.

Montebello, A.M., Bezerra, T., Rovira, R., Rago, L., Lafuente, J., and Gamisans, X. 2013. Operational aspects, pH transition and microbial shifts of a H2S desulfurizing biotrickling filter with random packing material. *Chemosphere* 93(11): 2675–2682.

Newman, S.A. 1985. *Acid and Sour Gas Treating Processes*. Gulf Publishing, Houston, Texas.

Nikolić, D., Kikkinides, E.S., and Georgiadis, M.C. 2009. Optimization of multibed pressure swing adsorption processes. *Ind. Eng. Chem. Res.* 48(11): 5388–5398.

Niesner, J., Jecha, D., and Stehlík, P. 2013. Biogas upgrading technologies: State of the art review in the European region. *Chem. Eng. Trans.* 35: 517–522.

Nishimura, S. and Yoda, M. 1997. Removal of hydrogen sulfide from an anaerobic biogas using a Bioscrubber. *Water Sci. Technol.* 36(6–7): 349–356.

Ozturk, B. and Demirciyeva, F. 2013. Comparison of biogas upgrading performances of different mixed matrix membranes. *Chem. Eng. J.* 222: 209–217.

Parkash, S. 2003. *Refining Processes Handbook*. Gulf Professional Publishing, Elsevier, Amsterdam, Netherlands.

Pitsinigos, V.D. and Lygeros, A.I. 1989. Predicting H_2S-MEA equilibria. *Hydrocarbon Processing* 58(4): 43–44.

Polasek, J. and Bullin, J. 1985. In *Acid and Sour Gas Treating Processes*, S.A. Newman (Editor). Gulf Publishing Company, Houston, Texas, Chapter 7.

Proell, T., Siefert, I.G., and Friedl, A. 2005. Removal of NH3 from biomass gasification producer gas by water condensing in an organic solvent scrubber. *Ind. Eng. Chem. Res.* 44: 1576–1584.

Ramírez, M., Gómez, J.M., and Cantero, D. 2015. Biogas: Sources, Purification and Uses. In *Hydrogen and Other Technologies*, U.C. Sharma, S. Kumar, R. Prasad (Editors). Studium Press LLC, New Delhi, India, Chapter 13, pp. 296–323.

Ranke, G. and Mohr, V.H. 1985. The Rectisol Wash New Developments in Acid Gas Removal from Synthesis Gas. In *Acid and Sour Gas Treating Processes*, S.A. Newman (Editor). Gulf Publishing Company, Houston, Texas.

Rapagna, S., Gallucci, K., and Marcello, M.D. 2010. Gas cleaning, gas conditioning and tar abatement by means of a catalytic filter candle in a biomass fluidized-bed gasifier. *Bioresour. Technol.* 101: 7123–7130.

Reddy, S. and Gilmartin, J. 2008. Fluor's Econamine Plus Technology for Post-Combustion CO2 Capture. *Proceedings of the GPA Gas Treatment Conference*, Amsterdam, Netherlands, February 20–22. https://www.fluor.com/SiteCollectionDocuments/FluorEFG-forPost-CombustionCO2CaptureGPAConf-Feb2008.pdf.

Rojey, A. and Larue, J. 1988. Integrated Process for the Treatment of a Methane-Containing Wet Gas in Order to Remove Water Therefrom, United States Patent 4,775,395.

Rojey, A., Procci, A., and Larue, J. 1990. Process and Apparatus for Dehydration, De-acidification, and Separation of Condensate from a Natural Gas, United States Patent 4,979,966.

Rongwong, W., Boributh, S., Assabumrungrat, S., Laosiripojana, N., and Jiraratananon, R. 2012. Simultaneous absorption of CO_2 and H_2S from biogas by capillary membrane contactor. *J. Membrane Sci.* 392/393: 38–47.

Ryckebosch, E., Drouillon, M., and Vervaeren, H. 2011. Techniques for transformation of biogas to biomethane. *Biomass Bioenergy* 35(5): 1633–1645.

Santos, M.P.S., Grande, C.A., and Rodrigues, A.E. 2013. Dynamic study of the pressure swing adsorption process for biogas upgrading and its responses to feed disturbances. *Ind. Eng. Chem. Res.* 52(15): 5445–5454.

Soreanu, G., Béland, M., Falletta, P., Edmonson, K., Svoboda, L., and Al-Jamal, M. 2011. Approaches concerning siloxane removal from biogas — A review. *Canadian Biosys. Eng.* 53: 8.1–8.18.

Soud, H. and Takeshita, M. 1994. *FGD Handbook.* No. IEACR/65. International Energy Agency Coal Research, London, United Kingdom.

Speight, J.G. 2014. *The Chemistry and Technology of Petroleum*, 5th Edition. CRC-Taylor and Francis Group, Boca Raton, Florida.

Speight, J.G. 2015. *Handbook of Petroleum Product Analysis*, 2nd Edition. John Wiley & Sons Inc., Hoboken, New Jersey.

Speight, J.G. 2017. *Handbook of Petroleum Refining.* CRC Press, Taylor & Francis Group, Boca Raton, Florida.

Speight, J.G. 2018. *Handbook of Natural Gas Analysis.* John Wiley & Sons Inc., Hoboken, New Jersey.

Speight, J.G. and El-Gendy, N. Sh. 2018. *Introduction to Petroleum Biotechnology.* Gulf Professional Publishing Company, Elsevier, Cambridge, Massachusetts.

Speight, J.G. 2019. *Natural Gas: A Basic Handbook*, 2nd Edition. Gulf Publishing Company, Elsevier, Cambridge, Massachusetts.

Syed, M., Soreanu, G., Falletta, P., and Béland, M. 2006. Removal of hydrogen sulfide from gas streams using biological processes — A review. *Canadian Biosys. Eng.* 48: 2.1–2.14.

Torres, W., Pansare, S.S., and Goodwin, J.G. 2007. Hot gas removal of tars, ammonia, and hydrogen sulfide from biomass gasification. *Gas. Catal. Rev.* 49: 407–56.

Yuan, S., Zhou, Z., and Li, J. 2010. HCN and NH3 Released from biomass and soybean cake under rapid pyrolysis. *Energy Fuels* 24: 6166–6171.

Chapter 10

The Fischer–Tropsch Process

10.1 Introduction

In Chapter 5, reference has been made to the use of the gasification process for converting carbonaceous feedstocks, such as crude oil residua, tar sand bitumen, coal, oil shale, and biomass, into starting chemicals for the production of petrochemicals. The chemistry of the gasification process is based on thermal decomposition of the feedstock and the reaction of the feedstock carbon and other pyrolysis products with oxygen, water, and fuel gases such as methane and can be represented by a sequence of simple chemical reactions (Table 10.1). However, the gasification process is often considered to involve two distinct chemical stages: (i) devolatilization of the feedstock to produce volatile matter and char, followed by (ii) char gasification, which is complex and specific to the conditions of the reaction — both processes contribute to the complex kinetics of the gasification process (Sundaresan and Amundson, 1978).

The Fischer–Tropsch process is a catalytic chemical reaction in which carbon monoxide (CO) and hydrogen (H_2) in the synthesis are converted into hydrocarbon derivatives of various molecular weights. The synthesis gas required as the feedstock for the Fischer–Tropsch process is typically produced through gasification of coal (Chapter 4), crude oil resid (Chapter 5), biomass (Chapter 6), or organic waste (Chapter 7), or through various reforming processes (Chapter 8) (Higman and Van der Burgt, 2008; Liu *et al.*, 2010; Rostrup-Nielsen and Christiansen, 2011;

Table 10.1: Reactions that occur during gasification of a carbonaceous feedstock.

Reaction
$2C + O_2 \rightarrow 2CO$
$C + O_2 \rightarrow CO_2$
$C + CO_2 \rightarrow 2CO$
$CO + H_2O \rightarrow CO_2 + H_2$ (shift reaction)
$C + H_2O \rightarrow CO + H_2$ (water–gas reaction)
$C + 2H_2 \rightarrow CH_4$
$2H_2 + O_2 \rightarrow 2H_2O$
$CO + 2H_2 \rightarrow CH_3OH$
$CO + 3H_2 \rightarrow CH_4 + H_2O$ (methanation reaction)
$CO_2 + 4H_2 \rightarrow CH_4 + 2H_2O$
$C + 2H_2O \rightarrow 2H_2 + CO_2$
$2C + H_2 \rightarrow C_2H_2$
$CH_4 + 2H_2O \rightarrow CO_2 + 4H_2$

Hu *et al.*, 2012). The Fischer–Tropsch process can be represented by a simple equation:

$$(2n+1)H_2 + nCO \rightarrow C_nH_{(2n+2)} + nH_2O$$

In this equation, n is an integer. Thus, for $n = 1$, the reaction represents the formation of methane, which in most gas-to-liquids (GTL) applications is considered to be an undesirable by-product.

The Fischer–Tropsch process conditions are usually chosen to maximize the formation of higher-molecular weight hydrocarbon liquid fuels, which are higher-value products. There are side reactions taking place in the process, among which the water–gas shift reaction is predominant:

$$CO + H_2O \rightarrow H_2 + CO_2$$

Depending on the catalyst, the temperature, and the type of process employed, hydrocarbon derivatives ranging from methane to higher-molecular weight paraffin derivatives and olefin derivatives can be obtained. Small amounts of low-molecular-weight oxygenated derivatives (such as alcohol derivatives and organic acid derivatives) are also formed. Typically, Fischer–Tropsch liquids are (unless the process is designed for the production of other products) hydrocarbon products (which range from naphtha-type liquids to wax) and non-hydrocarbon products. The production of non-hydrocarbon products requires the adjustments of the feedstock composition and the process parameters.

Briefly, synthesis gas is a gas mixture containing various amounts of carbon monoxide (CO) and hydrogen (H2) generated through the gasification of a carbonaceous material. Examples of methods include steam reforming of natural gas, crude oil residua, coal, and biomass. Synthesis gas is used as an intermediate in producing hydrocarbon derivatives via the Fischer–Tropsch process for use as gaseous and liquids fuels.

The synthesis gas is produced through the gasification conversion of carbonaceous feedstock such as crude oil residua, coal, and biomass and production of hydrocarbon products can be represented simply as follows:

$$CH_{\text{feedstock.}} + O_2 \rightarrow CO + H_2$$

$$nCO + nH_2 \rightarrow C_nH_{2n+2}$$

However, before the conversion of the carbon monoxide and hydrogen into hydrocarbon products, several reactions are employed to adjust the hydrogen/carbon monoxide ratio. Most important is the water–gas shift reaction in which additional hydrogen is produced (at the expense of carbon monoxide) to satisfy the hydrogen/carbon monoxide ratio necessary for the production of hydrocarbon derivatives:

$$H_2O + CO \rightarrow H_2 + CO_2$$

The boiling range of Fischer–Tropsch typically spans the naphtha and kerogen boiling ranges and is suitable for analysis by the application of standard test methods. By the suitable choice of a catalyst, the preference for products boiling in the naphtha range (<200°C, <390°F) or for products boiling in the diesel range (approximately 150–300°C, 300–570°F) can be realized.

The other product worthy of consideration is *bio-oil* (*pyrolysis oil*, or *bio-crude*), a liquid product produced by the thermal decomposition (destructive distillation) of biomass (Chapter 3) at temperatures of the order of 500°C (930°F). The product is a synthetic crude oil and is of interest as a possible complement (eventually a substitute) to crude oil. It can range from a light tarry material to a free-flowing liquid — both require further refining to produce specification grade fuels

(Parkash, 2003; Gary *et al.*, 2007; Speight, 2011, 2014a, 2017; Hsu and Robinson, 2017).

Hydrocarbon moieties are predominant in the product, but the presence of varying levels of oxygen (depending on the character of the feedstock) requires to be tested (using, for example, hydrotreating) during refining. On the other hand, the bio-oil can be used as a feedstock for the Fischer–Tropsch process for the production of lower boiling products, as is the case when naphtha and gas oil are used as feedstocks for the Fischer–Tropsch process. In summary, the Fischer–Tropsch process produces hydrocarbon products of different molecular weights from a gas mixture of carbon monoxide and hydrogen (synthesis gas), all of which can find use in various energy scenarios.

In the current context, the most valuable product is synthesis gas, which is the beginning of a wide range of chemicals (Figure 10.1). The production of synthesis gas, i.e. mixtures of carbon monoxide and hydrogen, has been known for several centuries. But it is only with the commercialization of the Fischer–Tropsch reaction that the importance of synthesis gas has been realized. The thermal cracking (pyrolysis) of crude oil or fractions thereof was an important method for producing gas in the years following its use for increasing the heat content of water gas. In addition to the gases obtained by distillation of crude oil, further gaseous products are produced during the processing of naphtha and middle

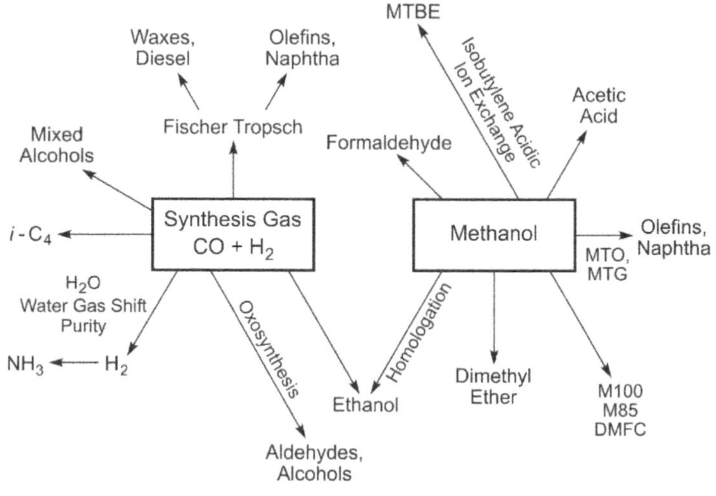

Figure 10.1: Routes to chemicals from synthesis gas and methanol.

distillates to produce gasoline. The hydrodesulfurization processes involving treatment of naphtha, distillates, and residual fuels and the coking or similar thermal treatments of vacuum gas oils and residual fuel oils also produce gaseous products (Parkash, 2003; Gary *et al.*, 2007; Speight, 2011, 2014a, 2017; Hsu and Robinson, 2017).

The chemistry of the oil-to-gas conversion was established several decades ago and can be described in general terms although the primary and secondary reactions can be truly complex. The composition of the gases produced from a wide variety of feedstocks depends not only on the severity of cracking but also to an equal or lesser extent on the feedstock type. In general terms, gas heating values are of the order of 950–1350 Btu/ft^3.

The second series of refining operations contributing to gas production are the catalytic cracking processes, such as fluid-bed catalytic cracking and other variants, in which heavy gas oils are converted into gas, naphtha, fuel oil, and coke (Parkash, 2003; Gary *et al.*, 2007; Speight, 2011, 2014a, 2017; Hsu and Robinson, 2017). The catalysts will promote steam reforming reactions that lead to a gas product that contains more hydrogen and carbon monoxide products and fewer unsaturated hydrocarbon products than the gas product from a non-catalytic process. The resulting gas is more suitable for use as a medium-heat-value gas than the rich gas produced by straight thermal cracking. The catalyst also influences the reaction rates in the thermal cracking reactions, which can lead to higher gas yields and lower tar and carbon yields.

Almost all crude oil fractions can be converted into gaseous fuels, although conversion processes for the heavier fractions require more elaborate technology to achieve the necessary purity and uniformity in the manufactured gas stream. In addition, the thermal yield from the gasification of heavier feedstocks is invariably lower than that from the gasification of light naphtha or liquefied crude oil gas, since in addition to the production of synthesis gas components (hydrogen and carbon monoxide) and various gaseous hydrocarbon derivatives, heavy feedstocks also yield some tar and coke.

Synthesis gas can be produced from heavy oil and other heavy crude feedstocks (such as residua, extra heavy oil, and tar sand bitumen) by a process known in the industry as partial oxidation:

$$[2CH]_{\text{crude oil}} + O_2 \rightarrow 2CO + H_2$$

In this process, the step consists of the reaction of the feedstock with a quantity of oxygen insufficient for burning the feedstock completely, yielding a mixture composed of carbon monoxide, carbon dioxide, hydrogen, and steam.

Success in partially oxidizing heavy feedstocks depends mainly on details of the burner design. The ratio of hydrogen to carbon monoxide in the product gas is a function of the reaction temperature and stoichiometry and can be adjusted, if desired, by varying the ratio of the carrier steam to the oil fed into the unit. The synthesis of hydrocarbon derivatives through the hydrogenation of carbon monoxide over transition metal catalysts involves the reaction of hydrogen with carbon dioxide at elevated temperatures (typically 300–400°C, 570–750°F) and elevated pressures in the presence of a nickel catalyst to produce methane and water.

The Fischer–Tropsch process (*Fischer–Tropsch synthesis*) is a series of catalyzed chemical reactions that convert a mixture of carbon monoxide and hydrogen into hydrocarbon derivatives. The process is a key component of GTL technology that produces liquid and solid hydrocarbon derivatives from coal, natural gas, biomass, or other carbonaceous feedstocks. Typical catalysts used are based on iron and cobalt and the hydrocarbon derivatives synthesized in the process are primarily liquid alkanes along with by-products, such as olefin derivatives, alcohols, and solid paraffin derivatives (waxes).

In the simplest terms, according to the following equation (Schulz, 1999; Chadeesingh, 2011):

$$(2n+1)\,H_2 + nCO \rightarrow C_nH_{2n+2} + nH_2O$$

For $n = 1$, the reaction represents the formation of methane, which in most coal-to-liquid or gas-to-liquid applications is considered an undesirable by-product. Thus, small amounts of low-molecular weight oxygenates (e.g. alcohol and organic acids) are also formed.

A major attraction of the use of synthesis gas is the very wide range of potential uses by converting synthesis gas into useful downstream products: (i) the Fischer–Tropsch synthesis of hydrocarbon derivatives, (ii) methanol synthesis, (iii) mixed alcohol synthesis, and (iv) synthesis gas fermentation. By choosing an appropriate catalyst (usually based on iron or cobalt) and appropriate reaction conditions (usually 200–350°C, 390–650°F, and pressures of the order of 300–600 psi), the process with its associated cracking and separation stages can be optimized to produce high-molecular weight wax for low-molecular weight olefin derivatives and naphtha for petrochemical production. The ideal feedstock for the Fischer–Tropsch process is synthesis gas consisting of a mixture of hydrogen and carbon monoxide with a molar ratio of 2:1 (Chadeesingh, 2011).

Methanol synthesis is another attractive conversion route because methanol is one of the top 10 petrochemical commodities insofar as, like synthesis gas, it can also be a source of chemicals (Figure 10.1). Synthesis gas can be converted into methanol over a copper–zinc oxide catalyst at 220–300°C (430–570°F) and 750–1500 psi:

$$CO + 2H_2 \rightarrow CH_3OH$$

Methanol can, in turn, be used to make acetic acid, formaldehyde for resins, gasoline additives and petrochemical building blocks such as ethylene and propylene.

Under slightly more severe process conditions (up to 425°C, 800°F, and 4500 psi), a wider range of mixed alcohols can be produced (Chadeesingh, 2011). The processes use catalysts modified from either Fischer–Tropsch synthesis or methanol synthesis, by addition of alkali metals. Finally, the fermentation route for the conversion of synthesis gas uses biochemical processes and reaction conditions that are close to ambient temperature and pressure to make ethanol or other alcohols. Biochemical processes are addressed in the following.

There are many options for converting synthesis gas into petrochemical feedstocks. For example, the olefin derivative conversion chain in which ethylene and propylene are converted into polymers (polyethylene, polypropylene, and PVC), glycol derivatives (ethylene glycol, $HOCH_2CH_2OH$, propylene glycol, and $CH_3CHOHCH_2OH$) and a range of familiar materials such as acetone (CH_3COCH_3), acetic acid (CH_3CO_2H), gasoline additives, and surfactants. The olefin derivatives can be produced by synthesizing naphtha in a Fischer–Tropsch process and then cracking it in a conventional naphtha cracker to make ethylene and propylene.

Depending on the source of biomass feedstock and the choice of gasifier technology, the raw synthesis gas can contain varying amounts of particulates (e.g. ash or char, which can lead to erosion, plugging, or fouling), alkali metals (which can cause hot corrosion and catalyst poisoning), water-soluble trace components (e.g. halides and ammonia), light oils or tars (e.g. benzene, toluene, xylene or naphthalene, which can lead to catalyst carbonization and fouling), polyaromatic compounds, sulfur components, phosphorus components as well as methane and carbon dioxide. Many of these can be removed (if required) using standard chemical-industry equipment such as cyclones, filters, electrostatic precipitators, water scrubbers, oil scrubbers, and activated carbon as adsorbents (Chapter 9).

Another important factor is the ratio of hydrogen to carbon dioxide in the synthesis gas. Different conversion routes require different ratios, e.g. 1.7:1 and 2.15:1 for producing Fischer–Tropsch naphtha/gasoline and diesel, respectively, or 3:1 for methanol synthesis. Because biomass feedstocks contain oxygen within the structure of the feedstock constituents, biomass-derived synthesis gas often requires the hydrogen/carbon monoxide ratio boosted. One method for achieving this is to react some of the synthesis gas with steam over a catalyst to produce hydrogen and carbon dioxide in the water–gas shift reaction (Chadeesingh, 2011; Speight, 2013a, 2013b) unless there is a by-product hydrogen source readily available:

$$CO + H_2O \rightarrow H_2 + CO_2 \qquad \text{(water–gas shift reaction)}$$

The extent to which gas clean-up is required depends on the choice of the synthesis gas conversion route. Generally, the level of particulates will need to be reduced considerably for any chemical synthesis process, but the precise extent to which (say) sulfur or halide levels need to be reduced depends on the catalysts that are going to be used. For the methanol synthesis process, for example, the sulfur content of the synthesis gas has to be below 100 ppb v/v. For the ammonia synthesis process, there is a similar sulfur constraint, and the carbon dioxide content must be below 10 ppm v/v.

10.2 Production of Synthesis Gas

Gasification for producing synthesis gas can proceed from almost any organic material, including biomass and plastic waste (Chapters 5 and 6) (Luque and Speight, 2015). The chemistry of the process for producing synthesis gas can be represented simply as follows:

$$\text{Carbonaceous feedstock} \rightarrow \text{(partial oxidation)} \rightarrow \text{synthesis gas}$$

Thus,

$$C_{\text{feedstock}} + O_2 \rightarrow CO_2$$

$$C_{feedstock} + CO_2 \rightarrow 2CO$$
$$C_{feedstock} + H_2O \rightarrow CO + H_2$$

The products designated as synthesis fuels include low-to-high-boiling hydrocarbon derivatives and methanol. Also, the high-boiling products (including wax products) can also be used as the feedstocks for gas production.

In addition, the actual process is composed of three components: (i) synthesis gas generation, (ii) waste heat recovery, and (iii) gas processing. Within each of the three above-mentioned systems, there are several options. For example, synthesis gas can be generated to yield a range of compositions ranging from high-purity hydrogen to high-purity carbon monoxide. Two major routes can be utilized for high-purity gas production: (i) pressure swing adsorption and (ii) the utilization of a cold box, where separation is achieved by distillation at low temperatures. In fact, both processes can also be used in combination as well. Unfortunately, both the processes require high capital expenditures. However, to address these concerns, research and development are ongoing and success can be measured by the demonstration and commercialization of technologies such as permeable membrane for the generation of high-purity hydrogen, which itself can be used to adjust the hydrogen/carbon monoxide ratio of the synthesis gas produced (Bayat *et al.*, 2010).

The resulting synthesis gas burns cleanly into water vapor and carbon dioxide. Alternatively, synthesis gas may be converted efficiently into either methane via the Sabatier reaction or a diesel-like synthetic fuel via the Fischer–Tropsch process. Inorganic components of the feedstock, such as metals and minerals, are trapped in an inert and environmentally safe form as char, which may be used as a fertilizer. When used as an intermediate in the large-scale, industrial synthesis of hydrogen and ammonia, it is produced from natural gas (via the steam reforming reaction) as follows:

$$CH_4 + H_2O \rightarrow CO + 3H_2$$

The synthesis gas produced in large waste-to-energy gasification facilities is used as the fuel to generate electricity.

A modified version of steam reforming called autothermal reforming (Hagh, 2004), which is a combination of partial oxidation near the reactor inlet with conventional steam reforming further along the reactor, improves the overall reactor efficiency and increases the flexibility of the process. Partial oxidation processes using oxygen instead of steam have also found wide application in synthesis gas manufacture, with the special feature that they could utilize low-value feedstocks such as heavy crude oil residua. In recent years, catalytic partial oxidation employing very short reaction times (milliseconds) at high temperatures (850–1000°C; 1560–1830°F) provides yet another approach for synthesis gas manufacture (Hickman and Schmidt, 1993). Nearly complete conversion of methane, with close about 100% selectivity to hydrogen and carbon dioxide, can be obtained with a rhenium monolith under well-controlled conditions. Experiments on the catalytic partial oxidation of *n*-hexane conducted with added steam give much higher yields of hydrogen than can be obtained in experiments without steam, a result of much interest in obtaining hydrogen-rich streams for fuel cell applications.

The route for converting a carbonaceous feedstock into synthetic automotive fuels, as practised by SASOL, is technically proven and a series of products with favorable environmental characteristics are produced (Luque and Speight, 2015). As is the case with essentially all carbonaceous feedstock conversion processes where air or oxygen is used for the utilization or partial conversion of the energy in the coal, the carbon dioxide burden is a drawback as compared to crude oil.

The uses of synthesis gas include use as a chemical feedstock and in GTL processes which employ Fisher–Tropsch chemistry to make liquid fuels as feedstocks for chemical synthesis. In addition to its use in the production of fuel additives, including diethyl ether and methyl *t*-butyl ether (MTBE), acetic acid and its anhydride, synthesis gas could also make an important contribution to chemical synthesis through conversion into methanol. There is also a method in which stranded natural gas is converted into synthesis gas followed by conversion into liquid fuels.

10.2.1 *Feedstocks*

In principle, synthesis gas can be produced from any hydrocarbon feedstocks, which include natural gas, naphtha, residual oil, crude oil coke, coal, biomass, and municipal or industrial waste (Chapter 1). The product gas stream is subsequently purified (to remove sulfur, nitrogen, and any particulate matter) after which it is catalytically converted into a mixture of liquid hydrocarbon products. In addition, synthesis gas may also be used to produce a variety of products, including ammonia and methanol.

Of all the carbonaceous materials used as feedstocks for gasification processes, coal represents the most widely used feedstock and, accordingly, the feedstock about which most is known. In fact, gasification of coal has been a commercially available proven technology (Speight, 2013a, 2013b). Modern gasification processes have evolved from three first-generation processing technologies: (i) Lurgi fixed-bed reactor, (ii) high-temperature Winkler fluidized-bed reactor, and (iii) Koppers–Totzek entrained flow reactor. In each case, steam/air/oxygen are passed through heated coal, which may either be a fixed bed or a fluidized bed or entrained in the gas. Exit gas temperatures from the reactor are 500°C (930°F), 900–1100°C (1650–2010°F), and 1300–1600°C (2370–2910°F), respectively. In addition to the steam/air/oxygen mixture being used as the feed gases, steam/oxygen mixtures can also be used for which membrane technology and a compressed oxygen-containing gas are employed.

In addition, low-value or negative-value materials and wastes, such as crude oil coke, refinery residua, refinery waste, municipal sewage sludge, biomass, hydrocarbon-contaminated soils, and chlorinated hydrocarbon by-products, have all been used successfully in gasification operations (Speight, 2008, 2013a, 2013b). In addition, synthesis gas is used as a source of hydrogen or as an intermediate for producing a variety of hydrocarbon products by means of the Fischer–Tropsch synthesis (Chadeesingh, 2011). In fact, gasification for producing synthesis gas can proceed from any carbonaceous material, including biomass and waste.

There are different sources for obtaining synthesis gas. It can be produced by steam reforming or partial oxidation of any hydrocarbon ranging from natural gas (methane) to crude oil residua. It can also be obtained by gasifying the carbonaceous feedstock into a medium-Btu gas, which consists of variable amounts of carbon monoxide, carbon dioxide, and hydrogen and is used principally as a fuel gas.

As already stated in Chapter 2, the first step in the process is feedstock preparation, which may include unit operations such as (i) size reduction, (ii) screening, and (iii) slurrying. The design of the feedstock preparation section depends on the properties of the raw material feed and on the requirements of the gasification technology that is selected. For example, preparation any form of crude oil resid, biomass, or waste for gasification is very different from the preparation of bituminous coal for the gasifier, and the type of gasifier suitable for the gasification may also (or more than likely) be feedstock-dependent (Chapter 3).

Once the feedstock is prepared for the gasifier, the major route for producing synthesis gas involves steam reforming of natural gas over a promoted nickel catalyst at temperatures on the order of 800°C (1470°F):

$$CH_4 + H_2O \rightarrow CO + 3H_2$$

In some countries, synthesis gas is mainly produced by steam reforming naphtha. Because naphtha is a mixture of hydrocarbon derivatives ranging approximately from C5 to C10, the steam reforming reaction may be represented using *n*-heptane:

$$CH_3(CH_2)_5CH_3 + 7H_2O \rightarrow 7CO + 15H_2$$

As the molecular weight of the hydrocarbon increases (lower H/C feed ratio), the hydrogen/carbon monoxide (H_2/CO) product ratio decreases. The hydrogen/carbon monoxide product ratio is approximately 3 for methane, 2.5 for ethane, 2.1 for heptane, and less than 2 for heavier hydrocarbon derivatives. The non-catalytic partial oxidation of hydrocarbon derivatives is also used to produce synthesis gas, but the hydrogen/carbon monoxide ratio is lower than from steam reforming. In practice, this ratio is even lower than what is shown by the stoichiometric equation, because part of the methane is oxidized to carbon dioxide and water. When resids are partially oxidized by oxygen and steam at 1400–1450°C (2550–2640°F) and pressures of the order of 11,500–13,000 psi, the gas consists of equal parts of hydrogen and carbon monoxide.

Synthesis gas is a mixture of carbon monoxide and hydrogen and an important intermediate used for producing methanol. It is also used to synthesize a wide variety of hydrocarbon derivatives, ranging from gases to naphtha to gas oil, using Fischer–Tropsch technology. This process may offer an alternative future route to obtain olefin derivatives and chemicals.

Synthesis gas is a major source of hydrogen and is used for producing ammonia. Ammonia is the host of many chemicals such as urea, ammonium nitrate, and hydrazine. Carbon dioxide, a by-product from synthesis gas production, reacts with ammonia to produce urea (H_2NCONH_2).

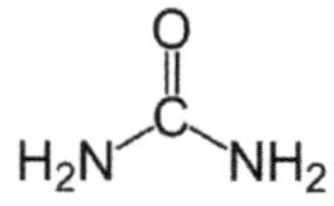

Urea

Urea (also known as carbamide) serves an important role in the metabolism of nitrogen-containing compounds by animals and is the main nitrogen-containing substance in the urine of mammals. It is a colorless, odorless solid, highly soluble in water, and, dissolved in water, it is neither an acid nor an alkali. It is formed in the liver by the combination of two ammonia molecules (NH_3) with a carbon dioxide (CO_2) molecule. It is widely used in fertilizers as a source of nitrogen and is an important raw material for the chemical industry.

Most of the production of hydrocarbon derivatives by the Fischer–Tropsch method uses synthesis gas produced from sources that yield a relatively low hydrogen/carbon monoxide ratio, such as is typical in coal gasifiers. This, however, does not limit this process to low hydrogen/carbon monoxide gas feeds. The only large-scale commercial process using this technology is in South Africa, where coal is an abundant energy source. The process of obtaining liquid hydrocarbon derivatives from coal through the Fischer–Tropsch process is termed indirect coal liquefaction, which was originally intended for obtaining liquid hydrocarbon derivatives from solid fuels. However, this method may well be applied in the future for the manufacture of chemicals through cracking the liquid products or by directing the reaction to produce more olefin derivatives.

The reactants in Fischer–Tropsch processes are carbon monoxide and hydrogen. The reaction may be considered a hydrogenative oligomerization of carbon monoxide in the presence of a heterogeneous catalyst. The main reactions occurring in Fischer–Tropsch processes are as follows:

(i) Olefin derivatives:

$$2n\mathrm{H_2} + n\mathrm{CO} \rightarrow \mathrm{C}_n\mathrm{H}_{2n} + n\mathrm{H_2O}$$

(ii) Paraffin derivatives:

$$(2n{+}1)\mathrm{H_2} + n\mathrm{CO} \rightarrow \mathrm{C}_n\mathrm{H}_{2n+2} + n\mathrm{H_2O}$$

(iii) Alcohol derivatives:

$$2n\mathrm{H_2} + n\mathrm{CO} \rightarrow \mathrm{C}_n\mathrm{H}_{2n+2}\mathrm{O} + (n{-}1)\mathrm{H_2O}$$

The coproduct water reacts with carbon monoxide (the shift reaction), yielding hydrogen and carbon dioxide:

$$CO + H_2O \rightarrow H_2 + CO_2$$

The hydrogen gained from the water shift reaction reduces the hydrogen demand for Fischer–Tropsch processes. Water–gas shift proceeds at about the same rate as the Fischer–Tropsch reaction. Another side reaction occurring in Fischer–Tropsch process reactors is the disproportionation of carbon monoxide to carbon dioxide and carbon:

$$2CO \rightarrow CO_2 + C$$

This reaction is responsible for the deposition of carbon in the reactor tubes in fixed-bed reactors and reducing the heat transfer efficiency.

Fischer–Tropsch technology is best exemplified by the SASOL projects in South Africa. After the carbonaceous feedstock is gasified to a synthesis gas mixture, it is purified in a Rectisol unit. The purified gas mixture is reacted in a Synthol unit over an iron-based catalyst. The main products are gasoline, diesel fuel, and jet fuels. By-products are ethylene, propylene, α-olefin derivatives, sulfur, phenol, and ammonia, which are used for the production of downstream chemicals. However, the exact mechanism is not fully established. One approach assumes a first-step adsorption of carbon monoxide on the catalyst surface followed by a transfer of an adsorbed hydrogen atom from an adjacent site to the metal carbonyl (M-CO). The polymerization continues (as in the last three steps shown above) until termination occurs and the hydrocarbon is desorbed. The last two steps mentioned above explain the presence of oxygenated derivatives in Fischer–Tropsch products.

Alternatively, an intermediate formation of an adsorbed methylene on the catalyst surface through the dissociative adsorption of carbon monoxide has been considered. The formed metal carbide (M–C) is then hydrogenated to a reactive methylene metal species. The methylene intermediate abstracts a hydrogen and is converted into an adsorbed methyl. Reaction of the methyl with the methylene produces an ethyl-metal species. Successive reactions of the methylene with the formed ethyl produces a long-chain adsorbed alkyl. The adsorbed alkyl species can either terminate to a paraffin by a hydrogenation step or to an olefin by a dehydrogenation step. The carbide mechanism, however, does not explain the formation of oxygenate derivatives in Fischer–Tropsch products.

10.3 The Fischer–Tropsch Process

The Fischer–Tropsch synthesis is based on the conversion of a mixture of carbon monoxide and hydrogen into liquid hydrocarbon derivatives. The Fischer–Tropsch synthesis can be represented by the following chemical reactions (Fischer and Tropsch, 1923):

$$(2n + 1)H_2 + nCO \rightarrow C_nH_{2n+2} + nH_2O$$
$$2nH_2 + nCO \rightarrow C_nH_{2n} + nH_2O$$

This process is a key route for the GTL technology, which produces synthetic fuels and chemicals from mainly natural gas, crude oil resids, coal, biomass, and carbonaceous wastes (Klerk and Furimsky, 2010; Chadeesingh, 2011; Luque and Speight, 2015).

10.3.1 *Chemistry*

The feedstock for the Fischer–Tropsch process, synthesis gas, consists primarily of carbon monoxide, carbon dioxide, and hydrogen and has less than half the energy density of natural gas. Synthesis gas is combustible and is often used as a fuel source or as an intermediate for the production of other chemicals. Synthesis gas for use as a fuel is most often produced by gasification of the carbonaceous feedstock or municipal waste mainly by the following paths:

$$C + O_2 \rightarrow CO_2$$
$$CO_2 + C \rightarrow 2CO$$
$$C + H_2O \rightarrow CO + H_2$$

Although the focus of this section is on the production of hydrocarbon derivatives from synthesis gas, it is worth noting that all or a part of clean synthesis gas can also be used (i) as chemical *building blocks* to produce a wide range of chemicals using processes well established in the chemical and petrochemical industries), (ii) as a fuel producer for highly efficient fuel cells (which run off the hydrogen made in a gasifier), or perhaps in the future, for hydrogen turbines and fuel cell–turbine hybrid systems, and (iii) as a source of hydrogen that can be separated from the gas stream and used as a fuel or as a feedstock for refineries (which use the hydrogen to upgrade crude oil products) (Parkash, 2003; Gary *et al.*, 2007;

Speight, 2011, 2014a; Hsu and Robinson, 2017; Speight, 2017). However, the decreasing availability and the increasing price of crude oil have renewed the worldwide interest in the production of liquid hydrocarbon derivatives from carbon monoxide and hydrogen using metal catalysts, also known as the Fischer–Tropsch synthesis or the Fischer–Tropsch process (Chapter 10) (Speight, 2019a).

The manufacture of gas mixtures of carbon monoxide and hydrogen has been an important part of chemical technology for about a century. Originally, such mixtures were obtained by the reaction of steam with incandescent coke and were known as *water gas*. First used as a fuel, water gas soon attracted attention as a source of hydrogen and carbon monoxide for the production of chemicals, at which time it gradually became known as synthesis gas. Eventually, steam reforming processes, in which steam is reacted with natural gas (methane) or crude oil naphtha over a nickel catalyst, found wide application in the production of synthesis gas.

The route for converting coal into synthetic automotive fuels, as practised by SASOL, is technically proven and products with favorable environmental characteristics are produced. As is the case with essentially all the carbonaceous feedstock conversion processes where air or oxygen is used for the utilization or partial conversion of the energy in the carbonaceous feedstock, the carbon dioxide burden is a drawback compared to crude oil.

The hydroformylation synthesis (also known as the oxo synthesis or the oxo process) is an industrial process for the production of aldehyde derivatives from alkene derivatives. This chemical reaction entails the addition of a formyl group (CHO) and a hydrogen atom to a carbon–carbon double bond. A key consideration of hydroformylation is the production of normal isomers or the production of *iso*-isomers in the product(s). For example, the hydroformylation of propylene can yield two isomeric products, butyraldehyde or *iso*-butyraldehyde:

$$H_2 + CO + CH_3CH{=}CH_2 \rightarrow CH_3CH_2CH_2CHO \text{ (}n\text{-butyraldehyde)}$$

$$H_2 + CO + CH_3CH{=}CH_2 \rightarrow (CH_3)_2CHCHO \text{ (}iso\text{-butyraldehyde)}$$

These isomers reflect the regio-chemistry (the preference of one direction of chemical bond formation or chemical bond scission over all other possible directions) of the insertion of the alkene into the M–H bond. Generally, both products are not equally desirable.

As an example of the hydroformylation process, the Exxon process, also known as Kuhlmann-oxo process or the PCUK-oxo process, is used for the hydroformylation of C_6–C_{12} olefin derivatives using cobalt-based catalysts. In order to recover the catalyst, an aqueous sodium hydroxide solution or sodium carbonate is added to the organic phase. By extraction with olefin and neutralization by addition of sulfuric acid solution under carbon monoxide pressure, the metal–carbonyl hydride can recovered. The recovered hydride is stripped out with synthesis gas, absorbed by the olefin, and returned to the reactor. The Exxon process, similar to the BASF process, is carried out at a temperature of the order of 160–180°C (320–355°F) and a pressure suitable for the reactants and products.

The Fischer–Tropsch reaction can be described as the synthesis of hydrocarbon derivatives via the hydrogenation of carbon monoxide using transition metal catalysts. The major catalysts used industrially are Fe and Co, but can also be Ru and Ni. From a mechanism perspective, the reactions can be regarded as a carbon chain building process where methylene (CH_2) groups are attached sequentially in a carbon chain (Table 10.2). Thus,

$$n\text{CO} + [n + m/2]\text{H}_2 \rightarrow \text{C}_n\text{H}_m + n\text{H}_2\text{O}$$

For example,

$$\text{CO} + 2\text{H}_2 \rightarrow -[\text{CH}_2]- + \text{H}_2\text{O}$$

Table 10.2: Carbon chain groups in the range of Fischer–Tropsch products.

Carbon number	Group name
C1–C2	Synthetic natural gas (SNG)
C3–C4	Liquefied petroleum gas (LPG)
C5–C7	Low boiling naphtha
C8–C10	High boiling naphtha
C11–C16	Middle distillate
C17–C30[a]	Low melting wax
C31–C60	High melting wax

Notes: [a]The C17 *n*-alkane (*n*-heptadecane) is the first member of the series that is not fully liquid under ambient conditions (melting point: 21°C (70°F)).

A common, salient feature of the reactions is the exothermicity of the reactions. As a general rule of thumb, those reactions that produce water and carbon dioxide as a product tend to be more exothermic on account of the very high heat of formation of these species. Some of the reactions proposed are as follows (Rauch, 2001):

$$CO_2 + 3H_2 \rightarrow -[CH_2]- + 2H_2O \qquad \Delta H = -125 \text{ kJ/mol}$$

$$CO + 2H_2 \rightarrow -[CH_2]- + H_2O \qquad \Delta H = -165 \text{ kJ/mol}$$

$$2CO + H_2 \rightarrow -[CH_2]- + CO_2 \qquad \Delta H = -204 \text{ kJ/mol}$$

$$3CO\ H_2 \rightarrow -[CH_2]- + 2CO_2 \qquad \Delta H = -244 \text{ kJ/mol}$$

There is also the water–gas shift reaction:

$$CO + H_2O \rightarrow H_2 + CO_2 \qquad \Delta H = -39 \text{ kJ/mol}$$

Due to the very high exothermic nature of the Fischer–Tropsch reactions, as illustrated in the reactions above, an important issue is, not surprisingly, the need to avoid an increase in temperature (Furnsinn *et al.*, 2005). The need for cooling is thus of critical importance in order to achieve the following: (i) maintain stable reaction conditions, (ii) avoid the tendency to produce lighter hydrocarbon derivatives, and (iii) prevent catalyst sintering and hence reduction in activity. Since the total heat of reaction is of the order of approximately 25% of the heat of combustion of the synthesis gas (i.e. reactants in the Fischer–Tropsch process), a theoretical limit on the maximum efficiency of the Fischer–Tropsch process is imposed (Rauch, 2001).

Two main chemical characteristics of Fischer–Tropsch synthesis are the unavoidable production of a wide range of hydrocarbon products (olefin derivatives, paraffin derivatives, and oxygenated products) and the liberation of a large amount of heat from the highly exothermic synthesis reactions. Product distributions are influenced by temperature, feed gas composition (hydrogen/carbon monoxide), pressure, catalyst type, and catalyst composition. Fischer–Tropsch products are produced in four main steps: synthesis gas generation, gas purification, Fischer–Tropsch synthesis, and product upgrading. Depending on the types and quantities of Fischer–Tropsch products desired, either low (200–240°C; 390–465°F) or high temperature (300–350°C; 570–660°F) synthesis is used with either an iron (Fe) or cobalt catalyst (Co) (Van Berge and Everson, 1997).

The required gas mixture of carbon monoxide and hydrogen (synthesis gas) is created through a reaction of coke or the carbonaceous feedstock with water steam and oxygen at temperatures over 900°C. In the past, town gas or gas for lamps was a carbon monoxide–hydrogen mixture, made by gasifying coke in gas works. In the 1970s, it was replaced with imported natural gas (methane). Gasification of carbonaceous feedstocks and Fischer–Tropsch hydrocarbon synthesis together bring about a two-stage sequence of reactions, which allows the production of liquid fuels like gasoline and diesel out of the solid combustible the carbonaceous feedstock.

The Fischer–Tropsch synthesis is, in principle, a carbon chain building process, where methylene groups are attached to the carbon chain. The actual reactions that occur have been, and remain, a matter of controversy as they have been in the last century since the 1930s:

$$(2n{+}1)H_2 + nCO \rightarrow C_nH_{2n+2} + nH_2O$$

Even though the overall Fischer–Tropsch process is described by the following chemical equation:

$$(2n{+}1)H_2 + nCO \rightarrow C_nH_{2n+2} + nH_2O$$

The initial reactants in the above reaction (i.e. carbon monoxide and H_2) can be produced either by other reactions such as the partial combustion of a hydrocarbon:

$$CnH_{2n+2} + \tfrac{1}{2}\,nO_2 \rightarrow (n{+}1)H_2 + nCO$$

for example (when $n = 1$), methane (in the case of GTL applications):

$$2CH_4 + O_2 \rightarrow 4H_2 + 2CO$$

or by the gasification of any carbonaceous source, such as biomass:

$$C + H_2O \rightarrow H_2 + CO$$

The energy needed for this endothermic reaction is usually provided by (exothermic) combustion with air or oxygen:

$$2C + O_2 \rightarrow 2CO$$

These reactions are highly exothermic, and to avoid an increase in temperature, which will result in lighter hydrocarbon derivatives, it is important to have sufficient cooling, to secure stable reaction conditions. The total heat of the reaction amounts to approximately 25% of the heat of combustion of synthesis gas, and puts thereby a theoretical limit on the maximal efficiency of the Fischer–Tropsch process.

The reaction is dependent on a catalyst, mostly iron or cobalt catalyst where the reaction takes place. There is either a low- or high-temperature process (low-temperature Fischer–Tropsch and high-temperature Fischer–Tropsch), with temperatures ranging between 200°C and 240°C (390°F and 465°F) for the low-temperature Fischer–Tropsch and 300–350°C (570–660°F) for the high-temperature Fischer–Tropsch process. The high-temperature Fischer–Tropsch process uses an iron catalyst and the low-temperature Fischer–Tropsch either iron or cobalt catalyst. Examples of the different catalysts include nickel-based and ruthenium-based catalysts, which also have enough activity for commercial use in the process. However, the availability of ruthenium is limited and the nickel-based catalyst has high activity but produces too much methane, and additionally the performance at high pressure is poor, due to the production of volatile carbonyls. This leaves only cobalt and iron as practical catalysts, and this study will only consider these two. Iron is cheap, but cobalt has the advantage of higher activity and longer life, though it is, on a metal basis, 1,000 times more expensive than iron catalyst.

10.3.2 *Process principles*

A Fischer–Tropsch plant incorporates three major process sections: (i) production of synthesis gas by a process such as steam reforming, (ii) conversion of the synthesis gas to aliphatic hydrocarbon derivatives, and (iii) hydrocracking the longer-chain, waxy synthetic hydrocarbon derivatives to fuel grade fractions. Of the above three steps, the production of synthesis gas is the most energy intensive as well as expensive.

Depending on the reaction temperature, the process can be divided into low-temperature Fischer–Tropsch synthesis (LTFT, at temperatures of the order of 200–260°C, 390–500°F) and high-temperature Fischer–Tropsch synthesis (HTFT, at temperatures of the order of 300–350°C, 570–660°F). During the LTFT process, higher boiling hydrocarbon derivatives (boiling above 360°C, 680°F) are produced. Also the total distillate yield is higher that the yield produced by the HTFT process.

Higher temperature leads to reactions that are more rapid than the reactions in the LTFT process and higher conversion rates also tend to favor shorter chains as well as the production of methane, olefin derivatives, and aromatic derivatives. Typical pressures range from 15 to 750 psi or higher. Increasing the pressure leads to (i) higher conversion rates and (ii) the formation of the much-desired long-chained alkane derivatives.

A variety of catalysts can be used for the Fischer–Tropsch process, but the most common are the transition metals such as cobalt, iron, and ruthenium. Nickel can also be used, but tends to favor the production of methane. In the LTFT process, a cobalt-based catalyst or an iron-based catalyst are the more common catalysts, whereas in the LTFT process iron-based catalysts are more common. The advantage of cobalt-based catalysts is a much longer lifetime and the yields of unsaturated hydrocarbon derivatives and alcohol derivatives are higher compared with the products from the use of iron-based catalysts. On the other hand, iron-based catalysts are more tolerant to sulfur and can be also used (by means of the water–gas shift reaction) to adjust the hydrogen/carbon monoxide ratio if it is lower than 2.

For large-scale commercial Fischer–Tropsch reactors, heat removal and temperature control are the most important design features for obtaining optimum product selectivity and long catalyst lifetimes. Over the years, basically four Fischer–Tropsch reactor designs have been used commercially. These are the multi-tubular fixed bed, the slurry reactor, and the fluidized-bed reactor (with either a fixed bed or a circulating bed). The fixed-bed reactor consists of thousands of small tubes with the catalyst as the surface-active agent in the tubes. Water surrounds the tubes and regulates the temperature by settling the pressure of evaporation. The slurry reactor is widely used and consists of fluid and solid elements, where the catalyst has no particular position, but flows around as small pieces of catalyst together with the reaction components. The slurry and fixed-bed reactors are used in the LTFT process. The fluidized-bed reactors are diverse but they are characterized by the fluid behavior of the catalyst.

The multi-tubular fixed-bed reactors (often referred to as Arge reactors) were developed jointly by Lurgi and Ruhrchemie and commissioned in 1955. They were used by Sasol to produce high-boiling Fischer–Tropsch liquid hydrocarbon derivatives and waxes in Sasolburg, in what Sasol called the LTFT synthesis process, aiming at liquid fuel production. Most, if not all, of these types of Arge reactors have now been replaced by slurry-bed reactors, which are considered the state-of-the-art technology for LTFT synthesis. Slurry-bed Fischer–Tropsch reactors offer better temperature control and higher conversion. Fluidized-bed Fischer–Tropsch reactors were developed for the HTFT synthesis to produce low molecular gaseous hydrocarbon derivatives and naphtha. This type of reactor was originally developed in a circulating mode (such as the Sasol synthol reactors) but has been replaced by a fixed fluidized-bed type of reactor (advanced synthol reactors), which are capable of a high throughput.

The Fischer–Tropsch process can be divided into two operating regimes: (i) the HTFT process and (ii) the LTFT process (Chadeesingh, 2011). The HTFT technology uses a fluidized catalyst at 300–330°C (570–635°F). Originally, circulating fluidized-bed units were used (Synthol reactors). Since 1989, a commercial scale classical fluidized-bed unit has been implemented and improved upon. The LTFT technology was originally used in tubular fixed-bed reactors at 200–230°C (390–260°F). This produces a more paraffin derivative and waxy product spectrum than the high-temperature technology. A new type of reactor (the Sasol slurry phase distillate reactor) has been developed and is now in commercial operation. This reactor uses a slurry phase system rather than a tubular fixed-bed configuration and is currently the preferred technology for the commercial production of synfuels.

The commercial Sasol Fischer–Tropsch reactors all use iron-based catalysts on the basis of the desired product spectrum and operating costs. Cobalt-based catalysts have also been known since the early days of this technology and have the advantage of higher conversion for low-temperature cases. Cobalt is not suitable for high-temperature use due to excessive methane formation at such temperatures. For once-through maximum diesel production, cobalt has, despite its high cost, advantages and Sasol has also developed cobalt catalysts that perform very well in the slurry phase process.

However, both iron and cobalt Fischer–Tropsch catalysts are sensitive to the presence of sulfur compounds in synthesis gas and can be poisoned by the sulfur compounds. In addition, the sensitivity of the catalyst to sulfur is higher for cobalt-based catalysts than that for iron-based catalysts. This is one reason why cobalt-based catalysts are preferred for Fischer–Tropsch synthesis with synthesis gas derived from natural gas, where the synthesis gas has a higher hydrogen/carbon monoxide ratio and is relatively lower in sulfur content. On the other hand, iron-based catalysts are preferred for lower quality feedstocks such as coal.

Under most circumstances the production of synthesis gas by reforming natural gas will be more economical than that from coal gasification, but site-specific factors need to be considered. In fact, any technological advances in this field (such as better energy integration or the oxygen transfer ceramic membrane reformer concept) will speed up the rate at which the synfuel technology will become common practice.

The product distribution is highly dependent on the chain growth probability, which again is dependent on (i) pressure and temperature, (ii) the composition of the feedstock gas, (iii) the catalyst type, (iv) the catalyst composition, and (v) the reactor design. The desire to increase the selectivity of some favorable products leads to the need to understand the relationship between reaction conditions and chain growth probability, which in turn requires a mathematical expression for the growth probability in order to make a suitable model of the process. The different attempts to model the growth probability, *a*, have resulted in some models that are regarded in the literature as appropriate to describe the product distribution. Two models will be presented here to show the influence of temperature and partial pressure.

10.3.3 *Product distribution*

The product distribution of hydrocarbon derivatives formed during the Fischer–Tropsch process follows an Anderson–Schulz–Flory distribution:

$$W_n/n = (1 - \alpha)^2 a^{n-1}$$

Where W_n is the weight fraction of hydrocarbon molecules containing n carbon atoms, α is the chain growth probability or the probability that a molecule will continue reacting to form a longer chain. In general, α is largely determined by the catalyst and the specific process conditions.

According to the above equation, methane will always be the largest single product; however, by increasing α close to 1, the total amount of methane formed can be minimized compared to the sum of all the various long-chain products. Increasing α increases the formation of long-chain hydrocarbon derivatives — waxes — which are solid at room temperature. Therefore, for production of liquid transportation fuels, it may be necessary to crack the Fischer–Tropsch longer chain products.

In order to avoid this, some researchers have proposed using zeolites or other catalyst substrates with fixed-size pores that can restrict the formation of hydrocarbon derivatives longer than some characteristic size (usually $n < 10$). This way they can drive the reaction so as to minimize methane formation without producing lots of long-chain hydrocarbon derivatives.

The synthesis of hydrocarbon derivatives from the hydrogenation of carbon monoxide over transition metal catalysts was discovered in 1902 when Sabatier and Sanderens produced methane from hydrogen and carbon monoxide mixtures passed over nickel, iron and cobalt catalysts. In 1923, Fischer and Tropsch reported the use of alkalized iron catalysts to produce liquid hydrocarbon derivatives rich in oxygenated compounds. The process is a catalyzed chemical reaction in which carbon monoxide and hydrogen are converted into liquid and solid hydrocarbon derivatives. The typical catalysts used are based on iron and cobalt — the hydrocarbon derivatives produced in the process are composed primarily of liquid alkanes; the by-products are olefins, alcohols, and solid paraffins (waxes).

Coal gasification and Fischer–Tropsch hydrocarbon synthesis together cause a two-stage sequence of reactions, which allows the production of liquid fuels like diesel and petrol out of the solid combustible coal.

The Fischer–Tropsch synthesis took its first serious place in industry in 1935 at Ruhrchemie in Oberhausen. By the beginning of the 1940s, about 600,000 tons of liquid hydrocarbon derivatives were produced per year in German facilities, made from coal using Fischer–Tropsch synthesis. Licensed by Ruhrchemie, four facilities in Japan, as well as a plant in France and in Manchuria, were in use. After World War II, competition from crude oil made petrol production from coal unprofitable. The only new production facilities were in South Africa, for political reasons, built starting in 1950 in Sasolburg.

The Fischer–Tropsch reaction is dependent on a catalyst, mostly iron or cobalt catalyst where the reaction takes place. There is either a LTFT process or a HTFT process, with temperatures ranging between 200°C and 240°C (390°F and 465°F) for the LTFT process and 300°C and 350°C (570°F and 650°F) for the HTFT process. The LTFT process uses either iron or cobalt catalyst while the HTFT process uses an iron catalyst. (Dry, 2001; Chadeesingh, 2011). The fluidized-bed reactor is used in the HTFT process.

The Fischer–Tropsch process is an established technology and has already been applied on a large scale, although its popularity is hampered by high capital costs, high operation and maintenance costs, and the uncertain and volatile price of crude oil, although this concept has been offset somewhat by the development of tight oil (i.e. crude oil in tight formations — low-permeability formations in the United States).

Furthermore, the combination of biomass gasification and Fischer–Tropsch synthesis (Chapter 10) is a very promising route to produce transportation fuels from renewable or green resources in the refinery of the future (Chapter 12).

Often a higher concentration of some types of hydrocarbon derivatives is wanted, which might be achieved by changed reaction conditions. Nevertheless, the product range is wide and infected with uncertainties, due to the lack of knowledge of the details of the process and of the kinetics of the reaction. Since the different products have quite different characteristics such as boiling point, physical state at ambient temperature and thereby different use and ways of distribution, often only a few of the carbon chains is wanted. As an example, the LTFT process is used when longer carbon chains are wanted, because lower temperature increases the proportion of longer chains. But too low temperature is not wanted because of reduced activity. When the desired products are shorter carbon chains, such as crude oil, the longer ones might be cracked into shorter chains.

The yield of diesel is therefore highly dependent on the chain growth probability, which again is dependent on pressure, temperature, feed gas composition, catalyst type, catalyst composition, and reactor design.

10.4 Reactors and Catalysts

Since its discovery, the Fischer–Tropsch synthesis has undergone periods of rapid development and periods of inaction. Within ten years of the discovery, German companies were building commercial plants. The construction of these plants stopped about 1940 but existing plants continued to operate during World War II.

10.4.1 *Reactors*

Currently, three reactor types are used commercially in the Fischer–Tropsch process: a fixed-bed reactor, a fluid-bed reactor, and a slurry-bed reactor. The fixed-bed reactors usually run at lower temperatures to avoid carbon deposition on the reactor tubes. Products from fixed-bed reactors are characterized by low olefin content, and they are generally heavier than products from fluid beds. Heat distribution in fluid beds, however, is better than fixed-bed reactors, and fluid beds are generally operated at higher temperatures. Products are characterized by (i) having more olefin derivatives, (ii) a high proportion of low-boiling hydrocarbon derivatives (gases), and (iii) lower molecular weight product slate than from fixed-bed types.

Originally, the Fischer–Tropsch synthesis was carried out in packed-bed reactors. However, gas-agitated multiphase reactors, also called "slurry reactors" or "slurry bubble columns," gained favor, because the circulation of the slurry makes it much easier to control the reaction temperature in a slurry-bed reactor than in a fixed-bed reactor. Gas-agitated multiphase reactors operate by suspending catalytic particles in a liquid and feeding the gas reactants into the bottom of the reactor through a gas distributor, which produces small gas bubbles. As the gas bubbles rise upward within the reactor, the reactants are absorbed into the liquid and diffuse toward the catalyst particles where, depending on the catalyst system used, they are typically converted into gaseous and liquid products.

A slurry-bed reactor is characterized by having the catalyst in the form of a slurry. The feed gas mixture is bubbled through the catalyst suspension. Temperature control is easier than in the other two reactor types. An added advantage of the slurry-bed reactor is that it can accept a synthesis gas with a lower hydrogen/carbon monoxide ratio than that in either fixed-bed or fluid-bed reactors.

In the slurry-phase reactor, preheated synthesis gas is fed into the bottom of the reactor, where it is distributed into the slurry consisting of liquid hydrocarbon and catalyst particles. As the gas bubbles rise upward within the slurry, they diffuse into the slurry and are converted into a range of hydrocarbon derivatives by the Fischer–Tropsch reaction. The heat generated from this reaction is removed through the cooling coils of the reactor, which generates steam. The heavier (wax) fraction is separated from the slurry containing the catalyst particles. The lower boiling fraction is extracted from a gas stream from the top of the reactor. The gas stream is cooled to recover the lighter hydrocarbon derivatives and water. The intermediate hydrocarbon streams are sent to the product upgrading unit, while the water stream is treated in a water recovery unit. The third step upgrades reactor products to diesel and naphtha. The reactor products are mainly paraffin derivatives, but the lighter products contain some olefin derivatives, and oxygenated derivatives that need to be removed for product stabilization. Hydrogen is added to hydrotreat (saturate) the olefin derivatives and oxygenated derivatives, thereby converting these derivatives to paraffin derivatives. Hydrogen is also added to the mild hydrocracker, which breaks the long-chain hydrocarbon derivatives into naphtha and kerosene.

As the process evolved, other types of reactors have been used and include: (i) the parallel plate reactors, (ii) a variety of fixed-bed tubular reactors, and (iii) gas-agitated multiphase reactors. For the parallel-plate type of reactor, the catalyst bed is located in tubes fixed between the plates, which were cooled by steam/water that passed around the tubes within the catalyst bed. In another version, the reactor may be regarded as finned-tube in which large fins are penetrated by a large number of parallel or connected catalyst-filled tubes. Various designs were utilized for the tubular fixed-bed reactor with the concentrically paced tubes being the preferred one. This type of reactor contained catalyst in the area between the two tubes with cooling water-steam flowing through the inner tube and on the exterior of the outer tube. The gaseous products formed enter the gas bubbles and are collected at the top of the reactor. Liquid products are recovered from the suspending liquid using different techniques, including filtration, settling, and hydrocyclones.

Because the Fischer–Tropsch reaction is exothermic, temperature control is an important aspect of Fischer–Tropsch reactor operation. Gas-agitated multiphase reactors or slurry bubble column reactors have very high heat transfer rates and therefore allow for a good thermal control of the reaction. On the other hand, because the desired liquid products are mixed in the suspending liquid, recovery of the liquid products can be relatively difficult. This difficulty is compounded by the tendency of the catalyst particles to erode in the slurry, forming fine catalyst particles that are also relatively difficult to separate from the liquid products. Fixed-bed reactors generally avoid the issues that arise from liquid separation and catalyst separation but they (the fixed-bed reactors) do not provide the mixing of phases that allows good thermal control in slurry bubble column reactors.

Furthermore, Fischer–Tropsch reactors are typically sized to achieve a desired volume of production. When a fixed-bed reactor is planned, economies of scale tend to result in the use of long (tall) reactors. Because the Fischer–Tropsch reaction is exothermic, however, a thermal gradient tends to form along the length of the reactor, with the temperature increasing with the distance from the reactor inlet. In addition, for most Fischer–Tropsch catalyst systems, every ten degree C rise in temperature increases the reaction rate approximately 60%, which in turn results in the generation of still more heat. To absorb the heat generated by the reaction and offset the rise in temperature, a cooling liquid is typically circulated through the reactor.

By shifting the equilibrium in the steam reforming process through removal of hydrogen from the reaction zone, membrane reactors can also be used to increase the equilibrium-limited methane conversion. Using palladium–silver (Pd–Ag) alloy membrane reactors, methane conversion can reach close to 100% (Shu *et al.*, 1995).

10.4.2 *Catalysts*

Of great importance to the Fischer–Tropsch process is the catalyst. The catalysts used in the latest generation of Fischer–Tropsch technologies are cobalt based, usually carried on alumina supports, sometimes with precious metal promoters. Coal-based processes, including a gasification step, use iron catalysts, which are better suited to high-temperature processes based on feedstocks containing impurities. But iron produces significant quantities of non-paraffin derivatives as by-products, while cobalt catalysts feature high selectivity and are more efficient for making paraffin derivatives from cleaner feedstocks.

Catalysts play a pivotal role in synthesis gas conversion reactions. In fact, fuel and chemical synthesis from synthesis gas does not occur in the absence of appropriate catalysts. The basic concept of a catalytic reaction is that reactants adsorb onto the catalyst surface and rearrange and combine into products that desorb from the surface. One of the fundamental functional differences between synthesis gas synthesis catalysts is whether or not the adsorbed carbon monoxide molecule dissociates on the catalyst surface. For the Fischer–Tropsch process and higher alcohol synthesis, carbon monoxide dissociation is a necessary reaction condition. For methanol synthesis, the carbon–oxygen bond remains intact. Hydrogen has two roles in catalytic synthesis gas synthesis reactions. In addition to serving as a reactant needed for carbon monoxide hydrogenation, it is commonly used to reduce the metalized synthesis catalysts and activate the metal surface.

Generally, the Fischer–Tropsch synthesis is catalyzed by a variety of transition metals such as iron, nickel, and cobalt. Iron-based catalysts are relatively low cost, have a higher water–gas shift activity and are therefore more suitable for a lower hydrogen/carbon monoxide ratio (H_2/CO) synthesis gas such as those derived from coal gasification. On the other hand, nickel-based catalysts tend to promote methane formation, as in a methanation process. Cobalt-based catalysts are more active and are generally preferred over ruthenium (Ru) and, in comparison to iron, Co has much less water–gas shift activity.

Thus, in many cases, an iron-containing catalyst is the preferred catalyst due to the higher activity, but a nickel-containing catalyst produces large amounts of methane while a cobalt-containing catalyst has a lower reaction rate and a lower selectivity than the iron-containing catalyst. While comparing cobalt and iron catalysts, it was found that cobalt promotes more middle-distillate products. In the Fischer–Tropsch process, a cobalt-containing catalyst produces hydrocarbon derivatives plus water while iron catalyst produces hydrocarbon derivatives and carbon dioxide. It appears that the iron catalyst promotes the shift reaction more than the cobalt catalyst.

Various metals, including but not limited to iron, cobalt, nickel, and ruthenium, alone and in conjunction with other metals, can serve as Fischer–Tropsch catalysts. Cobalt is particularly useful as a catalyst for converting natural gas to heavy hydrocarbon derivatives suitable for the production of diesel fuel. Iron has the advantage of being readily available and relatively inexpensive but it also has the disadvantage of greater water–gas shift activity. Ruthenium is highly active but quite expensive. Consequently, although ruthenium is not the economically preferred catalyst for commercial Fischer–Tropsch production, it is often used in low concentrations as a promoter with one of the other catalytic metals.

The most common catalysts are the transition metals cobalt, iron, and ruthenium. Nickel can also be used, but tends to favor methane formation (methanation). Cobalt seems to be the most active catalyst, although iron may be more suitable for low-hydrogen-content synthesis gases such as those derived from coal due to its promotion of the water–gas shift reaction. In addition to the active metal, the catalysts typically contain a number of promoters, including potassium and copper. Catalysts are supported on high-surface-area binders/supports such as silica (SiO_2), alumina (Al_2O_3), or the more complex zeolites. Cobalt catalysts are more active for Fischer–Tropsch synthesis when the feedstock is natural gas. Natural gas has a high hydrogen-to-carbon ratio, so the water–gas shift is not needed for cobalt catalysts. Iron catalysts are preferred for lower quality feedstocks such as crude oil residua, coal, or biomass.

Unlike the other metals used for this process (Co, Ni, and Ru), which remain in the metallic state during synthesis, iron catalysts tend to form a number of chemical phases, including various oxides and carbides during the reaction. Control of these phase transformations can be important in maintaining catalytic activity and preventing breakdown of the catalyst particles. For the synthesis of higher molecular weight alcohols, dissociation of carbon monoxide is a necessary reaction condition. For methanol synthesis, the carbon monoxide molecule remains intact. Hydrogen has two roles in catalytic synthesis gas synthesis reactions. In addition to serving as a reactant needed for the hydrogenation of carbon monoxide, it is commonly used to reduce metalized synthesis catalysts and activate the metal surface.

The (Group 1) alkali metals (including potassium) are poisons to cobalt catalysts but they are promoters of iron catalysts. Catalysts are supported on high-surface-area binders/supports such as silica, alumina, and zeolites. Cobalt catalysts are more active for Fischer–Tropsch synthesis when the feedstock is natural gas. Iron catalysts are preferred for lower quality feedstocks such as coal or biomass.

Although there are differences in the product distributions of cobalt-containing and iron-containing catalysts at similar temperatures and pressures (for example, at 240°C, 465°F, and 450 psi, a cobalt-containing catalyst has a somewhat higher selectivity for heavier hydrocarbon derivatives than that of an iron-containing catalyst), the product distribution is primarily driven by the choice of the operating temperature: a high temperature results in a naphtha/kerosene ratio of 2:1, while a low temperature results in a naphtha/kerosene ratio of the order of 1:2 more or less no matter whether the catalyst is an iron-containing catalyst or a cobalt-containing catalyst. Higher temperatures shift the selectivity toward fewer carbon number products and more hydrogenated products; the degree of branching increases and the number of secondary products such as ketones and aromatic derivatives also increases.

Fischer–Tropsch catalysts are sensitive to poisoning by sulfur-containing compounds. The sensitivity of the catalyst to sulfur is greater for cobalt-based catalysts than for their iron counterparts. Promoters also have an important influence on activity. Alkali metal oxides and copper are common promoters, but the formulation depends on the primary metal: iron or cobalt. Alkali oxides prepared on cobalt catalysts generally cause activity to drop severely even at very low alkali loadings. C_5+ and carbon dioxide selectivity increases, while methane and C_2–C_4 selectivity decrease. In addition, the olefin-to-paraffin ratio increases.

Fischer–Tropsch catalysts can lose activity as a result of the following: (i) conversion of the active metal site to an inactive oxide site, (ii) sintering, (iii) loss of active area by carbon deposition, and (iv) chemical poisoning. Some of these mechanisms are unavoidable and others can be prevented or minimized by controlling the impurity levels in the synthesis gas. By far, the most abundant, important, and most intensively studied catalyst poison for the Fischer–Tropsch process is sulfur. Other catalyst poisons include halides and nitrogen compounds (e.g. NH_3, NO_x and HCN).

The hydrocarbon derivatives formed are mainly aliphatic, and on a molar basis, methane is the most abundant; the amount of higher hydrocarbon derivatives usually decreases gradually with the increase in molecular weight. *Iso*-paraffin formation is more extensive over zinc oxide (ZnO) or thoria (ThO_2) at 400–500°C (750–930°F) and at higher pressure. Paraffin waxes are formed over ruthenium catalysts at relatively low temperatures (170–200°C, 340–390°F), at high pressures (1500 psi), and with a carbon monoxide/hydrogen ratio. The more highly branched product prepared over the iron catalyst is an important factor in the choice for the manufacture of automotive fuels. On the other hand, a high-quality diesel fuel (paraffin character) can be prepared over cobalt.

Secondary reactions play an important role in determining the final structure of the product. The olefin derivatives produced are subjected to both hydrogenation and double-bond shifting toward the center of the molecule; *cis* and *trans* isomers are formed in approximately equal amounts. The proportion of straight-chain molecules decreases with a rise in molecular weight, but even so they are more abundant than branched-chain compounds up to approximately C_{10}.

The small amount of aromatic hydrocarbon derivatives found in the product covers a wide range of isomer possibilities. In the C_6–C_9 range, benzene, toluene, ethylbenzene, xylene, *n*-propyl- and *iso*-propylbenzene, methyl ethyl benzene derivatives, and trimethylbenzene derivatives have been identified; naphthalene derivatives and anthracene derivatives are also present.

Alternatively, a methanol-to-olefin (MTO) option is available (Tian *et al.*, 2015). The methanol-to-olefin derivative reaction is one of the most important reactions in C_1 chemistry, which provides a viable option for producing basic petrochemicals from non-oil resources such as coal and natural gas. As olefin-based petrochemicals and relevant downstream processes have been well developed for many years, the MTO reaction provides a link between gasification chemistry and the modern petrochemical industry. In the process, olefin derivatives are produced from methanol using a zeolite catalyst. The methanol feedstocks are vaporized, mixed with recovered methanol, superheated and sent to the fluidized-bed reactor. In the reactor, methanol is first converted to a dimethyl ether (DME; CH_3OCH_3) intermediate and then converted to olefin derivatives with a very high selectivity for ethylene and propylene.

Dimethyl ether can be produced by any one of several routes but the most common route is using methanol produced from synthesis gas. In the process, water-free methanol is vaporized and sent to a reactor with an inlet temperature of the order of 220–250°C (430–480°F), and an outlet temperature of the order of 300–350°C (570–660°F). Thus,

$$2CH_3OH \rightarrow CH_3OCH_3 + H_2O.$$

The reactor effluents are sent to a distillation column where the dimethyl ether is separated from the top and condensed, after which the dimethyl ether is sent to storage. Water and methanol are discharged from the bottom and fed into a methanol column for methanol recovery. The purified methanol from this column is recycled to the reactor after mixing with feedstock methanol.

Generally, a low temperature favors the yield of high molecular-weight linear wax derivatives while a high temperature favors the production of naphtha and low molecular-weight olefin derivatives. In order to maximize the production of the naphtha fraction, it is best to use an iron-containing catalyst at a high temperature in a fixed fluid-bed reactor. On the other hand, in order to maximize the production of the kerosene fraction, a slurry reactor with a cobalt-containing catalyst is the more appropriate choice.

10.5 Products and Product Quality

The composition of products from the synthesis gas production processes is varied insofar as the gas composition varies with the type of feedstock and the gasification system employed (Speight, 2013, 2014a, 2014b; Luque and Speight, 2015). Furthermore, the quality of gaseous product(s) must be improved by removal of any pollutants such as particulate matter and sulfur compounds before further use (Chapter 4), particularly when the intended use is a water gas shift or methanation reaction.

10.5.1 *Products*

Low-Btu gas (low heat-content gas) is the product when the oxygen is not separated from the air and, as a result, the gas product invariably has a low heat content (150–300 Btu/ft^3). In *medium-Btu gas* (medium heat-content gas), the heating value is in the range of 300–550 Btu/ft^3 and the composition is much like that of low heat-content gas, except that there is virtually no nitrogen and the H_2/CO ratio varies from 2:3 to approximately 3:1 and the increased heating value correlates with higher methane and hydrogen contents as well as with lower carbon dioxide content. *High-Btu gas* (high heat-content gas) is essentially pure methane and is often referred to as synthetic natural gas or substitute natural gas. However, to qualify as a substitute natural gas, a product must contain at least 95% methane; the energy content of synthetic natural gas is 980–1080 Btu/ft^3. The commonly accepted approach to the synthesis of high heat-content gas is the catalytic reaction of hydrogen and carbon monoxide.

Hydrogen is also produced during gasification of carbonaceous feedstocks. Although several gasifier types exist (Chapter 2), entrained-flow gasifiers are considered most appropriate for producing both hydrogen and electricity from coal since they operate at temperatures high enough (approximately 1500°C, 2730°F) to enable high carbon conversion and prevent downstream fouling from tars and other residuals.

There is also a series of products that are called by older (even archaic) names that evolved from older coal gasification technologies and warrant mention: (i) producer gas, (ii) water gas, (iii) town gas, and (iv) synthetic natural gas. These products are typically low-to-medium-Btu gases that require a high degree of purification before use (Speight, 2013a, 2013b).

10.5.2 *Product quality*

Gas cleaning (also known as gas processing), although generally simple in chemical and/or physical principles, is often confusing because of the frequent changes in terminology and, often, the lack of cross-referencing

(Mokhatab *et al.*, 2006; Speight, 2008, 2013a, 2014a, 2019a) (Chapter 9). Although gas processing employs different process types, there is always overlap between the various concepts. And, with the variety of possible constituents and process operating conditions, a universal purification system cannot be specified for economic application in all cases.

The processes that have been developed for gas cleaning range from a simple once-through wash operation to complex multi-step systems with options for the recycle of the gases (Mokhatab *et al.*, 2006; Speight, 2008, 2014a, 2019a). In some cases, process complexities arise because of the need for recovery of the materials used to remove the contaminants or even recovery of the contaminants in the original, or altered, form.

In more general terms, gas cleaning is divided into removal of particulate impurities and removal of gaseous impurities. For the purposes of this chapter, the latter operation includes the removal of hydrogen sulfide, carbon dioxide, sulfur dioxide, and products that are not related to synthesis gas and hydrogen production. However, there is also a need for the subdivision of these two categories as dictated by needs and process capabilities: (i) coarse cleaning whereby substantial amounts of unwanted impurities are removed in the simplest, most convenient, manner, (ii) fine cleaning for the removal of residual impurities to a degree sufficient for the majority of normal chemical plant operations, such as catalysis or preparation of normal commercial products; or cleaning to a degree sufficient to discharge an effluent gas to the atmosphere through a chimney, (iii) ultra-fine cleaning where the extra step (as well as the extra expense) is justified by the nature of the subsequent operations or the need to produce a particularly pure product.

The products can range from (i) high-purity hydrogen, (ii) high-purity carbon monoxide, (iii) high-purity carbon dioxide, and (iv) a range of hydrogen/carbon monoxide mixtures. The plant is often times referred to as a HYCO if it is designed to produce both carbon monoxide and hydrogen at high purity, else it is referred to as a synthesis gas or a synthesis gas plant. In fact, the hydrogen/carbon monoxide ratio can be selected at will and the process scheme is chosen, in part, by the product composition required. The hydrogen/carbon monoxide ratio will likely vary between 1 and 3 for HYCO and synthesis gas plants. However, at one end of the scale, i.e. if hydrogen is the desired product, then the ratio can approach infinity by shifting all of the carbon monoxide to carbon dioxide. By contrast, on the other end, the ratio cannot be adjusted to zero because hydrogen and water are always produced. An interesting general rule of thumb exists in terms of the hydrogen/carbon monoxide ratio produced by the different gasification processes.

Gasification process	H_2/CO ratio
Steam methane reforming (SMR)	3.0–5.0
SMR + oxygen secondary reforming (O_2R)	2.5–4.0
Autothermal reforming (ATR)	1.6–2.65
Partial oxidation (POx)	1.6–1.9

It should be noted that in practice, however, the options are not limited to the ranges shown but extend to even greater hydrogen/carbon monoxide ratios, if adjustments are made, like the inclusion of a Shift converter to effect near-equilibrium water–gas shift conversion or adjusting the amount of steam.

Throughout the previous section, there has, of necessity, been frequent reference made to the production of hydrogen as an integral part of the production of carbon monoxide, since the two gases make up the mixture known as synthesis gas. Hydrogen is indeed an important commodity in the refining industry because of its use in hydrotreating processes, such as desulfurization, and in hydroconversion processes, such as hydrocracking. Part of the hydrogen is produced during reforming processes but that source, formerly sufficient, is now insufficient for the hydrogen needs of a modern refinery (Parkash, 2003; Gary *et al.*, 2007; Speight, 2011, 2014a; Hsu and Robinson, 2017; Speight, 2017).

In addition, optimum hydrogen purity at the reactor inlet extends catalyst life by maintaining desulphurization kinetics at lower operating temperatures and reducing carbon laydown. The typical purity increases resulting from hydrogen purification equipment and/or increased hydrogen sulfide removal as well as tuning hydrogen circulation and purge rates may extend catalyst life by as much as 25%. Indeed, since hydrogen use has become more widespread in refineries, hydrogen production has moved from the status of a high-tech specialty operation to an integral feature of most refineries (Raissi, 2001; Vauk *et al.*, 2008).

While the use of gasification of residua and coke to produce hydrogen and/or power may increase in refineries over the next two decades (Speight, 2011b), several other processes are available for the production of the additional hydrogen that is necessary for the various heavy feedstock hydroprocessing sequences (Speight, 2014a) and it is the aim of this section to present a general description of these processes.

Purities in excess of 99.5% of either the hydrogen or carbon monoxide produced from synthesis gas can be achieved if desired. Four of the major process technologies available are as follows: (i) cryogenics plus methanation, (ii) cryogenics plus pressure swing adsorption, (iii) methane-wash cryogenic process, and (iv) the Cosorb process.

(i) **Cryogenics plus methanation:** This process utilizes a cryogenic process (occurring in a cold box) whereby carbon monoxide is liquefied in a number of steps until hydrogen with a purity of ~98% is produced. The condensed carbon monoxide product, which would contain methane, is then distilled to produce essentially pure carbon monoxide and a mixture of carbon monoxide and methane. The hydrogen stream from the cold box is taken to a shift converter where the remaining carbon monoxide is converted to carbon dioxide and hydrogen. The carbon dioxide is then removed and any further carbon monoxide or carbon dioxide can be removed by methanation. The resulting hydrogen stream can be of a purity of ~99.7%.

(ii) **Cryogenics plus pressure swing adsorption (PSA):** This process utilizes the similar sequential liquefaction of carbon monoxide in a cold box until hydrogen of ~98% purity is achieved. Again, the carbon monoxide stream can be further distilled to remove methane until it is essentially pure. The hydrogen stream is then allowed to go through multiple swings of PSA cycles until the hydrogen purity is even as high as 99.999%.

(iii) **Methane-wash cryogenic process:** In this scheme, liquid carbon monoxide is absorbed into a liquid methane stream so that the hydrogen stream produced contains only ppm levels of carbon monoxide but higher amounts (approximately 5–8% v/v) of methane. Hence a hydrogen stream purity of the order of 95% v/v is possible. The liquid carbon monoxide/methane stream, however, can be distilled to produce an essentially pure carbon monoxide stream and a carbon monoxide/methane stream that can be used as a fuel.

(iv) **Cosorb process:** This process utilizes copper ions (cuprous aluminum chloride, $CuAlCl_4$) in toluene to form a chemical complex with the carbon monoxide, in effect separating it from the hydrogen, nitrogen, carbon dioxide, and methane. This process can capture approximately 96% v/v of the carbon monoxide to produce a stream of greater than 99% v/v purity. The downside of this process is that water, hydrogen sulfide and other trace chemicals which can poison the copper catalyst must be removed prior to entering the reactor. Furthermore, a hydrogen stream of up to 97% purity is obtained. However, while the efficiency of cryogenic separation decreases with low carbon monoxide content of the feed, the Cosorb process is able to process gases with a low carbon monoxide content more efficiently.

10.5.3 *Refining Fischer–Tropsch products*

The Fischer–Tropsch product stream typically contains hydrocarbon derivatives having a range of numbers of carbon atoms, including gases, liquids, and waxes. Depending on the molecular weight product distribution, different Fischer–Tropsch product mixtures are ideally suited to different uses. For example, Fischer–Tropsch product mixtures containing liquids may be processed to yield gasoline as well as heavier middle distillates. Hydrocarbon waxes may be subjected to additional processing steps for conversion to liquid and/or gaseous hydrocarbon derivatives. Thus, in the production of a Fischer–Tropsch product stream for processing into a fuel, it is desirable to obtain primarily hydrocarbon derivatives that are liquids and waxes (e.g. C_5+ hydrocarbon derivatives). Initially, the light gases in the raw product are separated and sent to a gas cleaning operation. The higher boiling product is distilled to produce separate streams of naphtha, distillate, and wax.

The naphtha stream is first hydrotreated, which produces a hydrogen-saturated liquid product (primarily paraffin derivatives), a portion of which is converted, by isomerization, from normal paraffin derivatives into *iso*-paraffin derivatives to boost its octane value. Another fraction of the hydrotreated naphtha is catalytically reformed to provide some aromatic content to (and further boost the octane value of) the final gasoline blend stock. The distillate stream is also hydrotreated, resulting directly in a finished diesel blend stock. The wax fraction is hydrocracked into a finished distillate stream and naphtha streams that augment the hydrotreated naphtha streams sent for isomerization and for catalytic cracking.

Thus, conventional refinery processes (Parkash, 2003; Gary *et al.*, 2007; Speight, 2011, 2014a, 2017; Hsu and Robinson, 2017) can be used for upgrading of Fischer–Tropsch liquid and wax products. Examples of possible processes for Fischer–Tropsch products are wax hydrocracking, distillate hydrotreating, catalytic reforming, naphtha hydrotreating, alkylation, and isomerization. Fuels produced using the Fischer–Tropsch synthesis are of a high quality due to very low aromaticity and zero sulfur content.

The diesel fraction has a high cetane number resulting in superior combustion properties and reduced emissions. New and stringent regulations may promote replacement or blending of conventional fuels by sulfur and aromatic-free products. Also, other products besides fuels can be manufactured with Fischer–Tropsch in combination with upgrading processes, for example, ethylene, propylene, α-olefin derivatives, alcohols, ketones, solvents, and waxes. These valuable by-products of the process have higher added values, resulting in an economically more attractive process economy (Parkash, 2003; Gary *et al.*, 2007; Speight, 2011, 2014a, 2017; Hsu and Robinson, 2017).

At this point, it is necessary to deal once again with the production of chemicals from the carbonaceous feedstock by gasification followed by conversion of the synthesis gas mixture (carbon monoxide, carbon monoxide, and hydrogen, H_2) to higher molecular weight liquid fuels and other chemicals (Penner, 1987; Speight, 2014a, 2019b).

The production of synthesis gas involves reaction of the carbonaceous feedstock with steam and oxygen.

In principle, synthesis gas can be produced from any hydrocarbon (or hydrocarbonaceous) feedstock. These include natural gas, naphtha, residual oil, crude oil coke, coal, and biomass. The lowest cost routes for synthesis gas production, however, are based on natural gas. The cheapest option is remote or stranded reserves.

The choice of technology for synthesis gas production also depends on the scale of the synthesis operation. Synthesis gas production from solid fuels can require an even greater capital investment with the addition of feedstock handling and more complex synthesis gas purification operations. The greatest impact on improving GTL plant economics is to decrease the capital costs associated with synthesis gas production and improve thermal efficiency through better heat integration and utilization. Improved thermal efficiency can be obtained by combining the GTL plant with a power generation plant to take advantage of the availability of low-pressure steam.

The synthesis of hydrocarbon derivatives from carbon monoxide and hydrogen (synthesis gas) (the Fischer–Tropsch synthesis) is a procedure for the indirect liquefaction of coal (Storch *et al.*, 1951; Batchelder, 1962). This process is the only coal liquefaction scheme currently in use on a relatively large commercial scale; South Africa is currently using the Fischer–Tropsch process on a commercial scale in their SASOL complex (Singh, 1981), although Germany produced roughly 156 million barrels of synthetic crude oil annually using the Fischer–Tropsch process during World War II.

Briefly, in the gasification process, the carbonaceous feedstock is converted into gaseous products at temperatures in excess of 800°C (1470°F) and at moderate pressures to produce synthesis gas:

$$C + H_2O \rightarrow CO + H_2$$

The gasification may be attained by means of any one of the several processes (Speight, 2013a; 2013b, 2014a, 2014b; Luque and Speight, 2015). The exothermic nature of the process and the decrease in the total gas volume in going from reactants to products suggest the most suitable experimental conditions to use in order to maximize product yields. The process should be favored by high pressure and relatively low reaction temperature.

In practice, the Fischer–Tropsch reaction is generally carried out at temperatures in the range 200–350°C (390–660°F) and at pressures in the range of 75–4000 psi; the hydrogen/carbon monoxide ratio is usually ca. 2.2:1 or 2.5:1. Since up to three volumes of hydrogen may be required to achieve the next stage of liquid production, the synthesis gas must then be converted (by means of the water–gas-shift reaction) to the desired level of hydrogen after which the gaseous mix is purified (acid gas removal, etc.) and converted to a wide variety of hydrocarbon derivatives.

$$CO + H_2O \rightarrow CO_2 + H_2,$$

$$CO + (2n{+}1)H_2 \rightarrow C_nH_{2n}{+}_2 + H_2O$$

These reactions result primarily in low- and medium-boiling aliphatic compounds; current commercial objectives are focused on the conditions that result in the production of *n*-hydrocarbon derivatives as well as olefin derivatives and oxygenated materials.

References

Batchelder, H.R. 1962. In *Advances in Petroleum Chemistry and Refining*, J.J. McKetta Jr. (Editor). Interscience Publishers Inc., New York, Volume V, Chapter 1.

Bayat, M., Rahmani, F., Mazinani, S., and Rahimpour, M.R. 2010. Enhancement of gasoline production in a novel optimized hydrogen-permselective membrane Fischer–Tropsch reactor in GTL technology. *Chem. Eng. Trans.* 21: 1477–1482.

Chadeesingh, R. 2011. The Fischer–Tropsch Process. In *The Biofuels Handbook*, J.G. Speight (Editor). The Royal Society of Chemistry, London, United Kingdom, Part 3, Chapter 5, pp. 476–517.

Fischer, F. and Tropsch, H.U. 1923. Űber die Herstellung synthetischer Őlgemische (syn-thol) durch Aufbau aus Kohlenoxyd und Wasserstoff. *Brennstoff Chemie* 4: 276–285.

Gary, J.G., Handwerk, G.E., and Kaiser, M.J. 2007. *Petroleum Refining: Technology and Economics*, 5th Edition. CRC Press, Taylor & Francis Group, Boca Raton, Florida.

Hagh, B.F. 2004. Comparison of Autothermal Reforming for Hydrocarbon Fuels. Preprints. *Div. Fuel Chem., Am. Chem. Soc.* 49 (1): 144–147.

Higman, C. and Van der Burgt, M. 2008. *Gasification*, 2nd Edition, Elsevier, Amsterdam, Netherlands.

Hsu, C.S. and Robinson, P.R. (Editors) 2017. *Handbook of Petroleum Technology*. Springer International Publishing AG, Cham, Switzerland.

Hu, J., Fei Yu, F., and Lu, Y. 2012. Application of Fischer–Tropsch Synthesis in Biomass to Liquid Conversion. *Catalysts* 2: 303–326.

Klerk, A. and Furimsky, E. 2010. *Catalysis in the Refining of Fischer–Tropsch Syncrude*, RSC Catalysis Series, No. 4. Royal Society of Chemistry, London, United Kingdom, pp. 1–294.

Liu, K. Song, C., and Subramani, V. (Editors) 2010. *Hydrogen and Syngas Production and Purification Technologies*. John Wiley & Sons Inc., Hoboken, New Jersey.

Luque, R. and Speight, J.G. (Editors) 2015. *Gasification for Synthetic Fuel Production: Fundamentals, Processes, and Applications*. Woodhead Publishing, Elsevier, Cambridge, United Kingdom.

Mokhatab, S., Poe, W.A., and Speight, J.G. 2006. *Handbook of Natural Gas Transmission and Processing*. Elsevier, Amsterdam, Netherlands.

Penner, S.S. 1987. *Proceedings of the Fourth Annual Pittsburgh Coal Conference*. University of Pittsburgh, Pittsburgh, Pennsylvania, p. 493.

Rauch, R. 2001. Biomass Gasification to produce Synthesis Gas for Fuel Cells, Liquid Fuels and Chemicals, IEA Bioenergy Agreement, Task 33: Thermal Gasification of Biomass.

Rostrup-Nielsen, J. and Christiansen, L.J. 2011. *Concepts in Syngas Manufacture*, Catalysis Science Series No. 10. Imperial College Press, London, United Kingdom.

Schulz, H. 1999. Short history and present trends of Fischer–Tropsch synthesis. *Applied Catalysis A: General* 86(1–2): 3–12.

Shu, J., Grandjean, B.P.A., and Kaliaguine, S. 1995. *Catalysis Today* 25: 327–332.

Speight, J.G. 2008. *Synthetic Fuels Handbook: Properties, Processes, and Performance*. McGraw-Hill, New York.

Speight, J.G. 2011. *The Refinery of the Future*. Gulf Professional Publishing, Elsevier, Oxford, United Kingdom.

Speight, J.G. 2013a. *The Chemistry and Technology of Coal*, 3rd Edition. CRC Press, Taylor & Francis Group, Boca Raton, Florida.

Speight, J.G. 2013b. *Coal-Fired Power Generation Handbook*. Scrivener Publishing, Beverly, Massachusetts.

Speight, J.G. 2014a. *The Chemistry and Technology of Petroleum*, 5th Edition. CRC Press, Taylor & Francis Group, Boca Raton, Florida.

Speight, J.G. 2014b. *Gasification of Unconventional Feedstocks*. Gulf Professional Publishing, Elsevier, Oxford, United Kingdom.

Speight, J.G. 2017. *Handbook of Petroleum Refining Processes*. CRC Press, Taylor & Francis Group, Boca Raton, Florida.

Speight, J.G. 2019a. *Natural Gas: A Basic Handbook*, 2nd Edition. Gulf Publishing Company, Elsevier, Cambridge, Massachusetts.

Speight, J.G. 2019b. *Handbook of Petrochemical Processes*. CRC Press, Taylor & Francis Group, Boca Raton, Florida.

Storch, H.H., Golumbic, N., and Anderson, R.B. 1951. *The Fischer Tropsch and Related* Syntheses. John Wiley & Sons Inc., New York.

Tian, P., Wei, Y., Ye, M., and Liu, Z., 2015. Methanol to Olefins (MTO): From fundamentals to commercialization. *ACS Catalysts* 5(3): 1922–1938.

Van Berge, P.J. and Everson, R.C. 1997. Cobalt as an Alternative Fischer–Tropsch Catalyst to Iron for the Production of Middle Distillates. In *Stud. Surf. Sci. Catal.* 107 Natural Gas Conversion IV, M. de Pontes, R.L. Espinoza, C.P. Nicolaides, J.H. Scholtz, and M.S. Scurrell (Editors). Elsevier B.V., Amsterdam, Netherlands, pp. 207–212.

Part 3

Properties of Alternate Fuels and the Future

Chapter 11

Properties of Alternate Fuels

11.1 Introduction

Currently, ever since the oil crises of the 1970s and despite crude oil from tight formations, there is a growing concern about the future of energy supply and the depletion of natural gas and crude oil reserves, which can lead to a future shortage of these resources. For these reasons, alternative fuels are being used and developed to replace the traditional fuels. One application of these fuels is in spark-ignited gaseous-fueled internal combustion engines (ICEs). These alternative gaseous fuels are derived from a variety of feedstock such as biomass, waste products, and coal. Different feed stocks introduce variations in fuel composition, properties, and performance, the effects of which have not been fully investigated. One key property yet to be explored for these fuels is knock tendency.

Alternative fuels, also known as non-conventional fuels and advanced fuels, are any materials or substances that can be used as fuels, other than conventional fuels like the fossil fuels natural gas, crude oil, coal, extra heavy oil, tar sand bitumen, and oil shale. Some well-known alternative fuels include bio-diesel, bioalcohol (methanol, ethanol, propanol, and butanol), refuse-derived fuel, biohydrogen, vegetable oil, and other biomass sources. In order to be appropriate for the near future, the properties of various fuels (Table 11.1) will require modification to match the specifications by (i) blending with suitable materials, (ii) the choice of appropriate feedstocks, and/or (iii) engine modification.

Table 11.1: Illustration of the properties of different fuels.

Property	Gasoline	No. 2 diesel	Methanol	Ethanol	Hydrogen	Methane	Propane
Chemical formula	C4 to C12	C12 to C25	CH_3OH	C_2H_5OH	H_2	CH_4	C_3H_8
Molecular weight	100–105	200–300	32.04	46.07	2.02	16.04	44.1
Composition (w/w%)							
Carbon	85–88	84–87	37.5	52.2	0	75	82
Hydrogen	12–15	33–16	12.6	13.1	100	25	18
Oxygen	0	0	49.9	34.7	0	—	—
Specific gravity (60°F/60°F)	0.72–0.78	0.81–0.89	0.796	0.796	0.07	0.424	0.508
Density (lb/gal at 60°F)	6.0–6.5	6.7–7.4	6.63	6.61	—	1.07	4.22
Boiling point (°F)	80–43[a]	370–650[a]	149	172	–4,230	–259	–44
Reid vapor pressure, psi	8–15	0.2	4.6 (0)	2.3 (0)	—	2,400	208
Octane number							
Research octane number	90–100	—	107	108	130+	—	112
Motor octane number	81–90	—	92	92	—	—	97
Cetane number	5–20	40–55	—	—	—	—	—
Water solubility (at 70°F)							
Fuel in water (v/v%)	Negligible	Negligible	100	100	—	—	—
Water in fuel (v/v%)	Negligible	Negligible	100	100	—	—	—
Freezing point (°F)	–40	–40–30	–143.5	–173.2	–435	–296	–305.8
Viscosity							
Centipoise (at 60°F)	0.37–0.44	2.6–4.1	0.59	1.19		—	—
Flash point, closed cup (°F)	–45	165	52	55	—	–300	–100 to –150
Autoignition temperature (°F)	495	600	867	793	1,050–1,080	1,004	850–950
Flammability limits (v/v%)							
Lower	1.4	1	7.3	4.3	4.1	5.3	2.2
Higher	7.6	6	36	19	74	15	9.5
Latent heat of vaporization							
Btu/gal (at 60°F)	900	700	3,340	2,378	—	—	775
Btu/lb (at 60°F)	150	100	506	396	192.1	219	193.1

Note: [a]Boiling range.

However, the prefix *bio* often used to refer to alternative fuels from biosources may cause confusion. Methanol and bio-methanol are the same in terms of properties and behavior — it is only the source that makes a difference. There is no difference between the properties of methanol (from, say, non-biological sources) and biomethanol (from biological sources). In fact, to allay any possible confusion and erroneous thoughts about property differences, there may be a strong case for omitting the prefix *bio* in such cases. However, in some cases — such as bio-diesel and crude oil-based diesel — there are vast chemical differences in the constituents, properties, and behavior between these two diesel fuels. Bio-diesel is made from non-hydrocarbon starting materials such as animal fats or vegetable oils, which are renewable resources that come from plants such as atrophy, soybean, sunflower, corn, olive, peanut, palm, coconut, safflower, canola, sesame, and cottonseed by a process known as transesterification.

11.2 Gaseous Fuels

By definition, gaseous fuels are those fuels that are in the gaseous phase under ambient conditions (room temperature and pressure). They are produced by a variety of processes (Tables 11.2 and 11.3) (Speight, 2013, 2014) and are the most convenient fuels, requiring the least amount of handling and the simplest and most maintenance-free burner systems. Gas is delivered on tap via a distribution network and is therefore well suited to a high-population or industrial density area. However, large-scale consumers do use gas holders and some produce their own gas.

11.2.1 *Natural gas*

Natural gas (also called *marsh gas* and *swamp gas* in the older literature and more recently *landfill gas*) is a naturally occurring gaseous fossil fuel that is found in conjunction with crude oil in coal seams, as well as in reservoirs that contain natural gas without any crude oil.

Briefly, for the sake of clarity, natural gas is not the same as town gas, manufactured from coal, and coal gas, manufactured gas, producer gas, and synthetic natural gas (SNG), also produced from coal. Depending on the production process, synthesis gas produced from a carbonaceous feedstock is a mixture of hydrogen, carbon monoxide, carbon dioxide, methane, and volatile hydrocarbon derivatives in varying amounts as well as small amounts of nitrogen.

Prior to the development of resources, virtually all fuel and lighting gases were manufactured from coal. The by-product coal tar produced during the manufacture of gas from coal was an important feedstock for the chemical industry. The development of manufactured gas parallels the onset and evolution of the Industrial Revolution and urbanization.

Table 11.2: Nomenclature of relevant properties of the various gas products.[a]

Gas	Relevant property
Biogas	Gas produced from a biological source; also referred to as biogas
Blue water gas	Higher heat content (calorific value) gas produced from coal
Casinghead gas	Gas extracted from an oil well by separation at the surface.
Carbureted water gas	Higher heat content (calorific value) gas produced from coal
Coalbed methane	Natural gas from coal seams
Dry gas	High methane content
High-Btu gas	High-energy gas; also called high-heat-content gas
Landfill gas	Gas produced by the decay of organic matter in a landfill
Lean gas	High methane content
Low-Btu gas	Low-energy gas; also called low-heat-content gas
Medium-Btu gas	Medium-energy gas; also called medium-heat-content gas
Producer gas	Very low heat content (calorific value) gas produced from coal
Residue gas	Gas remaining after the condensing process; predominantly methane
Sour gas	High concentration of hydrogen sulfide (H_2S)
Sweet gas	Low concentration of hydrogen sulfide or no hydrogen sulfide is present
Synthesis gas	A mixture of carbon monoxide and hydrogen
Wet gas	High concentration of higher-boiling hydrocarbon derivatives (C_5–C_{10})
Wood gas	Gas produced from wood

Note: [a]Listed alphabetically.

11.2.1.1 *Composition*

Natural gas is a combustible mixture of hydrocarbon gases (Mokhatab *et al.*, 2006; Speight, 2018, 2019a). The principal constituents of natural gas, which are present in different amounts depending on the source, are methane (CH_4), ethane (C_2H_6), propane (C_3H_8), butanes (C_4H_{10}), pentanes (C_5H_{12}), hexane (C_6H_{14}), heptane (C_7H_{16}), sometimes trace amounts of octane (C_8H_{18}), and higher molecular weight hydrocarbon derivatives (Table 11.4). Some aromatics (BTX — namely benzene (C_6H_6), toluene ($C_6H_5CH_3$), and xylenes ($CH_3C_6H_4CH_3$)) can also be present, raising

Table 11.3: Gaseous fuels from different sources.

(1) Fuels naturally found in Nature:

Natural gas: (1) from crude oil reservoirs and (2) from natural gas reservoirs; the most volatile fraction from the distillation of oil, petroleum gas (mainly butane, propane)

Methane from coal mines (Firedamp)

Methane from the digestion of biomass by bacteria (landfill gas, digester gas)

(2) Fuel gases produced from solid fuel

- (a) Gases derived from coal
  - (i) Pyrolysis, carbonization: the product of the heating of a solid or liquid to devolatilize the substance to form a char, coke, or charcoal
  - (ii) Coke oven gas: liberated in the high-temperature carbonization (coking) of coal.
  - (iii) Gas from charcoal production
  - (iv) Gasification — The product of complete gasification of a solid or liquid feedstock — i.e. including the char. The only residue is ash.

    Producer gas: low calorific value (low heat content) gas from partial combustion of coal in air

    Water gas: medium calorific value (medium heat content) gas from gasification of coal with steam.

    Carbureted water gas: medium-to-high calorific value (medium-to-high heat content) gas.

    Blue gas or water gas: a higher calorific value gas produced by intermittently blasting the incandescent bed with air and steam such that the overall heat balance is maintained; often used as a synthesis gas in the chemical industry.
- (b) Gases derived from waste and biomass

(3) Gases derived from crude oil

Oil gas: also called refinery gas; formed by the thermal cracking of crude oil.

Liquefied petroleum gas (LPG)

Refinery gas: formed by the cracking of higher boiling fractions

Gas from heavy oil gasification or from resid gasification

(4) Gas from some fermentation process

safety issues due to their toxicity. The non-hydrocarbon gas portion of the natural gas contains nitrogen (N_2), carbon dioxide (CO_2), helium (He), hydrogen sulfide (H_2S), water vapor (H_2O), and other sulfur compounds, such as carbonyl sulfide (COS) and mercaptans (e.g. CH_3SH) and trace amounts of other gases. Carbon dioxide and hydrogen sulfide are

Table 11.4: Typical range of compositions of natural gas.

Name	Formula	v/v%
Methane	CH_4	85–95
Ethane	C_2H_6	3–8
Propane	C_3H_8	1–5
Butane	C_4H_{10}	1–2
Pentane[a]	C_5H_{12}	1–5
Carbon dioxide	CO_2	1–2
Hydrogen sulfide	H_2S	1–2
Nitrogen	N_2	1–5
Helium	He	<0.5

Note: [a]Pentane and higher molecular weight hydrocarbon derivatives up to octane as well as benzene (C_6H_6), toluene ($C_6H_5CH_3$), and in some cases the xylene isomers (1,2-, 1,3- and 1,4-$CH_3C_6H_4CH_3$).

commonly referred to as *acid gases* since they form corrosive compounds in the presence of water. Nitrogen, helium, and carbon dioxide are also referred to as *diluents* since none of these burn, and thus they have no heating value. Mercury can also be present, either as a metal in vapor phase or as an organo-metallic compound in liquid fractions. Concentration levels are generally very small, but even at these very small concentration levels, mercury can be detrimental due to its toxicity and its corrosive properties (reaction with aluminum alloys).

Several general definitions have so far been applied to natural gas depending on its composition. For example, *lean* gas is gas in which methane is the major constituent. On the other hand, w*et* gas contains considerable amounts of higher molecular weight hydrocarbon derivatives. To further define the terms *dry* and *wet* in quantitative measures, the term *dry* natural gas indicates that there is less than 0.1 gal (1 US gal = 264.2 m^3) of gasoline vapor (higher molecular weight paraffin derivatives) per 1,000 ft^3 (1 ft^3 = 0.028 m^3). The term *wet natural gas* indicates that there are such paraffin derivatives present in the gas, in fact, more than 0.1 gal/1,000 ft^3. Natural gas is considered to be *dry* when it is almost pure methane, having had most of the other commonly associated hydrocarbon derivatives removed. When other hydrocarbon derivatives are present, the natural gas is *wet*.

Sour gas contains hydrogen sulfide whereas *sweet* gas contains very little, if any, hydrogen sulfide. *Residue gas* is natural gas from which the higher molecular weight hydrocarbon derivatives have been extracted and *casing head gas* is derived from petroleum but is separated at the separation facility at the well-head.

However, in its purest form, the natural gas that is delivered to the consumer is almost pure methane and the remaining hydrocarbon derivatives and non-hydrocarbon derivatives have been removed through refining. The non-hydrocarbon constituents include, but are not limited to, carbon dioxide (CO_2), hydrogen sulfide (H_2S), nitrogen (N_2), and helium (He). Because natural gas is colorless, shapeless, odorless, and tasteless in its pure form, it is not possible to see or smell natural gas. Therefore, an odorant (e.g. mercaptan, also called a thiol, which is a sulfur-containing compound with the general formula R-SH) is added to natural gas for safety reasons so that it can be smelled if there is a gas leak. Mercaptan is a chemical odorant that smells a little like rotten eggs or skunk spray.

11.2.1.2 *Properties*

The properties of unrefined natural gas are variable because the composition of natural gas is never constant. Therefore, the properties and behavior of natural gas — often equated to the properties of methane (Table 11.5) — are best understood by investigating the properties and behavior of its constituents (Table 11.4) (Speight, 2019a). Thus, assuming that natural gas has been cleaned (insofar as any constituents such as carbon dioxide, hydrogen sulfide, and other contaminants have been removed and the only constituents remaining are hydrocarbon derivatives), the properties and behavior of natural gas become a study of the properties and behavior of the relevant hydrocarbon derivatives (Speight, 2019a).

The composition of natural gas varies depending on the field, the formation, or the reservoir from which it is extracted and these are an artifact

Table 11.5: Boiling point and density of methane relative to air and water.

- Boiling point (760 mmHg): –161.5°C (–258.7°F)
- Vapor density: 0.55 to 0.64 (air = 1.00)
- Gas Density (varies slightly): 0.0438 lb/scf
- Specific gravity of liquefied natural gas: 0.42 to 0.46 (water = 1.00)

of its formation. The different hydrocarbon derivatives that form natural gas can be separated using their different physical properties such as weight, boiling point, or vapor pressure. Depending on its content of higher molecular weight hydrocarbon components, natural gas can be considered to be rich (5 or 6 gal or more of recoverable hydrocarbon components per cubic feet) or lean (less than 1 gal of recoverable hydrocarbon components per cubic foot).

11.2.1.2.1 Density

Density, a physical property of matter, is a measure of the relative *heaviness* of hydrocarbon derivatives and other chemicals at a constant volume, and each constituent of natural gas has a unique density associated with it. Another term, *specific gravity* is commonly used in relation to the properties of hydrocarbon derivatives. The specific gravity of a substance is a comparison of its density to that of water. For most chemical compounds (i.e. those that are solid or liquid), the density is measured relative to that of water (1.00). For gases, the density is more likely to be compared to the density of air (also given the number 1.00 but this is arbitrary and bears no relationship to the density of water). As an example, the density of liquefied natural gas (LNG) is approximately 0.41–0.5 kg/L, depending on the temperature, pressure and composition; in comparison, the density of water is 1.0 kg/L. *Density* values (including those of natural gas hydrocarbon derivatives) are given at room temperature unless otherwise indicated by a superscript number — for example, 2.487^{15} indicates a density of 2.487 g/cm at 15°C (59°F).

The density of any gas compared to the density of air is the *vapor density*, which is a very important characteristic of the constituents of natural gas (Speight, 2019a). Furthermore, a statement is often made that "*Natural gas is lighter than air*." This statement often arises because of the continued insistence by engineers and scientists that the properties of a mixture are determined by the mathematical average of the properties of individual constituents of the mixture. Such mathematical bravado and inconsistency of thought are *detrimental to safety* and need to be qualified.

Relative to air, methane is less dense (Speight, 2019a) but the other hydrocarbon constituents of unrefined natural gas (i.e. ethane, propane, butane, etc.) are denser than air. Therefore, should a natural gas leak occur in field operations, especially where the natural gas contains constituents other than methane, only methane dissipates readily into the air and the higher molecular weight hydrocarbon constituents pose a considerable risk if the ambient conditions are such that they accumulate at the ground level.

11.2.1.2.2 Heat of combustion

All of the hydrocarbon constituents of natural gas are combustible, but non-flammable non-hydrocarbon components (carbon dioxide, hydrogen sulfide, nitrogen, and helium) detract slightly from the heating value of natural gas. The *heat of combustion* (*energy content*) of natural gas is the amount of energy that is obtained from the burning of a volume of natural gas and is measured in British thermal units (Btu). The value of natural gas is calculated by its Btu content. One Btu is the quantity of heat required to raise the temperature of one pound of water of 1°F at atmospheric pressure. A cubic foot of natural gas has an energy content of approximately 1,031 Btu, but the range of values is between 500 and 1,500 Btu, depending on the composition of the gas.

Thus, the energy content of natural gas is variable, because natural gas has variations in the amount and type of energy gases (methane, ethane, propane, and butane) it contains: the more non-combustible gases in the natural gas, the lower the energy (Btu). In addition, the volume mass of energy gases that are present in a natural gas accumulation also influences the Btu value of natural gas. The more carbon atoms in a hydrocarbon gas, the higher the Btu value (heat content) of the gas. It is necessary to conduct the Btu analysis of natural gas at each stage of the supply chain. Gas chromatographic process analyzers are used in order to conduct fractional analysis of the natural gas streams, separating natural gas into identifiable components. The components and their concentrations are converted into a gross heating value in Btu per cubic foot.

In the USA, at retail, natural gas is often sold in units of *therms* (th) (1 therm = 100,000 Btu. Wholesale transactions are generally done in *decatherms* (Dth) or thousand decatherms (MDth) or million decatherms (MMDth). A million decatherms is roughly one billion cubic feet of natural gas.

The gross heats of combustion of crude oil and its products are given with fair accuracy by the equation:

$$Q = 12{,}400 - 2100d^2.$$

In this equation, *d* is the specific gravity (60/60°F). Deviation from the formula is generally less than 1%.

11.2.1.2.3 Volatility, flammability, and explosive properties

The boiling point (boiling temperature) of a substance is the temperature at which the vapor pressure of the substance is equal to atmospheric pressure.

At the boiling point, a substance changes from the liquid to gas state. A stricter definition of boiling point is the temperature at which the liquid and vapor (gas) phases of a substance can exist in equilibrium. When heat is applied to a liquid, the temperature of the liquid rises until the *vapor pressure* of the liquid equals the pressure of the surrounding atmosphere (gases). At this point there is no further rise in temperature, and the additional heat energy supplied is absorbed as *latent heat* of vaporization to transform the liquid into gas. This transformation occurs not only at the surface of the liquid (as in the case of *evaporation*) but also throughout the volume of the liquid, where bubbles of gas are formed. The boiling point of a liquid is lowered if the pressure of the surrounding atmosphere (gases) is decreased. On the other hand, if the pressure of the surrounding atmosphere (gases) is increased, the boiling point is raised. For this reason, it is customary when the boiling point of a substance is given to include the pressure at which it is observed, if that pressure is other than standard, i.e. 760 mm of mercury or 1 atmosphere [standard temperature and pressure (STP)].

The boiling points of petroleum fractions are rarely, if ever, distinct temperatures. It is, in fact, more correct to refer to boiling ranges of various fractions; the same is true for natural gas. To determine these ranges, the material in question is tested in various methods of distillation, either at atmospheric pressure or at reduced pressure. Thus, the boiling points of the hydrocarbon constituents of natural gas increase with molecular weight and the initial boiling point of natural gas corresponds to the boiling point of the most volatile constituent (i.e. methane) (Speight, 2019a).

Purified natural gas is neither corrosive nor toxic, its ignition temperature is high, and it has a narrow flammability range, making it an apparently safe fossil fuel compared to other fuel sources. In addition, purified natural gas (i.e. methane) having a specific gravity (0.60) lower than that of air (1.00) rises if escaping and dissipates from the site of any leak. However, methane is highly flammable, and burns easily and almost completely. Therefore, natural gas can also be hazardous to life and property through an explosion. When natural gas is confined, such as within a house or in a coal mine, a high concentration of the gas can form explosive mixtures that, if ignited, results in blasts that could destroy buildings.

The *flash point* of petroleum or a petroleum product, including natural gas, is the temperature to which the product must be heated under specified conditions to give off sufficient vapor to form a mixture with air that can be ignited momentarily by a specified flame (ASTM D56, D92, and D93). As is the case with other properties, the flash point is dependent on the composition of the gas and the presence of other hydrocarbon constituents (Speight, 2019a).

The *fire point* is the temperature to which the gas must be heated under the prescribed conditions of the method to burn continuously when the mixture of vapor and air is ignited by a specified flame (ASTM D92). In addition to the fire point, and from the viewpoint of safety, information related to the flash point is of most significance at or slightly above the maximum temperatures (30–60°C, 86–140°F)) that may be encountered in storage, transportation, and use of liquid petroleum products, in either closed or open containers. In this temperature range, the relative fire and explosion risk can be estimated from the flash point. For products with flash point below 40°C (104°F), special precautions are necessary for safe handling. Flash points above 60°C (140°F) gradually lose their safety significance until they become indirect measures of some other quality.

The flash point of a petroleum product is also used to detect any contamination. A substantially lower flash point than expected for a product is a reliable indicator that the product has become contaminated with a more volatile product, such as gasoline. The flash point is also an aid in establishing the identity of a particular petroleum product.

A further aspect of volatility that received considerable attention is the vapor pressure of petroleum and its constituent fractions. The *vapor pressure* is the force exerted on the walls of a closed container by the vaporized portion of a liquid. Conversely, it is the force that must be exerted on the liquid to prevent it from vaporizing further (ASTM D323). The vapor pressure increases with temperature for a given gasoline, liquefied petroleum gas, or other product. The temperature at which the vapor pressure of a liquid, either a pure compound or a mixture of many compounds, equals 1 atmosphere (14.7 psi, absolute) is designated as the boiling point of the liquid.

The flammable range is expressed by the lower explosive limit (LEL) and the upper explosive limit (UEL). The lower explosive limit is the concentration of natural gas in air below which the propagation of a flame will not occur on contact with an ignition source. The LEL for natural gas is 5% by volume in air and, in most cases, the smell of the gas would be detected well before combustion conditions are met. The UEL is the concentration of natural gas in air above which the propagation of a flame will not occur on contact with an ignition source. The natural gas UEL is 15% by volume in air.

In any form, a minute amount of an odorant with an obvious smell is added to the otherwise colorless and odorless gas so that leaks can be detected well before a fire or explosion occurs. Odorants are considered to be non-toxic in the extremely low concentrations occurring in natural gas delivered to the end user.

11.2.1.2.4 Environmental properties

The environmental issues related to the use of natural gas were discussed in detail elsewhere (Chapter 8) but a brief mention of such properties is necessary here. However, in order to evaluate fully the environmental effects of natural gas, the general properties of the constituents must also be considered in addition to the effects of the combustion properties.

Currently, natural gas constitutes approximately one-quarter of the energy consumed in the United States, with further increases in use projected over the next decade. These increases are expected because emissions of greenhouse gases with the consumption of natural gas are much lower relative to the consumption of other fossil fuels. For example, natural gas, when burned, emits smaller amounts of greenhouse gases and criteria pollutants per unit of energy produced than other fossil fuels; this is partly because natural gas is fully combusted more easily and partly because natural gas contains fewer impurities than any other fossil fuel. However, the major constituent of natural gas, methane, contributes directly to the greenhouse effect through the venting or leaking of natural gas into the atmosphere (Speight, 2005).

Purified natural gas (methane) is the cleanest among all fossil fuels. The main products of the combustion of natural gas are carbon dioxide and water vapor. Coal and petroleum, on the other hand, release large amounts of harmful emissions, including a higher proportion of carbon dioxide emissions, nitrogen oxides (NO_x), and sulfur dioxide (SO_2). Coal and fuel oil release into the environment ash particles, which are substances that do not burn but instead are carried into the atmosphere and contribute to pollution. The combustion of purified natural gas, on the other hand, releases very small amounts of sulfur dioxide and nitrogen oxides, virtually no ash or particulate matter, and smaller amounts of carbon dioxide, carbon monoxide, and other reactive hydrocarbon derivatives.

Natural gas has no known toxic or chronic physiological effects (i.e. it is not poisonous) but it is dangerous insofar as an atmosphere rich in natural gas would result in death to humans and animals. Even though exposure to a moderate concentration of natural gas may result in a headache or similar symptoms owing to oxygen deprivation, it is likely that the smell (due to the presence of an odorant) would be detected, by as well before the concentrations are high enough for this symptom to occur.

In fact, in the natural gas refining industries (Speight, 2005), as in other industries, air emissions include point and non-point sources. Point sources are those emissions that exit stacks and flares and, thus, can be monitored and treated. Non-point sources are *fugitive emissions* that are difficult to locate and capture. Fugitive emissions occur throughout refineries and arise from, for example, the thousands of valves, pipe connections, seals in pumps and compressors, storage tanks, pressure relief valves, and flanged joints.

Although individual leaks are typically small, the sum of all fugitive leaks at a gas processing plant can constitute one of its largest emission sources. These leaks can release methane and volatile constituents of natural gas into the air. Companies can minimize *fugitive emissions* by designing facilities with the fewest possible components and connections and avoiding components known to cause significant fugitive emissions. Once companies quantify fugitive emissions, this provides important information that can be used to design the most effective leak repair program for the companies. Inspection and maintenance programs are designed to identify the source of these leaks and to prioritize and plan their repair in a timely fashion. A reliable and effective directed inspection and maintenance program for an individual facility will be composed of a variety of components, including methods for leak detection, a definition of a leak, planned schedules and targeted devices for leak surveys, and allowable repair time.

A directed inspection and maintenance program begins with a baseline survey to identify and quantify leaks. Quantification of the leaks is critical because this information is used to determine which leaks are serious enough to justify their repair costs. Repairs are then made only to those leaking components that are cost-effective to fix. Subsequent surveys are then scheduled and designed based on the information collected from previous surveys, thus permitting operators to focus on those components that are more likely to leak. Some natural gas companies have demonstrated that directed inspection and maintenance programs can profitably eliminate as much as 95% of gas losses due to equipment leaks.

11.2.2 *Synthetic gases*

In the context of this book, a synthetic gas is a gas that has been produced by a process from a carbonaceous feedstock (Table 11.6) and there is at present an increased interest in power generation by the gasification potential of waste and biomass.

11.2.2.1 *Producer gas*

This gas is produced by blowing air and sometimes steam through an incandescent fuel bed (this process is self-heating). The reaction with air is exothermic but an insufficient amount of air is added; hence, carbon monoxide is produced. Steam addition results in the formation of hydrogen by the water–gas reaction. This reaction is endothermic and hence balances out the exothermic air reaction. Producer gas has a low calorific value and hence is typically used as an on-site fuel gas.

In this process — using coal as an example of the carbonaceous feedstock — air is blown through an incandescent bed of coal. The amount of air is sufficient to maintain the process temperature but insufficient to complete the combustion reaction. The product gas is a mixture containing approximately 50% v/v nitrogen from the air, 29% v/v carbon monoxide due to incomplete combustion of the feedstock, and 4% v/v carbon dioxide, as well as sundry other products, including hydrocarbon derivatives. The heat content of the gas is low and, although not suitable for distribution to consumers, it can be used on-site as a fuel gas. In order to modify the product distribution, steam can be used in the process and endothermic reactions (such as the water–gas reaction) result in the formation of hydrogen.

Initially, oxidation of the feedstock (coal) in the presence of available oxygen occurs, after which there is competition for oxygen between the carbon and the water and the water–gas reactions occur:

Table 11.6: Types of gases produced by gasification of carbonaceous feedstocks.[a]

Producer Gas

The simplest method of gasification; available for well over 100 years.

Air is blown through an incandescent bed of the feedstock.

Enough air to maintain the temperature; insufficient air to complete the combustion reaction.

Composition: a mixture of the order of 50% v/v nitrogen (from air), 29% v/v CO (incomplete combustion), and 4% v/v CO_2

Very low calorific value; not suitable for distribution but can be used on site as fuel gas.

Steam is added to modify the products. Endothermic reactions (water gas) result in the formation of hydrogen.

Chemistry:

Initially, oxidation of feedstock in the presence of available oxygen:

$$2C + O_2 \rightarrow 2CO \text{ (exothermic)}$$

Then competition for oxygen between the carbon and the water. The water–gas reactions.

$$C + H_2O \rightarrow CO + H_2 \text{ (endothermic)}$$
$$C + 2H_2O \rightarrow CO_2 + 2H_2$$
$$CO + H_2O \rightarrow CO_2 + H_2$$

The ratio of steam to air may be used to modify the composition of the gas produced.

Blue Water Gas

The feedstock bed is simultaneously blasted with air followed by steam.

Air reaction is exothermic; steam reaction is endothermic.

The mixture of gas produced is higher in quality than the produced gas.

Composition: a mixture of the order of 5% v/v CO_2, 40% v/v CO, 49% H_2, 1% v/v methane, and <5% v/v nitrogen.

If oxygen is used instead of air, the process can be continuous.

Carbureted Water Gas

Low-quality gas is enriched by adding a carburetor.

Fuel oil is sprayed into a brick lined chamber during the blow period of the blue water gas plant with air. This heats the bricks to around 1,000°C (1,830°F)

During the make period, the air is turned off and the oil is cracked into smaller hydrocarbon derivatives in the now heated chamber.

Composition: mainly methane.

Note: [a]Using coal as an example of the feedstock.

$$2C_{\text{feedstock}} + O_2 \rightarrow 2CO \text{ (exothermic)},$$
$$C_{\text{feedstock}} + H_2O \rightarrow CO + H_2 \text{ (endothermic)},$$
$$C_{\text{feedstock}} + 2H_2O \rightarrow CO_2 + 2H_2 \text{ (endothermic)},$$
$$CO + H_2O \rightarrow CO_2 + H_2 \text{ (exothermic)}.$$

The ratio of steam to air may be used to modify the composition of the gas produced.

11.2.2.2 *Blue water gas*

Blue water gas (sometimes referred to as blue gas) is manufactured in a similar manner to producer gas but allows the production of a higher-heat content gas by intermittently blasting the incandescent bed with air and steam such that the overall heat balance is maintained. The products of the air blast contain nitrogen, which reduces the heat content of the product gas.

This process follows similar principles to the production of producer gas except that the problem of nitrogen dilution is overcome. The feedstock bed is simultaneously blasted with air followed by steam — the air reaction is exothermic, causing the bed to increase in temperature and this is balanced by the endothermic reaction of the steam. The typical composition of the product gas is carbon dioxide 5% v/v, carbon monoxide 41% v/v, hydrogen 49%, v/v, methane 1% v/v, and nitrogen 4% v/v. If oxygen is used instead of air, the process can be continuous and coke is preferred to coal because coal can continue to devolatilize in the blow period, thereby reducing process efficiency — there are also issues related to process efficiency that arise from the use of coking coal as the feedstock.

The blue water gas may be enriched by adding a carburetor in which heavy (high–density, high-boiling) fuel oil is sprayed into a brick-lined chamber during the blow period of the blue water gas plant with air. During the make period, the air is turned off and the oil is cracked into smaller hydrocarbon derivatives in the now heated chamber. This produces a mixture of hydrocarbon derivatives (predominantly methane), which increases the heat content and enriches the product gas.

11.2.2.3 *Carbureted water gas*

Carbureted water gas is the result of combining the water gas and oil gas methods. Briefly, oil gas is a gas formed by thermal cracking of crude oil. If oil is sprayed onto heated checker work (refractory), it cracks to form lower gaseous hydrocarbon derivatives. These types of hydrocarbon derivatives depend entirely on the type of feedstock but do cause an increase in the heat content of the product gas.

In the oil gas method, an oil product (typically heavy fuel oil) is sprayed into the hot water gas chamber to produce a better quality (higher heat content) gas. The amount of heavy fuel oil is a determinant of the quality of the gas. In the 19th century and early 20th century, this was the method used to produce town gas but it has largely been superseded by natural gas in countries that have an abundant supply. As supplies of natural gas are diminishing and various carbonaceous feedstocks are being used for gas production, the process may once again rise in importance.

11.2.2.4 *Coal and coke oven gas*

As mentioned above, gases are liberated during the high-temperature carbonization (coking) of coal and other carbonaceous feedstocks. These gases are cleaned (Chapter 9) and used as fuel or as petrochemical feedstocks (Speight, 2019a, 2019b). If coke is not required and another carbonaceous feedstock is used, steam injected at the end of the cycle reacts with the coke to form blue water gas. This reduces the heat content of the product gas, but the thermal efficiency of conversion rises.

11.2.2.5 *Biogas*

Gasification technology is an attractive route for the production of fuel gases from various types of biomass (Van Doorn, 1995; Faaij *et al.*, 1997). By gasification, solid biomass is converted into a combustible gas mixture normally called *producer gas*, consisting primarily of hydrogen (H_2) and carbon monoxide (CO), with lesser amounts of carbon dioxide (CO_2), water (H_2O), methane (CH_4), and higher boiling (higher molecular weight) hydrocarbon derivatives (C_xH_y), as well as nitrogen (N_2) and particulates.

A variety of biomass gasifiers have been developed and they can be grouped into four major classes: (i) a fixed-bed updraft or counter-current gasifier, (ii) a fixed-bed downdraft or co-current gasifier, (iii) a bubbling fluidized-bed gasifier, and (iv) a circulating fluidized-bed gasifier. This differentiation is based on the means of supporting the biomass in the reactor vessel, the direction of flow of both the biomass and the oxidant, and the way that heat is supplied to the reactor. The processes occurring in any gasifier include drying, pyrolysis, reduction, and oxidation. A unique feature of the updraft gasifier is the sequential occurrence of these processes: they are separated spatially and therefore temporally.

As an illustration, using the updraft gasifier, biomass and air are fed in an opposite direction. In the highest zone, biomass is heated and releases its moisture. In the pyrolysis zone, biomass undergoes a further increase in temperature and decomposes into hydrocarbon derivatives, gas products, and char in the temperature range of 150–500°C (300–930°F). The major reactions are as follows:

$$\text{Biomass} \rightarrow C_xH_y + C_xH_yO_z + H_2O + CO_2 + CO + H_2.$$

The hydrocarbon fraction consists of methane to high-boiling tar. The composition of this fraction can be influenced by many parameters, such as the particle size of the biomass, temperature, pressure, heating rate, residence time, and catalysts. The operative reactions are as follows:

$$C + CO_2 \rightarrow 2CO,$$

$$C + H_2O \rightarrow H_2 + CO,$$

$$C + 2H_2O \rightarrow H_2 + CO_2,$$

$$C + 2H_2 \rightarrow CH_4,$$

$$CO + H_2O \rightarrow CO_2 + H_2,$$

$$CO + 3H_2 \rightarrow CH_4 + H_2O.$$

The composition of producer gases varies widely depending on the properties of the biomass, the gasifying agent, and the process conditions. Depending on the nature of the raw solid feedstock and the process conditions, the char formed from pyrolysis contains 20–60% of the energy input. Therefore, the gasification of char is an important step for the complete conversion of solid biomass into gaseous products and for the efficient utilization of energy in the biomass.

The producer gases from the reduction zone rise beyond the reduction zone. When they come into contact with the cooler biomass, the temperature drops down and the aforementioned reactions are frozen. The unreacted char further undergoes oxidation with air in the lowest zone:

$$C + O_2 \rightarrow CO_2.$$

As a result, ash is left at the bottom of the reactor. The produced carbon dioxide flows upward and is involved in the reactions in the reduction zone. The heat released in the oxidation zone drives both the reduction and pyrolysis processes.

The greenhouse gas production associated with lignocellulosic-based feedstocks is anticipated to be much lower than that associated with conventional fuels. The environmental performance depends very much on the specific life cycle of the fuel, including the country in which the life cycle assessment (LCA) was conducted, the feedstock on which the fuel is based, the vehicle used, the propulsion system, and the overall state of technology (Quirin *et al.*, 2004).

In general, fuels (and chemicals) made from lignocellulosic materials are characterized by reduced carbon dioxide emissions when compared to similar products derived from crude oil. Most bio-fuels can reduce greenhouse gas emissions significantly and substituting emissions by utilizing biobased energy can create an overall negative emission for the fuel (Braune, 1998).

Fischer–Tropsch (FT) fuels based on bioresidues are likely to have the lowest possible emissions; this is typical of diesel propulsion systems that have better energy recovery. If energy crops are utilized as a feedstock, the overall emissions rise slightly, because the benefit of residue disposal is lost. Ethanol from residues or from energy crops also have relatively low emissions, particularly compared to conventional fuels including gasoline and diesel fuel.

For ethanol from lignocellulose, there is a potential to reduce greenhouse gas emissions with improved technology. This reflects the close-to-commercial status of the technology and the anticipated improvements that will be seen as this technology advances.

Biomass-based fuels (many components of domestic and industrial waste can be included in this category) are non-fossil fuels. Vegetative biomass can be classified into woody, herbaceous, agricultural by-products, energy crops, and black liquor — a by-product of the forest industry (Grace *et al.*, 1989). Because of the high concentration of organic materials in industrial waste and domestic waste (including discarded food), many constituents of waste are actually biomass. But the issues arise from the other constituents of the waste that make the waste truly heterogeneous and, in many cases, of indeterminate composition.

11.2.2.6 *Wood gas*

Wood can be used to make both liquid and gaseous fuels. When wood is heated in the absence of air or with a reduced air supply, it is possible to produce a liquid fuel that can be used in a similar way to conventional oil fuels. It can be used to run internal combustion engines in vehicles or electric generators. The gas produced from wood is a mixture of hydrogen and carbon monoxide, which is similar to the coal gas that was made before the arrival of natural gas from the North Sea. This wood gas can be used in internal combustion engine or in gas turbine that can be used to power a generator.

Gasification of wood has the potential to meet all fuel needs for small oil- or gas-fired boilers and supplemental fuel for large boilers. Gasification is carried out at elevated temperatures (500–1,500°C, 930–2,730°F) and at atmospheric or elevated pressures. The process involves the conversion of biomass, which is carried out in the absence of air or with less air than the stoichiometric amount required for complete combustion. Partial combustion produces carbon monoxide as well as hydrogen, which are both combustible gases. The producer gas can be used in different ways: it can be burned directly to produce heat or used as a fuel for gas engines and gas turbines to generate electricity; in addition, it can also be used as a feedstock (synthesis gas) in the production of chemicals, e.g. methanol.

11.2.2.7 *Refuse gas*

There are two ways of producing gaseous fuel from refuse or waste: (i) gasification of the refuse at high temperatures by partial oxidation and then conversion of materials containing carbon into synthesis gas (mainly hydrogen and carbon monoxide) and (ii) production of bio-gas, mainly methane, by the anaerobic digestion of refuse in a landfill (Chapters 6 and 7).

The principle behind waste gasification and the production of gaseous fuels is that waste contains carbon and it is this carbon that is converted to gaseous products via the usual gasification chemistry. Thus, when fuel is fed to a gasifier, water and volatile matter are released fast and a char residue is left to react further. The char gasification is what mainly controls the conversion achieved in the process. From solid carbon, product gas is formed according to the following main reactions:

$$C + H_2O \rightarrow CO + H_2,$$

$$C + CO_2 \rightarrow 2CO,$$

$$C + 2H_2 \rightarrow CH_4.$$

In addition to the reactions of solid carbon, the most important reaction is the water–gas shift reaction, which takes place in the gas phase:

$$CO + H_2O \leftrightarrow H_2 + CO_2.$$

The product gas generally contains large amounts of hydrogen and carbon monoxide and a small amount of methane, as well as carbon dioxide and steam, and in air gasification, nitrogen. In addition, a significant amount of other organic components in the gas, known as tar, is formed. Tar formation is a well-known phenomenon in the thermal reactions of coal (Speight, 2013) and can also be anticipated to form in the gasification or pyrolysis of any complex carbonaceous material.

11.3 Liquid Fuels

By definition, liquid fuels are combustible or energy-generating molecules that can be harnessed to create mechanical energy, usually producing kinetic energy; they must also take the shape of their container. It is the fumes of liquid fuels that are flammable instead of the fluid. Most liquid fuels in widespread use are derived from fossil fuels. However, there are several types, such as hydrogen fuel (for automotive uses), ethanol, and bio-diesel, that are also categorized as a liquid fuel. Many liquid fuels play a primary role in transportation and the economy.

The main advantage of liquid fuels over their gaseous alternatives is the extremely high thermal energy/volume ratio. Hence, they have become the conventional gas for transport applications. Liquid fuels (light and heavy fuel oils) are also used in furnaces and boilers. Not surprisingly, the most heavily specified of all fuels are transport fuels which have a variety of physical parameters that must be met and are described as specifications (Table 11.7) (Speight, 2014, 2015a).

11.3.1 *Methanol*

Methanol (Table 11.8) is another possible replacement for conventional motor fuels. In fact, it has been considered as a possible large-volume motor fuel substitute at various times during gasoline shortages. In fact, it was often used in the early 20th century to power automobiles before the introduction of inexpensive gasoline. Later, synthetically produced methanol was widely used as a motor fuel in Germany during World War II.

Table 11.7: Examples of the properties used in the specifications for liquid fuels.

Property	
Density	
Cloud point	The temperature at which a sample has visible cloudiness
Pour point	The temperature at which the fuel will no longer pour
Freezing point	
Flash point	The temperature above which the vapor above a fuel in air becomes flammable
Viscosity	
Gum content	The quantity of solid residue remaining after a sample has been heated and evaporated for a prescribed period of time in air
Ash content	
Sulfur content	
Trace metal content	Va, Ni, Cu, and Fe
Aromatic content	The percentage of aromatic hydrocarbon derivatives in the fuel
Thermal stability	The resistance of the fuel to thermal degradation

Table 11.8: Properties of methanol.

Chemical formula	CH_3OH or CH_4O
Molecular weight	32.04 g mol^{-1}
Appearance	Colorless liquid
Density	0.792 g/cm^3
Melting point	–97.6°C (–143.7°F; 175.6°K)
Boiling point	64.7°C (148.5°F; 337.8°K)
Solubility in water	Miscible
Vapor pressure	13.02 kPa (at 20°C)
Acidity (pK_a)	15.5[3]
Refractive index (n_D)	1.33141[6]
Viscosity	0.545 mPa·s (at 25°C)[7]
Dipole moment	1.69 D

Again, during the oil crisis of the 1970s, blending methanol with motor fuel received attention due to its availability and low cost.

Methanol can be produced from biomass, essentially the primary energy source. Thus, the choice of fuel in the transportation sector is, to some extent, determined by the availability of biomass. As regards the difference between hydrogen and methanol production costs, conversion of natural gas, biomass, and coal into hydrogen is generally more energy-efficient and less expensive than conversion into methanol.

Methanol is poisonous and burns with an invisible flame. Similar to ethanol, methanol has a high octane rating and hence is suitable for Otto engines. Methanol is a colorless, odorless, and nearly tasteless alcohol; it is produced from crops and is used as a fuel. Methanol, like ethanol, burns more completely but releases as much or more carbon dioxide than its gasoline counterpart. The balance is often seen as the various bi-processes that draw carbon dioxide from the atmosphere, so there is no net modern release as there is for fossil fuels.

Methanol is also known as *wood alcohol*. Generally, methanol is easier to find than ethanol. Methanol is produced from synthetic gas or biogas and evaluated as a fuel for internal combustion engines. The production of methanol is a cost-intensive chemical process. It is commonly used in bio-diesel production for its reactivity with vegetable oils. Methanol can be used as one possible replacement for conventional motor fuels. Many tests have shown promising results by using 85–100% v/v methanol as a transportation fuel in automobiles, trucks, and buses.

11.3.2 *Ethanol*

Ethanol (Table 11.9) is produced from organic feedstocks, such as biomass, by a fermentation process. However, recent efforts have focused on the production of ethanol from waste materials using a gasification process. The overall process consists of a thermo-chemical conversion of synthesis gas, which is then converted to higher molecular weight alcohols by a catalytic process after which high-purity ethanol is separated by distillation. The key differentiating factors are the catalysts and their operating parameters.

Most current research efforts have focused on bacteria, enzyme, and acid/solvent hydrolysis–fermentation based ethanol production. However, the action of bacteria and enzymes is often strongly dependent on the feedstock. In addition, these processes generate a small quantity of ethanol over a period of days and the dilute aqueous ethanol product must be distilled to recover the ethanol. In the gasification–synthesis process, various carbonaceous feedstocks can be used, with appropriate modifications in the synthesis gas production step.

Table 11.9: Properties of ethanol.

Chemical formula	C_2H_6O
Molecular weight	46.069 g·mol^{-1}
Appearance	Colorless liquid
Density	0.7893 g/cm^3 (at 20°C, 68°F)
Melting point	−114.14 ± 0.03°C (−173.45 ± 0.05°F
Boiling point	78.24 ± 0.09°C (172.83 ± 0.16°F
Solubility in water	Miscible
Vapor pressure	5.95 kPa (at 20°C, 68°F)
Acidity (pK_a)	15.9
Refractive index (n_D)	1.3611[2]
Viscosity	1.2 mPa·s (at 20°C, 68°F), 1.074 mPa·s (at 25°C, 77°F)
Dipole moment	1.69 D

11.3.3 *Propanol and butanol*

Propanol and butanol exist in various isomeric (structural) forms — isomers are ions or molecules with identical formulas but distinct structures. Thus, for propanol and butanol, the isomers are *n*-propanol, isopropanol (or *iso*-propanol), *n*-butanol, *iso*-butanol (more often referred to as *sec*-butanol or secondary butanol), and tertiary-butanol. The various structures are as follows:

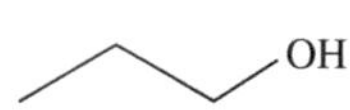

n-propanol

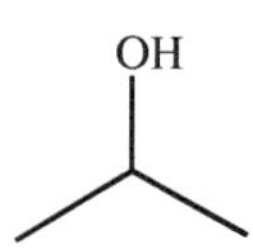

iso-propanol

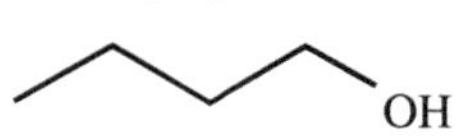

n-butanol

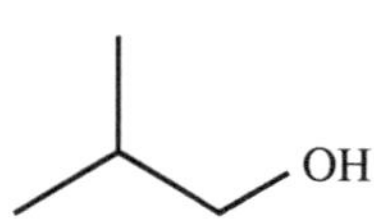

iso-butanol

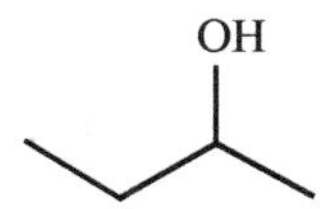

sec-butanol

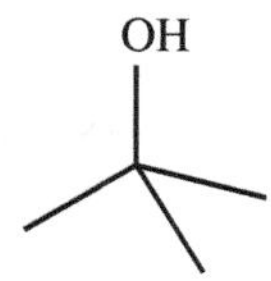

tert-butanol

The unmodified term *butanol* usually refers to the straight-chain isomer with the alcohol functional groups at the terminal carbon, and is also known as *n*-butanol or 1-butanol. The straight-chain isomer with the alcohol at an internal carbon is *sec*-butanol or 2-butanol. The branched isomer with the alcohol at a terminal carbon is *iso*-butanol (isobutanol) or 2-methyl-1-propanol, while the branched isomer with the alcohol at the internal carbon is *tert*-butanol or 2-methyl-2-propanol. These two alcohols are considerably less toxic and less volatile than methanol. In particular, butanol has a high flash point of 35°C (95°F), which is beneficial for fire safety but may be a difficulty for starting engines in cold weather.

Also, the different isomer forms can cause subtle-to-significant differences in the properties of the respective isomers (see Tables 11.10, 11.11, and Table 11.12 for the propanol isomers and Tables 11.13 and 11.14 for the butanol isomers). For example, the butanol isomers have different melting and boiling points. *n*-Butanol and *iso*-butanol have limited solubility in water (73 g/L at 25°C), *sec*-butanol has substantially greater solubility in water (290 g/L), and *tert*-butanol is miscible with water above the melting point of *tert*-butanol. The hydroxyl group renders the molecule polar, promoting solubility in water, whereas the longer hydrocarbon chain mitigates the polarity and reduces solubility.

11.3.4 *Bio-diesel*

Bio-diesel fuels are attracting increasing attention worldwide as blending components or direct replacements for diesel fuel in vehicle engines (Table 11.15). Bio-diesel is a diesel-equivalent, processed fuel derived from biological sources. Bio-diesel fuel typically consists of lower alkyl fatty acid (chain length C14–C22) esters of short-chain alcohols,

Table 11.10: Properties of 1-propanol.

Chemical formula	C_3H_8, $CH_3CH_2CH_2OH$
Molecular weight	60.096 g·mol^{-1}
Appearance	Colorless liquid
Odor	mild, alcohol-like
Density	0.803 g/mL
Melting point	–126°C (–195°F)
Boiling point	97–98°C (206–208°F)
Solubility in water	Miscible
Vapor pressure	1.99 kPa (at 20°C, 68°F)
Acidity (pK_a)	16
Basicity (pK_b)	–2
Refractive index (n_D)	1.387
Viscosity	1.959 mPa·s (at 25°C, 77°F)
Dipole moment	1.68 D

Table 11.11: Properties of 2-propanol.

Chemical formula	C_3H_8O, $CH_3CH(OH)CH_3$
Molecular weight	60.096 g·mol^{-1}
Appearance	Colorless liquid
Density	0.786 g/cm^3 (20°C, 68°)
Melting point	–89°C (–128°F)
Boiling point	82.6°C (180.7°F)
Solubility in water	Miscible
Solubility	Miscible with benzene, chloroform, ethanol, ether, glycerin, acetone
Acidity (pK_a)	16.5
Refractive index (n_D)	1.3776
Viscosity	2.86 cP at 15°C (59°F); 1.96 cP at 25°C (77°F); 1.77 cP at 30°C (86°F)
Dipole moment	1.66 D (gas)

Table 11.12: Properties of *n*-Butanol.

Chemical formula	$C_4H_{10}O$, $CH_3CH_2CH_2CH_2OH$
Molecular weight	74.123 g·mol^{-1}
Appearance	Colorless
Odor	Harsh, alcoholic and sweet
Density	0.81 g/cm^3
Melting point	–89.8°C; –129.6°F
Boiling point	117.7°C; 243.9°F
Solubility in water	73 g/L at 25°C
Solubility	Very soluble in acetone; miscible with ethanol, ethyl ether
Vapor pressure	6 mmHg (20°C, 68°F)
Acidity (pK_a)	16.10
Refractive index (n_D)	1.3993 (20°C)
Viscosity	2.573 mPa·s (at 25°C, 77°)
Dipole moment	1.66 D

Table 11.13: Properties of 2-butanol.

Chemical formula	$C_4H_{10}O$, $CH_3CH_2CH(OH)CH_3$
Molecular weight	74.123 g·mol^{-1}
Density	0.808 g cm^{-3}
Melting point	–115°C; –175°F
Boiling point	98–100°C; 208–212°F
Solubility in water	290 g/L
Vapor pressure	1.67 kPa (at 20°C, 68°F)
Acidity (pK_a)	17.6
Refractive index (n_D)	1.3978 (at 20°C, 68°F)

primarily methanol or ethanol. Various methods have been reported for the production of bio-diesel from vegetable oil, for examples direct use and blending, micro-emulsification, pyrolysis, and transesterification. Among these, transesterification is an attractive and widely accepted technique. The purpose of the transesterification process is to lower the viscosity of the oil. The most important variables affecting methyl ester yield

Table 11.14: Properties of *tert*-butanol.

Chemical formula	$C_4H_{10}O$, $(CH_3)_3COH$
Molecular weight	74.123 g·mol^{-1}
Appearance	Colorless solid
Odor	Camphor-like
Density	0.775 g/mL
Melting point	25–26°C; 77–79°F
Boiling point	82–83°C; 179–181°F
Solubility in water	Miscible
Vapor pressure	4.1 kPa (at 20°C)
Acidity (pK_a)	16.54
Refractive index (n_D)	1.387
Dipole moment	1.31 D

Table 11.15: Examples of the specifications for crude oil-derived diesel and bio-diesel fuels.

Property	Diesel	Bio-diesel
Standard	ASTM D975	ASTM D6751
Composition	HC[a] (C10–C21)	FAME[b] (C12–C22)
Kin. viscosity (mm^2/s) at 313 K	1.9–4.1 0.85	1.9–6.0 0.88
Specific gravity (g/mL)	333–353	373–443
Flash point (K)	258–278	270–285
Cloud point (K)	238 to 258	258–289
Cetane number	87	77

Notes: [a]Hydrocarbon derivatives.
[b]Fatty acid methyl esters.

during the transesterification reaction are the molar ratio of alcohol to vegetable oil and the reaction temperature. Methanol is the commonly used alcohol in this process, due in part to its low cost.

Methyl esters of vegetable oils (bio-diesel) have several outstanding advantages over other renewable and clean engine fuel alternatives. Bio-diesel fuel is a renewable, substitute fuel (for crude oil diesel or petrodiesel fuel) made from vegetable or animal fats; it can be used in any mixture with petrodiesel fuel, as it has very similar characteristics, but it has lower exhaust emissions. Bio-diesel fuels are attracting increasing attention worldwide as a blending component or a direct replacement for diesel fuel in vehicle engines. Bio-diesel, as an alternative fuel for internal combustion engines, is defined as a mixture of mono-alkyl esters of long chain fatty acids (FAME) derived from a renewable lipid feedstock, such as vegetable oil or animal fat.

Bio-diesel is a liquid that varies in color — between golden and dark brown — depending on the production feedstock. It is practically immiscible with water and has a high boiling point and a low vapor pressure. A typical methyl ester bio-diesel has a flash point of the order of 150°C (300°F), making it rather non-flammable. Bio-diesel has a density of 0.88 g/cm³, less than that of water. Bio-diesel uncontaminated with starting material can be regarded as non-toxic. Bio-diesel has a similar viscosity to the diesel derived from crude oil (often referred to as petrodiesel) and can be used as an additive in formulations of diesel to increase the lubricity of ultra-low-sulfur diesel (ULSD) fuel, which is advantageous because it has virtually no sulfur content. Much of the world uses a system known as the "B" factor to state the amount of bio-diesel in any fuel mix, in contrast to the "BA" or "E" system used for fuels containing ethanol. For example, pure bio-diesel is referred to as B100 while a fuel containing 20% bio-diesel is labeled B20.

Bio-diesel is a renewable fuel that is safe and bio-degradable and that reduces air pollutants, such as particulates, carbon monoxide, and hydrocarbon derivatives. Blends of 20% bio-diesel with 80% crude oil diesel (B20) can generally be used in unmodified diesel engines. Bio-diesel can also be used in its pure form (B100), but may require certain engine modifications to avoid maintenance and performance problems. Bio-diesel has approximately 5–8% lower energy density, but better lubricity and more complete combustion can make the energy output of a diesel engine only 2% lower per volume when compared to petrodiesel.

A series of standards applied to bio-diesel ensure that the following important factors in the fuel production process are satisfied: (i) complete reaction, (ii) removal of glycerin, (iii) removal of catalyst, (iv) removal of alcohol, (v) absence of free fatty acids, (vi) low-to-no low sulfur. Pure, non-blended bio-diesel can be poured straight into the tank of any diesel vehicle. As with normal diesel, low-temperature bio-diesel is sold during winter months to prevent viscosity problems. Some older diesel engines still have natural rubber parts, which will be affected by bio-diesel, but in practice these rubber parts should have been replaced long ago.

The temperature at which pure (B100) bio-diesel starts to gel varies significantly and depends on the mixture of esters and therefore the feedstock oil used to produce the bio-diesel. For example, bio-diesel produced from varieties of canola seed starts to gel at approximately –10°C (50°F). Bio-diesel produced from tallow tends to gel at around 16°C (61°F). Some users modify the vehicles to permit the use of bio-diesel without mixing and without the possibility of gelling at low temperatures, which is a similar practice to the one used for running straight vegetable oil. A second fuel tank is installed, which is insulated, and a heating coil — a temperature sensor — is installed to notify the driver when the fuel is warm enough to burn; the driver then switches the tank the engine is drawing from. Bio-diesel may contain small but problematic quantities of water. Although bio-diesel is hydrophobic (non-miscible with water), it is hygroscopic to the point of attracting water from the atmosphere. In addition, there may be water that is residual to processing or resulting from storage tank condensation.

Bio-diesel fuel has several advantages over fossil fuel. The most celebrated advantage is that bio-diesel fuel is less costly to make than gasoline. As an added bonus, bio-diesel is an environmentally friendly fuel. The use of bio-diesel in an unmodified diesel engine will substantially reduce the emissions of harmful unburned hydrocarbon derivatives, sulfates, polycyclic aromatic hydrocarbon derivatives, and carbon monoxide.

Bio-diesel fuel can be used in almost any diesel engine found in cars, trucks, tractors, planes, trains, boats, buses, and almost all agricultural equipment, and most diesel engines can use bio-diesel without having to be modified in any way. Bio-diesel fuel reduces carbon dioxide emissions by up to 80%, thereby reducing the formation of acid rain because it produces no sulfur dioxide, which is the key element in creating acid rain. Furthermore, bio-diesel burns much cleaner and also takes considerably less time to degrade in the event of a spill, usually of the order of 28 days, which is approximately four times faster than gasoline or regular diesel fuel. In fact, the extra lubrication provided by bio-diesel fuel helps to improve the longevity of the engine, in addition to boosting engine performance, and helping to eliminate engine knocks and noise. In addition, bio-diesel fuel can be stored in any type of tank and has a much higher flash point of 300°C (570°F) than that of regular diesel of 150°C (300°F).

Bio-diesel is a light-to-dark yellow liquid that is immiscible with water and that has a high boiling point and a low vapor pressure. A typical methyl ester bio-diesel has a flash point of the order of 150°C (300°F), making it non-flammable. Bio-diesel has a density of 0.88 g/cm³. Bio-diesel has a similar viscosity to diesel fuel derived from crude oil (the current term to differentiate the different forms of diesel) and can be used as an additive in formulations of diesel to increase the lubricity of ULSD fuel. Much of the world uses a system known as the "B" factor to state the amount of bio-diesel in any fuel mix, in contrast to the "BA" or "E" system used for ethanol mixtures. For example, a fuel containing 20% v/v bio-diesel is labeled B20 whereas pure bio-diesel (100% v/v) is referred to as B100.

Another source of liquid fuels is waste cooking oil; usually bio-diesel is the desired product (Zhang *et al.*, 2003). In fact, waste cooking oil is a valuable asset for conversion into liquid fuels, and the quantity of bio-diesel that can be produced from waste cooking oil can be determined in a similar manner to that of bio-diesel produced directly from agriculture. The only difference is that rather than calculating crop yields it is necessary to quantify the amount of waste oil available.

11.3.5 *Hydrocarbon fuels*

Hydrocarbon derivatives, as such, are not usually produced from crops, there being an insufficient amount of hydrocarbon derivatives present in the plant tissue to make the process economical. However, bio-diesel is produced from crops, thereby offering an excellent renewable fuel for diesel engines.

The distillable oil obtained from the fast pyrolysis of biomass has high oxygen content. Ketone derivatives and aldehyde derivatives, carboxylic acid derivatives, ester derivatives, aliphatic and aromatic alcohol derivatives, and ether derivative have been detected in significant quantities. Because of the reactivity of oxygenated groups, the main problems of the oil are instability. The hydrodeoxygenation (HDO) of bio-oil in the presence of a cobalt molybdate catalyst was studied (Zhang *et al.*, 2003), which is the most important route for upgrading the oil.

$$-(CH_2O)- + H_2 \rightarrow -(CH_2)- + H_2O. \quad (11.1)$$

The reaction has strong analogies with typical refinery processes, such as the hydrodesulfurization and hydrodenitrogenation processes

(Parkash, 2003; Gary *et al.*, 2007; Speight, 2011a, 2014, 2017; Hsu and Robinson, 2017). In general, most of the hydrodeoxygenation studies have been performed using existing hydrodesulfurization catalysts (NiMo and CoMo on suitable carriers). However, these types of catalysts need activation using a suitable sulfur source and this is a major drawback when using low-to-no-sulfur resources.

11.3.5.1 *Naphtha and gasoline*

The naphtha–gasoline fraction is a crude oil-derived liquid mixture (Table 11.16), primarily used as fuel in internal combustion engines, specifically in spark-ignition engines. However, gasoline is not typically produced directly from crude oil but is a blend of various fraction — the basic fraction is the low-boiling naphtha fraction.

The amount of naphtha that can be distilled from oil crude depends on where the crude is obtained. For example, heavy crude oil, such as Venezuelan heavy (high-density) crude oil, yields very little naphtha (approximately 5% v/v), whereas West Texas crude oil or Arabian light (low-density) crude oil yields approximately 30% v/v naphtha. The density of gasoline is of the order of 0.71–0.77 kg/L. An important characteristic of the finished gasoline is the octane number or octane rating, which is a measure of how resistant the gasoline is to the abnormal combustion phenomenon known as pre-detonation (also known as knocking, pinging, spark knock, etc.). The octane number is measured relative to a mixture of 2,2,4-trimethylpentane and *n*-heptane. Octane number is a measure of the gasoline quality for the prevention of early ignition, which leads to cylinder knocks. Higher octane numbers are preferred in internal combustion engines. For gasoline production, aromatic derivatives, naphthene (cycloalkane) derivatives and *iso*-alkane derivatives are highly desirable, whereas olefin derivatives and *n*-paraffin derivatives are less desirable.

The typical composition of gasoline hydrocarbon derivatives (% volume) is as follows: 4–8% v/v alkane derivatives, 2–5% v/v alkene derivatives, 25–40% *iso*-alkane derivatives, 3–7% v/v cycloalkane derivatives, 1–4% cycloalkene derivatives, and 20–50% aromatic derivatives (including 0.5–2.5% v/v benzene) (Table 4.2). Additives and blending agents are added to the hydrocarbon mixture to improve the performance and stability of gasoline. These compounds include anti-knock agents, antioxidants, metal deactivators, lead scavengers, anti-rust agents, anti-icing agents, upper-cylinder lubricants, detergents, and dyes.

Table 11.16: Typical constituents of gasoline.

Component	Composition (w/w%)[a]
n-Alkane derivatives	
C5	3
C6	12
C7	2
C9	1
C10–C13	1
Total of *n*-alkane derivatives	19
Branched alkane derivatives	
C4	2
C5	15
C6	8
C7	2
C8	2
C9	2
C10–C13	1
Total of branched alkane derivatives	32
Cycloalkane derivatives	
C6	3
C7	1
C8	1
Total of cycloalkane derivatives	5
Olefin derivatives	
C6	2
Total of olefin derivatives	2
Aromatic derivatives	
Benzene	3
Toluene	5
Xylene isomers	7
Ethylbenzene	1
C3-benzene derivatives	4
C4-benzene derivative	8
Others	3
Total aromatic derivatives	31
Other constituents	11

Note: [a]Approximate; varies across different crude oils.

Table 11.17: Physical and chemical properties of methanol, ethanol, and gasoline.

Property	Methanol CH_3OH	Ethanol C_2H_5OH	Gasoline C4–C12
Molecular weight (g/mol)	32	46	~114
Specific gravity	0.789 (298 K)	0.788 (298 K)	0.739 (288.5 K)
Vapor density rel. to air	1.10	1.59	3.0–4.0
Liquid density (g cm^{-3} at 298 K)	0.79	0.79	0.74
Boiling point (K)	338	351	300–518
Melting point (K)	175	129	
Vapor pressure at 311 K (psia)	4.6	2.5	8–10
Heat of evaporation (Btu/lb)	472	410	135
Heating value (kBtu gal^{-1})			
• Lower	58	74	111
• Upper	65	85	122
Tank design pressure (psig)	15	15	15
Viscosity (cp)	0.54	1.20	0.56
Flash point (K)	284	287	228
Flammability/explosion limits			
• (%) Lower (LFL)	6.7	3.3	1.3
• (%) Upper (UFL)	36	19	7.6
Autoignition temperature (K)	733	636	523–733
Solubility in H_2O (%)	Miscible (100%)	Miscible (100%)	Negl. (~0.01)
Azeotrope with H_2O	None	95% EtOH	Immiscible

Methanol and ethanol have a long history as alternative transportation fuels. The properties allow for a higher compression ratio and a shorter burn time, which lead to theoretical efficiency advantages over gasoline in an internal combustion engine. Both methanol and ethanol fuels are technically and economically suitable for internal combustion engines (Table 11.17).

The hydrothermal liquefaction process (HTL process) or direct liquefaction is a promising technology to treat waste streams from various sources and to produce valuable bioproducts such as biocrude oil. A major problem with commercializing the hydrothermal liquefaction processes for biomass conversion is that it remains uneconomical when compared to the costs of diesel or gasoline production. The high transportation costs of large quantities of biomass increase production costs, and poor conversion efficiency coupled with a lack of understanding of complex reaction mechanisms inhibits the growth of the process commercially.

In the HTL process, biomass is reacted in liquid water at elevated temperature and pressure. The phase equilibria in the HTL process are very complicated due to the presence of water, supercritical carbon dioxide, alcohols, and the so-called biocrude. The biocrude is a mixture with a wide molecular weight distribution and consists of various kinds of molecules. Bio-crude oil contains 10–13% w/w oxygen and is upgraded by catalytic hydrodeoxygenation in a central facility. Preliminary process studies on the conversion of various biomass types into liquid fuels have indicated that hydrothermal liquefaction is more attractive than pyrolysis or gasification.

In the HTL process, biomass, typically as a 25% w/w slurry in water, is treated at temperatures of 300–430°C (570–805°F) and at varying pressures in the presence of liquid water for 5–20 min to yield a mixture of gas (mainly carbon dioxide) and synthetic crude oil (sometimes referred to as bio-crude). Subsequent processing is necessary to turn the synthetic crude oil to a usable fuel (Goudriaan and Peferoen, 1990).

Biomass, such as wood, having a lower energy density is converted to bio-crude having a higher energy density, organic compounds including mainly alcohols and acids, and gases including mainly CO_2. Water is also a by-product. In the products, CO_2, the main component of the gas product, can be used to represent all gases produced, and methanol and ethanol represent organic compounds. The weight fraction of each component is assigned on the basis of the data of the vacuum flash of bio-crude (Feng *et al.*, 2004).

11.3.5.2 *Kerosene and diesel*

Gasification followed by Fischer–Tropsch synthesis (FTS) is currently the most promising method for upgrading low-value biomass to high-value liquid fuels and chemicals (Speight, 2013, 2014, 2019a, 2019b; Tijmensen *et al.*, 2002; Jin and Datye, 2000). To produce synthesis gas from a biomass fuel, the following procedures are necessary: (i) gasification of the biomass, (ii) cleaning of the product gas, and (iii) the use of synthesis gas to produce liquid fuel via FTS.

Synthesis gas (a mixture of carbon monoxide and hydrogen), produced by the gasification of fossil fuels or biomass, can be converted into a large number of organic compounds that can be used as chemical feedstocks, fuels, and solvents. Many of the conversion technologies were first developed for coal gasification but the focus on process economic aspects has resulted in a shift to natural-gas-derived synthesis gas. These conversion technologies successfully apply to biomass-derived synthesis gas as well (Spath and Mann, 2000).

Bio-fuels are considered to be relevant technologies by both developing and industrialized countries due to energy security reasons, environmental concerns, foreign exchange savings, and socio-economic issues related to the rural sector. Catalytic conversion will be a primary tool for industry to produce valuable fuels, chemicals, and materials from biomass platform chemicals. Catalytic conversion of biomass is best suitable for synthesis gas. Economic considerations dictate that the current production of liquid fuels from synthesis gas translates into the use of natural gas as the hydrocarbon source. Biomass is the only renewable fuel that can meet our demand for carbon-based liquid fuels and chemicals. Bio-fuels as well as green motor fuels produced from biomass by FTS are the most modern biomass-based transportation fuels. Green motor fuels are the renewable replacement of crude oil-based diesel. Biomass energy conversion facilities are important for obtaining bio-oil by pyrolysis. The aim of the Fischer–Tropsch process is synthesis of long-chain hydrocarbon derivatives from the carbon monoxide and hydrogen gas mixture.

The products from the Fischer–Tropsch process are mainly aliphatic straight-chain hydrocarbon derivatives (C_xH_y). Besides C_xH_y, also branched hydrocarbon derivatives, unsaturated hydrocarbon derivatives, and primary alcohols are formed in minor quantities. The Fischer–Tropsch process is capable of producing liquid hydrocarbon fuels from synthesis gas. The large hydrocarbon derivatives can be hydrocracked to form mainly diesel of excellent quality. The process for producing liquid fuels from biomass, which integrates biomass gasification with the Fischer–Tropsch process, converts a renewable feedstock into a clean fuel.

The Fischer–Tropsch process is a process for producing mainly straight-chain hydrocarbon derivatives from a synthesis gas rich in CO and H_2. Catalysts are usually employed. Typical operating conditions for FTS are: temperatures of 200–350°C and very high pressures, depending on the desired products. The product range includes low-boiling hydrocarbon derivatives such as methane (CH_4) and ethane (C_2H_6), propane (C_3H_8)

butane (C_4H_{10}), gasoline (C_5–C_{12}), diesel (C_{13}–C_{22}), and waxes (C_{23}–C_{33}). The distribution of products depends on the catalyst and the process conditions (temperature, pressure, and residence time). Synthesis gas must have very low tar and particulate matter content (Jun *et al.*, 2004; Demirbaş 2007).

The Fischer–Tropsch process was established in 1923 by German scientists Franz Fischer and Hans Tropsch. The main aim of the Fischer–Tropsch process is synthesis of long-chain hydrocarbon derivatives from the carbon monoxide and hydrogen gas mixture (synthesis gas). The process chemistry is described by simple equations (although it is much more complex than what the simple equations indicate) (Schulz, 1999):

$$nCO + (n + m/2)\,H_2 \rightarrow C_nH_m + nH_2O,$$

where n is the average length of the hydrocarbon chain and m is the number of hydrogen atoms per carbon. All reactions are exothermic and the product is a mixture of different hydrocarbon derivatives in which paraffin derivatives and olefin derivatives are the main constituents.

Typical operation conditions for the Fischer–Tropsch process are a temperature range of 475–625 K and pressures of 15–40 bar, depending on the process. The kind and quantity of liquid product obtained is determined by the reaction temperature, pressure and residence time, the type of reactor, and the catalyst used. Iron catalysts have a higher tolerance for sulfur, are cheaper, and produce more olefin products and alcohols. However, the lifetime of the Fe catalysts is short and in commercial installations is generally limited to eight weeks (Davis, 2002). Bulk iron (Fe) catalysts are the catalysts of choice for converting low-H_2/CO ratio synthesis gas produced by gasification of biomass or coal into fuels via the Fischer–Tropsch process. These relatively low-cost catalysts have low methane selectivity and high water–gas shift activity. However, the development of a bulk Fe catalyst that combines high FT activity, low methane selectivity, high attrition resistance, and long-term stability is still elusive and presents a widely recognized barrier to the commercial deployment of the Fischer–Tropsch process for biomass conversion. The critical property determining the activity and deactivation of Fe catalysts for the Fischer–Tropsch process appears not to be Fe in the metallic state but the carburized Fe surface.

The design of a biomass gasifier integrated with a Fischer–Tropsch process reactor must aim at achieving a high yield of liquid hydrocarbon derivatives. For the gasifier, it is important to avoid methane formation as much as possible, and convert all of the carbon in the biomass into mainly carbon monoxide and carbon dioxide (Prins *et al.*, 2004). Gas cleaning is an important process before the Fischer–Tropsch process. Gas cleaning is even more important for the integration of a biomass gasifier with a catalytic reactor. To prevent poisoning of the process catalyst, tar, hydrogen sulfide, carbonyl sulfide, ammonia, hydrogen cyanide, alkali, and dust particles must be removed thoroughly (Stelmachowski and Nowicki, 2003).

Synthetic diesel fuels from the Fischer–Tropsch process can show excellent auto-ignition characteristics. This diesel fuel is composed of only straight-chain hydrocarbon derivatives and has no aromatics or sulfur. Reaction parameters are temperature, pressure and the H_2/CO ratio. The product composition from the Fischer–Tropsch process is strongly influenced by the catalyst composition — for example, the product from the use of a cobalt catalyst has a higher content of paraffin derivatives and the product from the use of an iron catalyst has a higher content of olefin derivatives and oxygenate derivatives (Demirbaş, 2007). The design of a biomass gasifier integrated with a Fischer–Tropsch process reactor must aim at achieving a high yield of liquid hydrocarbon derivatives.

11.3.5.3 *Other fuels*

The implementation of bio-fuel into the transportation market has already commenced but the continued implementation into the hydrocarbon-based transport infrastructure faces serious energy-density and compatibility issues. As an example, the bio-oil obtained from the fast pyrolysis of biomass has high oxygen content, especially when produced from wood (Table 11.18). Because of the reactivity of oxygenated groups, the main problem of bio-oil is related to its instability and incompatibility with hydrocarbon-type products and there is a need for hydrodeoxygenation (HDO) of bio-oil, which can be accomplished using a cobalt molybdate catalyst (Zhang *et al.*, 2003).

The main HDO reaction is represented by the following simple equation:

$$-(CH_2O)- + H_2 \rightarrow -(CH_2)- + H_2O.$$

Table 11.18: Properties of pyrolysis oil from various wood feedstocks.

Property	Birch	Pine	Poplar	Various
Solids (wt.%)	0.06	0.03	0.045	0.01–1
pH	2.5	2.4	2.8	2.0–3.7
Water (wt.%)	18.9	17.0	16.8	15–30
Density	1.25	1.24	1.20	1.2–1.3
Viscosity cSt/50°C	28	28	13.5	13–80
Ash (wt.%)	0.004	0.03	0.007	0.004–0.3
Carbon residue (wt.%)	20	16	—	14–23
C (wt.%)	44.0	45.7	48.1	32–49
H (wt.%)	6.9	7.0	5.3	6.9–8.6
N (wt.%)	<0.1	<0.1	0.14	0.0–0.2
S (wt.%)	0.00	0.02	0.04	0.0–0.05
O (wt.%)	49.0	47.0	46.1	44–60
Na+K (ppm)	29	22	2	5–500
Ca (ppm)	50	23	1	4–600
Flash point (°C)	62	95	64	50–100
Pour point (°C)	–24	–19	—	–36 to –9

This is the most important route for chemical upgrading. The reaction has strong analogies with typical refinery hydrogenations like hydrodesulfurization and hydrodenitrogenation. In general, most of the HDO studies have been performed using existing hydrodesulfurization catalysts (NiMo and CoMo on suitable carriers). Such catalysts need activation using a suitable sulfur source and this is a major drawback when using nearly sulfur-free resources like bio-oil.

The primary objective of catalytic partial HDO is to increase the energetic value of the oil by removing bound oxygen in the form of water. HDO of bio-oils involves treating the oils at moderate temperatures with high-pressure hydrogen in the presence of heterogeneous catalysts. The process is carried out in two distinct stages: a first stage at relatively low temperatures (250–300°C, 480–570°F), aimed at stabilizing the bio-oil, and a second stage at higher temperatures (300–400°C, 570–750°F) to deoxygenate the intermediate product. Different types of catalysts are screened, ranging from the conventional sulfide-promoted catalysts used in the hydrodesulfurization process (i.e. $NiMo/Al_2O_3$ and $CoMo/Al_2O_3$) to novel non-sulfide-promoted catalysts based on noble metal catalysts (i.e. Ru/Al_2O_3). Operating conditions were optimized to obtain the highest yield of a hydrocarbon-like liquid product (Huber *et al.*, 2005; Metzger, 2006).

It is possible to produce low-boiling alkanes by aqueous-phase reforming of biomass-derived oxygenates such as sorbitol, which can be obtained from glucose by hydrogenation. The production of alkanes from aqueous carbohydrate solutions would be advantageous because of the easy separation of the alkanes from water. Much hydrogen is needed to reduce biomass-derived oxygenates to alkanes as shown in Eq. (11.2).

$$C_6H_{14}O_6 + 6H_2 \rightarrow C_6H_{14} + 6H_2O \tag{11.2}$$

It was shown in earlier work that hydrogen can be produced in a catalytic process at 500 K in water from biomass-derived polyols having a C/O stoichiometry of 1:1 such as sorbitol (Metzger, 2006).

$$C_6H_{14}O_6 + 6H_2O \rightarrow 13H_2 + 6CO_2$$

The overall reaction is an exothermic process and converts approximately 1.5 mol of sorbitol into 1 mol of hexane:

$$19C_6O_6H_{14} \rightarrow 13C_6H_{14} + 36CO_2 + 42H_2O$$

Approximately 95% of the heating value and only 30% of the mass of the biomass-derived reactant are retained in the product. The remaining 70% of the biomass is found as CO_2 and water. A maximum yield of 61% of hydrogen with respect to sorbitol was obtained over Pt/Al_2O_3 under optimized reaction conditions (Metzger, 2006).

Destructive reaction of cellulose is characterized by a decreasing polymerization degree and, thus, thermal degradation of cellulose proceeds through a gradual degradation, decomposition and charring on heating at lower temperatures, and a rapid volatilization accompanied by the formation of levoglucosan on pyrolysis at higher temperatures. The glucose chains in cellulose are first cleaved to glucose and, in a second stage, glucosan is formed by the splitting off of one molecule of water. Initial degradation reactions include depolymerization, hydrolysis, oxidation, dehydration, and decarboxylation (LeVan, 1989; Demirbaş, 2000; Hwang and Obst (2003).

11.4 Solid Fuels

In the present context, the term *solid fuel* is the term applied to wood or any other non-gaseous or non-liquid fuel produced from wood.

Burning wood as a fuel is the largest current use of biomass-derived energy and, as such, wood can be used in many forms: as a solid fuel for cooking or heating, occasionally for steam engines (locomotives), and for steam turbines for the generation of electricity. The particular form of wood fuel used depends on (among other things) its source, quantity, and quality. The available forms include logs, bolts, blocks, firewood, stovewood (often from split blocks), charcoal, chips, sheets, pellets, and sawdust. Wastes from saw mills and construction industry by-products include various forms of lumber tailings.

The main components of wood cells are cellulose, hemicellulose, and lignin, forming approximately 99% w/w of the wood material. Cellulose and hemicellulose are composed of long chains of carbohydrate moieties whereas lignin is a complicated component of polymeric phenol constituents. Lignin is rich in carbon and hydrogen, which are the main heat-producing elements. Thus, the calorific value of lignin is higher than that of cellulose and hemicellulose (carbohydrates). Wood and bark also contain the so-called extractives, such as terpenes, fats, and phenols. The amount of extractives from wood is relatively small compared to the amount of extractives from bark and foliage. The nitrogen (N) content of wood is approximately 0.75% w/w but does vary from one tree species to another. The occurrence of sulfur-containing species in wood is minimal and usually not more than 0.05% w/w.

Wood fuel is a general term applied to wood and wood products (including coppices, scrubs, branches, and forest waste) as well as industrial waste that is burned primarily for heating, cooking, or lighting.

The use of wood as a fuel for fires to produce heat and light is the largest current use of biomass as an energy source. The particular form of wood fuel used depends on (among other things) its source, quantity, and quality. Sawmill waste and construction industry by-products include various forms of lumber tailings.

Fuel wood (i.e. wood itself) is the most common solid fuel and continues to be widely used as a major source of energy for households, especially in developing countries. Charcoal is also increasingly used in many African countries by urban dwellers as a result of a relentless process of migration of people from rural areas towards urban centers.

Firewood can be sold by weight but freshly cut (*green*) wood can contain from 40% to 60% moisture by weight, whereas properly seasoned wood contains only 15–20%. Select the driest wood when buying by weight. Green wood will shrink approximately 8% in volume (i.e. approximately 10 cubic feet per cord) when properly seasoned.

Wood should be dried as much as possible before burning. Properly seasoned wood has approximately 7,700 Btu/lb vs. only approximately 5,000 Btu/lb available from green wood. For best results, season or air-dry wood for at least 6–8 months after cutting, which should reduce the moisture content down to 15–20% w/w.

The combustion of wood is a chemical reaction. If the burning process is complete, the carbon and hydrogen in the wood are converted to carbon dioxide and water:

$$C + O_2 \rightarrow CO_2$$

$$2H_2 + O_2 \rightarrow 2H_2O$$

In practice, the burning process is never complete. Combustion always produces small amounts of unburned hydrocarbon derivatives (CO and C_xH_y).

Depending on the combustion efficiency the ash may contain combustibles. The *ash deformation temperature* is the temperature at which first signs of rounding due to melting of the tip or edges of the test piece occur. *The ash flow temperature* is the temperature at which the *ash* is spread out over the supporting tile in a layer, the height of which is one-third of the height of the test piece at the ash hemisphere temperature. The *ash fusibility temperature* is the temperature at which the ash melts and is the characteristic physical state of the ash obtained by heating under specific conditions.

11.4.1 *Fuelwood*

The most common form of wood fuel at the moment is logs. These will usually come from local sources and can be brought from a variety of outlets — e.g. coal merchants, farmers, and tree surgeons. It is important that the logs are dry and well-seasoned. Burning wet or unseasoned wood is less efficient and can cause harmful buildup of deposits in the chimney over a very short time. Thick coatings of creosote or resinous material can cause chimney fires, or prevent the chimney functioning properly. This can allow harmful fumes to escape into the dwelling. Efficient burning is achieved by setting the appliance to burn fast after stoking. This will ensure that all the gases are fully burned. Just set to slow burn when all the wood has been reduced to charcoal and ash. Newly added wood set to burn slowly creates smoke and creosote in the chimney. The stove should not be banked up with logs for overnight burning. A bright fire which has turned the wood into charcoal should be left with the ash, no secondary air and minimal primary air.

11.4.2 *Logs and wood chips*

Log-fired heating conjures up images of open fires and log stoves for many people, but there are also sophisticated, controllable log-boilers available, which can provide central heating and hot water. Logs are readily available in many areas where there are existing markets for domestic fuel wood. Logs should be well seasoned before burning to ensure the most efficient combustion, and should ideally be stored under cover in a well-ventilated log store for a year or more.

Logs are ideal for providing heat for domestic buildings and are suitable for heating loads of up to around 50 kW. Log heating systems do require manual stoking once a day, which makes logs less convenient than more automated heating systems such as woodchips and pellets. Log-fired systems are available with outputs greater than 50 kw to heat larger buildings such as village halls; however, these larger systems will require more frequent fueling.

Woodchips are made from whole trees, branch wood, or coppice products which have been mechanically shredded by a chipping machine. For some types of boilers, the wood should have been air-dried before chipping, or the chips dried before burning. Woodchips are a bulky fuel and sufficient storage and delivery access needs to be considered when designing a heating system. Transport costs can be high, but if wood chips are sourced within 20 miles they can be a very cost-effective fuel.

The potential for woodchip heating in many countries is high. The use of a timber resource for local woodchip heating would provide a valuable economic return and stimulate the rural economy. Wood chip systems can provide automated, clean, and convenient heating for larger domestic properties with outputs of 20–30 kW, up to large-scale systems for hospitals, factories, schools, and district heating schemes with heating loads in the megawatt range. Wood chips can also be used to fuel combined heat and power plants in which the heat produced during electricity generation is used to provide hot water, and is not lost as in conventional power stations.

Logs that have not been seasoned should be stored under cover but open to allow free air circulation for at least a year — some logs may take 3–4 years to fully season. In practice, the fuel should be allowed to dry before use. Wood from different trees have different heat values Wood fuel has typically less than half the calorific value of coal and is a smokeless fuel and, as a result, a greater volume of wood (than the required volume of coal) is required to produce the same amount of heat.

Woodchips are generally a by-product of the arboriculture industry or are created from waste timber. This fuel is used in screw-fed boilers that have the capacity to fuel local micro-chip systems and larger buildings such as leisure centers, where commercial-sized boilers are needed. It is important to ensure that stored woodchips do not contain excessive quantities of moisture.

These products are manufactured from sawdust bound or pressed together into the shape of a log. They incorporate an inflammable substance that renders them easy to light. Some of them are suitable for open fires. They are relatively expensive but can give a fire lasting for 2 or 3 h. They are clean and handy to use and are ideal for a quick fire on a cold night. These products can be purchased from supermarkets, garages, and other shops, as well as from some coal merchants.

11.4.3 *Pellets and briquettes*

Pellets, made of compressed sawdust or wood shavings, have been available in many parts of the world for several decades. Because of being compressed, pellets offer a more concentrated form of fuel than wood chips. Consequently, they need lower storage space and are easier to handle. The manufacture of wood pellets requires more energy than that of wood chips, and the capital cost for a production plant is high; pellets are therefore more expensive than chips.

Wood pellets are usually made of compressed sawdust, which is a waste product from some industries (e.g. sawmills). They are held together by the natural lignin present in wood — typically no binder glue is needed. Lignin constitutes approximately one-quarter to one-third of dry wood. It strengthens wood in addition to showing waterproofing properties.

Wood pellets have a very low moisture content, typically 6–10%, and thus are a very good source of energy because very little of the energy in the pellets is used to evaporate moisture. The general properties of wood pellets are as follows:

Moisture: 3.5–5.5%
Ash: 0.2–0.5%
Calorific value: 8,800 Btu/lb
Bulk density: 600–750 kg/m

Wood pellets are extremely dense, and their low humidity content (below 10%) allows them to burn with a very high combustion efficiency. The regular geometry of these pellets and their small size allow automatic feeding with very fine calibration. Their high density enables compact storage and rational transport over a long distance.

Woodchips are a solid fuel made from wood that is often used as a raw material for technical wood processing. In industry, processing of bark chips is often separated after peeling the logs due to different chemical properties. Woodchips can also be used as a wood fuel.

Pellets can be used to fuel a variety of appliances and heating systems, the smallest being pellet stoves with an output of up to around 9 kW, which are suitable for heating individual rooms and are electronically controlled to deliver regulated heat output and need fueling only once every few days. Pellet boilers are available in a wide range of outputs from the small domestic scale to the large industrial scale to heat schools or hospitals. For the domestic user, pellets offer the most user-friendly form of wood heating. In Scandinavia, wood pellets are delivered by tanker and are pumped into storage silos, which feed automatically into the boiler.

Wood pellets are used on specially designed stoves. Pellets are typically made from the timber waste from sawmills. This wood goes through a fairly lengthy process of transformation before it is finally extruded as hard pellets. In spite of this processing, the fuel is still carbon-neutral and is therefore less harmful than other fossil fuels to the environment. For commercial undertakings, as is the case with any wood fuel, wood pellets do not attract climate change levy tax. At present, there are a limited number of pellet-producing plants in the UK although there is some encouragement to produce them from local sources. In the meantime, some pellets are being imported. Pellets for domestic room heaters and boilers are being supplied in prepacked bags, usually 10 kg in weight, which renders them a handy size to store and load into the hopper. For larger boiler models, pellets can be delivered in bulk, but an appropriate silo must be available, connected to the boiler. At the moment most of the pellets are supplied by the stove shops that market the boilers.

Fuel briquettes (sometime called fuel briquets), especially briquettes manufactured from coal (Speight, 2013), emerged as a significant business enterprise in the 20th century. In the 1950s, several economic methods were developed to make briquettes without a binder. A multitude of factories throughout the world produced literally tens of millions of tons of usable and economic materials that met the household and industrial energy needs. During the two World Wars, households in many European countries made their own briquettes from soaked waste paper and other combustible domestic waste using simple lever-operated presses (Lardinois and Klundert, 1993). Modern industrial briquetting machines, although much larger and more complex, operate on the same principle although the marketed briquettes are now sold at a premium for occasional backyard barbeques rather than for everyday use.

For over 100 years, informal waste collectors in Cairo have separated and dried organic waste products for sale as solid fuel for domestic use. This process faded somewhat when fossil fuel sources became available. Switching to conventional fuels may prove advantageous to those who can afford them, but given the economic and energy conditions in many cities, urban and agricultural wastes remain a viable alternative fuel.

Briquetting is undergoing resurgence, principally due to the convergence of three critical factors. First, recent developments in briquette processing and binding have dramatically changed the economic aspects of using fuel briquettes as an energy resource. Secondly, the shortage of fuel wood has become increasingly severe in most of the developing countries. Finally, there has been a steady increase of environmental concerns to address the problem of domestic and urban waste disposal, a dilemma that briquetting can help to remedy.

11.4.3.1 *Briquette manufacture*

Generally, briquette manufacture (*briquetting*) involves the collection of combustible materials that are not usable as such because of their low density, and compressing them into a solid fuel product of any convenient shape that can be burned like wood or charcoal. Thus, the material is compressed to form a product of higher bulk density, lower moisture content, and uniform size, shape, and material properties. Briquettes are easier to package and store, cheaper to transport, and more convenient to use, and their burning characteristics are better than those of the original organic waste material.

The raw material of a briquette must bind during compression; otherwise, when the briquette is removed from the mold, it will crumble. Improved cohesion can be obtained with a binder and also without one, since under high temperature and pressure, some materials such as wood bind naturally. A binder must not cause smoke or gummy deposits, and the creation of excess dust must also be avoided. Two different sorts of binders may be employed. Combustible binders are prepared from natural or synthetic resins, animal manure or treated, dewatered sewage sludge. Non-combustible binders include clay, cement, and other adhesive minerals. Although combustible binders are preferable, non-combustible binders may be suitable if used in sufficiently low concentrations. For example, if organic waste is mixed with too much clay, the briquettes will not easily ignite or burn uniformly. Suitable binders include starch (5–10% w/w) or molasses (15–25% w/w) although their use can prove expensive (Lardinois and Klundert, 1993). It is important to identify additional, inexpensive materials to serve as briquette binders in Kenya and find their optimum concentrations. The exact method of preparation depends on the material being briquetted, as illustrated by the following three cases of compressing sugar bagasse, sawdust, and urban waste into cooking fuel briquettes.

11.4.3.2 *Bagasse briquettes*

Surplus bagasse presents a disposal problem for many sugar factories. Briquetting technology remains simple and involves the following steps: (i) size reduction, in which the bagasse is chopped, rolled or hammered, (ii) drying, in which moisture is removed by open air-drying or by using forced, heated air in a large rotating drum, (iii) carbonization, in which the bagasse is combusted in a limited supply of oxygen in a buried pit or trench until it carbonizes into charcoal, (iv) feedstock preparation, in which the carbonized bagasse is mixed with a binder such as clay or molasses, (v) compaction and extrusion, in which the material is passed through a machine-operated or manually operated extruder to form *rolls* of charcoal, (vi) drying in which the rolls are air-dried for 1–3 days, causing them to break into chunks, and (vii) packaging in which the briquettes are made ready for sales.

11.4.3.3 *Sawdust briquettes*

Sawdust is the waste material from all types of primary and secondary wood processing. In milling operations 10–13% w/w of a log is reduced to sawdust. Sawdust is bulky and is therefore expensive to store and transport. Also, the calorific value of sawdust is quite low making briquetting an ideal way of reducing the bulk, increasing the density, and thus increasing the calorific value. The equipment required to produce sawdust briquettes consists of a drier, a press and an extruder with a tapered screw, and a large revolving disk.

Sawdust briquettes are formed under a pressure sufficiently high to produce cohesion between the wood particles. In the process, lignin softens and then binds the briquettes, so no additional binder is needed. The use of sawdust briquettes has several advantages, including (i) they have usually approximately the same price as fuelwood but are much more convenient to use as they do not require further cutting and chopping, (ii) they have good burning characteristics in any kind of solid fuel stove or boiler, (iii) quick ignition followed by clean burning, leaving only 1%–6% w/w mineral ash, (iv) they are sulfur-free and burn without producing an odor, and (v) they have a heat content almost equivalent to that of medium-quality coal.

11.4.3.4 *Urban waste briquettes*

Disposal of solid waste (refuse) is one of the most serious urban environmental problems in developing countries. Many municipal authorities collect and adequately dispose of in places other than landfill sites less than half of these wastes. This failure is attributed to (i) the inadequate number of landfill sites, (ii) a variety of environmental regulations, (iii) the absence of a sufficient capacity for processing and recycling waste, and last but not least, (iv) the planned obsolescence of packaging and many of the items that form the basis of the waste.

Open or crude dumping is the method most commonly used by municipal authorities. Waste poses a health hazard when it lies scattered in streets and at dumping sites. It is now an accepted environmental philosophy that wastes have value and should be utilized based on principles of *reduce*, *reuse*, *recover*, and *recycle*. Through recycling, urban wastes can be transformed into useful products. Waste paper and leaves, in particular, provide a potentially important alternative source of cooking fuel.

Recycling organic combustible materials into fuel briquettes contributes to solving urban needs such as income generation, insufficient land for waste disposal, and maintaining environmental quality. Since the resources of the Earth are finite, greater resource recovery and utilization is essential to achieve an acceptable level of organic waste management. Enhancing the recovery of organic waste can restore various natural cycles, thus preventing the loss of raw materials, energy, and nutrients.

On the other hand, the demand for energy in many countries is expected to add to the emissions of greenhouse gases through burning of fossil fuels. There is an urgent need to promote climate-friendly technologies and fuel briquetting appears to be one such technology that addresses the multiple needs of society and the environment.

Current needs are focused on finding better binders for bagasse briquettes, achieving improved calorific values and combustion by producing higher density briquettes, designing more efficient extrusion methods, and reducing the production costs.

11.4.4 *Charcoal*

The most well-known solid fuel produced from wood is charcoal, but there are other sources such as coconut shells and crop residues. Charcoal is produced by heating wood under conditions that severely restrict the amount of oxygen. When wood is heated slowly to approximately 280°C (535°F), an exothermic reaction occurs. The usual carbonization procedure is prolonged, heating to a final temperature of 400–500°C (750–930°F) in the absence of air. The yield of charcoal is approximately 25% w/w on a dry basis.

Thus, charcoal is produced in kilns by pyrolysis, i.e. breaking down the chemical structure of wood under high temperature in the absence of air. During the process, first the water is driven from the wood (drying), and next, pyrolysis starts when the temperature in the kiln is high enough. When pyrolysis is complete, the kiln gradually cools down after which charcoal can be removed from the kiln. Because some of the wood is burned to drive off the water, dry wood produces better charcoal and at a higher efficiency. Typically, approximately two-thirds of the energy is lost in the process, but charcoal has advantages over fuel wood such as stoves with higher efficiency, higher convenience, and easier distribution.

Commercial charcoal is found either as (i) lump charcoal, (ii) briquetted charcoal, or (iii) extruded charcoal. *Lump charcoal* is made directly from hardwood material and usually produces far less ash than briquettes. *Briquettes* are made by compressing charcoal, typically made from sawdust and other wood by-products, with a binder (usually starch). *Extruded charcoal* is made by extruding either raw ground wood or carbonized wood into logs without the use of a binder — the heat and pressure of the extruding process hold the charcoal together. If the extrusion is made from raw wood material, the extruded logs must be subsequently carbonized.

Charcoal quality can be specified and measured in various ways, depending on the end use. Efficiency to use normally means the ability to transfer the maximum amount of heat content of charcoal to the object to be heated and is dependent on the use of properly designed charcoal burning equipment.

Charcoal burns by reacting with oxygen in the air at a glowing red heat to form colorless carbon monoxide gas, which then burns with a blue flame with more oxygen from the air to produce carbon dioxide gas. Due to the heat liberated by both reactions, charcoal reaches a glowing red and radiates heat energy, and hot carbon dioxide gas leaves the combustion zone. The gas temperature falls as it transfers heat and it passes into the room. Carbon monoxide gas may be released by burning charcoal. It is very poisonous and therefore ventilation of any rooms where charcoal is burning is necessary.

Char is one of the thermal degradation products of wood. These products consist of moisture, volatiles, char, and ash. Volatiles are further subdivided into gases and liquid (including tarry products). Some properties vary with the species, location within the tree, and growth conditions. Other properties depend on the combustion environment (Ragland *et al.*, 1991).

Char is typically a fine-grained charcoal that is high in organic carbon and largely resistant to decomposition. It is also produced by the pyrolysis of other (non-woody) plant and waste feedstocks. During the pyrolysis, several changes take place in wood as the temperature is increased and the different chemical components undergo thermal degradation, which affects the performance of wood. The extent of the changes depends on the temperature level and length of time under exposure conditions.

For example, permanent changes in the properties of the wood start occurring at temperatures above 65°C (150°F), with the amount depending on the temperature, pH of wood, moisture content, heating medium, exposure period, and species of wood. This is likely to be due to depolymerization reactions involving no significant carbohydrate weight loss. Chemical bonds begin to break at temperatures in excess of 100°C (212°F) as carbohydrate and lignin weight losses of various types increase with the temperature (wood pyrolysis) (Shafizadeh, 1984). Between 100°C (212°F) and 200°C (390°F), wood dehydrates as it loses water vapor and other noncombustible gases and liquids (including carbon dioxide). With prolonged exposure to higher temperatures, wood starts to form char. From 200°C to 300°C (390°F to 570°F), wood components begin to undergo significant pyrolysis, and significant amounts of carbon monoxide and high-boiling-point tar are evolved — hemicellulose and lignin constituents are pyrolyzed in the range 200–450°C (390–840°F).

Dehydration reactions are primarily responsible for the pyrolysis of hemicellulose and lignin and result in a high char yield for wood. Although cellulose remains mostly unpyrolyzed, its thermal degradation can be accelerated in the presence of water, acids, and oxygen. As the temperature increases, the degree of polymerization of cellulose decreases further and any tar formed undergoes cracking to lighter gases and repolymerization to char while streaming through the hot char residue. All wood components end their volatile emissions at approximately 450°C (840°F), and at higher temperatures the remaining wood residue is char, which undergoes further degradation (if oxygen is absent) to a higher-carbon residue or (if oxygen is present) to carbon dioxide, carbon monoxide, and water.

Upon completion of pyrolysis, the char has a porosity of the order of 0.8–0.9 (for wood heated under combustion conditions) and the initial char density is 10–20% that of the dry wood and as the char continues to react the density decreases further (Ragland *et al.*, 1991).

The oldest and, probably, even now the most widely used method for charcoal production is the earth kiln. Two varieties exist: earth pit kilns and earth mound kilns. An earth pit kiln is constructed by first digging a small pit in the ground. Then the wood is placed in the pit and lit from the bottom, after which the pit is covered first with green leaves or metal sheets and then with earth to prevent complete burning of the wood. An earth mound kiln is built by covering a mound or a pile of wood on the ground with earth. The mound is preferred over the pit where the soil is rocky, hard, or shallow, or the water table is close to the surface. Mounds can also be built over a long period, by stacking gathered wood in position and allowing it to dry before covering and burning.

Earth kilns can be made at minimal cost, and are often used near wood resources, since they can then be made entirely from local materials. Earth kilns can be made in any size, with the duration of the process ranging from three days to two months. Gross variations in the quality of the charcoal can occur, because in a batch, some of the wood is burned, while some of the wood is only partly carbonized. Efficiencies are generally low, around 10 to 20% by weight and 20–40% in energy terms. The efficiency and quality vary depending on the construction of the kiln (e.g. walls can be lined with rocks or bricks and external chimneys can be used) and the monitoring of the carbonization process.

Several other types of charcoal kilns have been developed; they generally have higher efficiencies but require higher investments than the earth kiln. Two most often-used types are fixed kilns made of mud, clay, and bricks and portable steel kilns. Fixed kilns usually have a beehive shape. Smaller beehive kilns are usually made of mud and so are not very durable, while larger beehive kilns are made of bricks and have external chimneys. Beehive kilns have an opening for loading the wood and for unloading the charcoal; it is closed after loading.

Portable steel kilns can be made from oil drums, and can be used in both the horizontal and vertical positions. They generally have a short lifetime. When used in the horizontal position, an opening is made on the side, through which the wood is loaded. For the vertical kiln the top is cut out and used as a lid.

The production of briquettes (briquetting or densification) is used to improve the characteristics of materials destined for transport and use as an energy source. Raw materials include sawdust, loose crop residues, and charcoal fines. The material is compacted under pressure, and depending on the material, pressure, and speed of densification, additional binders may be needed to bind the material. The two main briquetting technologies are the piston press and the screw press. In the piston press the material is punched into a die by a ram with a high pressure. In the screw press, the material is compacted continuously by a screw. Using the screw press, generally briquettes of higher quality can be produced.

Charcoal is produced by heating wood under conditions that severely restrict the amount of oxygen. When wood is heated slowly to approximately 280°C (535°F), an exothermic reaction occurs. The usual carbonization procedure is prolonged, heating to a final temperature of 400–500°C (750–930°F) in the absence of air. The term *charcoal* includes charcoal made from bark.

Charcoal is produced commercially in either kilns or continuous furnaces from primary wood processing residues and from low-quality round wood. A kiln is used if the raw material is in the form of round wood, sawmill slabs, or edgings. The most common kilns in the United States are constructed from poured concrete. They have a capacity of 40–100 cords of wood and operate on an approximately 7–12-day cycle. If the raw material is in the form of sawdust, shavings, or milled wood and bark, it is converted to charcoal in a continuous multiple-hearth furnace, commonly referred to as a Herreshoff furnace. The capacity is usually at least one ton of charcoal per hour. The yield of charcoal is approximately 25% by weight on a dry basis.

The proximate analysis of charcoal is approximately 20–25% volatile matter, 70–75% fixed carbon, and 5% ash. The higher heating value is approximately 12,000 Btu/lb (28,000 J/kg). Charcoal briquettes have a higher heating value of only approximately 10,000–11,000 Btu/lb (23,000–25,000 J/kg) due to added ingredients. To alleviate the air pollution problem associated with charcoal kilns and furnaces, the gases from the kilns and furnaces are burned. The gases can be burned with additional fossil fuel to recover heat and steam, or they can be burned in afterburners to nearly eliminate visible air pollution and odors.

Charcoal was an important industrial raw material in the United States for iron ore reduction until it was replaced by coal in the early 1880s. Charcoal production did continue to increase, however, due to the demand for the by-products acetic acid, methanol, and acetone.

The increase in production of charcoal since the 1940s reflects the use of charcoal briquettes for home recreational cooking. Nearly all the charcoal presently produced is consumed for cooking using charcoal briquettes. Some charcoal is consumed in certain metallurgical and filtration processes and horticultural uses. In Brazil, charcoal is produced (in beehive-type kilns from naturally and plantation-grown trees) for use as an iron ore-reducing agent because Brazil does not have abundant supplies of coking coal. In many developing countries, charcoal is the preferred fuel for domestic cooking. Very often, it is produced in pit-type kilns, but beehive and portable sheet metal kilns are also used where the construction materials are available.

11.4.5 *Coke*

The carbonization of coal is essentially a process for the production of a coke (the carbonaceous residue) by thermal decomposition (with simultaneous removal of distillate) of organic substances (Speight, 2013):

$$C_{organic} \rightarrow C_{coke/char/carbon} + \text{liquids} + \text{gases}$$

The process may also be referred to as *destructive distillation* and has been applied to a whole range of organic materials, but more particularly to natural products such as wood, sugar, and vegetable matter to produce charcoal. Coal usually yields "coke", which is physically dissimilar to charcoal and has the more familiar honeycomb-like structure.

The thermal decomposition of coal varies strongly as a function of the coal type or rank. Low-rank coals, such as lignite and subbituminous coal, produce relatively large amounts of light gases and very little tar. Bituminous coal produces much more tar than the low-rank coals and moderate amounts of light gases. Higher rank coal produces relatively small amounts of both low-boiling gases and tar. In fact, only a limited number of coals produce acceptable metallurgical cokes. These coals are in the bituminous rank range but not all bituminous coals are coking coals. Prime coking coals are expensive and not always available nationally. It is quite possible that the remaining indigenous coals available for coke making are poorer in coking quality.

The thermal decomposition of coal is a complex sequence of events which can be described in terms of several important physico-chemical changes, such as the tendency of the coal to soften and flow when heated (plastic properties) or the relationship to carbon-type in the coal. In fact, some coals become quite fluid at temperatures of the order of 400–500°C (750–930°F) and there is a considerable variation in the degree of maximum plasticity, the temperature of maximum plasticity, as well as the plasticity temperature range for various coals. Indeed, significant changes also occur in the structure of the char during the various stages of devolatilization.

At temperatures of just over 100°C (212°F) the coal residual moisture evolves. The evolution of low-boiling gases begins at temperatures of

200–500°C (390–930°F); these early gases consist mainly of carbon monoxide (CO) and carbon dioxide (CO_2) and low-boiling hydrocarbons. Tar formation was seen in low-heating rate experiments to begin at around 330°C (625°F) and increases to temperatures above 530°C (985°F). Cross-linking reactions are thought to occur at different temperatures — depending on the coal and heating rates — due to different kinetics of the competing processes of bond breaking, vaporization, and bond formation. Pyrolysis experiments conducted with a number of coals indicate that early cross-linking begins in the range of 400–500°C (750–930°F), and later cross-linking continues as temperatures increase. The exact temperatures at which these pyrolysis steps occur are dependent on many factors and it is likely that the heating rate and coal type have the largest effect on the temperatures at which these steps occur.

On the other hand, coke is also produced by the destructive distillation (thermal decomposition with simultaneous removal of distillate) of crude oil residua (Parkash, 2003; Gary *et al.*, 2007; Speight, 2011a, 2014, 2017; Hsu and Robinson, 2017). That formed in catalytic cracking operations is usually non-recoverable, as it is often employed as fuel for the process. The composition of the coke varies with the source of the crude oil, but in general, large amounts of high-molecular-weight complex hydrocarbon derivatives (rich in carbon but correspondingly poor in hydrogen) make up a high proportion. The solubility of petroleum *coke* in carbon disulfide has been reported to be as high as 50–80%, but this is in fact a misnomer, since the coke is the insoluble, honeycomb material that is the end product of thermal processes.

Three physical structures of coke can be produced by delayed coking: (1) shot coke, (2) sponge coke, or (3) needle coke.

Shot coke is an abnormal type of coke resembling small balls. Due to mechanisms not well understood, the coke from some coker feedstocks forms into small, tight, non-attached clusters that look like pellets, marbles, or ball bearings. It is usually a very hard coke, i.e. low Hardgrove grindability index (Speight, 2013). Such a coke is less desirable to the end users because of difficulties in handling and grinding. It is believed that feedstocks that are high in asphaltene constituents and low in API favor shot coke formation. Blending aromatic materials with the feedstock and/or increasing the recycle ratio reduces the yield of shot coke. Fluidization in the coke drums may cause the formation of shot coke.

Occasionally, the smaller *shot coke* may agglomerate into ostrich egg-sized pieces. While shot coke may look like it is entirely made up of shot, most shot coke is not 100% shot. Interestingly, even sponge coke may have some quantity of embedded shot coke. The test for such a measurement is not precise but a low shot coke percentage is sometimes specified for anode grades of petroleum coke. In the case of fuel coke production, there is a general belief that the shot structure of coke means that the coker is being run under the most economical conditions.

Sponge coke is the common type of coke produced by delayed coking units. It is in a form that resembles a sponge and has been called honeycombed. Sponge coke, mostly used for anode grade, is dull and black, having a porous, amorphous structure.

Needle coke (*acicular coke*) is a special quality coke produced from aromatic feedstocks, is silver-gray, has a crystalline broken needle structure, and is believed to be chemically produced through cross-linking of condensed aromatic hydrocarbon derivatives during coking reactions. It has a crystalline structure with more unidirectional pores and is used in the production of electrodes for the steel and aluminum industries, where it is particularly valuable because the electrodes must be replaced regularly.

Petroleum coke is employed for a number of purposes, but its chief use is in the manufacture of carbon electrodes for aluminum refining, which requires a high-purity carbon — low in ash and sulfur-free; the volatile matter must be removed by calcining. In addition to its use as a metallurgical reducing agent, petroleum coke is employed in the manufacture of carbon brushes, silicon carbide abrasives, and structural carbon (e.g. pipes and Rashig rings), as well as calcium carbide manufacture from which acetylene is produced:

$$\text{Coke} \rightarrow \text{CaC}_2$$

$$\text{CaC}_2 + \text{H}_2\text{O} \rightarrow \text{HC}{\equiv}\text{CH}$$

11.5 Fuel Quality

Fuel quality is dictated by the extent of refining required to convert the synthetic oil into specification-grade transport fuels, which, in turn, is determined by the feedstock quality in relation to the fuel specifications. A central theme in the refining literature of synthetic oil (oil obtained from non-fossil fuel sources) is catalytic hydroprocessing. Heteroatoms in general and sulfur in particular must be removed to very low levels in transport fuels.

It is possible to meet the specifications related to the heteroatoms content of transport fuels by hydroprocessing the refinery products at high temperature and pressure. However, heteroatom content is not the only fuel specification to consider when producing transport fuels and hydroprocessing alone may be insufficient to produce transport fuels and may even cause some fuel properties to deteriorate. Thus, in order to produce transport fuels, all of the fuel specifications must be met. While hydroprocessing of any synthetic oil may be an important pretreatment step, the hydrotreated oil must still be refined and the implication for the production of specification-grade fuel production is that hydroprocessing must take place at the highest practical pressure of hydrogen and the lowest practical temperature.

After the fuel is produced, the general aspects of fuel quality (whatever the source of the fuel) are assessed by the measurement of physical properties such as relative density (specific gravity), refractive index, or viscosity, or by empirical tests such as pour point or oxidation stability that are intended to relate to behavior in service. In some cases the evaluation may include tests in mechanical rigs and engines either in the laboratory or under actual operating conditions.

11.5.1 *Bulk properties*

Measurements of bulk properties are generally easy to perform and, therefore, quick and economical. Several properties may correlate well with certain compositional characteristics and are widely used as a quick and inexpensive means to determine these. The most important properties of a whole crude oil are its boiling-point distribution, its density (or API gravity), and its viscosity. Density and viscosity are measured for secondary reasons. The former helps to estimate the paraffinic character of the fuel, and the latter permits the assessment of its undesirable residual material that causes resistance to flow. Sulfur content, another crucial and primary property of a crude oil, is also readily determined. Certain composite characterization values, calculated from density and mid-boiling point correlate better with molecular composition than density alone.

However, as a preliminary step, the value of any product is judged by the characteristics of the sample as determined by laboratory tests. The sample used for the test(s) must be representative of the bulk material, otherwise data will be produced that are not representative of the material and will, to be blunt, incorrect no matter how accurate or precise the test method. In addition, the type and cleanliness of sample containers are important; if the container is contaminated or is made of material that either reacts with the product or is a catalyst, the test results may be wrong.

Thus, the importance of the correct sampling of any fuel destined for analysis should always be overemphasized. Incorrect sampling protocols can lead to erroneous analytical data from which the performance of the product in service cannot be accurately deduced. In addition, adequate records of the circumstances and conditions during sampling have to be made; for example, in sampling from storage tanks, the temperatures and pressures of the separation plant and the atmospheric temperature would be noted.

Gaseous fuels derived from the gasification and pyrolysis of MSW, crop and wood waste, biomass, scrap tyres (pyrolysis gas and synthesis gas), biogas from anaerobic digestion (AD), and land fill gas (LFG) were compared to natural gas (NG).

Liquid waste-derived fuels such as used lubricant engine oils, diesel and gasoline-like fuel, straight and bio-diesel blends, biocrude, and pyrolysis derived liquids were compared to regular petroleum-derived fuels (diesel fuel, fuel oil, gasoline, etc.) using common fuel characteristics such as cetane number, octane number, carbon residue, cloud, flash, pour point, acid number (mg KOH/g), viscosity, ash, iodine value (oxidative stability), and heating value as an indication of its suitability for application in combustion ignition (CI) and spark ignition (SI) engines. Brief explanations of these important fuel characteristics are given.

Cetane number is a dimensionless descriptor related to the ignition delay time a fuel experiences upon injection into the combustion chamber of a diesel engine. The higher the cetane number, the shorter the ignition delay time and vice versa. A cetane scale using hydrocarbon derivatives has been established and is commonly used, with hexadecane being the high cetane number reference compound on this scale (CN = 100).

Octane number is a measure of the gasoline quality for the prevention of early ignition, which leads to cylinder knocks. Higher octane numbers are preferred in internal combustion engines. For gasoline production, aromatics, naphthenes, and iso-alkane derivatives are highly desirable, whereas olefin derivatives and *n*-paraffin derivatives are less desired.

Viscosity is the thickness of the oil, that is, the resistance the oil gives to motion while being deformed by stress (either shear stress or tensile stress). Viscosity describes the internal resistance of a fluid to flow and may be thought of as a measure of fluid friction. The less viscous a fluid is, the greater its ease of movement.

The *flash point* of a volatile liquid is the temperature at which the oil will vaporize and form an ignitable mixture in air. The flash requires an ignition source, and the vapor may cease to burn if the source of ignition is removed. The flash point should not be confused with the autoignition temperature, which does not require an ignition source.

The *cloud point* is the temperature at which the first solids appear in a fuel but the fuel can still flow; these solids can lead to fuel filter plugging. Several other methods exist for determining the low-temperature properties of bio-diesel. These are the cold filter plugging point (CFPP) and the low-temperature flow test (LTFT). The low-temperature properties of bio-diesel are again influenced by the properties of the individual components.

The *pour point* is the temperature at which the oil stops flowing like a liquid. The pour point is usually a few degrees below the cloud point and is the temperature at which the fuel can no longer be freely poured.

Mixtures such as bio-diesel do not possess melting points, but rather melting ranges; this is reflected in the specifications used in bio-diesel standards (*cold flow*). The melting points of fatty esters, which are the main constituents of bio-diesel, generally increase with chain length (although chains with odd numbers of carbons have slightly lower melting points than the preceding even-number chain) and increasing saturation.

The *iodine value* (IV) is a measure of the total unsaturation of a sample and has often been used in connection with issues such as oxidative stability, with the implication that materials with a high iodine value are less stable to oxidation than those with a low iodine value. The iodine value is also molecular weight dependent; for example, ethyl esters have a lower iodine value than methyl esters although the kind of ester is highly unlikely to affect the oxidative stability of the double bonds in a fatty acid chain. Therefore, the iodide value should not be considered in the choice of optimized bio-diesel components.

The *heating value* (*heat content*, *calorific value*, and *Btu value*) of a fuel may be expressed as the *higher heating value* (HHV) or the *lower heating value* (LHV). The lower heating value is the total quantity of sensible heat released during combustion of a specified quantity of fuel. The higher heating value is the lower heating value plus the enthalpy of evaporation of water vapor resulting from combustion (can be substantial for hydrogen-rich fuels, typical for renewable energy applications). The higher heating value is usually favored because its value allows easier checking of energy calculations when examining technologies for implementation.

Biomass pyrolysis liquids are of significant interest because of their potential as biofuels for combustion systems (engines, boilers, and turbines). The quality of pyrolysis liquids is determined by several physical and chemical properties, namely, *phase separation, solid and water content of the liquid product and higher heating values of the organic phase and liquid product*. Phase separation is one of the main problems of pyrolysis liquids in terms of their application and commercialization. Low solids content is required because the solids can contain metals with catalytic activity that will cause the liquid to become unstable during storage. Solids are capable of causing mechanical problems in turbines and engines, because of its erosion power. Moderate water content (<27% w/w) appears beneficial from the point of view of viscosity and atomization, although higher water contents can lead to a low heating value and to a phase-separated pyrolysis liquid. The heating value of the liquid product is necessary for evaluating the possibility of using pyrolysis liquids as fuels and for checking the energy balance of the process.

Finally, *waste-derived solid fuels* (briquettes, coal supplements, granulated fuels, and pyrolysis solids) are compared using proximate analysis and heating value. The concept of proximate analysis was developed for coal as a convenient and effective means for determining the distribution of products obtained by heating coal under a set of standard conditions (Table 11.19) (Speight, 2013, 2015b). This particular group that falls within the proximate analysis group of tests has been used widely as the basis for coal characterization in connection with coal utilization. The proximate analysis (ASTM D3172) of coal may also be considered for the determination of the general properties of coal and is, in reality, the determination of the moisture content, volatile matter content, ash yield, and (by difference) fixed carbon yield, in contrast to the ultimate analysis of coal, which provides the elemental composition (Speight, 2013). Similarly, these test methods can be adapted to determine the properties of solid fuels.

The moisture, volatile matter, and ash results are typically among the primary parameters used to assess the quality of the fuel. The moisture result is utilized for calculating the dry basis results of other analytical results. The ash from biomass fuels is typically richer in calcium, silicon,

Table 11.19: Outline of the data retrieved from the ultimate analysis and the proximate analysis of solid fuels.

	w/w%	As received[a]	Dry[b]	Dry-ash-free[c]
Ultimate analysis	Carbon			
	Hydrogen			
	Nitrogen			
	Oxygen			
	Sulfur			
Proximate analysis	w/w%			
	Moisture	√	0	0
	Ash	√	√	0
	Volatile matter	√	√	√
	Fixed carbon	√	√	√

Notes: [a]Often designated as AR.
[b]Often designated as MF (moisture-free).
[c]Often designated as MMF (mineral-moisture-free) or DAF (dry ash-free).

and alkali minerals and micronutrients such as Zn, Cu, and Mn, in comparison with the ash from coal combustion (Vamvuka and Kakaras, 2011). Some could be useful for soil amendment or the cement industry. Slagging/fouling problems should be expected in boilers operating above 1000°C, especially those firing cotton residue, vine shoots, and bituminous coal without pre-treatment. However, the environmental impact of either biomass or coal ashes after their disposal is expected to be very low, as leaching tests have shown. For coal/biomass blends, the composition and fusibility of the ashes varied with those of the individual components.

The ash result is utilized in the ultimate analysis for calculation of oxygen by difference (ASTM D3176) and for calculating material balance and ash load purposes in industrial boiler systems. The volatile matter result indicates the coke/char yield from thermal processes and provides additional information relating to the combustion characteristics of the solid fuel. Fixed carbon is a calculated value of the difference between 100 and the sum of the moisture, ash, and volatile matter if all values are on the same moisture reference base.

Thus, the objective of the proximate analysis is to determine the amount of moisture, volatile matter yield, ash yield, and fixed carbon from the fuel sample. Mineral matter is not measured directly but may be obtained by one of a number of empirical formulae either from the yield of mineral ash or from data derived from the ultimate analysis.

The variables are measured in percent by weight (w/w%) and calculated on several different bases. The (as-received AR) basis is the most widely used basis in industrial applications and it puts all variables into consideration and uses the total weight as the basis of measurement. The (air-dried AD) basis neglects the presence of moistures other than inherent moisture. The (dry basis DB) leaves out all moistures, including surface moisture, inherent moisture, and other moistures. The (dry, ash-free DAF) basis neglects all moisture and ash constituents in the fuel. The (dry, mineral-matter-free DMMF) basis leaves out the presence of moisture and mineral matter in coal, such as quartz, pyrite, calcite, and clay. An example of these bases is given below (Speight, 2013, 2015b):

Proximate analysis	AR	AD	DB	DAF
Moisture (w/w%)	3.3	2.7		
Ash (w/w%)	22.1	22.2	22.8	
Volatile matter (w/w%)	27.3	27.5	28.3	36.6
Fixed carbon (w/w%)	47.3	47.6	48.9	63.4

The proximate analysis of the solid fuel covers the determination of moisture, volatile matter, ash content, and fixed carbon (initial sample mass – (volatile matter + ash content)). Volatile matter and fixed carbon are the dominant components to produce heat. Density, compressive strength, and ignition temperature are important for storage, transportation, and combustion safety.

11.5.2 *Specifications*

Presenting the specifications for the various fuels is beyond the scope of this book but selected items should be considered when placing a fuel on the market. Nevertheless, it is essential that there is compliance with national, regional, and state fuel specifications to avoid the risk of costly fines.

In the United States (as is the case with various other countries), fuel specifications (i.e. the properties of the fuel) vary from state to state or from region to region. Thus, it is especially important to know the correct fuel type to pull and mention the right one in supplier contracts. Among the many issues to be considered are (i) state-to-state or region-to-region fuel taxes, (ii) renewable fuel standards, (iii) ultra-low-sulfur fuel standards, and (iv) the Reid vapor pressure (RVP) standards, by which in the summer months, the reference limit is 7.00 psi.

Furthermore, with the increasingly stringent legislation with respect to transportation fuel specifications, the growing demand for high-quality fuel is likely to be met by synthetic production via the proposed gas-to-liquid (GTL) plants based on the Fischer–Tropsch technology synthesis of hydrocarbon derivatives — especially high-molecular-weight hydrocarbon derivatives — followed by hydrocracking. Since this feedstock differs substantially from traditional crude oil refinery feedstocks, existing hydrocracking catalysts and processes are not necessarily optimized for use in gas-to-liquid operations. Base metal catalysts appear to be principally suitable for hydrocracking high-molecular-weight Fischer–Tropsch products and have the advantage of producing less-branched, high-cetane number diesel.

Thus, no matter what processes have been applied to fuel production, it is necessary to ensure that the fuel specifications contain the right requirements. Different parameters have different significance — some will affect plant operation and availability, while others will influence properties such as emissions and ash quality. Moreover, specifying the property limits for all possible properties may not be the correct method — it is necessary to identify the relevant properties and try to find the appropriate limits for those properties. It is also necessary to be aware of changes in the fuel source and changes in the processes by which the fuels are produced. Changes in the fuel source and the production processes can be strong indicators of changes in fuel properties and, hence, in fuel quality (Chapter 6). In fact, any fuel that is clearly outside of the specifications should be rejected.

References

Faaij, A., van Doorn, J., Curvers, A., Waldheim, L., Olsson, E., van Wijk, A., Daey-Ouwens, C., and Turkenburg, W. 1997. Characterization and availability of biomass waste and residues in the Netherlands for gasification. *Biomass and Bioenerg.* 12(4): 225–240.

Gary, J.H., Handwerk, G.E., and Kaiser, M.J. 2007. *Petroleum Refining: Technology and Economics*, 5th Edition. CRC Press, Taylor & Francis Group, Boca Raton, Florida.

Hsu, C.S. and Robinson, P.R. (Editors) 2017. *Handbook of Petroleum Technology*. Springer International Publishing AG, Chaim, Switzerland.

Mokhatab, S., Poe, W.A., and Speight, J.G. 2006. *Handbook of Natural Gas Transmission and Processing*. Elsevier, Amsterdam, Netherlands.

Parkash, S. 2003. *Refining Processes Handbook*. Gulf Professional Publishing, Elsevier, Amsterdam, Netherlands.

Speight, J.G. 2013. *The Chemistry and Technology of Coal*, 3rd Edition. CRC Press, Taylor & Francis Group, Boca Raton, Florida.

Speight, J.G. 2014. *The Chemistry and Technology of Petroleum*, 5th Edition. CRC Press, Taylor & Francis Group, Boca Raton, Florida.

Speight, J.G. 2015a. *Handbook of Petroleum Product Analysis*, 2nd Edition. John Wiley & Sons Inc., Hoboken, New Jersey.

Speight, J.G. 2015b. *Handbook of Coal Analysis*, 2nd Edition. John Wiley & Sons Inc., Hoboken, New Jersey.

Speight, J.G. 2017. *Handbook of Petroleum Refining*. CRC Press, Taylor & Francis Group, Boca Raton, Florida.

Speight, J.G. 2018. *Handbook of Natural Gas Analysis*. John Wiley & Sons Inc., Hoboken, New Jersey. 2018.

Speight, J.G. 2019a. *Natural Gas: A Basic Handbook*, 2nd Edition. Gulf Publishing Company, Elsevier, Cambridge, Massachusetts. 2019.

Speight, J.G. 2019b. *Handbook of Petrochemical Processes*. CRC Press, Taylor & Francis Group, Boca Raton, Florida.

Vamvuka, D. and Kakaras, E. 2011. Ash properties and environmental impact of various biomass and coal fuels and their blends. *Fuel Process. Technol.* 92(3): 570–581.

Van Doorn, J. 1995. Characterization of energy crops and biomass and waste streams, Netherlands Energy Research Foundation. Report ECN-C-95-047, Petten, Netherlands, July.

Chapter 12

Production of Fuels in a Biorefinery

12.1 Introduction

The previous chapters have focused on the production of fuels from fossil fuel and non-fossil fuel feedstocks by a variety of processes. The latter aspect leads to the concept in which fuels might be produced from non-fossil fuel feedstocks in a central location such as in tandem (on-site) with the production of crude oil-based products or, as may be the case in the future, in a refinery solely dedicated to the conversion of, say, biomass-type feedstocks only (Taylor, 2008).

An important aspect to be considered is the nature of the feedstocks. The current refinery will evolve into a redesigned version that is designed to facilitate an evolution of product slates and through the move to heavier feedstocks, such as extra heavy oil and tar sand bitumen, whatever the current case against coal and oil shale, there could well be a resurgence of interest in these two fossil fuels as the sources of liquid fuels. Thus, the refinery of the future may also require flexibility of design to accommodate feedstocks based on the various types of biomass (Chapter 5), either as mixed feedstocks in which the biomass is a co-feedstock with extra heavy oil and/or tar sand bitumen or as a feedstock that includes only the various forms of biomass. In either case, it is likely that biomass feedstocks will make serious inroads into refineries.

Gaseous, liquid, and solid biofuels may be made from a variety of biomass sources, including waste organic materials, through a number of conversion routes. Several types of products suitable for road and other forms of transportation fuels can be produced including, in the future, aviation fuel, perhaps using micro-algae for feedstock supply. Current products include bioethanol, biodiesel, and other bio-based products such as biobutanol. The feedstocks include oil-based crops such as oilseed rape, used for the production of biodiesel, and sugar- and starch-based crops such as sugar beet and sugarcane, used for the production of bioethanol. Such crops that have a food use are often referred to as *first-generation* bioenergy crops. All lignocellulosic biomass (biomass that is mostly plant cell walls, which have a high carbon content) can be converted to liquid fuels through biological processes (esterification and fermentation) or through thermochemical routes such as pyrolysis. Such dedicated bioenergy crops with no food use are often referred to as *second-generation* bioenergy crops. Hydrogen may also be produced from biomass, but this is still at the research stage.

The supply of crude oil, the basic feedstock for refineries and for the petrochemical industry, is finite and its dominant position will become unsustainable as supply/demand issues erode its economic advantage over other alternative feedstocks. This situation will be mitigated to some extent by the exploitation of more technically challenging fossil resources and the introduction of new technologies for fuel and chemical production from natural gas, crude oil, and coal (Speight, 2008, 2013a, 2014, 2017).

Reducing the dependence of any country on imported crude oil is of critical importance for long-term security and continued economic growth. Supplementing crude oil consumption with renewable biomass resources is a first step towards this goal. The realignment of the chemical industry from one of petrochemical refining to a biorefinery concept is, given time, feasible and has become a national goal of many oil-importing countries. However, clearly defined goals are necessary for the construction of a biorefinery and with the increasing use of biomass-derived feedstocks in industrial chemical production, it is important to keep this goal in perspective. In this context, the increased use of biofuels should be viewed as *one* of a range of possible measures for achieving self-sufficiency in energy rather than a panacea (Crocker and Crofcheck, 2006).

A *biorefinery* is the means by which biomass can be converted to other products — in the current context the other products are *biofuels*, which have the potential to replace certain crude oil-derived fuels. *Biomass* refers to (i) energy crops grown specifically to be used as fuel, such as fast-growing trees or switch grass, (ii) agricultural residues and by-products, such as straw, sugarcane fiber, and rice hulls, and (iii) residues from forestry, construction, and other wood-processing industries (NREL, 2003; Wright *et al.*, 2006). Biomass is a renewable energy source, unlike the fossil fuel resources (natural gas, crude oil, and coal) and, like the fossil fuels, biomass is a form of stored solar energy but is produced in lesser time than the fossil fuels (Speight, 2008). Thus, a biorefinery is a facility that integrates biomass conversion processes with equipment to produce fuels, power, and chemicals from biomass. The concept is based on the modern crude oil refinery, which produces multiple fuels and products from crude oil (Parkash, 2003; Gary *et al.*, 2007; Speight, 2014, 2017, 2019a; Hsu and Robinson, 2017). Industrial biorefineries have been identified as the most promising route to the creation of a new domestic bio-based industry.

By producing multiple products, a biorefinery can take advantage of the differences in biomass components and intermediates and maximize the value derived from the biomass feedstock. A biorefinery might, for example, produce one or several low-volume, but high-value, chemical products and a low-value, but high-volume, liquid transportation fuel, while generating electricity and process heat for its own use and perhaps enough electricity for sale. The high-value products enhance profitability, the high-volume fuel helps meet national energy needs, and the power generation reduces costs and avoids greenhouse-gas emissions.

In theory, a biorefinery can use all kinds of *biomass,* including wood and dedicated agricultural crops, plant- and animal-derived waste, municipal waste, and aquatic biomass (algae, seaweeds). A biorefinery produces a spectrum of marketable products and energy including intermediate and final products: food, feed, materials, chemicals, fuels, power, and/or heat. However, the differences in the various biomass feedstocks may dictate that a biorefinery be constructed and operated on the basis of the chemical composition of the feedstock and the means by which the feedstock is to be processed.

The aim of this chapter is to present the concept of a biorefinery and the different biomass feedstocks that might be used in the biorefinery.

12.2 A Biorefinery

The concept of biorefining is not new when activities such as the production of vegetable oils, beer, and wine requiring pretreatment are considered. Many of these activities are known to have been in practice for millennia. Biorefining offers a method to access the integrated production of chemicals, materials, and fuels. Although the concept of a biorefinery is analogous to that of an oil refinery, the differences in the various biomass feedstocks require a divergence in the methods used to convert the feedstocks to fuels and chemicals (Ruth, 2004; Speight, 2014, 2017, 2019a). Thus, a biorefinery, like a crude oil refinery, may need to be a facility that integrates biomass conversion processes with equipment to produce fuels, power, and chemicals from biomass. In a manner similar to the crude oil refinery, a biorefinery would integrate a variety of conversion processes to produce multiple product streams such as motor fuels and other chemicals from biomass such as the inclusion of gasification processes and fermentation processes to name only two possible process options (Speight, 2011b).

In short, a biorefinery should combine the essential technologies to transform biological raw materials into a range of industrially useful intermediates. However, the type of biorefinery would have to be differentiated by the nature of the feedstock. For example, the *crop biorefinery* would use raw materials such as cereals or maize and the *lignocellulose biorefinery* would use raw materials with high cellulose content, such as straw, wood, and paper waste. However, the different compositional nature of the biomass feedstock, compared to crude oil, will require the application of a wider variety of processing tools in the biorefinery. Processing the individual components will utilize conventional thermochemical operations and state-of-the-art bioprocessing techniques.

The production of synthesis gas through pyrolysis is accompanied by the generation of char, which can then be gasified to provide process heat and energy for the thermochemical platform. A variety of commercial-scale processes exist to transform fossil fuels such as natural gas, crude oil, and coal into liquid fuels, including Fischer–Tropsch fuels. However, the use of biomass instead of fossil fuels changes the composition of synthesis gas, creating a more heterogeneous intermediate product and increasing the difficulty in downstream catalysis.

The bioconversion platform typically uses a combination of physical or chemical pretreatment and enzymatic hydrolysis to convert lignocellulose into its component monomers. Once liberated, the carbohydrate components of wood may be processed into a number of chemical and fuel products. A number of US-led projects are paving the way for new chemical products from the lignocellulose-based biorefinery, including bioethanol, lactic acid and polylactide, propanediol, and succinic acid.

Cellulosic-based bioethanol from a demonstration-scale plant in Ottawa is already being produced and blended as an oxygenate in fuels. Other chemical products can be used to create consumer products such as bioplastics, or used as platform chemicals in a number of industrial applications. The development of better ways to separate lignin from the lignocellulose matrix during bioconversion has created the possibility of developing value-added lignin-based products as well. The bioconversion platform therefore has the ability to serve as the basis for full-fledged wood-based biorefining operations, generating value-added bioproducts as well as fuel and energy for the forest sector.

Using components of these platforms, biomass can provide a sustainable, renewable source of bioenergy. This chapter illustrates how evolutions in technology may be combined to create truly revolutionary processes that can transform the energy sector. It is also shown how each technological platform might be used to generate other, valuable chemical products or energy, thus creating a biorefinery. An important conclusion is that the development of the biorefinery should take precedence over specific biofuel and bioenergy projects.

A key requirement for the biorefinery is the ability of the refinery to develop process technology that can economically access and convert the five- and six-membered ring sugars present in the cellulose and hemicellulose fractions of the lignocellulosic feedstock. Although engineering technology exists to effectively separate the sugar-containing fractions from the lignocellulose, the enzyme technology to economically convert the five-ring sugars to useful products requires further development.

Plants are very effective chemical mini-factories or refineries insofar as they produce chemicals by specific pathways. The chemicals they produce include sugars and amino acids that are essential for the growth of the plant, as well as more complex compounds. Unlike crude oil-derived petrochemicals where most chemicals are built from the bottom up, biofeedstocks already have some valuable products to skim off the top before being broken down and used to build new molecules.

As a feedstock, biomass can be converted by thermal or biological routes to a wide range of useful forms of energy including process heat, steam, electricity, as well as liquid fuels, chemicals, and synthesis gas. As a raw material, biomass is a nearly universal feedstock due to its versatility, domestic availability, and renewable nature. At the same time, it also has its limitations. For example, the energy density of biomass is low compared to that of coal, liquid crude oil, or crude oil-derived fuels.

The heat content of biomass, on a dry basis (7,000–9,000 Btu/lb), is at best comparable with that of a low-rank coal or lignite, and substantially (50–100%) lower than that of anthracite, most bituminous coals, and crude oil. Most biomass, as received, has a high burden of physically adsorbed moisture, up to 50% by weight. Thus, without substantial drying, the energy content of a biomass feed per unit mass is even lower.

These inherent characteristics and limitations of biomass feedstocks have led to a focus on the development of efficient methods for chemically transforming and upgrading biomass feedstocks in a refinery. By analogy with crude oil, every element of the plant feedstock will be utilized including the low-value lignin components. Processing of the individual components will utilize conventional thermochemical operations and state-of-the-art bioprocessing techniques. The production of biofuels in the biorefinery complex will serve existing high-volume markets, providing economy-of-scale benefits and large volumes of by-product streams at minimal cost for upgrading to valuable chemicals. A pertinent example of this is the production of glycerol (glycerin) as a by-product in biodiesel plants (Figures 12.3 and 12.4). Glycerol has high functionality and is a potential platform chemical for conversion into a range of higher value chemicals. The high-volume product streams in a biorefinery need not necessarily be a fuel but could also be a large-volume chemical intermediate such as ethylene or lactic acid.

If the biorefinery is truly analogous to an oil refinery in which crude oil is separated into a series of products, such as gasoline, heating oil, jet fuel, and petrochemicals.

12.3 Feedstocks

Biomass is a term used to describe any material of recent biological origin, including plant materials such as trees, grasses, agricultural crops, and even animal manure (Chapter 5). Other biomass components, which are generally present in minor amounts, include triglycerides, sterols, alkaloids, resins, terpenes, terpenoids and waxes. This includes everything from *primary sources* of crops and residues harvested/collected directly from the land to *secondary sources* such as sawmill residuals, and *tertiary sources* of post-consumer residuals that often end up in landfills. A *fourth source*, although not usually categorized as such, includes the gases that result from anaerobic digestion of animal manures or organic materials in landfills (Wright *et al.*, 2006).

More generally, biomass feedstocks are recognized or classified by the specific plant content of the feedstock or the manner in which the feedstocks is produced. For example, *primary biomass feedstocks* are thus primary biomass that is harvested or collected from the field or forest where it is grown. Examples of primary biomass feedstocks currently being used for bioenergy include grains and oilseed crops used for transportation fuel production, plus some crop residues (such as orchard trimmings and nut hulls) and some residues from logging and forest operations that are currently used for heat and power production.

Secondary biomass feedstocks differ from *primary biomass feedstocks* in that the secondary feedstocks are a by-product from processing of the primary feedstocks. By *processing*, it is meant that there is substantial physical or chemical breakdown of the primary biomass and the production of by-products; *processors* may be factories or animals. Field processes such as harvesting, bundling, chipping, or pressing do not cause a biomass resource that was produced by photosynthesis (e.g. tree tops and limbs) to be classified as secondary biomass. Specific examples of secondary biomass include sawdust from sawmills, black liquor (which is a by-product of paper making), and cheese whey (which is a by-product of cheese making processes). Manures from concentrated animal feeding operations are collectable under secondary biomass resources. Vegetable oils used for biodiesel that are derived directly from the processing of oil seeds for various uses are also a secondary biomass resource.

Tertiary biomass feedstocks include post-consumer residues and wastes, such as fats, greases, oils, construction and demolition wood debris, and other waste wood from the urban environments, as well as packaging wastes, municipal solid wastes, and landfill gases. A category called *other wood waste from the urban environment* includes trimmings from urban trees, which technically fits the definition of primary biomass. However, because this material is normally handled as a waste stream along with other post-consumer wastes from urban environments (and included in those statistics), it makes the most sense to consider it to be part of the tertiary biomass stream.

Tertiary biomass often includes *fats and greases*, which are by-products of the reduction of animal biomass into component parts, since most fats and greases, and some oils, are not available for bioenergy use until after they become a post-consumer waste stream. Vegetable oils derived from processing of plant components and used directly for bioenergy (e.g. soybean oil used in biodiesel) would be a secondary biomass resource, though amounts being used for bioenergy are most likely to be tracked together with fats, greases and waste oils.

One aspect of designing a refinery for any feedstocks is the composition of the feedstocks. For example, a heavy oil refinery would differ somewhat from a conventional refinery and a refinery for extra heavy oil and tar sand bitumen would be significantly different from both (Speight, 2008, 2014, 2017). Furthermore, the composition of biomass is highly variable (Speight, 2008), which is reflected in the range of heat value (heat content, calorific value) of biomass, which is somewhat lower than that for coal and much lower than the heat value for crude oil, generally falling in the range 6,000–8,500 Btu/lb (Table 12.1). Moisture content is probably the most important determinant of heating value. Air-dried biomass typically has approximately 15–20% moisture, whereas the moisture content for oven-dried biomass is around 0%. Moisture content is also an important characteristic of coals, varying in the range of 2–30%. However, the bulk density (and hence energy density) of most biomass feedstocks is generally low, even after densification, approximately 10% and 40% of the bulk density of most fossil fuels.

Plants (Table 12.2) offer a unique and diverse feedstock for chemicals and the production of *biofuels* from biomass requires some knowledge of the chemistry of biomass, the chemistry of the individual constituents of

Table 12.1: Heating value of selected fuels.

Fuel	Btu/lb
Natural gas	23,000
Gasoline	20,000
Crude oil	18,000
Heavy oil	16,000
Coal (anthracite)	14,000
Coal (bituminous	11,000
Wood (farmed trees, dry)	8,400
Coal (lignite)	8,000
Biomass (herbaceous, dry)	7,400
Biomass (corn stover, dry)	7,000
Wood (forest residue, dry)	6,600
Bagasse (sugar cane)	6,500
Wood	6,000

Table 12.2: Typical plants used as a source of energy.

Type of biomass	Plant species	Predominant use
Wood	Butea monosperma, Casuarina equisetifolia, Eucalyptus globulus, Leucaena leucocephala, Melia azadirachta, Tamarix dioica	Firewood
Starch	Cereals, millets, root and tuber crops, e.g. potato	Bioethanol
Sugar	Sugarcane, sugar beet	Bioethanol
Hydrocarbons	Euphorbia lathyris, Aslepia speciosa, Copaifera multijuga, algae	Biodiesel
Wastes	Crop residues, animal/human refuge, sewage	Biogas

biomass, and the chemical means by which the biomass can be converted to fuel. It is widely recognized that further significant production of plant-based chemicals will only be economically viable in highly integrated and efficient production complexes producing a diverse range of chemical products. This biorefinery concept is analogous to conventional oil refineries and petrochemical complexes that have evolved over many years to maximize process synergies, energy integration, and feedstock utilization to drive down production costs.

In addition, the specific components of plants such as (i) carbohydrate derivatives, (ii) vegetable oils, (iii) plant fibers, and (iv) complex organic molecules, such as the organic constituents of waste, can be utilized to produce a range of valuable monomers, chemical intermediates, and pharmaceuticals.

12.3.1 *Carbohydrates*

Plants capture solar energy as fixed carbon during which carbon dioxide is converted to water and sugars $(CH_2O)_x$:

$$CO_2 + H_2O \rightarrow (CH_2O)_x + O_2$$

The sugars produced are stored in three types of polymeric macromolecules: (i) starch, (ii) cellulose, and (iii) hemicellulose.

In general, sugar polymers such as cellulose (Figure 12.1) and starch can be readily broken down to their constituent monomers by hydrolysis, preparatory to conversion to ethanol or other chemicals. In contrast, lignin has an unknown complex structure containing aromatic groups that is

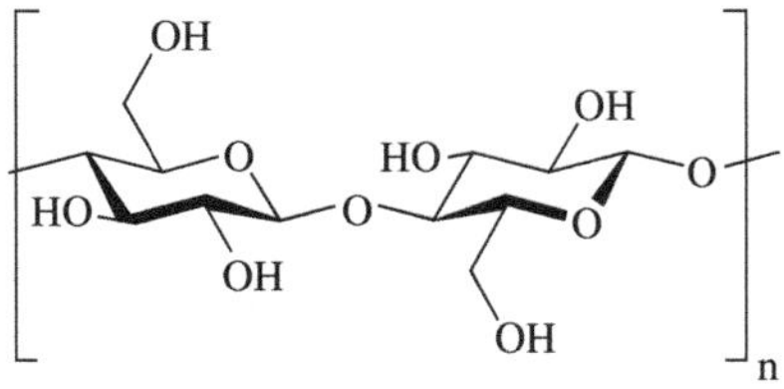

Figure 12.1: Generalized structure of cellulose.

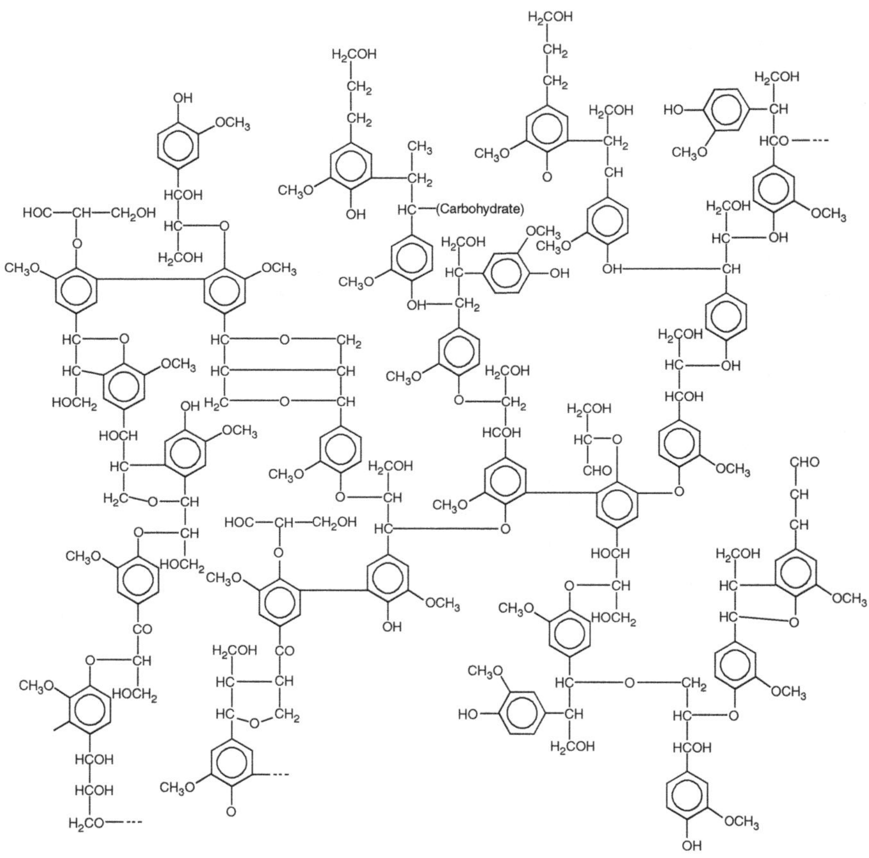

Figure 12.2: Hypothetical structure of lignin to illustrate the complexity of the molecule.

totally hypothetical (Figure 12.2) and is less readily degraded than starch or cellulose.

Although lignocellulose is one of the cheapest and most abundant forms of biomass, it is difficult to convert this relatively unreactive

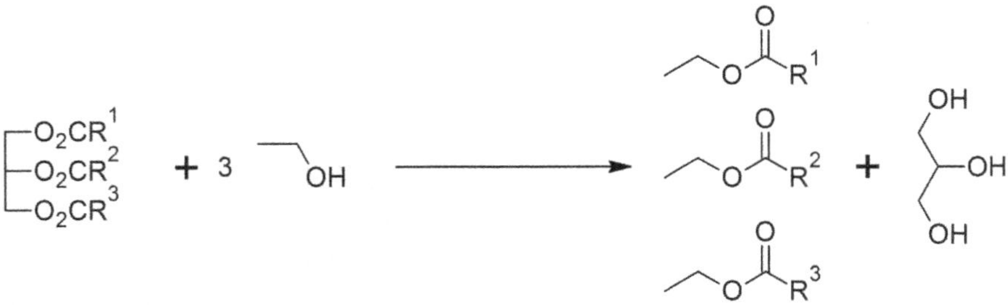

Figure 12.3: The chemistry of biodiesel production.

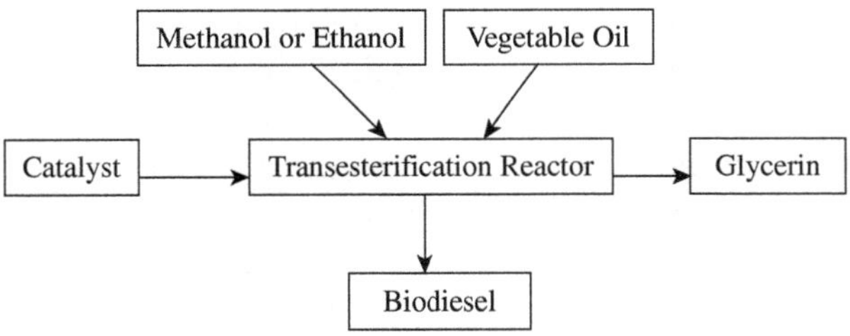

Figure 12.4: Catalytic production of biodiesel.

material into sugars. Among other factors, the walls of lignocellulose are composed of lignin, which must be broken down in order to render the cellulose and hemicellulose accessible to acid hydrolysis. For this reason, many efforts focused on ethanol production from biomass are based almost entirely on the fermentation of sugars derived from the starch in corn grain (Figures 12.3 and 12.4).

Starch is readily obtained from wheat and potato, whilst cellulose is obtained from wood pulp. The structures of these polysaccharides can be readily manipulated to produce a range of biodegradable polymers with properties similar to those of conventional plastics such as polystyrene foams and polyethylene film. In addition, these polysaccharides can be hydrolyzed catalytically or enzymatically to produce sugars, a valuable fermentation feedstock for the production of ethanol, citric acid, lactic acid, and dibasic acids such as succinic acid.

12.3.2 *Vegetable oils*

Vegetable oils (sometimes referred to as vegetable fats) are oils extracted from seeds, or less often, from other parts of fruits. Similar to animal fats, vegetable fats are mixtures of triglycerides. Soybean oil and rape seed oil are examples of fats from seeds, while olive oil and palm oil are examples of fats from other parts of fruits. In common usage, vegetable oil may refer exclusively to vegetable fats, which are liquid at room temperature. Vegetable oils are usually edible; non-edible oils derived mainly from petroleum are termed mineral oils. The predominant source of vegetable oils in many countries is rapeseed oil. Vegetable oils are a major feedstock for the oleochemicals industry (surfactants, dispersants, and personal care products) and are now successfully entering new markets such as diesel fuel, lubricants, polyurethane monomers, functional polymer additives, and solvents.

Unsaturated vegetable oils can be transformed through partial or complete hydrogenation into oils of higher melting point. The hydrogenation process involves sparging the oil at high temperature and pressure with hydrogen in the presence of a catalyst, typically a nickel compound. As each carbon–carbon double-bond is chemically reduced to a single bond, two hydrogen atoms each form single bonds with the two carbon atoms. An oil may be hydrogenated to increase resistance to oxidation (which can turn the oil rancid) or to change its physical characteristics. As the degree of saturation increases, the viscosity and the melting point of the oil increase.

Vegetable oils are used as an ingredient or component in many manufactured products. They are used to make soaps, skin products, candles, perfumes and other personal care and cosmetic products. Some oils are particularly suitable for use as drying oils, and are used in making paints and other wood treatment products. Vegetable oils are increasingly being used in the electrical industry as insulators since vegetable oils are bridgeable if spilled and are not toxic to the environment — they also have high flash and fire points. However, vegetable oils are less stable chemically, so they are generally used in systems where they are not exposed to oxygen and they are more expensive than crude oil distillate.

More important in the present context, vegetable oils are also used to make biodiesel, which can be used like conventional diesel. Some vegetable oils are used in unmodified vehicles but straight vegetable oil, also known as pure plant oil, needs specially designed vehicles that have a means for heating the oil to reduce the viscosity.

12.3.3 *Plant fibers*

Lignocellulosic fibers extracted from plants such as hemp and flax can replace cotton and polyester fibers in textile materials and glass fibers in insulation products.

Lignin is a complex chemical compound that is most commonly derived from wood and is an integral part of the cell walls of plants, especially in tracheids, xylem fibers, and sclereids. The chemical structure of lignin is unknown and, at best, can only be represented by hypothetical formulas. However, it is one of the most abundant organic compounds on the Earth after cellulose and chitin.

By way of clarification, chitin ($(C_8H_{13}O_5N)_n$) is a long-chain polymeric polysaccharide of β-glucose that forms a hard, semitransparent material found throughout the natural world. Chitin is the main component of the cell walls of fungi and is also a major component of the exoskeletons of arthropods, such as the crustaceans (e.g. crab, lobster, and shrimp) and the insects (e.g. ants, beetles, and butterflies), and the beaks of cephalopods (e.g. squids and octopuses).

Lignin makes up approximately one-quarter to one-thirds of the dry mass of wood and is generally considered to be a large, cross-linked hydrophobic, aromatic macromolecule with a molecular mass that is *estimated* to be in excess of 10,000. Degradation studies indicate that the molecule consists of various types of substructures that appear to repeat in a random manner.

Lignin fills the spaces in the cell wall between cellulose, hemicellulose, and pectin components and is covalently linked to hemicellulose. Lignin also forms covalent bonds to polysaccharides, which enables cross-linking to different plant polysaccharides. Lignin confers mechanical strength to the cell wall (stabilizing the mature cell wall) and therefore the entire plant.

12.3.4 *Waste*

As already stated (Chapter 5), waste (refuse) comes in a variety of forms and must be dealt with on a day-to-day basis. Industry produces huge amounts of *industrial waste* and *domestic waste* makes a large contribution to the general waste problem. In spite of the fact, many insidious waste products escape (inadvertently or deliberately) into the surrounding environment. Thus, there have been numerous pollution incidents. On the other hand, there are technologies available for the treatment of most of the wastes we produce. The level of treatment is largely a matter of cost but the conversion of waste to new products is a concept that has long been ready to hatch.

Most important in the context of this book is organic waste (which occurs in both domestic waste and industrial waste), defined as any material that is biodegradable and comes from either a plant or an animal. Biodegradable waste is organic material that can be broken into carbon dioxide, methane, or simple organic molecules. Examples of organic waste include green waste, food waste, food-soiled paper, non-hazardous wood waste, green waste, and landscape and pruning waste.

Domestic waste (also known as rubbish, garbage, trash, or junk) is unwanted or undesired material (although the old adage *One man's waste is another man's treasure* sometimes applies). *Waste* is the general term; though there are other terms treated loosely as its synonyms, they have more specific meanings. Thus, (i) rubbish or trash is mixed household waste including paper and packaging; (ii) food waste or garbage (North America) is kitchen and table waste, and (iii) junk or scrap is metallic or industrial material. There are other categories of waste as well: sewage, ash, manure, and plant materials from garden operations, including *grass* cuttings, fallen leaves, and pruned branches. On the other hand, *industrial waste* is waste produced by industrial operations such as factories, mills, and mines and has existed since the onset of the Industrial Revolution. Chemical waste and toxic waste are two additional (but specific) designations of industrial waste (Chapters 5 and 6).

Municipal solid waste (MSW) is a waste type that includes predominantly household waste (domestic waste) with sometimes the addition of commercial wastes collected by a municipality within a given area. They are in either solid or semisolid form and generally exclude industrial hazardous waste. The term *residual waste* relates to waste left from household sources containing materials that have not been separated out or sent for reprocessing.

To some observers, organic waste is not considered to be a source of environmental pollution. However, when organic waste is dumped in landfills, it undergoes anaerobic decomposition (due to the lack of oxygen) and produces methane (Chapter 7). Methane is 20–25 times more potent as a greenhouse gas than carbon dioxide, which is also produced from organic waste during the combustion of the waste.

12.4 Process Options

A crude oil refinery is a series of integrated unit processes by which crude oil can be converted to a slate of useful (salable) products and, as currently configured, is unsuitable for processing raw, or even partially processed, biomass. However, a typical refinery might be suitable for

Table 12.3: Simplified representation of the potential products from a biorefinery.

Feedstock[a]				
Pretreatment[b]				
	Gasification/pyrolysis			
		Gas conditioning		
		Gas separation		
			Hydrogen	
			Methane	Chemicals
			Ethane	Chemicals
			Propane	Chemicals
			Butane	Chemicals
			Pentane plus	
				Gasoline
			Synthesis gas	
				Methanol
				Ethanol
				Propanol
				Butanol
				Hydrocarbon fuels
	Enzymatic hydrolysis			
		Sugar products		
			Fermentation	
				Ethanol
		Lignin intermediates		
			Various products	Chemicals

Notes: [a]Biomass, waste.
[b]Whenever necessary.

processing products such as gas, liquid or solid products from biomass processing (Table 12.3). These products from biomass might be acceptable as a single feedstock to a specific unit or, more likely, as a feedstock to be blended with refinery streams to be co-processed in various refinery units. Thus, a biorefinery might, in the early stages of development, be a series of unit processes that covert biomass to a primary product, which requires further processing to become the final salable product. This is analogous to the processing of bitumen from tar sand, which is fist processed to a synthetic crude oil (primary processing) and then sent to a refinery for conversion to salable fuel products (Speight, 2008, 2014, 2017).

The basic types of processes used to generate energy from biomass as might be incorporated into a biorefinery are: (i) anaerobic digestion, (ii) direct combustion, (iii) fermentation, (iv) gasification, (v) pyrolysis, and (vi) transesterification. These options are described in the following paragraphs but are listed alphabetically and not in any order of presence; the choice of any of the options is feedstock and product slate dependent.

12.4.1 *Direct combustion*

The simplest, cheapest, and most common method for obtaining energy from biomass is *direct combustion*. Any organic material that has a water content low enough to allow for sustained combustion can be burned to produce energy. The heat of combustion can be used to provide space or process heat, water heating or, through the use of a steam turbine, electricity. In the developing world, many types of biomass such as dung and agricultural wastes are burned for cooking and heating. In fact, biofuels such as wood or dried dung have been used since man learned to control fire — there were no attempts to refine these feedstocks other than drying to reduce the water content.

Direct combustion involves burning the energy crop and then using the resulting hot combustion gases to raise steam. The steam is, in turn, used to drive a steam turbine, which drives a generator to produce electricity. Combustion facilities can burn many types of biomass fuel, including wood, agricultural residues, wood pulping liquor, MSW, and refuse-derived fuel. Combustion technologies convert biomass fuels into several forms of useful energy for commercial or industrial uses: hot air, hot water, steam, and electricity.

A furnace is the simplest combustion technology. In a furnace, biomass fuel burns in a combustion chamber, converting biomass into heat energy. As the biomass burns, hot gases are released. These hot gases contain approximately 85% of the fuel's potential energy. Commercial and industrial facilities use furnaces for heat, either directly or indirectly through a heat exchanger in the form of hot air or water. In fact, a boiler fired by biomass is a more adaptable direct combustion technology because a boiler transfers the heat of combustion into steam. Steam can be used for electricity, mechanical energy or heat. Biomass boilers supply energy at low cost for many industrial and commercial uses.

Pile burners consist of cells, each having an upper and a lower combustion chamber. Biomass fuel burns on a grate in the lower chamber, releasing volatile gases. The gases burn in the upper (secondary) combustion chamber and pile burners must be shut down periodically to remove ash. Although capable of handling high-moisture fuels and fuels mixed with dirt, pile burners have become obsolete with the development of more efficient combustion designs with automated ash removal systems.

In a stationary or traveling grate combustor, an automatic feeder distributes the fuel onto a grate, where the fuel burns. Combustion air enters from below the grate. In the stationary grate design, ashes fall into a pit for collection. In contrast, a traveling grate system has a moving grate that drops the ash into a hopper. On the other hand, fluidized-bed combustors burn biomass fuel in a hot bed of granular material, such as sand. Injection of air into the bed creates turbulence resembling a boiling liquid. The turbulence distributes and suspends the fuel. This design increases heat transfer and allows for operating temperatures below 972°C (1700°F), reducing nitrogen oxide (NO_x) emissions. Fluidized-bed combustors can handle high-ash fuels and agricultural biomass residue.

Conventional combustion equipment is not designed for burning agricultural residues. Straw and grass contain alkali (potassium and sodium) compounds, which are also present in all annual crops and crop residues and in the annual growth of trees and plants. During combustion, alkali combines with silica, which is also present in agricultural residues. This reaction causes slagging and fouling problems in conventional combustion equipment designed for burning wood at higher temperatures.

Volatile alkali lowers the fusion temperature of ash. In conventional combustion equipment having furnace gas exit temperatures above 1450°F, combustion of agricultural residue causes slagging and deposits on heat transfer surfaces. Specially designed boilers with lower furnace exit temperatures could reduce slagging and fouling from combustion of these fuels. Low-temperature gasification may be another method of using these fuels for efficient energy production while avoiding the slagging and fouling problems encountered in direct combustion.

Combustion facilities that produce electricity from steam-driven turbine generators have a conversion efficiency of 17–25%. Using a boiler to produce both heat and electricity (cogeneration) improves overall system efficiency to as much as 85 percent. That is, cogeneration converts 85% of the fuel's potential energy into useful energy in two forms: electricity and steam heat. Two cogeneration arrangements, or cycles, are possible for combining electric power generation with industrial steam production. Steam can be used in an industrial process first and then routed through a turbine to generate electricity. This arrangement is called a bottoming cycle. In the alternative arrangement, steam from the boiler passes first through a turbine to produce electric power. The steam exhaust from the turbine is then used for industrial processes or for space and water heating. This arrangement is called a topping cycle.

The direct-fired gas turbine is another combustion technology for converting biomass to electricity. In this technology, fuel pretreatment reduces biomass to a particle size of less than 2 mm and a moisture content of less than 25%. Then the fuel is burned with compressed air. Cleanup of the combustion gas reduces particulate matter before the gas expands through the turbine stage. The turbine drives a generator to produce electricity.

Co-firing biomass as a secondary fuel in a coal-burning power plant using high-sulfur coal could help reduce sulfur dioxide and nitrogen oxide emissions. Also, co-firing decreases the net carbon dioxide emissions from the power plant (if the biomass fuel comes from a sustainable source). Co-firing may require wood fuel preparation or boiler modifications to maintain boiler efficiency.

12.4.2 *Fermentation and hydrolysis*

A number of processes allow biomass to be transformed into gaseous fuels such as methane or hydrogen (Sørensen *et al.*, 2006). One pathway uses algae and bacteria that have been genetically modified to produce hydrogen directly instead of the conventional biological energy carriers. A second pathway uses plant material such as agricultural residues in a fermentation process leading to biogas from which the desired fuels can be isolated. This technology is established and in widespread use for waste treatment, but often with the energy produced only for onsite use, which often implies less than maximum energy yields. Finally, high-temperature gasification supplies a crude gas, which may be transformed into hydrogen by a second reaction step. In addition to biogas, there is also the possibility of using the solid by-product as a biofuel.

Traditional fermentation plants producing biogas are in routine use, ranging from farms to large municipal plants. As feedstock they use manure, agricultural residues, urban sewage, and waste from households, and the output gas is typically 64% methane. The biomass conversion process is accomplished by a large number of different agents, from the microbes decomposing and hydrolyzing plant material, to the acidophilic bacteria dissolving the biomass in aquatic solution, to the strictly anaerobic methane bacteria responsible for the gas formation. Operating a biogas plant for a period of some months usually makes the bacterial composition stabilize in a way suitable for obtaining high conversion efficiency (typically above 60%, the theoretical limit being nearly 100%), and it is important not to vary the feedstock compositions abruptly, if optimal operation has to be maintained. The operating temperatures for the bacterial processes are only slightly above ambient temperatures, e.g. in the mesophilic region around 30°C.

The production of ethanol from corn is a mature technology that holds much potential (Nichols *et al.*, 2006). Substantial cost reductions may be possible, however, if cellulose-based feedstocks are used instead of corn. The feed for all ethanol fermentations is sugar — traditionally a hexose (a six-carbon or "C6" sugar) such as those present naturally in sugar cane, sugar beet and molasses. Sugar for fermentation can also be recovered from starch, which is actually a polymer of hexose sugars (*polysaccharide*).

Biomass, in the form of wood and agricultural residues such as wheat straw, is viewed as a low cost alternative feed to sugar and starch. It is also potentially available in far greater quantities than sugar and starch feeds. As such it receives significant attention as a feed material for ethanol production. Like starch, wood and agricultural residues contain polysaccharides. However, unlike starch, while the cellulose fraction of biomass is principally a polymer of easily fermented C6 sugars, the hemicellulose fraction is principally a polymer of C5 sugars, with quite different characteristics for recovery and fermentation the cellulose and hemicellulose in biomass are bound together in a complex framework of crystalline organic material known as lignin.

There are several different methods of hydrolysis: (i) concentrated sulfuric acid, (ii) dilute sulfuric acid, (iii) nitric acid, and acid pretreatment followed by enzymatic hydrolysis. The greatest potential for ethanol production from biomass, however, lies in enzymatic hydrolysis of cellulose. The enzyme cellulase, now used in the textile industry to stone wash denim and in detergents, simply replaces the sulfuric acid in the hydrolysis step. The cellulase can be used at lower temperatures, 30–50°C, which reduces the degradation of the sugar. In addition, process improvements now allow simultaneous saccharification and fermentation (SSF). In the saccharification and fermentation process, cellulase and fermenting yeast are combined, so that as sugars are produced, the fermentative organisms convert them to ethanol in the same step.

Once the hydrolysis of the cellulose is achieved, the resulting sugars must be fermented to produce ethanol. In addition to glucose, hydrolysis produces other six-carbon sugars from cellulose and five-carbon sugars from hemicellulose that are not readily fermented to ethanol by naturally occurring organisms. They can be converted to ethanol by genetically engineered yeasts that are currently available, but the ethanol yields are not sufficient to make the process economically attractive. It also remains to be seen whether the yeasts can be made hardy enough for production of ethanol on a commercial scale.

The fermentation processes to produce propanol and butanol from cellulose are fairly tricky to execute, and the *Clostridium acetobutylicum* currently used to perform these conversions produces an extremely unpleasant smell, and this must be taken into consideration when designing and locating a fermentation plant. This organism also dies when the butanol content of whatever it is fermenting rises to 7%. For comparison, yeast dies when the ethanol content of its feedstock hits 14%. Specialized strains can tolerate even greater ethanol concentrations — so-called turbo yeast can withstand up to 16% ethanol. However, if ordinary Saccharomyces yeast can be modified to improve its ethanol resistance, scientists may yet 1 day produce a strain of the Weizmann organism with a butanol resistance higher than the natural boundary of 7%. This would be useful because butanol has a higher energy density than ethanol, and because waste fiber left over from sugar crops used to make ethanol could be made into butanol, raising the alcohol yield of fuel crops without there being a need for more crops to be planted.

Wet milling and dry milling are the means by which grain and straw fractions are processed into a variety of end products. The processes encompass fermentation and distilling of grains (wheat, rye or maize). Wet milling starts with water-soaking the grain adding sulphur dioxide to soften the kernels and loosen the hulls, after which it is ground. It uses well-known technologies and allows separation of starch, cellulose, oil and proteins. Dry milling grinds whole grains (including germ and bran). After grinding, the flour is mixed with water to be treated with liquefying enzymes and, further, cooking the mash to breakdown the starch. This hydrolysis step can be eliminated by simultaneously adding saccharifying enzymes and fermenting yeast to the fermenter (simultaneous saccharification and fermentation).

After fermentation, the mash (called *beer*) is sent through a multi-column distillation system, followed by concentration, purification and dehydration of the alcohol. The residue mash (stillage) is separated into a solid (wet grains) and liquid (syrup) phase that can be combined and dried to produce distiller's dried grains with soluble constituents, to be used as cattle feed. Its nutritional characteristics and high vegetable fiber content make distiller's dried grains with soluble constituents unsuitable for other animal species.

12.4.3 *Gasification*

Gasification is a high temperature process which produces gas which can then be used in an internal combustion engine or fuel cell (Luque and Speight, 2015). Thus, gasification is the thermal degradation in the presence of an externally supplied oxidizing (oxygen containing) agent e.g. air, steam, oxygen. Various gasification concepts have been developed over the years, mainly for the purposes of power generation. However, efficient biomass-to-liquids production imposes completely different requirements for the composition of the gas. The reason is that in power generation the gas is used as a fuel, while in biomass-to-liquids processing it is used as a chemical feedstock to obtain other products. This difference has implications with respect to the purity and composition of the gas.

12.4.3.1 *Gasification chemistry*

In the gasifier, biomass is converted into a gaseous mixture of hydrogen, carbon monoxide, carbon dioxide, and other compounds by applying heat under pressure in the presence of steam and a controlled amount of oxygen. The biomass produces synthesis gas.

$$C_6H_{12}O_6 + O_2 + H_2O \rightarrow CO + CO_2 + H_2 + \text{other products}$$

The above reaction uses glucose as a surrogate for cellulose. Biomass has highly variable composition and complexity, with cellulose as one major component.

Gasification occurs through the thermal decomposition of biomass with the help of an oxidant such as pure oxygen or oxygen enriched air to yield a combustible gas such as synthesis gas rich in carbon monoxide and hydrogen. The synthesis gas is post-treated, by steam reforming or partial oxidation, to convert the hydrocarbon derivatives produced by gasification into hydrogen and carbon monoxide. The carbon monoxide is then put through the shift process to obtain a higher fraction of hydrogen, by carbon dioxide-removal and methanation or by pressure swing adsorption (Mokhatab *et al.*, 2006; Speight, 2019b).

The basic gasification of biomass (such as, e.g. wood scrap) is a well-known process, taking place in pyrolysis (oxygen supply far below what is required for complete combustion, the fraction called "equivalence ratio") or fluidized-bed type of reactors. Conditions such as operating temperature determine whether hydrogen is consumed or produced in the process. Hydrogen evolution is largest for near-zero equivalence ratios, but the energy conversion efficiency is highest at an equivalence ratio around 0.25. The hydrogen fraction (in this case typically some 30%) must be separated for most fuel-cell applications, as well as for long-distance pipeline-transmission. In the pyrolysis-type application, gas production is low and most energy is in the oily substances that must be subsequently reformed in order to produce significant amounts of hydrogen. Typical operating temperatures are around 850°C (1,560°). An overall energy conversion efficiency of around 50% is attainable, with considerable variations. Alternative concepts use membranes to separate the gases produced, and many reactor types uses catalysts to help the processes to proceed in the desired direction, notably at a lower temperature (down to some 500°C, 930°F).

Environmental concerns include disposal of associated tars and ashes, particularly for the fluidized bed reactors, where these substances must be separated from the flue gas stream (in contrast to the pyrolysis plants, where most tar and ash deposits at the bottom of the reactor). Biomass ash has also the potential to be used as a clarifying agent in water treatment, as a wastewater adsorbent, as a liquid waste adsorbent, as a hazardous waste solidification agent, as a lightweight fill for roadways, parking areas, and structures, as asphalt mineral filler, or as a mine spoil amendment.

Methanol production from natural gas feed consists of three steps: (i) synthesis gas generation — in the case of natural gas feed, synthesis gas production consists of converting methane (CH_4) into carbon monoxide (CO) and hydrogen (H_2) via steam reforming, (ii) synthesis gas upgrading — primarily removal of carbon dioxide, plus any contaminants such as sulfur, and (iii) methanol synthesis and purification — reacting the carbon monoxide, hydrogen, and steam over a catalyst in the presence of a small amount of carbon dioxide and at elevated temperature and pressure. The methanol synthesis is an equilibrium reaction and excess reactants must be recycled to optimize yields.

Methods proposed for the production of methanol from biomass involve the conversion of the biomass to a suitable synthesis gas, after which processing steps are very similar to those developed for methanol from natural gas. However, the gasification techniques proposed are still at a relatively early stage of development using biomass feed and the methods are based on similar techniques used widely already with natural gas as feed.

12.4.3.2 *Gasifiers*

The gasifier is the heart of the gasification process (Luque and Speight, 2015). Gasifiers are designed to process the fuel in a variety of ways consistent with the type of fuel, the end use of the gas, the size of the process, and the source of oxygen. The oxygen may be introduced as a pure gas or may come from air or steam. Some gasifiers operate under pressure; others do not.

The simplest type of gasifier is the *fixed-bed counter current gasifier.* The biomass is fed at the top of the reactor and moves downwards as a result of the conversion of the biomass and the removal of ashes. The air intake is at the bottom and the gas leaves at the top. The biomass moves in counter current to the gas flow, and passes through the drying zone, the distillation zone, reduction zone and the oxidation zone.

The major advantages of this type of gasifier are its simplicity, high charcoal burn-out and internal heat exchange leading to low gas exit temperatures and high gasification efficiency. In this way also fuels with high moisture content (up to 50% w/w) can be used. Major drawbacks are the high amounts of tar and pyrolysis products, because the pyrolysis gas is not lead through the oxidation zone. This is of minor importance if the gas is used for direct heat applications, in which the tar is burnt. In case the gas is used for engines, gas cleaning is required, resulting in problems of tar-containing condensates.

In the conventional *downdraft gasifier* (sometimes called the *co-flow* gasifier), biomass is fed at the top of the reactor and air intake is at the top or from the sides. The gas leaves at the bottom of the reactor, so the fuel and the gas move in the same direction. The pyrolysis gasses are lead through the oxidation zone (with high temperatures) and or more or less burnt or cracked. Therefore the producer gas has low tar content and is suitable for engine applications. In practice however, a tar-free gas is seldom if ever achieved over the whole operating range of the equipment. Because of the lower level of organic components in the condensate, downdraft gasifiers suffer less from environmental objections than updraft gasifiers.

Successful operation of a downdraft gasifier requires drying the biomass fuel to a moisture content of less than 20% w/w. The advantage of the downdraft design is the very low tar content of the producer gas. However, disadvantages of the downdraft gasifier are: (i) the high amounts of ash and dust particles in the gas, (ii) the inability to operate on a number of unprocessed fuels, often pelletization or briquetting of the biomass is necessary, (iii) the outlet gas has a high temperature leading to a lower gasification efficiency, and (iv) the moisture content of the biomass must be less than 25% w/w.

A more recent development is the *open core* gasifier design for gasification of small sized biomass with high ash content but the producer gas is not tar-free, which must be removed (Chapter 9). In the open core gasifier the air is sucked over the whole cross section from the top of the bed. This facilitates better oxygen distribution since the oxygen will be consumed over the whole cross section, so that the solid-bed temperature will not reach the local extremes (hot spots) observed in the oxidation zone of conventional gasifiers due to poor heat transfer. Moreover, the air nozzles in conventional gasifiers generate caves and create obstacles that may obstruct solid flow especially for solids of low bulk like rice husk. On the other hand, the entry of air through the top of the bed creates a downward flow of the pyrolysis gases, and transports the tars products to the combustion zone. Thus, flow problems due to the caking of rice husk caused by back mixing of tar are avoided.

The gasification process is amenable to a variety of biomass feedstocks such as waste rice hulls, wood waste, grass, and the dedicated energy crops. Gasification is a clean process with few air emissions and, when crops are used as the feedstock, little or no ash. However, before biomass can be gasified it must be pre-treated to meet the processing constraints of the gasifier. This typically involves size reduction, and drying to keep moisture contents below specific levels. Thereafter, biomass gasification involves heating biomass in the presence of low levels of oxygen (i.e. less than required for complete combustion to carbon dioxide and water). Above certain temperatures the biomass will break down into a gas stream and a solid residue. The composition of the gas stream is influenced by the operating conditions for the gasifier, with some gasification processes more suited than others to producing a gas for methanol production. In particular, simple gasification with air creates a synthesis gas stream that is diluted with large quantities of nitrogen. This nitrogen is detrimental to subsequent processing to methanol and so techniques using indirect gasification or an oxygen feed are preferred. For large scale gasification, pressurized systems are considered to be more economic than atmospheric systems.

The gasification process could play a significant role in meeting the goal of greenhouse gas mitigation. It is likely that both in the transition phase to a hydrogen economy and in the steady state, a significant fraction of hydrogen might be derived from domestically abundant crops. In addition, the co-firing applications of crops (and other biomass) biomass with coal, biomass can provide up to 15% of the total energy input of the fuel mixture. Such concepts address greenhouse gas mitigation by co-firing biomass and coal to offset the losses of carbon dioxide to the atmosphere that are inherent in coal combustion processes (even with the best-engineered capture and storage of carbon). Since growth of biomass fixes atmospheric carbon, its combustion leads to no net addition of atmospheric carbon dioxide even if vented. Thus, co-firing of crops or crop residues (or other biomass with coal) in an efficient coal gasification process, affording the opportunity for capture and storage of carbon dioxide, could lead to a net reduction of atmospheric carbon dioxide. Cheaper, though less plentiful, biomass residue could supplant crops as gasifier feedstock leading to a less significant impact on the environment than would farming bioenergy crops.

Along similar lines, willow biomass crops have been shown to be a good fuel for farm-based power production using advanced gasification technology (Pian *et al.*, 2006). The fuel gas can be used for generating electricity, using micro-turbines modified to operate on low-Btu gas, or for other energy needs. Willow biomass was found to make a fuel for ash-rejection gasifiers with a predicted net gasification efficiency of approximately 85%. Analysis showed that developing a method to co-gasify willow with various amounts of low-cost wastes, such as dairy farm animal waste, can be an excellent way to reduce the fuel cost, to increase the overall fuel availability and help work around problems resulting from seasonal availability of bioenergy crops. Co-gasification of dairy farm wastes along with willow offers an economical way to dispose of the wastes and manage nutrient flows on a dairy farm.

Agricultural residues can be divided into two groups: (i) crop residues and (ii) agro-industrial residues. Crop residues are plant materials left behind in the farm after removal of the main crop produce. The remaining materials could be of different sizes, shapes, forms, and densities like straw, stalks, sticks, leaves, haulms, fibrous materials, roots, branches, and twigs.

Due to high energy content, straw is one of the best crop residues for solid biofuels. However, straw has several disadvantages — it has a higher ash content, which results in lower calorific value. In order to improve its bulk density, the straw is generally baled before transportation. Straw burning requires a specific technology. There are four basic types of straw burners: those that accept shredded, loose straw; burners that use densified straw products such as pellets, briquettes or cubes and straw logs; small, square bale burners and round bale burners. To be suitable for heat and electricity production straw should not have a large content of moisture, preferably not more than 20% as the moisture reduces the boiler efficiency. Also straw color as well as straw chemistry should be considered before burning as it indicates the quality of the straw.

The agro-industrial residues are by-products of the post-harvest processes of crops such as cleaning, threshing, sieving, and crushing. These could be in the form of husk, dust, and straw. Furthermore, the quantity of agricultural residues produced differs from crop to crop and is affected by soil type and irrigation conditions. Production of agricultural residues is directly related to the corresponding crop production and ratio between the main crop produce and the residues, which varies from crop to crop and, at times, with the variety of the seeds in one crop itself. Thus, for known amounts of crop production, it may be possible to estimate the amounts of agricultural residues produced using the residue to crop ratio.

Most crop or agricultural residues are not found throughout the year but are available only at the time of harvest. The amount available depends upon the harvesting time, storage-related characteristics, and the storage facility.

12.4.3.3 *Synthesis gas*

Synthesis gas, as generated from biomass, is not the same as biogas (Chapter 8). Biogas is a clean and renewable form of energy generated from biomass that could very well substitute for conventional sources of energy (Speight, 2019c). The gas is generally composes of methane (55–65%), carbon dioxide (35–45%), nitrogen (0–3%), hydrogen (0–1%), and hydrogen sulfide (0–1%).

The calorific value of the gas is the prime factor for power generation — the higher the value, the better. Hence, the availability in the gas of any compounds that increase calorific value is generally welcomed — product gas, which contains carbon monoxide (CO), hydrogen (H_2), various hydrocarbon derivatives (methane (CH_4), ethylene (C_2H_4), ethane (C_2H_6), tar, and char). The presence of inert components — such as water (H_2O), carbon dioxide (CO_2), and nitrogen (N_2) — is also acceptable, provided the amounts of these components are maintained at a low level.

Synthesis gas, a product of biomass gasification, is composed primarily of hydrogen and carbon monoxide, also called *biosyngas* (Cobb, 2007). The production of high-quality synthesis gas from biomass is the feedstock for BTL production (Figure 12.5). The production of synthesis

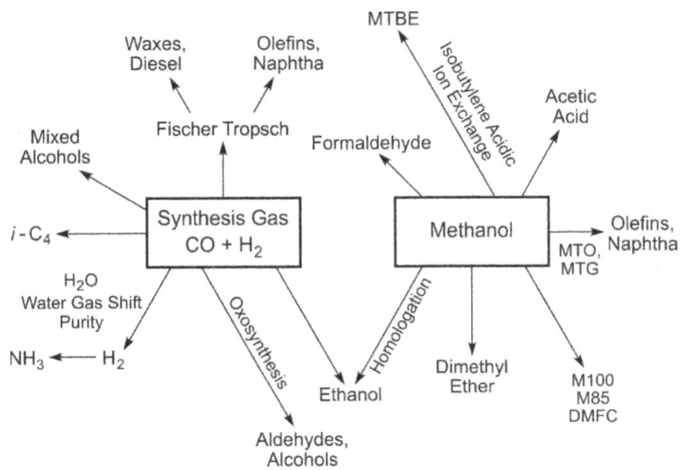

Figure 12.5: Products from synthesis gas.

gas from biomass is indeed the novel component in the gas-to-liquids concept — obtaining synthesis gas from fossil raw materials (natural gas, crude oil, and coal) is a relatively mature technology. For example, the Fischer–Tropsch process can be used to convert synthesis gas-to-liquid fuels needed for transportation. The water–gas shift process converts synthesis gas to more concentrated hydrogen for fuel cells. A variety of other catalytic processes can turn synthesis gas into a myriad of chemicals or other potential fuels or products.

12.4.4 *Pyrolysis*

Pyrolysis (thermal decomposition, thermal cracking) is a medium temperature method (typically using temperatures less than 500°C, 930°F) which produces gas, oil and char from crops which can then be further processed into useful fuels or feedstock. Wood and many other similar types of biomass which contain lignin and cellulose, (agricultural wastes, cotton gin waste, wood wastes, peanut hulls, etc.) can be converted through a thermochemical process, such as pyrolysis, into solid, liquid or gaseous fuels. Pyrolysis, used to produce charcoal since the dawn of civilization, is still the most common thermochemical conversion of biomass to commercial fuel.

During pyrolysis, biomass is heated in the absence of air and breaks down into a complex mixture of liquids, gases, and a residual char. If wood is used as the feedstock, the residual char is what is commonly known as charcoal. With more modern technologies, pyrolysis can be carried out under a variety of conditions to capture all the components, and to maximize the output of the desired product be it char, liquid or gas.

Pyrolysis is often considered to be the gasification of biomass in the absence of oxygen. However, the chemistry of each process may differ significantly. In general, biomass does not gasify as easily as coal, and it produces other hydrocarbon compounds in the gas mixture exiting the gasifier; this is especially true when no oxygen is used. As a result, typically an extra step must be taken to reform these hydrocarbon derivatives with a catalyst to yield a clean synthesis gas mixture of hydrogen, carbon monoxide, and carbon dioxide.

Fast pyrolysis is a thermal decomposition process that occurs at moderate temperatures with a high heat transfer rate to the biomass particles and a short hot vapor residence time in the reaction zone. Several reactor configurations have been shown to assure this condition and to achieve yields of liquid product as high as 75% based on the starting dry biomass weight. They include bubbling fluid beds, circulating and transported beds, cyclonic reactors, and ablative reactors.

Fast pyrolysis of biomass produces a liquid product, pyrolysis oil or bio-oil that can be readily stored and transported. Pyrolysis oil is a renewable liquid fuel and can also be used for production of chemicals. Fast pyrolysis has now achieved a commercial success for production of chemicals and is being actively developed for producing liquid fuels. Pyrolysis oil has been successfully tested in engines, turbines and boilers, and been upgraded to high quality hydrocarbon fuels.

Direct hydrothermal liquefaction involves converting biomass to an oily liquid by contacting the biomass with water at elevated temperatures (300–350°C, 570–660°F) with sufficient pressure to maintain the water primarily in the liquid phase for residence times up to 30 min. Alkali may be added to promote organic conversion. The primary product is an organic liquid with reduced oxygen content (approximately 10%) and the primary by-product is water containing soluble organic compounds. (http://www1.eere.energy.gov/biomass/pyrolysis.html).

The importance of the provisions for the supply of feedstocks as crops and other biomass are often underestimated since it is assumed that the supplies are inexhaustible. While this may be true over the long term, short term supply of feedstocks can be as much as risk as any venture.

12.4.5 *Anaerobic digestion*

Anaerobic digestion is a natural process and is the microbiological conversion of organic matter to methane in the absence of oxygen. The *biochemical conversion* of biomass is completed through alcoholic fermentation to produce liquid fuels and anaerobic digestion or fermentation, resulting in biogas (hydrogen, carbon dioxide, ammonia, and methane) usually through four steps (hydrolysis, acidogenesis, acetogenesis, and methanogenesis) (Table 12.4). The decomposition is caused by natural bacterial action in various stages and occurs in a variety of natural anaerobic environments, including water sediment, water-logged soils, natural hot springs, ocean thermal vents and the stomach of various animals (e.g. cows). The digested organic matter resulting from the anaerobic digestion process is usually called *digestate*.

Symbiotic groups of bacteria perform different functions at different stages of the digestion process. There are four basic types of

Table 12.4: Stages in the anaerobic digestion process.

Stage	Reactions
Hydrolysis:	Carbohydrates → sugars Fats → fatty acids Proteins → amino acids
Acidogenesis:	Sugars → carbon acids + alcohols + hydrogen + carbon dioxide + ammonia Fatty acids → carbon acids + alcohols + hydrogen, carbon dioxide + ammonia Amino acids → carbon acids + alcohols + hydrogen, carbon dioxide + ammonia
Acetogenesis:	Carbon acids + alcohols → acetic acid + carbon dioxide + hydrogen
Methanogenesis:	Acetic acid → methane + carbon dioxide

microorganisms involved. Hydrolytic bacteria break down complex organic wastes into sugars and amino acids. Fermentative bacteria then convert those products into organic acids. Acidogenic microorganisms convert the acids into hydrogen, carbon dioxide and acetate. Finally, the methanogenic bacteria produce biogas from acetic acid, hydrogen and carbon dioxide.

The process of anaerobic digestion occurs in a sequence of stages involving distinct types of bacteria. Hydrolytic and fermentative bacteria first break down the carbohydrates, proteins and fats present in biomass feedstock into fatty acids, alcohol, carbon dioxide, hydrogen, ammonia and sulfides. This stage is the hydrolysis stage (also known as the liquefaction stage). In the next stage, acetogenic (acid-forming) bacteria further digest the products of hydrolysis into acetic acid, hydrogen and carbon dioxide. Methanogenic (methane-forming) bacteria then convert these products into biogas.

The combustion of digester gas can supply useful energy in the form of hot air, hot water or steam. After filtering and drying, digester gas is suitable as fuel for an internal combustion engine, which, combined with a generator, can produce electricity. Future applications of digester gas may include electric power production from gas turbines or fuel cells. Digester gas can substitute for natural gas (methane) or propane (and butane) in space heaters, refrigeration equipment, cooking stoves or other equipment. Furthermore, compressed digester gas can be used as an alternative transportation fuel. Also, there are three principal by-products of anaerobic digestion: (i) biogas, (ii) acidogenic digestate, and (iii) methanogenic digestate.

Biogas is a gaseous mixture comprising mostly of methane and carbon dioxide, but also containing a small amount of hydrogen and occasionally trace levels of hydrogen sulfide (Speight, 2019c). Biogas can be burned to produce electricity, usually with a reciprocating engine or microturbine. The gas is often used in a cogeneration arrangement, to generate electricity and use waste heat to warm the digesters or to heat buildings. Since the gas is not released directly into the atmosphere and the carbon dioxide comes from an organic source with a short carbon cycle biogas does not contribute to increasing atmospheric carbon dioxide concentrations; because of this, it is considered to be an environmentally friendly energy source. The production of biogas is not a steady stream; it is highest during the middle of the reaction. In the early stages of the reaction, little gas is produced because the number of bacteria is still small. Toward the end of the reaction, only the hardest to digest materials remain, leading to a decrease in the amount of biogas produced.

The second by-product (*acidogenic* digestate) is a stable organic material comprised largely of lignin and chitin, but also of a variety of mineral components in a matrix of dead bacterial cells; some plastic may be present. This resembles domestic compost and can be used as compost or to make low grade building products such as fiberboard. The third by-product is a liquid (*methanogenic digestate*) that is rich in nutrients and can be an excellent fertilizer dependent on the quality of the material being digested. If the digested materials include low levels of toxic heavy metals or synthetic organic materials such as pesticides or polychlorobiphenyls, the effect of digestion is to significantly concentrate such materials in the digester liquor. In such cases further treatment will be required in order to dispose of this liquid properly. In extreme cases, the disposal costs and the environmental risks posed by such materials can offset any environmental gains provided by the use of biogas. This is a significant risk when treating sewage from industrialized catchments.

Nearly all digestion plants have ancillary processes to treat and manage all of the by-products. The gas stream is dried and sometimes sweetened before storage and use. The sludge liquor mixture has to be separated by one of a variety of ways, the most common of which is filtration. Excess water is also sometimes treated in sequencing batch reactors (SBRs) for discharge into sewers or for irrigation.

Digestion can be either *wet* or *dry*. Dry digestion refers to mixtures which have a solid content of 30% or greater, whereas wet digestion refers to mixtures of 15% or less.

Anaerobic digester systems can reduce fecal coliform bacteria in manure by more than 99%, virtually eliminating a major source of water pollution. Separation of the solids during the digester process removes approximately 25% of the nutrients from manure, and the solids can be sold out of the drainage basin where nutrient loading may be a problem. In addition, the digester´s ability to produce and capture methane from the manure reduces the amount of methane that otherwise would enter the atmosphere. Scientists have targeted methane gas in the atmosphere as a contributor to global climate change.

Controlled anaerobic digestion requires an airtight chamber, called a digester. To promote bacterial activity, the digester must maintain a temperature of at least 68°F. Using higher temperatures, up to 150°F, shortens processing time and reduces the required volume of the tank by 25–40%. However, there are more species of anaerobic bacteria that thrive in the temperature range of a standard design (mesophilic bacteria) than there are species that thrive at higher temperatures (thermophilic bacteria). High-temperature digesters also are more prone to upset because of temperature fluctuations and their successful operation requires close monitoring and diligent maintenance.

The biogas produced in a digester (*digester gas*) is actually a mixture of gases, with methane and carbon dioxide making up more than 90% v/v of the total (Speight, 2019c). Biogas typically contains smaller amounts of hydrogen sulfide, nitrogen, hydrogen, methyl mercaptans and oxygen. Methane is a combustible gas. The energy content of digester gas depends on the amount of methane it contains. Methane content varies from approximately 55% to 80%. Typical digester gas, with a methane concentration of 65%, contains approximately 600 Btu of energy per cubic foot.

There are three basic digester designs and all of them can trap methane and reduce fecal coliform bacteria, but they differ in cost, climate suitability and the concentration of manure solids they can digest.

A *covered lagoon digester*, as the name suggests, consists of a manure storage lagoon with a cover. The cover traps gas produced during decomposition of the manure. This type of digester is the least expensive of the three. Covering a manure storage lagoon is a simple form of digester technology suitable for liquid manure with less than 3% solids. For this type of digester, an impermeable floating cover of industrial fabric covers all or part of the lagoon. A concrete footing along the edge of the lagoon holds the cover in place with an airtight seal. Methane produced in the lagoon collects under the cover. A suction pipe extracts the gas for use. Covered lagoon digesters require large lagoon volumes and a warm climate. Covered lagoons have low capital cost, but these systems are not suitable for locations in cooler climates or locations where a high water table exists.

A *complete mix digester* converts organic waste to biogas in a heated tank above or below ground. A mechanical or gas mixer keeps the solids in suspension. Complete mix digesters are expensive to construct and cost more than plug-flow digesters to operate and maintain. Complete mix digesters are suitable for larger manure volumes having solids concentration of 3% to 10%. The reactor is a circular steel or poured concrete container. During the digestion process, the manure slurry is continuously mixed to keep the solids in suspension. Biogas accumulates at the top of the digester. The biogas can be used as fuel for an engine-generator to produce electricity or as boiler fuel to produce steam. Using waste heat from the engine or boiler to warm the slurry in the digester reduces retention time to less than 20 days.

Plug-flow digesters are suitable for ruminant animal manure that has a solids concentration of 11–13%. A typical design for a plug-flow system includes a manure collection system, a mixing pit and the digester itself. In the mixing pit, the addition of water adjusts the proportion of solids in the manure slurry to the optimal consistency. The digester is a long, rectangular container, usually built below-grade, with an airtight, expandable cover.

New material added to the tank at one end pushes older material to the opposite end. Coarse solids in ruminant manure form a viscous material as they are digested, limiting solids separation in the digester tank. As a result, the material flows through the tank in a *plug*. Average retention time (the time a manure plug remains in the digester) is 20–30 days. Anaerobic digestion of the manure slurry releases biogas as the material flows through the digester. A flexible, impermeable cover on the digester traps the gas. Pipes beneath the cover carry the biogas from the digester to an engine-generator set.

A plug-flow digester requires minimal maintenance. Waste heat from the engine-generator can be used to heat the digester. Inside the digester, suspended heating pipes allow hot water to circulate. The hot water heats the digester to keep the slurry at 25–40°C (77–104°F), a temperature range suitable for methane-producing bacteria. The hot water can come from recovered waste heat from an engine generator fueled with digester gas or from burning digester gas directly in a boiler.

12.4.6 *Transesterification*

The transesterification process is a means of biodiesel production (Marchetti *et al.*, 2005) in which glycerin is separated from the fat or vegetable oil. The process leaves behind two products: (i) methyl esters, which are the components of biodiesel and (ii) glycerin, which is a valuable by-product usually sold to be used in soaps and other products).

Biodiesel (fatty acid methyl esters, FAME) is a notable alternative to the widely used crude oil derived diesel fuel since it can be generated by domestic natural sources such as soybeans, rapeseeds, coconuts, and even recycled cooking oil, and thus reduces dependence on diminishing crude oil fuel from foreign sources. In addition, because biodiesel is largely made from vegetable oils, it reduces lifecycle greenhouse gas emissions by as much as 78% (Iso *et al.*, 2001; Ban-Weiss *et al.*, 2007).

Vegetable oils and animal fats belong to an extensive family of chemicals called lipids. Lipids are bioproducts from the metabolism of living creatures. As a result, they can be found widely distributed in nature. Their biofunctions are diverse, but they are most known for their energy storage capacity. Most lipids can easily dissolve in common organic solvents, meaning that they are hydrophobic. If a lipid is a solid at 25°C, it is classified as a fat; otherwise, it is oil. Typically, fats are produced by animals and oils by plants, but both are mainly made of triglyceride (TG) molecules, which are tri-esters of glycerol (a triol) and free fatty acids (long alkyl chain carboxylic acids). Other glyceride species, such as di-glycerides and mono-glycerides, are obtained from triglycerides by the substitution of one and two fatty acid moieties, respectively, with hydroxyl groups (Lotero *et al.*, 2006).

12.4.6.1 *Feedstocks*

Soybeans are widely used as a source of biodiesel. However, a variety of oils can be used to produce biodiesel. These include: (i) virgin oil, (ii) waste vegetable oil. And (iii) animal fats. In terms of A Virgin oil feedstock, rapeseed oil and soybean oil are most commonly used, soybean oil alone accounting for approximately 90% of all fuel stocks; It also can be obtained from field pennycress and Jatropha while other crops such as mustard, flax, sunflower, canola, pal oil, hemp, and even algae have been used. Waste vegetable oil is any oi produced from the source that has been used and cannot be used again for the original purpose. Animal fats, including tallow, lard, yellow grease, chicken fat, and the by-products of the production of omega-3 fatty acids from fish oil. Many advocates suggest that waste vegetable oil is the best source of oil to produce biodiesel. However, the available supply is drastically less than the amount of crude oil-based fuel that is burned for transportation and home heating in the world.

12.4.6.2 *Transesterification*

Transesterification (sometime referred to as alcoholysis) is the conversion of triacylglycerol lipids by alcohols to alkyl esters without first isolating the free fatty acids (May, 2004). Transesterification of a vegetable oil was conducted as early as 1853 many years before the first diesel engine became functional. The prime model of Rudolf Diesel, a single 10 ft (3 m) iron cylinder with a flywheel at its base, ran on its own power for the first time in Augsburg, Germany, on August 10, 1893. Diesel later demonstrated his engine and received the *Grand Prix* (highest prize) at the World Fair in Paris, France in 1900. This engine stood as an example of the vision of Rudolf Diesel because it was powered by peanut oil, which is a biofuel, though not *biodiesel*, since it was not transesterified. He believed that the utilization of biomass fuel was the real future of his engine and that the use of vegetable oils for engine fuels would become as important as crude oil and the coal-tar products of the present time.

The purpose of transesterification of vegetable oils to their methyl esters (biodiesel) process is to lower the viscosity of the oil. The transesterification reaction is affected by alcohol type, molar ratio of glycerides to alcohol, type and amount of catalyst, reaction temperature, reaction time and free fatty acids and water content of vegetable oils or animal fats. The transesterification reaction proceeds with or without a catalyst by using primary or secondary monohydric aliphatic alcohols having 1–8 carbon atoms as follows (Demirbaş, 2006a, 2006b):

$$\text{Triglycerides} + \text{Monohydric alcohol} \leftrightarrow \text{Glycerin} + \text{Mono-alkyl esters}$$

Generally, the reaction temperature near the boiling point of the alcohol is recommended (Çanakçı and Özsezen, 2005). The reactions take place at low temperatures (approximately 65°C) and at modest pressures (2 atmospheres, 1 atmosphere = 14.7 psi = 101.325 kPa). Biodiesel is further purified by washing and evaporation to remove any remaining methanol. The oil (87%), alcohol (9%), and catalyst (1%) are the inputs in the production of biodiesel (86%), the main output (Lucia *et al.*, 2006). Pretreatment is not required if the reaction is carried out under high pressure (9,000 kPa) and high temperature (240°C), where simultaneous esterification and transesterification take place with maximum yield obtained at temperatures ranging from 60°C to 80°C at a molar ratio of 6:1 (Barnwal and Sharma, 2005). The alcohols employed in the transesterification are generally short chain alcohols such as methanol, ethanol, propanol, and butanol. It was reported that when transesterification of soybean oil using methanol, ethanol and butanol was performed, 96–98% of ester could be obtained after 1 h (Dmytryshyn *et al.*, 2004).

12.4.6.3 *Catalytic transesterification*

Transesterification reactions can be catalyzed by *alkalis* (Korytkowska *et al.*, 2001; Dmytryshyn *et al.*, 2004; Stavarache *et al.*, 2005; Varghaa and Truterb, 2005; Meher *et al.*, 2006a), *acids* (Lee *et al.*, 2000; Goff *et al.*, 2004; Lopez *et al.*, 2005; Liu *et al.*, 2006) or *enzymes* (Watanabe *et al.*, 2000, 2002; Ghanem, 2003; Reyes-Duarte *et al.*, 2005; Royon *et al.*, 2007; Shah and Gupta, 2007; Bernardes *et al.*, 2007). The catalytic transesterification of vegetable oils with methanol is an important industrial method used in biodiesel synthesis. Also known as methanolysis, this reaction is well studied and established using acids or alkalis, such as sulfuric acid or sodium hydroxide as catalysts. However, these catalytic systems are less active or completely inactive for long chain alcohols. Usually, industries use sodium or potassium hydroxide or sodium or potassium methoxide as catalyst, since they are relatively cheap and quite active for this reaction (Macedo *et al.*, 2006). Enzyme-catalyzed procedures, using lipase as catalyst, do not produce side reactions, but the lipases are very expensive for industrial scale production and a three-step process was required to achieve a 95% conversion. The acid-catalyzed process is useful when a high amount of free acids are present in the vegetable oil, but the reaction time is very long (48–96 h), even at the boiling point of the alcohol, and a high molar ratio of alcohol was needed (20:1 w/w to the oil) (Stavarache *et al.*, 2005).

The transesterification process is catalyzed by alkaline metal alkoxides, and hydroxides, as well as sodium or potassium carbonates. Alkali-catalyzed transesterification with short-chain alcohols, for example, generates high yields of methyl esters in short reaction times (Jeong and Park, 1996). The alkaline catalysts show high performance for obtaining vegetable oils with high quality, but a question often arises; that is, the oils contain significant amounts of free fatty acids which cannot be converted into biodiesels but to a lot of soap (Furuta *et al.*, 2004). These free fatty acids react with the alkaline catalyst to produce soaps that inhibit the separation of the biodiesel, glycerin, and wash water (Çanakcı and Van Gerpen, 2003). Triglycerides are readily transesterified in a batch operation in the presence of alkaline catalyst at atmospheric pressure and at a temperature of approximately 60–70°C with an excess of methanol (Srivastava and Prasad, 2000). It often takes at least several hours to ensure the alkali (NaOH or KOH) catalytic transesterification reaction is complete. Moreover, removal of these catalysts is technically difficult and brings extra cost to the final product (Demirbaş, 2002, 2003). Nevertheless, they are a good alternative since they can give the same high conversions of vegetable oils just by increasing the catalyst concentration to 1 or 2 mol%. Alkaline metal alkoxides (as CH_3ONa for the methanolysis) are the most active catalysts, since they give very high yields (> 98%) in short reaction times (30 min) even if they are applied at low molar concentrations (0.5 mol%) (Schuchardta *et al.*, 1998).

The transesterification process is catalyzed by sulfuric, hydrochloric, and organic sulfonic acids. In general, acid-catalyzed reactions are performed at high alcohol-to-oil molar ratios, low-to-moderate temperatures and pressures, and high acid catalyst concentrations (Lotero *et al.*, 2006). These catalysts give very high yields in alkyl esters but these reactions are slow, requiring typically temperature above 100°C and more than 3 h to complete the conversion (Meher *et al.*, 2006b). Studies of the acid-catalyzed system have been very limited in number. No commercial biodiesel plants to date have been reported to use the acid-catalyzed process. Despite its relatively slow reaction rate, the acid-catalyzed process offers benefits with respect to its independence from free fatty acid content and the consequent absence of a pretreatment step. These advantages favor the use of the acid-catalyzed process when using waste cooking oil as the raw material (Zhang *et al.*, 2003).

Enzyme (such as lipase)-catalyzed reactions have advantages over traditional chemical-catalyzed reactions: the generation of no by-products, easy product recovery, mild reaction conditions, and catalyst recycling. Also, enzymatic reactions are insensitive to free fatty acids and water content in waste cooking oil (Kulkarni and Dalai, 2006). As for the enzyme-catalyzed system, it requires a much longer reaction time than the other two systems (Zhang *et al.*, 2003). The enzyme reactions are highly specific and chemically clean. Because the alcohol can be inhibitory to the enzyme, a typical strategy is to feed the alcohol into the reactor in three steps of 1:1 mole ratio each. The reactions are very slow, with a three step sequence requiring from 4 to 40 h, or more. The reaction conditions are modest, from 35°C to 45°C (Van Gerpen *et al.*, 2004). The main problem of the enzyme-catalyzed process is the high cost of the lipases used as catalyst (Royon *et al.*, 2007).

Synthesis of biodiesel using enzymes such as *Candida antarctica, Candida rugasa, Pseudomonas cepacia, immobilized lipase* (*Lipozyme RMIM*), *Pseudomonas* sp., and *Rhizomucor miehei* is well reported in the literature. The best yield (98% w/w) was obtained by using *Pseudomonas cepacia* lipase immobilized on celite at 50°C in the presence of 4–5% (w/w) water in 8 h (Shah and Gupta (2007).

12.4.6.4 *Transesterification using supercritical methanol*

The transesterification of triglycerides by supercritical methanol (SCM), ethanol, propanol and butanol has proved to be the most promising process. Recently, a catalysts-free method was developed for biodiesel production by employing supercritical methanol (Saka and Kusdiana, 2001). The supercritical treatment at 350°C, 43 MPa and 240 s with a molar ratio of 42 in methanol is the optimum condition for transesterification of rapeseed oil to biodiesel fuel (Kusdiana and Saka, 2004a).

To achieve more moderate reaction conditions, further effort was made through the two-step preparation (Kusdiana and Saka, 2004b). In this method, oils/fats are, first, treated in subcritical water for hydrolysis reaction to produce fatty acids. After hydrolysis, the reaction mixture is separated into oil phase and water phase by decantation. The oil phase (upper portion) is mainly fatty acids, while the water phase (lower portion) contains glycerol in water. The separated oil phase is then mixed with methanol and treated at supercritical condition to produce FAME thorough methyl esterification. After removing unreacted methanol and water produced in reaction, FAME can be obtained as biodiesel.

Therefore, in this process, methyl esterification is the main reaction for FAME formation, while in the one-step method, transesterification is the major one (Saka and Minami, 2006).

Reaction by supercritical methanol has some advantages (i) glycerides and free fatty acids are reacted with equivalent rates, (ii) the homogeneous phase eliminates diffusive problems, (iii) the process tolerates great percentages of water in the feedstock catalytic process require the periodical removal of water in the feedstock or in intermediate stage to prevent catalyst deactivation, (iv) the catalyst removal step is eliminated, and (v) if high methanol/oil ratios are used, total conversion of the oil can be achieved in a few minutes (Vera *et al.*, 2005).

Some disadvantages of the one-stage supercritical method are clear: (i) it operates at very high pressures, (ii) the high temperatures bring along proportionally high heating and cooling costs, (iii) high methanol/oil ratios — usually on the order of 42 — involve high costs for the evaporation of the unreacted methanol, and (iv) the process as posed to date does not explain how to reduce free glycerol to less than 0.02% as established in the ASTM D6584 or other equivalent international standards.

12.4.6.5 *Effect of process parameters*

The main factors affecting transesterification are the molar ratio of glycerides to alcohol, catalyst, reaction temperature and pressure, reaction time and the contents of free fatty acids and water in oils.

The free fatty acids and moisture content are key parameters for determining the viability of the vegetable oil transesterification process (Meher *et al.*, 2006b). In the transesterification, free fatty acids and water always produce negative effects, since the presence of free fatty acids and water causes soap formation, consumes catalyst and reduces catalyst effectiveness, all of which result in a low conversion (Demirbaş and Karslıoğlu, 2007). To carry the base catalyzed reaction to completion; a free fatty acid value lower than 3% is needed (Meher *et al.*, 2006b).

The presence of water has a greater negative effect on transesterification than that of the free fatty acids. In the transesterification of beef tallow catalyzed by sodium hydroxide (NaOH) in presence of free fatty acids and water, the water and free fatty acid contents must be maintained at specified levels (Ma *et al.*, 1998, 1999; Kusdiana and Saka, 2004a; Bala, 2005).

Transesterification can occur in different temperatures depending on the type of oil employed (Ma and Hanna, 1999) — the reaction at room temperature has also been reported (Marinetti, 1962, 1966; Graboski and McCormick, 1998; Encinar *et al.*, 2002). Generally, the conversion rate increases with reaction time. The transesterification of Rice bran oil with methanol was studied at molar ratios of 4:1, 5:1, and 6:1 (Gupta *et al.*, 2007). At molar ratios of 4:1 and 5:1, there was significant increase in yield when the reaction time was increased from 4 to 6 h. Among the three molar ratios studied, ratio 6:1 gave the best results.

One of the most important factors that affect the yield of ester is the molar ratio of alcohol to triglyceride. Although the stoichiometric molar ratio of methanol to triglyceride for transesterification is 3:1, higher molar ratios are used to enhance the solubility and to increase the contact between the triglyceride and alcohol molecules (Noureddini *et al.*, 1998). In addition, investigation of the effect of molar ratio on the transesterification of sunflower oil with methanol showed that when the molar ratio varied from 6:1 to 1:1 and concluded that 98% conversion to ester was obtained at a molar ratio of 6.1 (Freedman *et al.*, 1986).

Another important variable affecting the yield of methyl ester is the type of alcohol to triglyceride. In general, short chain alcohols such as methanol, ethanol, propanol, and butanol can be used in the transesterification reaction to obtain high methyl ester yields. Çanakcı and Van Gerpen (1999) investigated the effect of different alcohol types on acid-catalyzed transesterification of pure soybean oil. They obtained yields from 87.8% to 95.8% after 48 and 96 h of reaction.

Catalysts used for the transesterification of triglycerides are classified as alkali, acid, enzyme. Alkali-catalyzed transesterification is much faster than acid-catalyzed transesterification and is most often used commercially (Ma and Hanna, 1999) and, quite often, for the base-catalyzed transesterification the best yields were obtained when the catalyst was used in small concentration, i.e. 0.5% w/w of oil (Stavarache *et al.*, 2005). On the other hand, data show that during the production of free and bound ethyl ester (FAEE) from castor oil, hydrochloric acid is much more effective than sodium hydroxide at higher reaction temperatures (Meneghetti *et al.*, 2006).

12.5 Alternate Fuels

Alternate fuels (fuels) are primarily used to fuel vehicles, but can also fuel engines or fuel cells for electricity generation. A brief introduction is

Table 12.5: Options for fuels production via gasification.

Option	Use
Fischer–Tropsch Liquids	Blend with crude oil-derived liquids[a] Use as substitute for crude oil-derived liquids
Methanol	Convert to naphtha/gasoline Use directly as fuel Convert to dimethyl ether
Dimethyl ether	Use directly as fuel

Note: [a]Subject to no issues arising from incompatibility of the liquids.

warranted here with a more complete description IS available elsewhere this text (Chapter 3).

The options for the production of biofuels to replace natural gas and crude oil is in active development (Table 12.5). The focus is on the use of cheap organic matter (usually cellulose, agricultural and sewage waste) in the efficient production of liquid and gas biofuels which yield a high net energy gain. The carbon in biofuels was recently extracted from atmospheric carbon dioxide by growing plants, so burning it does not result in a net increase of carbon dioxide in the atmosphere of the Earth. As a result, biofuels are seen by many as a way to reduce the amount of carbon dioxide released into the atmosphere by using them to replace non-renewable sources of energy.

12.5.1 *Gaseous fuels*

Most biomass materials are easier to gasify than coal because they are more reactive with higher ignition stability. This characteristic also makes them easier to process thermochemically into higher-value fuels such as methanol or hydrogen. Ash content is typically lower than for most coals, and sulfur content is much lower than for many fossil fuels. Unlike coal ash, which may contain toxic metals and other trace contaminants, biomass ash may be used as a soil amendment to help replenish nutrients removed by harvest. A few biomass feedstocks stand out for their peculiar properties, such as high silicon or alkali metal contents — these may require special precautions for harvesting, processing and combustion equipment. Note also that mineral content can vary as a function of soil type and the timing of feedstock harvest. In contrast to their fairly uniform physical properties, biomass fuels are rather heterogeneous with respect to their chemical elemental composition.

Biogas contains methane and can be recovered in industrial anaerobic digesters and mechanical biological treatment systems. Landfill gas is a less clean form of biogas which is produced in landfill sites through naturally occurring anaerobic digestion. Paradoxically if this gas is allowed to escape into the atmosphere it is a potent greenhouse gas.

When biomass is heated with no oxygen or only approximately one-third the oxygen needed for efficient combustion (amount of oxygen and other conditions determine if biomass gasifies or is thermally decomposed by pyrolysis), it produces a mixture of carbon monoxide and hydrogen (synthesis gas).

Combustion is a function of the mixture of oxygen with the hydrocarbon fuel. Gaseous fuels mix with oxygen more easily than liquid fuels, which in turn mix more easily than solid fuels. Synthesis gas therefore inherently burns more efficiently and cleanly than the solid biomass from which it was made. Biomass gasification can thus improve the efficiency of large-scale biomass-based power facilities, such as those for forest industry residues and specialized facilities such as black liquor recovery boilers of the pulp and paper industry, both major sources of biomass power. Like natural gas, synthesis gas can also be burned in gas turbines, a more efficient electrical generation technology than steam boilers to which solid biomass and fossil fuels are limited.

12.5.2 *Liquid fuels*

Ethanol is the predominant fuel produced form crops and has been used as fuel in the many countries United States since at least 1908. Currently the production of ethanol by fermentation of corn-derived carbohydrates is the main technology used to produce liquid fuels from biomass resources. Furthermore, amongst different biofuels, suitable for application in transport, bioethanol and biodiesel seem to be the most feasible ones at present. The key advantage of bioethanol and biodiesel is that they can be mixed with conventional petrol and diesel respectively, which allows using the same handling and distribution infrastructure. Another important strong point of bioethanol and biodiesel is that when they are mixed at low concentrations (≤10% bioethanol in petrol and ≤20% biodiesel in diesel, no engine modifications are necessary.

Alternatively, biomass can be converted into fuels and chemicals indirectly (by gasification to synthesis gas followed by catalytic conversion to liquid fuels) or directly to a liquid product by thermochemical means. Direct thermochemical conversion processes include pyrolysis, liquefaction, and solvolysis (Kavalov and Peteves, 2005). Biologically produced alcohols, most commonly ethanol and methanol, and less commonly propanol and butanol are produced by the action of microbes and enzymes through fermentation.

Methanol is a colorless, odorless and nearly tasteless alcohol and is also produced from crops and is also used as a fuel. Methanol, like ethanol, burns more completely but releases as much or more carbon dioxide than its gasoline counterpart. Propanol and butanol are considerably less toxic and less volatile than methanol. In particular, butanol has a high flashpoint of 35°C (95°F), which is a benefit for fire safety, but may be a difficult for starting engines in cold weather (Ezeji *et al.*, 2006).

Biodiesel is a diesel-equivalent fuel derived from biological sources (such as vegetable oils) which can be used in unmodified diesel-engine vehicles. It is thus distinguished from the straight (unused) vegetable oil or waste vegetable oil that are used as fuel in some diesel vehicles. In the current context, biodiesel refers to the alkyl ester products produced by the transesterification of vegetable oil or animal fat. Biodiesel fuel is a fuel made from the oil of certain oilseed crops such as soybean, canola, palm kernel, coconut, sunflower, safflower, corn and a hundreds of other oil producing crops (Bourne, 2007). The oil is extracted by the use of a press and then mixed in specific proportions with other agents, which causes a chemical reaction. The results of this reaction are two products, biodiesel and soap. After a final filtration, the biodiesel is ready for use. After curing, the glycerin soap which is produced as a by-product can be used as is, or can have scented oils added before use. In general, biodiesel compares well to crude oil-based diesel (Lotero *et al.*, 2006). Pure biodiesel fuel (100% esters of fatty acids) is called B100. When blended with diesel fuel the designation indicates the amount of B100 in the blend, e.g. B20 is 20% B100 and 80% diesel, and B5 used in Europe is contains 5% B100 in diesel (Pinto *et al.*, 2005).

Hydrocarbon derivatives are products a various plant species belonging to different families which convert a substantial amount of photosynthetic products into latex. The latex of such plants contains liquid hydrocarbon derivatives of high molecular weight (10,000). These hydrocarbon derivatives can be converted into high grade transportation fuel (i.e. crude oil). Therefore, hydrocarbon producing plants are called crude oil plants or petroplants and their crop as petrocrop. Natural gas is also one of the products obtained from hydrocarbon derivatives. Thus, crude oil plants can be an alternative source for obtaining crude oil to be used in diesel engines. Normally, some of the latex-producing plants of families Euphorbiaceae, Apocynaceae, Asclepiadaceae, Sapotaceae, Moraceae, Dipterocarpaceae, etc. are petroplants. Similarly, sunflower (family Composiae), *Hardwickia pinnata* (family Leguminosae) are also petroplants. Some algae also produce hydrocarbon derivatives.

However, hydrocarbon derivatives, as such, are not usually produced from crops, there being insufficient amount of the hydrocarbon derivatives present in the plant tissue to make the process economical. However, biodiesel is produced from crops thereby offering an excellent renewable fuel for diesel engines.

Bio-oil is a product that is produced by a totally different process than that used for biodiesel production. The process (*fast pyrolysis*, *flash pyrolysis*) occurs when solid fuels are heated at temperatures between 350°C and 500°C (570°F and 930°F) for a very short period of time (<2 s). The bio-oils currently produced are suitable for use in boilers for electricity generation. In another process, the feedstock is fed into a fluidized bed (at 450–500°C (840–930°F) and the feedstock flashes and vaporizes. The resulting vapors pass into a cyclone where solid particles, char, are extracted. The gas from the cyclone enters a quench tower where they are quickly cooled by heat transfer using bio-oil already made in the process. The bio-oil condenses into a product receiver and any non-condensable gases are returned to the reactor to maintain process heating. The entire reaction from injection to quenching takes only 2 s.

12.5.3 *Solid fuels*

Examples of solid fuels from biofuel feedstocks include wood and wood-derived charcoal and dried dung, particularly cow dung. One widespread use of such fuels is in home cooking and heating. The biofuel may be burned on an open fireplace or in a suitably-constructed stove. The efficiency of this process may vary widely, from 10% for a well-made fire (even less if the fire is not made carefully) up to 40% for a custom designed charcoal stove. Inefficient use of fuel is a cause of deforestation (though this is negligible compared to deliberate destruction to clear land for agricultural use) but more importantly it means that more work has to be put into gathering fuel, thus the quality of cooking stoves has a direct influence on the viability of biofuels.

12.6 Environmental Benefits and the Future

The projections for the continued use of fossil fuels indicate that there will be at least another five decades of fossil fuel use (especially in the cases of coal and crude oil) before biomass and other forms of alternate energy take hold and are able to provide sufficient energy not only to subsidize the production of energy from fossil fuels but to replace the fossil fuels. Furthermore, estimations that the era of fossil fuels (natural gas, crude oil, coal, and biomass, and carbonaceous wastes) will be almost over when the cumulative production of the fossil resources reaches 85% of their initial total reserves may or may not have some merit. In fact, the relative scarcity (compared to a few decades ago) of crude oil was real but it seems that the remaining reserves make it likely that there will be an adequate supply of energy for several decades. The environmental issues are very real and require serious and continuous attention.

Technologies which ameliorate the effects of fossil fuels combustion on acid rain deposition, urban air pollution, and global warming must be pursued vigorously (Vallero, 2008). There is a challenge that must not be ignored and the effects of acid rain in soil and water leave no doubt in relation to the need to control its causes. Indeed, recognition of the need to address these issues is the driving force behind recent energy strategies as well as a variety of research and development programs.

While regulations on the greenhouse gas (GHG) carbon dioxide (CO_2) would be an immediate hurdle to deployment of coal plants, gasification plants are in the best position compared to other coal-based alternatives to capture carbon dioxide. However, with the continued uncertainty of carbon dioxide regulation, there is industry reluctance to make large investments in projects with high emissions of carbon dioxide since a cost-effective solution for reducing such emissions is not yet available. Nevertheless, the reduction in greenhouse gas emissions can be an enhancing factor for gasification in the long run because the carbon dioxide from a gasification plant is more amenable to capture.

As new technology is developed, emissions may be reduced by repowering in which aging equipment is replaced by more advanced and efficient substitutes. Such repowering might, for example, involve an exchange in which an aging unit is exchanged for a newer combustion chamber, such as the atmospheric fluidized-bed combustor (AFBC) or the pressurized fluidized-bed combustor (PFBC).

Indeed, recognition of the production of these atmospheric pollutants in considerable quantities every year has led to the institution of national emission standards for many pollutants. Using sulfur dioxide as the example, the various standards are not only very specific but will become more stringent with the passage of time. Atmospheric pollution is being taken very seriously and there is also the threat, or promise, of heavy fines and/or jail terms for any pollution-minded miscreants who seek to flaunt the laws (Vallero, 2008). Nevertheless, a trend to the increased use of fossil fuels will require more stringent approaches to environmental protection issues than we have ever known at any time in the past. The need to protect the environment is strong.

The careless combustion of fossil fuels can account for the large majority of the sulfur oxides and nitrogen oxides released to the atmosphere. Whichever technologies succeed in reducing the amounts of these gases in the atmosphere should also succeed in reducing the amounts of urban smog, those notorious brown and grey clouds that are easily recognizable at some considerable distances from urban areas, not only by their appearance but also by their odor.

$$SO_2 + H_2O \rightarrow H_2SO_4 \text{ (sulfurous acid)}$$

$$2SO_2 + O_2 \rightarrow 2SO_3$$

$$SO_3 + H_2O \rightarrow H_2SO_4 \text{ (sulfuric acid)}$$

$$2NO + H_2O \rightarrow HNO_2 + HNO_3 \text{ (nitrous acid + nitric acid)}$$

$$2NO + O_2 \rightarrow 2NO_2$$

$$NO_2 + H_2O \rightarrow HNO_3 \text{ (nitric acid)}$$

Current awareness of these issues by a variety of levels of government has resulted, in the United States, of the institution of the Clean Fossil fuels Program to facilitate the development of pollution abatement technologies. And it has led to successful partnerships between government and industry (United States Department of Energy, 1993). In addition, there is the potential that new laws, such as the passage in 1990 of the Clean Air Act Amendments in the United States (United States Congress, 1990; Stensvaag, 1991) will be a positive factor and supportive of the controlled clean use of fossil fuels. However, there will be a cost but industry is supportive of the measure and confident that the goals can be met.

Besides fuel and product flexibility, gasification-based systems offer significant environmental advantages over competing technologies, particularly coal-to-electricity combustion systems. Gasification plants can readily capture carbon dioxide, the leading greenhouse gas, much more easily and efficiently than coal-fired power plants. In many instances, this carbon dioxide can be sold, creating additional value from the gasification process.

Carbon dioxide captured during the gasification process can be used to help recover oil from otherwise depleted oil fields. The Dakota Gasification plant in Beulah, North Dakota, captures its carbon dioxide while making substitute natural gas and sells it for enhanced oil recovery.

It very likely that a biorefinery, by producing multiple products, can take advantage of the differences in biomass components and intermediates and maximize the value derived from the biomass feedstock.

Limitations in optimal plant size are caused by feedstock transportation needs: larger plants demanding larger distances to fulfill feedstock requirements year round. Long transportation distances are especially harmful for feedstocks with high concentrations of water (transport of which is expensive but not effective), minerals or organic components (required to maintain local soil quality). In contrast to fossil feedstocks, that can generally be recovered following the exact timing of its demand most biomass types (wood being the exception) are harvested only during a short period of the year. Year round biomass availability requires expensive storage facilities, while crops with high water concentrations cannot be stored over long periods.

In many traditional biorefineries the very large volumes that are processed often result in the duplication of equipment because larger equipment cannot be built because of physical limitations. The choice may be to use unit operations that show only small economy of scale benefits such as the usage of membrane processes, instead of evaporation using heat for concentration purposes. Another strategy could be to convert the desired components in intermediates that can be conveniently recovered or even to leave the component in the process water and subsequently convert these components to biogas that can be used on site or fed to the grid.

Until sufficient quantities of bio-feedstocks are guaranteed to operate a large biorefinery, constructing a modular biorefinery based on two (or more) different conversion platforms to accommodate different feedstocks and to promote different product slates and the association of such a facility with a crude oil-based refining operation may be the most economical near-term reality.

Finally, the perceived conflict between energy and food production can be allayed developing technologies based on lignocellulosic materials. Biorefining requires further innovation but offers opportunities to all economic sectors. Building a bio-based economy can help to overcome present difficulties while laying the foundation of an sustainable and environmentally benign industry.

References

Bala, B.K. 2005. Studies on biodiesels from transformation of vegetable oils for Diesel engines. *Energy Edu. Sci. Technol.* 15: 1–43.

Ban-Weiss, G.A., J.Y. Chen, B.A. Buchholz, and R.W. Dibble. 2007. A numerical investigation into the anomalous slight NOx increase when burning biodiesel — A new (old) theory. *Fuel Process. Technol.* 88: 659–667.

Barnwal, B.K. and Sharma, M.P. 2005. Prospects of biodiesel production from vegetable oils in India. *Renew. Sust. Energ. Rev.* 9: 363–378.

Bernardes, O.L. and Bevilaqua, J.V., Leal, M.C.M.R., Freire, D.M.G., and Langone, M.A.P. 2007. Biodiesel fuel production by the transesterification reaction of soybean oil using immobilized lipase. *Appl. Biochem. & Biotechnol.* 137–140: 105–114.

Çanakçı, M., and Van Gerpen, J. 2003. A pilot plant to produce biodiesel from high free fatty acid feedstocks. *Transactions ASAE.* 46: 945–955.

Çanakçı, M. and Özsezen, A.N. 2005. Evaluating waste cooking oils as alternative diesel fuel. G.U. *J. Sci.* 18: 81–91.

Cobb, J.T. Jr. 2007. Production of Synthesis Gas by Biomass Gasification. *Proceedings. Spring National Meeting AIChE.* Houston, Texas, April 22–26, 2007.

Crocker, M. and Crofcheck, C. 2006. Reducing national dependence on imported oil. Energeia Vol. 17, No. 6. Center for Applied Energy Research, University of Kentucky, Lexington, Kentucky.

Demirbaş, A. 2002. Biodiesel from vegetable oils via transesterification in supercritical methanol. *Energy Convers. Mgmt.* 43: 2349–2356.

Demirbaş, A. 2003. Biodiesel fuels from vegetable oils via catalytic and non-catalytic supercritical alcohol transesterification and other methods: A survey. *Energy Convers. Mgmt.* 44: 2093–2109.

Demirbaş, A. 2006a. Biodiesel from sunflower oil in supercritical methanol with calcium oxide. *Energy Convers. Mgmt.* 48: 937–941.

Demirbaş, A. 2006b. Biodiesel production via non-catalytic SCF method and biodiesel fuel characteristics. *Energy Convers Mgmt.* 47: 2271–2282.

Demirbaş, A. and S. Karslıoğlu. 2007. Biodiesel production facilities from vegetable oils and animal fats. *Energy* Sources, *Part A* 29: 133–141.

Dmytryshyn, S.L., Dalai, A.K., Chaudhari, S.T., Mishra, H.K., and Reaney, M.J. 2004. Synthesis and characterization of vegetable oil derived esters: evaluation for their diesel additive properties. *Bioresour. Technol.* 92: 55–64.

Encinar, J.M., Gonzalez, J.F., Rodriguez, J.J., and Tejedor, A. 2002. Biodiesel fuels from vegetable oils: Transesterification of Cynara cardunculus L. oils with ethanol. *Energy and Fuel* 16: 443–450.

Freedman, B., Butterfield, R.O., and Pryde, E.H. 1986. Transesterification kinetics of soybean oil. *J. Am. Oil. Chem. Soc.* 63: 1375–1380.

Furuta, S., Matsuhasbi, H., and Arata, K. 2004. Biodiesel fuel production with solid superacid catalysis in fixed bed reactor under atmospheric pressure. *Catal. Commun.* 5: 721–723.

Gary, J.H., Handwerk, G.E., and Kaiser, M.J. 2007. *Petroleum Refining: Technology and Economics*, 5th Edition. CRC Press, Taylor & Francis Group, Boca Raton, Florida.

Ghanem, A. 2003. The utility of cyclodextrins in lipase-catalyzed transesterification in organic solvents: Enhanced reaction rate and enantioselectivity. *Org. Biomol. Chem.* 1: 1282–1291.

Goff, M.J., Bauer, N.S., Lopes, S., Sutterlin, W.R., and Suppes, G.J. 2004. Acid-catalyzed alcoholysis of soybean oil. *J. Am. Oil. Chem. Soc.* 81: 415–420.

Graboski, M.S. and McCormick, R.L. 1998. Combustion of fat and vegetable oil derived fuels in diesel engines. *Prog. Energy Combust. Sci.* 24: 125–164.

Gupta, P.K., Kumar, R., Panesar, P.S., and Thapar, V.K. 2007. Parametric Studies on Bio-diesel prepared from Rice Bran Oil. Agricultural Engineering International: The CIGR Journal, Manuscript EE 06 007. Vol. IX. April.

Hsu, C.S. and Robinson, P.R. (Editors). 2017. *Handbook of Petroleum Technology*. Springer International Publishing AG, Cham, Switzerland.

Iso, M., Chen, B., Eguchi, M., Kudo, T., and Shrestha, S. 2001. Production of biodiesel fuel from triglycerides and alcohol using immobilized lipase. *J. Mol. Catal. B: Enzym.* 16: 53–58.

Jeong, G.T. and Park, D.H. 1996. Batch (one- and two-stage) production of biodiesel fuel from rapeseed oil. *Appl. Biochem. Biotechnol.* 131: 668–679.

Kavalov, B. and Peteves, S.D. 2005. Status and Perspectives of Biomass-to-Liquid Fuels in the European Union. European Commission. Directorate

General Joint Research Centre (DG JRC). Institute for Energy, Petten, The Netherlands.

Knothe, G. 2005. Dependence of biodiesel fuel properties on the structure of fatty acid alkyl esters. *Fuel Process. Technol.* 86: 1059–1070.

Korytkowska, A., Barszczewska-Rybarek, I., and Gibas, M. 2001. Side-reactions in the transesterification of oligoethylene glycols by methacrylates. *Des. Monomers Poly.* 4: 27–37.

Kulkarni, M.G. and Dalai, A.K. 2006. Waste Cooking Oils — An Economical Source for Biodiesel: A Review. *Ind. Eng. Chem. Res.* 45: 2901–2913.

Kusdiana, D. and Saka, S. 2004a. Effects of water on biodiesel fuel production by supercritical methanol treatment. *Bioresour. Technol.* 91: 289–295.

Kusdiana, D. and Saka, S. 2004b. Two-step preparation for catalyst-free biodiesel fuel production: Hydrolysis and methyl esterification. *Appl. Biochem. Biotechnol.* 115: 781–791.

Liu, Y., Lotero, E., and Goodwin, J.G Jr. 2006. Effect of water on sulfuric acid catalyzed esterification. *J. Mol. Catal. A: Chem.* 245: 132–140.

Lopez, D.E., Goodwin, J.G. Jr, Bruce, D.A., and Lotero, E. 2005. Transesterification of triacetin with methanol on solid acid and base catalysts. *App. Catal. A: Gen.* 295: 97–105.

Lotero, E., Goodwin, J.G. Jr., Bruce, D.A., Suwannakarn, K., Liu, Y., and Lopez, D.E. 2006. The catalysis of biodiesel synthesis. *Catalysis* 19: 41–83.

Lucia, L.A., Argyropoulos, D.S., Adamopoulos, L., and Gaspar, A.R. 2006. Chemicals and energy from biomass. *Can. J. Chem.* 8: 960–970.

Ma, F., Clements, L.D., and Hanna, M.A. 1998. The effect of catalyst, free fatty acids, and water on transesterification of beef tallow. *Trans. ASAE* 41: 1261–1264.

Ma, F. and Hanna, M.A. 1999. Biodiesel production: A review. *Bioresour. Technol.* 70: 1–15.

Macedo, C.C.S., Abreu, F.R., Tavares, A.P., Alves, M.P., Zara, L.F., Rubim, J.C., and Suarez, P.A.Z. 2006. New heterogeneous metal-oxides based catalyst for vegetable oil trans-esterification. *J. Braz. Chem. Soc.* 17: 1291–1296.

Marchetti, J.M., Miguel, V.U., and Errazu, A.F. 2005. Possible methods for biodiesel production. *Renew. Sust. Energ. Rev.* 11: 1300–1311.

Marinetti, G.V. 1962. Hydrolysis of lecithin with sodium methoxide. *Biochemistry* 1: 350–353.

Marinetti, G.V. 1966. Low temperature partial alcoholysis of triglycerides. *J. Lipid Res.* 7: 786–788.

Meher, L.C., Kulkarni, M.G., Dalai, A.K., and Naik, S.N. 2006a. Transesterification of karanja (Pongamia pinnata) oil by solid basic catalysts. *Eur. J. Lipid Sci. Tech.* 108: 389–397.

Meher, L.C., Sagar, D.V., and Naik, S.N. 2006b. Technical aspects of biodiesel production by transesterification — A review. *Renew. Sust. Energ. Rev.* 10: 248–268.

Meneghetti, P.S.M., Meneghetti, M.R., Wolf, C.R., Silva, E.C., Lima, G.E.S., Coimbra, D.A., Soletti, J.I., and Carvalho, S.H.V. 2006. Ethanolysis of Castor and Cottonseed Oil: A Systematic Study Using Classical Catalysts. *JAOCS* 83: 819–822.

Mokhatab, S., Poe, W.A., and Speight, J.G. 2006. *Handbook of Natural Gas Transmission and Processing*. Elsevier, Amsterdam, Netherlands.

Nichols, N.N., Dien, B.S., Bothast, R.J., and Cotta, M.A. 2006. The Corn Ethanol Industry. In *Alcoholic Fuels*, S. Minteer (Editor). CRC-Taylor & Francis, Boca Raton, Florida. Chapter 4.

Noureddini, H., Harkey, D., and Medikonduru, V. 1998. A Continuous Process for the Conversion of Vegetable Oils into Methyl Esters of Fatty Acids. *JAOCS* 75: 1775–1783.

NREL. 2003. Dollars from Sense. National Renewable Energy Laboratory, Golden, Colorado. http://www.nrel.gov/docs/legosti/fy97/20505.pdf.

Parkash, S. 2003. *Refining Processes Handbook*. Gulf Professional Publishing, Elsevier, Amsterdam, Netherlands.

Pian, C.C.P., Volk, T.A., Abrahamson, L.P., White, E.H., and Jarnefeld. J. 2006. Biomass Gasification for Farm-Based Power Generation Applications. *Transactions on Ecology and the Environment*, Wessex Institute of Technology, Press. 92: 267.

Pinto, A.C., Guarieiro, L.N.N., Rezende, M.J.C., Ribeiro, N.M., Torres, E.A., Lopes, W.A., Pereira, P.A.P., and De Andrade, J.B. 2005. Biodiesel: An overview. *J. Braz. Chem. Soc.* 16: 1313–1330.

Reyes-Duarte, D., N. Lopez-Cortes, M. Ferrer, F. Plou, and Ballesteros, A. 2005. Parameters affecting productivity in the lipase-catalysed synthesis of sucrose palmitate. *Biocatal. Biotransformation* 23: 19–27.

Royon, D., Daz, M., Ellenrieder, G., and Locatelli, S. 2007. Enzymatic production of biodiesel from cotton seed oil using *t*-butanol as a solvent. *Bioresour. Technol.* 98: 648–653.

Ruth, M. 2004. Development of a Biorefinery Optimization Model. Renewable Energy Modeling Series Forecasting the Growth of Wind and Biomass. National Bioenergy Centre, National Renewable Energy Laboratory, Golden, Colorado. (http://www.epa.gov/cleanenergy/pdf/ruth2_apr20.pdf).

Saka, S. and Kusdiana, D. 2001. Biodiesel fuel from rapeseed oil as prepared in supercritical methanol. *Fuel* 80: 225–231.

Saka, S. and Minami, E. 2006. A Novel Non-catalytic Biodiesel Production Process by Supercritical Methanol as NEDO "High Efficiency Bioenergy Conversion Project". *The 2nd Joint International Conference on "Sustainable Energy and Environment (SEE 2006)*", Bangkok, Thailand, November 21–23.

Sørensen, B.E., Njakou, S., and Blumberga, D. 2006. Gaseous Fuels Biomass. Proceedings. World Renewable Energy Congress IX. WREN, London.

Schuchardta, U., Serchelia, R., and Vargas, R.M. 1998. Transesterification of Vegetable Oils: A Review. *J. Braz. Chem. Soc.* 9: 199–210.

Shah, S. and Gupta, M.N. 2007. Lipase catalyzed preparation of biodiesel from Jatropha oil in a solvent free system. *Process Biochem.* 42: 409–414.

Speight, J.G. 2008. *Synthetic Fuels Handbook: Properties, Processes, and Performance*. McGraw-Hill, New York.

Speight, J.G. 2011a. *The Refinery of the Future*. Gulf Professional Publishing, Elsevier, Oxford, United Kingdom.

Speight, J.G. (Editor). 2011b. *The Biofuels Handbook*. Royal Society of Chemistry, London, United Kingdom.

Speight, J.G. 2011c. *An Introduction to Petroleum Technology, Economics, and Politics*. Scrivener Publishing, Salem, Massachusetts.

Speight, J.G. 2013a. *The Chemistry and Technology of Coal*, 3rd Edition. CRC Press, Taylor & Francis Group, Boca Raton, Florida.

Speight, J.G. 2013b. *Coal-Fired Power Generation Handbook*. Scrivener Publishing, Salem, Massachusetts.

Speight, J.G. 2014. *The Chemistry and Technology of Petroleum*, 5th Edition. CRC Press, Taylor & Francis Group, Boca Raton, Florida.

Speight, J.G. 2017. *Handbook of Petroleum Refining*. CRC Press, Taylor & Francis Group, Boca Raton, Florida.

Speight, J.G. 2019a. *Handbook of Petrochemical Processes*. CRC Press, Taylor & Francis Group, Boca Raton, Florida.

Speight, J.G. 2019b. *Natural Gas: A Basic Handbook*, 2nd Edition. Gulf Publishing Company, Elsevier, Cambridge, Massachusetts.

Speight, J.G. 2019c. *Biogas: Production and Properties*. Nova Science Publishers, New York.

Srivastava, A. and Prasad, R. 2000. Triglycerides-based diesel fuels. *Renew. Sust. Energ. Rev.* 4: 111–133.

Stavarache, C., Vinatoru, M. Nishimura, R., and Maed, Y. 2005. Fatty acids methyl esters from vegetable oil by means of ultrasonic energy. *Ultrason. Sonochem.* 12: 367–372.

Stensvaag, J.-M. 1991. *Clean Air Act Amendments: Law and Practice*. John Wiley and Sons Inc., New York.

Taylor, G. 2008. Biofuels and the refinery concept. *Energy Policy* 36: 4406–4409.

United States Congress, 1990. Public Law 101–549. An Act to Amend the Clean Air Act to Provide for Attainment and Maintenance of Health protective National Ambient Air Quality Standards, and for Other Purposes. November 15.

United States Department of Energy, 1990. Gas Research Program Implementation Plan. DOE/FE-0187P. United States Department of Energy, Washington, D.C. April.

United States Department of Energy. 1993. Clean Fossil Fuels Technology Demonstration Program. DOE/FE-0272. United States Department of Energy, Washington, D.C. February.

United States General Accounting Office, 1990. Energy Policy: Developing Strategies for Energy Policies in the 1990s. Report to Congressional Committees. GAO/RCED-90–85. United States General Accounting Office, Washington, D.C. June.

Vallero, D. 2008. *Fundamentals of Air Pollution*, 4th Edition. Elsevier, London, United Kingdom.

Van Gerpen, J., Shanks, B., Pruszko, R., Clements, D., and Knothe, G. 2004. Biodiesel Analytical Methods: August 2002–January 2004. National Renewable Energy Laboratory, NREL/SR-510-36240, Colorado, July.

Varghaa, V. and Truterb, P. 2005. Biodegradable polymers by reactive blending trans-esterification of thermoplastic starch with poly(vinyl acetate) and poly(vinyl acetate-co-butyl acrylate. *Eur. Polym. J.* 41: 715–726.

Vera, C.R., D'Ippolito, S.A., Pieck, C.L., and Parera, J.M. 2005. Production of biodiesel by a two-step supercritical reaction process with adsorption refining. In *2nd Mercosur Congress on Chemical Engineering and 4th Mercosur Congress on Process Systems Engineering* (*ENPROMER-2005*), Rio de Janeiro. Brasil, August 14–18.

Watanabe, Y., Shimada, Y., and Sugihara, A. 2000. Continuous production of biodiesel fuel from vegetable oil using immobilized Candida antarctica lipase. *J. Am. Oil. Chem. Soc.* 77: 355–360.

Watanabe, Y., Shimada, Y., Sugihara, A., and Tominaga, T. 2002. Conversion of degummed soybean oil to biodiesel fuel with immobilized Candida antarctica lipase. *J. Mol. Catal. B: Enzym.* 17: 151–155.

Wright, L., Boundy, R, Perlack, R., Davis, S., and Saulsbury. B. 2006. Biomass Energy Data Book: Edition 1. Office of Planning, Budget and Analysis, Energy Efficiency and Renewable Energy, United States Department of Energy. Contract No. DE-AC05-00OR22725. Oak Ridge National Laboratory, Oak Ridge, Tennessee.

Zhang, Y., Dube, M.A., McLean, D.D., and Kates, M. 2003. Biodiesel production from waste cooking oil: 1. Process design and technological assessment. *Bioresour. Technol.* 89: 1–16.

Conversion Factors

1. General

1 acre = 43,560 sq ft
1 acre foot = 7758.0 bbl
1 atmosphere = 760 mm Hg = 14.696 psia = 29.91 in. Hg
1 atmosphere = 1.0133 bars = 33.899 ft. H_2O
1 barrel (oil) = 42 gal = 5.6146 *cu* ft
1 barrel (water) = 350 lb at 60°F
1 barrel per day = 1.84 cu cm per second
1 Btu = 778.26 ft-lb
1 centipoise × 2.42 = lb mass/(ft) (h), viscosity
1 centipoise × 0.000672 = lb mass/(ft) (s), viscosity
1 cubic foot = 28,317 cu cm = 7.4805 gal
Density of water at 60°F = 0.999 gram/cu cm = 62.367 lb/cu ft = 8.337 lb/gal
1 gallon = 231 cu in. = 3,785.4 cu cm = 0.13368 cu ft
1 horsepower-hour − 0.7457 kWh = 2544.5 Btu
1 horsepower = 550 ft-lb/s = 745.7 W
1 inch = 2.54 cm
1 meter = 100 cm = 1,000 mm = 10 microns = 10 angstroms (Å)
1 ounce = 28.35 g
1 pound = 453.59 g = 7,000 grains
1 square mile = 640 acres

2. Concentration Conversions

1 part per million (1 ppm) = 1 microgram per liter (1 μg/L)
1 microgram per liter (1 μg/L) = 1 milligram per kilogram (1 mg/kg)
1 microgram per liter (μg/L) × 6.243 × 10^8 = 1 lb per cubic foot (1 lb/ft^3)
1 microgram per liter (1 μg/L) × 1^{-3} = 1 milligram per liter (1 mg/L)
1 milligram per liter (1 mg/L) × 6.243 × 10^5 = 1 pound per cubic foot (1 lb/ft^3)
I gram mole per cubic meter (1 g mol/m^3) × 6.243 × 10^5 = 1 pound per cubic foot (1 lb/ft^3)
10,000 ppm = 1% w/w
1 ppm hydrocarbon in soil × 0.002 = 1 lb of hydrocarbons per ton of contaminated soil

3. Weight Conversion

1 ounce (1 oz) = 28.3495 grams (18.2495 g)
1 pound (1 lb) = 0.454 kg
1 pound (1 lb) = 454 grams (454 g)
1 kilogram (1 kg) = 2.20462 pounds (2.20462 lb)
1 stone (English, 1 st) = 14 pounds (14 lb)
1 ton (US; 1 short ton) = 2,000 lbs
1 ton (English; 1 long ton) = 2,240 lbs
1 metric ton = 2204.62262 pounds
1 tonne = 2204.62262 pounds

4. Temperature Conversions

°F = (°C × 1.8) + 32
°C = (°F – 32)/1.8
(°F – 32) × 0.555 = °C
Absolute zero = –273.15°C
Absolute zero = –459.67°F

5. Area

1 square centimeter (1 cm^2) = 0.1550 square inches
1 square meter 1 (m^2) = 1.1960 square yards

1 hectare = 2.4711 acres
1 square kilometer (1 km^2) = 0.3861 square miles
1 square inch (1 $inch^2$) = 6.4516 square centimeters
1 square foot (1 ft^2) = 0.0929 square meters
1 square yard (1 yd^2) = 0.8361 square meters
1 acre = 4046.9 square meters
1 square mile (1 mi^2) = 2.59 square kilometers

6. Other Approximations

14.7 pounds per square inch (14.7 psi) = 1 atmosphere (1 atm)
1 kiloPascal (kPa) × 9.8692 × 10^{-3} = 14.7 pounds per square inch (14.7 psi)
1 yd^3 = 27 ft^3
1 US gallon of water = 8.34 lbs
1 imperial gallon of water – 10 lbs
1 yd^3 = 0.765 m^3
1 acre-inch of liquid = 27,150 gallons = 3.630 ft^3
1 ft depth in 1 acre (in-situ) = 1,613 × (20 to 25% excavation factor) = ~2,000 yd^3
1 yd^3 (clayey soils-excavated) = 1.1 to 1.2 tons (US)
1 yd^3 (sandy soils-excavated) = 1.2 to 1.3 tons (US)

7. SI Metric Conversion Factors

(E = exponent; i.e. E + 03 = 10^3 and E – 03 = 10^{-3}

acre-foot × 1.233482	E + 03 = meters cubed
barrels × 1.589873	E – 01 = meters cubed
centipoise × 1.000000	E – 03 = pascal seconds
Darcy × 9.869233	E – 01 = micro meters squared
feet × 3.048000	E – 01 = meters
pounds/acre-foot × 3.677332	E – 04 = kilograms/meters cubed
pounds/square inch × 6.894757	E + 00 = kilo pascals
dyne/cm × 1.000000	E + 00 = mN/m
parts per million × 1.000000	E + 00 = milligrams/kilograms

8. Converting Volumes of Gas

Multiply flow of	By	To obtain flow of
Natural Gas	0.625	Propane
	0.547	Butane
	0.775	Air
Propane	1.598	Natural Gas
	0.874	Butane
	1.237	Air
Butane	1.826	Natural Gas
	1.140	Propane
	1.414	Air
Air	1.290	Natural gas
	0.808	Propane
	0.707	Butane

9. Natural Gas Conversion Table*

ft^3	Ccf	Mcf	MMcf	therm	dekatherm	BTU	MMBTU	kJ	kWh
0.0009	0.00001	0.000001	0	0.00001	0.0001	0.9482	0.000001	1	0
0.001	0.00001	0.000001	0.000000001	0.00001	0.0001	1	0.000001	1.055	0
1	0.01	0.001	0.000001	0.01	0.001	1000	0.001	1054	0.293
100	1	0.1	0.0001	1	0.1	100000	0.1	105461.5	29
1000	10	1	0.001	10	1	1000000	1	1054615	293
1000000	10000	1000	1	10000	1000	1.00E+09	1000	1054615000	293071

Note: *Assumed 1 cf = 1000 BTU, for approximate reference only.

10. Heat Content for Various Fuels

Automotive gasoline	125,000 Btu/gal(gross)	115,400 Btu/gal(net)
Hydrogen	134,200 Btu/kg(gross)	113,400 Btu/kg(net)
Diesel motor fuel	138,700 Btu/gal (gross)	128,700 Btu/gal (net)
Biodiesel	126,206 Btu/gal (gross)	117,093 Btu/gal (net)
Methanol	64,600 Btu/gal (gross)	56,560 Btu/gal (net)
Ethanol	84,600 Btu/gal (gross)	75,670 Btu/gal (net)
Gasohol	120,900 Btu/gal (gross)	112,417 Btu/gal (net)
Aviation gasoline	120,200 Btu/gal (gross)	112,000 Btu/gal (net)

(*Continued*)

Propane	91,300 Btu/gal (gross)	83,500 Btu/gal (net)
Butane	103,000 Btu/gal (gross)	93,000 Btu/gal (net)
Jet fuel (naphtha)	127,500 Btu/gal (gross)	118,700 Btu/gal (net)
Jet fuel (kerosene)	135,000 Btu/gal (gross)	128,100 Btu/gal (net)
Lubricants	144,400 Btu/gal (gross)	130,900 Btu/gal (net)
Waxes	131,800 Btu/gal (gross)	120,200 Btu/gal (net)
Asphalt and road oil	158,000 Btu/gal (gross)	157,700 Btu/gal (net)
Petroleum coke	143,400 Btu/gal (gross)	
Natural gas		
Wet	1,109 Btu/ft^3	
Dry	1,027 Btu/ft^3	
Compressed	960 Btu/ft^3	20,551 Btu/pound
Liquid	90,800 Btu/gal (gross)	87,600 Btu/gal (net)
Crude petroleum	138,100 Btu/gal (gross)	131,800 Btu/gal (net)
Fuel oil		
Residual	149,700 Btu/gal (gross) = 138,400 Btu/gal (net)	
Distillate	138,700 Btu/gal (gross) = 131,800 Btu/gal (net)	
Coal		
Anthracite	21.711 × 10^6 Btu/short ton (2,000 lbs)	
Bituminous and lignite	21.012 × 10^6 Btu/short ton (2,000 lbs)	

11. Biomass Energy Conversions

Cord: a stack of wood comprising 128 cubic feet (3.62 m^3); standard dimensions are 4 × 4 × 8 feet, including air space and bark. One cord contains approx. 1.2 US tons (oven-dry) = 2400 pounds = 1089 kg

1.0 metric tonne wood = 1.4 cubic meters (solid wood, not stacked)

Energy content of wood fuel (HHV, bone dry) = 18–22 GJ/t (7,600–9,600 Btu/lb)

Energy content of wood fuel (air dry, 20% moisture) = about 15 GJ/t (6,400 Btu/lb)

Energy content of agricultural residues (range due to moisture content) = 10–17 GJ/t (4,300–7,300 Btu/lb)

Metric tonne charcoal = 30 GJ (= 12,800 Btu/lb) (but usually derived from 6–12 t air-dry wood, i.e. 90–180 GJ original energy content)

(*continued*)

(*Continued*)

Metric tonne ethanol = 7.94 petroleum barrels = 1262 liters
Ethanol energy content = 11,500 Btu/lb = 75,700 Btu/gallon = 26.7 GJ/t = 21.1 MJ/L
Ethanol HHV = 84,000 Btu/gallon = 89 MJ/gallon = 23.4 MJ/L
Ethanol density (average) = 0.79 g/mL (= metric tonnes/m^3)
Metric tonne biodiesel = 37.8 GJ (33.3–35.7 MJ/L)
biodiesel density (average) = 0.88 g/mL (= metric tonnes/m^3)

12. Mass Conversions

	To				
From	Pound	Kilogram	Short ton	Long ton	Metric ton
Pound	1	0.4536	5.0×10^{-4}	4.4643×10^{-4}	4.5362×10^{-4}
Kilogram	2.205	1	1.1023×10^{-3}	9.8425×10^{-4}	1.0×10^{-3}
Short ton	2000	907.2	1	0.8929	0.9072
Long ton	2240	1016	1.12	1	1.016
Metric ton	2205	1000	1.102	0.9842	1

13. Volume and Flow Rate Conversions

1 U.S. gal = 231 $inch^3$	1 liter = 61.02 $inch^3$
= 0.1337 ft3	= $3.531 \times 10^{-2} ft^3$
= 3.785 liters	= 0.2624 US gal
= 0.8321 imperial gal	= 0.2200 imperial gal
= 0.0238 bbl	= 6.29×10^{-3} bbl
= 0.003785 m^3	= 0.001 m^3

14. Composition of Selected Feedstocks

Feedstock	Cellulose	Hemicellulose	Lignin
Corn stover	35	28	16–21
Hardwood	45	30	20
Softwood	42	21	26

(*Continued*)

Feedstock	Cellulose	Hemicellulose	Lignin
Bamboo	41–49	24–28	24–26
Switchgrass	44–51	42–50?	13–20
Miscanthus	44	24	17
Arundo donax	31	30	21

15. Biomass Characteristics

Biomass feedstocks and fuels exhibit a wide range of physical, chemical, and agricultural/process engineering properties. Despite their wide range of possible sources, biomass feedstocks are remarkably uniform in many of their fuel properties, compared with competing feedstocks such as coal or petroleum. For example, there are many kinds of coals whose gross heating value ranges from 20 to 30 GJ/tonne (gigajoules per metric tonne; 8,600–12,900 Btu/lb). However, nearly all kinds of biomass feedstocks destined for combustion fall in the range 15–19 GJ/tonne (6,450–8,200 Btu/lb). For most agricultural residues, the heating values are even more uniform — about 15–17 GJ/tonne (6,450–7,300 Btu/lb); the values for most woody materials are 18–19 GJ/tonne (7,750–8,200 Btu/lb). Moisture content is probably the most important determinant of heating value. Air-dried biomass typically has about 15–20% moisture, whereas the moisture content for oven-dried biomass is around 0%. Moisture content is also an important characteristic of coals, varying in the range of 2–30%. However, the bulk density (and hence energy density) of most biomass feedstocks is generally low, even after densification, about 10 and 40% of the bulk density of most fossil fuels. Liquid biofuels have comparable bulk densities to fossil fuels.

Most biomass materials are easier to gasify than coal because they are more reactive with higher ignition stability. This characteristic also makes them easier to process thermochemically into higher-value fuels such as methanol or hydrogen. Ash content is typically lower than for most coals, and sulphur content is much lower than for many fossil fuels. Unlike coal ash, which may contain toxic metals and other trace contaminants, biomass ash may be used as a soil amendment to help replenish nutrients removed by harvest. A few biomass feedstocks stand out for their peculiar properties, such as high silicon or alkali metal contents — these may require special precautions for harvesting, processing and combustion equipment. Note also that mineral content can vary as a function of soil type and the timing of feedstock harvest. In contrast to their fairly uniform physical properties, biomass fuels are rather heterogeneous with respect to their chemical elemental composition.

Among the liquid biomass fuels, biodiesel (vegetable oil ester) is noteworthy for its similarity to petroleum-derived diesel fuel, apart from its negligible sulfur and ash content. Bioethanol has only about 70% the heating value of petroleum distillates such as gasoline, but its sulfur and ash contents are also very low. Both of these liquid fuels have lower vapor pressure and flammability than their petroleum-based competitors — an advantage in some cases (e.g. use in confined spaces such as mines) but a disadvantage in others (e.g. engine starting at cold temperatures).

Fire hazard properties of flammable liquids, gases and volatile solids

Substance	Lower flammable limit %v/v	Upper flammable limit %v/v	Flash point °C	Auto ignition temperature °C
Acetylene	2	83	—	305
Benzene	1	8	−11	560
Butane	2	9	—	365
Cyclohexane	1	8	−18	260
Ethane	3	12	—	515
Ethanol	3	19	12	363
Ethylene	3	34	—	305
Heptane	1	7	−4	204
Hexane	1	8	−22	223
Hydrogen	4	76	—	—
Methanol	6	37	11	385
Octane	1	7	13	206
Pentane	1	8	−40	285
Propane	3	12	—	515
Propylene	2	11	−108	460
Toluene	1	7	4	530
o-Xylene	1	6	17	464
m-Xylene	1	7	25	528
p-Xylene	1	7	25	529

The main parameters which describe the ability of mixtures of substances in air to propagate a flame and/or explode are flammable limits, flash points and auto-ignition temperatures. None of these parameters are basic physico-chemical properties of a substance or mixture but are defined by standard tests and, for this reason, values are given to the nearest integer. If the test conditions are varied, then the values of the parameters will change. In particular, values will change considerably if the concentration of oxygen in the mixture is changed.

The flammable limits, lower flammable limit and upper flammable limit, are the values, by volume, of the concentrations in air, at normal temperature and pressure, between which a mixture can propagate a flame.

The flash point is the lowest temperature, at one atmosphere pressure, at which a liquid gives off flammable vapour in sufficient quantity to ignite when mixed with air at or near the surface of the liquid on application of a flame or spark. Substances with a flash point <21°C are termed "extremely flammable". Substances with flash points between 21°C and 55°C are termed "flammable" while the term "combustible" applies to all other combustible solids and liquids, including those with flash points above 100°C.

The auto-ignition temperature is the lowest temperature of a glass surface at which droplets of a combustible liquid falling onto the surface will undergo spontaneous combustion. A glass surface is assumed to be chemically inert and in practice, because of catalytic effects, the values for other types of surfaces can be lower.

Aerosols of combustible solids will also propagate a flame in the appropriate concentration range. However, the flammable limits of aerosols are affected by the size of the particles although, in general, the smaller the particle size, the closer the behaviour of the aerosol resembles that of a vapor.

Glossary

Acetic acid: An acid with the structure of CH_3COOH; acetyl groups are bound through an ester linkage to hemicellulose chains, especially xylan derivatives, in wood and other plants; the natural moisture present in plants hydrolyzes the acetyl groups to acetic acid, particularly at elevated temperatures.

Acid: Any of a class of substances whose aqueous solutions are characterized by a sour taste, an ability to turn blue litmus red, and an ability to react with bases and certain metals to form salts; a substance that yields hydrogen ions when dissolved in water and can act as a proton (H^+) donor.

Acid detergent fiber (ADF): The organic matter that is not solubilized after 1 h of refluxing in an acid detergent of cetyltrimethylammonium bromide in 1N sulfuric acid; includes cellulose and lignin; this analytical method is commonly used in the feed and fiber industries.

Acid hydrolysis: A chemical process in which acid is used to convert cellulose or starch to sugar.

Acid-insoluble lignin: Lignin is mostly insoluble in mineral acids and therefore can be analyzed gravimetrically after hydrolyzing the cellulose and hemicellulose fractions of the biomass with sulfuric acid; ASTM E1721 is the standard method for determining acid-insoluble lignin in biomass; see ASTM International.

Acid-soluble lignin: A small fraction of the lignin in a biomass sample is solubilized during the hydrolysis process of the acid-insoluble lignin method. This lignin fraction is referred to as acid-soluble lignin and may be quantified by ultraviolet spectroscopy; see lignin and acid-insoluble lignin.

Aerosol: A dispersion of a liquid or solid in a gas.

Agitator: A device such as a stirrer that provides complete mixing and uniform dispersion of all the components in a mixture; generally used continuously during thermal processes and intermittently during fermentation.

Agricultural residue: Agricultural crop residues are the plant parts, primarily stalks and leaves, not removed from the fields containing the primary food or fiber product; examples include corn stover (stalks, leaves, husks, and cobs), wheat straw, and rice straw.

Agricultural waste: Waste produced at agricultural premises as a result of an agricultural activity.

Air quality maintenance area: A specific populated area where air quality is a problem because of one or more pollutants.

Alcohol: The family name for a group of organic-chemical compounds composed of carbon, hydrogen, and oxygen. The molecules in the series vary in chain length and are composed of a hydrocarbon plus a hydroxyl group. Alcohol includes methanol and ethanol.

Aldoses: Aldoses occur when the carbonyl group of a monosaccharide is an aldehyde.

Algae: Primitive plants, usually aquatic, that are capable of synthesizing their own food by photosynthesis; currently being investigated as a possible feedstock for producing biodiesel.

Aliphatic: Any non-aromatic organic compound having an open chain structure.

Alkali: A soluble mineral salt.

Alkali lignin: Lignin obtained by acidification of an alkaline extract of wood.

Alkylation: a process for manufacturing high-octane blending components used in unleaded petrol or gasoline.

Annual removals: The net volume of growing stock trees removed from the inventory during a specified year by harvesting, cultural operations such as timber stand improvement, or land clearing.

Alternative fuel: As defined in the United States Energy Policy Act of 1992 (EPACT); alternative fuels are methanol, denatured ethanol, and other alcohols, separately or in blends of at least 10% by volume with gasoline and other fuels: compressed natural gas, liquefied natural gas, liquefied propane gas, hydrogen, coal-derived liquid fuels, fuels other than alcohols derived from biological materials, electricity, biodiesel, and any other fuel deemed to be substantially not petroleum and offering potential energy security benefits and substantial environmental benefits.

Ambient air quality: The condition of the air in the surrounding environment.

American Society for Testing and Materials (ASTM): Now known as ASTM International; an international voluntary standards organization that develops and produces technical standards for materials, products, systems, and services.

Anaerobic: Biological processes that occur in the absence of oxygen.

Anaerobic digestion: Decomposition of biological wastes by microorganisms, usually under wet conditions, in the absence of air (oxygen), to produce a gas composed mostly of methane and carbon dioxide.

Anhydrous: without water; transesterification of biodiesel needs to be an anhydrous process; the water in vegetable oil causes either no reaction or a cloudy biodiesel, and the water in lye or methanol renders it less useful or even useless, depending on how much water is present.

API gravity: a measure of the lightness or heaviness of petroleum that is related to density and specific gravity; °API = (141.5/sp. gr. at 60°F) – 131.5.

Aquatic plants: The wide variety of aquatic biomass resources, such as algae, giant kelp, other seaweed, and water hyacinth; certain microalgae can produce hydrogen and oxygen, while others produce hydrocarbons and a host of other products; examples of microalgae include Chlorella, Dunaliella, and Euglena.

Arabinan: A polymer of arabinose; can be hydrolyzed to arabinose.

Arabinose: A five-carbon sugar; a product of the hydrolysis of arabinan found in the hemicellulose fraction of biomass.

Aromatics: a range of hydrocarbons with a distinctive sweet smell; they include benzene and toluene, occur naturally in petroleum, and are extracted for use as a petrochemical feedstock, as well as for use as solvents.

Asphaltene (asphaltenes): the brown-to-black powdery material produced upon treating of petroleum, heavy oil, bitumen, or residuum with a low-boiling liquid hydrocarbon.

ASTM International: Formerly the American Society for Testing and Materials (ASTM); an international voluntary standards organization that develops and produces technical standards for materials, products, systems, and services.

Atmospheric pressure: Pressure of the air and atmosphere surrounding us; changes from day to day; equal to 14.7 psia.

Attainment area: A geographic region where the concentration of a specific air pollutant does not exceed federal standards.

Auger: A rotating, screw-type device that moves material through a cylinder.

Available production capacity: The biodiesel-production capacity of those refining facilities that are not specifically designed to produce biodiesel.

Average megawatt (MWa or aMW): One megawatt of capacity produced continuously over a period of 1 year; 1 aMW = 1 MW × 8760 h/year = 8,760 MWh = 8,760,000 kWh.

B100: Another name for pure biodiesel.

Background level: The average amount of a substance present in the environment; it originally refined to naturally occurring phenomena; it is used in toxic-substance monitoring.

Backup rate: A utility charge for providing occasional electricity service to replace on-site generation.

Backup electricity: Power or services needed occasionally, for example, when on-site generation equipment fails.

Baffle Chamber: In an incinerator design, a chamber designed to settle fly ash and coarse particulate matter by changing the direction and reducing the velocity of the combustion gases.

Bagasse: Sugarcane waste.

Bark: The outer protective layer of a tree outside the cambium comprising the inner bark and the outer bark; the inner bark is a layer of living bark that separates the outer bark from the cambium and, in a living tree, is generally soft and moist; the outer bark is a layer of dead bark that forms the exterior surface of the tree stem; the outer bark is frequently dry and corky.

Barrel (bbl): the unit of measure used by the petroleum industry; equivalent to approximately 42 US gallons or approximately 34 (33.6) Imperial gallons or 159 L; 7.2 barrels are equivalent to one metric tonne of oil.

Barrel of oil equivalent (boe): The amount of energy contained in one barrel of crude oil, i.e. approximately 6.1 GJ (5.8 million Btu), equivalent to 1,700 kWh.

Base: A classification of those substances that when combined with an acid, will form a salt plus water, usually producing hydroxide ions when dissolved.

Baseload capacity: The power output that generating equipment can continuously produce.

Baseload demand: The minimum demand experienced by an electric utility, usually 30–40% of the utility's peak demand.

Batch distillation: A process in which the liquid feed is placed in a single container and the entire volume is heated, in contrast to continuous distillation in which the liquid is fed continuously through the still.

Batch fermentation: Fermentation as conducted from the start to finish in a single vessel; see fermentation.

Batch process: Unit of operation where 1 cycle of feedstock preparation, cooking, fermentation, and distillation is completed before the next cycle is started.

Beer: A general term for all fermented malt beverages flavored with hops; a low-level (6–12%) alcohol solution derived from the fermentation of mash by microorganisms.

Beer still: The stripping section of a distillation column for concentrating ethanol.

Benzene: A toxic, six-carbon aromatic component of gasoline; a known carcinogen.

Billion: 1×10^9.

Biobutanol: Alcohol containing four carbon atoms per molecule, produced from the same feedstocks as ethanol, but with a modified fermentation and distillation process; less water-soluble than ethanol, biobutanol has a higher energy density and can be transported by pipeline more easily.

Biochemical conversion: The use of fermentation or anaerobic digestion to produce fuels and chemicals from organic sources.

Bioconversion platform: Typically uses a combination of physical or chemical pretreatment and enzymatic hydrolysis to convert lignocellulose into its component monomers.

Biochemical conversion process: The use of living organisms or their products to convert organic materials to fuels, chemicals, or other products.

Biochemical oxygen demand (BOD): A standard means for estimating the degree of water pollution, especially of water bodies that receive contamination from sewage and industrial waste; the amount of oxygen needed by bacteria and other microorganisms to decompose organic matter in water — the greater the BOD, the greater the degree of pollution; BOD is a process that occurs over a period of time and is commonly measured for a five-day period, referred to as BOD5.

Bioconcentration (bioaccumulation): The accumulation of a chemical in tissues of an organism to levels greater than those in the environment in which the organism lives.

Biodegradable: Capable of decomposing rapidly under natural conditions.

Biodiesel: A fuel that is derived from biological sources and that can be used in diesel engines instead of petroleum-derived diesel; through the process of transesterification, the triglycerides in biologically derived oils are separated from glycerin, creating a clean burning, renewable fuel.

Biodiesel Blend: A blend of biodiesel and diesel fuels — the blend can be with Diesel #1, Diesel #2, or JP8; one standard blend meeting the minimum requirements of the federal EPA Clean Air Act criteria is B20. The number after "B" indicates the percentage of biodiesel included in the blend — in B20, there would be 20% biodiesel and 80% diesel in the fuel blend; a biodiesel blend can come with any mixture percentage, such as B2, B5, B50, and B85.

Biodiesel recipe: The most common recipe uses waste vegetable oil (WVO), methanol (wood alcohol), and sodium hydroxide (caustic soda/lye) to produce biodiesel and glycerin; the steps are as follows: (1) cleaning/heating the waste vegetable oil, (2) titration of the waste vegetable oil sample, (3) combining methanol and sodium hydroxide in exact amounts, (4) combining step (3) with (1) and mixing at 50°C, (5) settling, (6) separating the biodiesel from the wastes, (7) washing and drying the biodiesel, and (8) disposing of the wastes.

Biodegradable waste: Waste that is capable of being broken down by plants (including fungi) and animals (including worms and microorganisms).

Bioenergy: Useful, renewable energy produced from organic matter — by the conversion of the complex carbohydrates present in organic matter into energy; organic matter may be used directly as a fuel or processed into liquids and gases, or be a residual of processing and conversion.

Bioethanol: The ethanol produced from biomass feedstocks; includes ethanol produced by fermentation of crops such as corn, as well as cellulosic ethanol produced from woody plants or grasses.

Biofuels: A generic name for those liquid or gaseous fuels that are not derived from petroleum-based fossil fuels or that contain a proportion of non-fossil fuel; fuels produced from plants and crops such as sugar beet, rape seed oil or re-processed vegetable oils or fuels made from gasified biomass; fuels that are made from renewable biological sources and that include ethanol, methanol, and biodiesel; such sources include but are not limited to corn, soybeans, flaxseed, rapeseed, sugarcane, palm oil, raw sewage, food scraps, animal parts, and rice.

Biogas: A combustible gas derived from decomposing biological waste under anaerobic conditions. Biogas normally consists of 50–60% methane. See also landfill gas.

Biofuels: A generic name for liquid or gaseous fuels that are not derived from petroleum.

Bioheat: A name applied sometimes to biodiesel when it is used for heating purposes.

Biological assessment: A specific process required to be performed during an environmental assessment; an evaluation of the potential effects of a proposed project on proposed, endangered, threatened, and sensitive animal and plant species and their habitats.

Biological oxidation: Decomposition of organic materials by microorganisms.

Biomass: Any organic matter that is available on a renewable or recurring basis, including agricultural crops and trees, wood and wood residues, plants (including aquatic plants), grasses, animal manure, municipal residues, and other residue materials. Biomass is generally produced, in a sustainable manner, from water and carbon dioxide by photosynthesis. There are three main categories of biomass: primary, secondary, and tertiary.

Biomass fuel: Liquid, solid, or gaseous fuel produced by conversion of biomass.

Biomass processing residues: By-products from the processing of all forms of biomass that have significant energy potential; the residues are typically collected at the point of processing; they can be convenient and relatively inexpensive sources of biomass for energy.

Bionaphtha (bio-naphtha): A term used in some eastern European nations for biodiesel.

Biopower: The use of biomass feedstock to produce electric power or heat through direct combustion of the feedstock, through gasification and then combustion of the resultant gas, or through other thermal conversion processes. Power is generated using engines, turbines, fuel cells, or other equipment.

Bioreactor: A vessel in which a chemical process occurs, which usually involves organisms or biochemically active substances derived from such organisms.

Biorefinery: A facility that processes and converts biomass into value-added products. These products can range from biomaterials to fuels such as ethanol or important feedstocks for the production of chemicals and other materials.

Biomass to liquid (BTL): The process of converting biomass to liquid fuels.

Bitumen: also, on occasion, referred to as native asphalt, and extra heavy oil; a naturally occurring material that has little or no mobility under reservoir conditions and that cannot be recovered through a well by conventional oil well production methods including currently used enhanced recovery techniques; current methods involve mining for bitumen recovery.

Boiler: Any device used to burn biomass fuel in order to heat water for the purpose of generating steam.

Boiler horsepower: A measure of the maximum rate of heat energy output of a steam generator; one boiler horsepower equals 33,480 Btu/hour output in steam.

Bone-dry: Having zero percent moisture content. Wood that has been heated in an oven at a constant temperature of 100°C (212°F) or above until its weight stabilizes is considered bone-dry or oven-dry.

Bottoming cycle: A cogeneration system in which steam is used first for process heat production and then for electric power production.

Black liquor: Solution of lignin residue and the pulping chemicals used to extract lignin during the manufacture of paper.

Brewing: Generically, the entire beer-making process, but technically only the part of the process during which the beer wort is cooked in a brew kettle and during which the hops are added; after brewing, the beer is fermented.

British thermal unit (Btu): A non-metric unit of heat, still widely used by engineers; 1 Btu is the heat energy needed to raise the temperature of one pound of water from 60°F to 61°F at 1 atmospheric pressure. 1 Btu = 1055 joules (1.055 kJ).

Brown grease: Waste grease, which is the least expensive among the various grades of waste grease.

Bubble cap trays: Cross-flow trays, usually installed in rectifying columns handling liquids free of suspended solids; the bubble caps consist of circular cups inverted over small vapor pipes — the vapor from the tray below passes through the vapor pipes into the caps and curves downward to escape below the rim into the liquid; the rim of each cap is slotted or serrated to break up the escaping vapor into small bubbles, thereby increasing the surface area of the vapor as it passes through the liquid.

Bubble wash: A method for final washing of biodiesel through air agitation. Biodiesel floats above a quantity of water; bubbles from an aquarium air pump and air stone are injected into the water, causing the bubbles to rise — at the water/biodiesel interface, the air bubbles carry water upward through the biodiesel by surface tension; simple diffusion causes water-soluble impurities in the biodiesel to be extracted into the water; as the bubbles reach the surface and break, the water is freed and percolates back downward through the biodiesel again.

Bunker: A storage tank.

Butanol: Although generally produced from fossil fuels, this four-carbon alcohol can also be produced through bacterial fermentation of alcohol.

Byproduct (by-product): A substance, other than the principal product, generated as a result of creating a biofuel.

Canola: A member of the *Brassica* family, which includes broccoli, cabbage, cauliflower, mustard, radish, and turnip; it is a variant of the crop rapeseed, containing less erucic acid and glucosinolate derivatives than rapeseed; it is grown for its seed; the seed is crushed for the oil contained within, and after the oil has been extracted, the by-product is in turn a protein-rich meal used by the intensive livestock industry.

Capacity: The maximum power that a machine or system can produce or carry safely; the maximum instantaneous output of a resource under specified conditions; the capacity of electricity generating equipment is generally expressed in kilowatts or megawatts.

Capacity factor: The amount of energy that a power plant actually generates as compared to its maximum rated output, expressed as a percentage.

Capital cost: The total investment needed to complete a project and bring it to a commercially operatable status; the cost of construction of a new plant; the expenditures for the purchase or acquisition of existing facilities.

Carbohydrate: A chemical compound made up of carbon, hydrogen, and oxygen; it includes sugars, cellulose, and starches.

Carbon chain: The atomic structure of hydrocarbons in which a series of carbon atoms, saturated by hydrogen atoms, form a chain; volatile oils have shorter chains while fats have longer chain lengths and waxes, have extremely long carbon chains.

Carbon dioxide (CO_2): A product of combustion; acts as a greenhouse gas for the atmosphere of the Earth, thereby trapping heat and contributing to climate change.

Carbon monoxide (CO): A lethal gas produced by the incomplete combustion of carbon-containing fuels in internal combustion engines. It is colorless, odorless, and tasteless.

Carbon sequestration: The absorption and storage of carbon dioxide from the atmosphere; naturally occurs in plants.

Carbon sink: A geographical area whose vegetation and/or soil soaks up significant carbon dioxide from the atmosphere. Such areas, typically in tropical regions, are increasingly being sacrificed for energy crop production.

Catalyst: A substance that accelerates a chemical reaction without itself being affected. In refining, catalysts are used in the cracking process to produce blending components for fuels.

Cellulose: Fiber contained in the leaves, stems, and stalks of plants and trees; the most abundant organic compound on the Earth; it is a polymer of glucose with a repeating unit of $C_6H_{10}O_5$ strung together by β-glycosidic linkages — the β-linkages in cellulose form linear chains that are highly stable and resistant to chemical attack because of the high degree of hydrogen bonding that can occur between chains of cellulose; hydrogen bonding between cellulose chains makes the polymers more rigid, inhibiting the flexing of the molecules that may occur during the hydrolytic breaking of the glycosidic linkages — hydrolysis can reduce cellulose to a cellobiose repeating unit, $C_{12}H_{22}O_{11}$, and ultimately to the six-carbon sugar glucose, $C_6H_{12}O_6$.

Cetane number: A measure of the ignition quality of diesel fuel; the higher the cetane number, the more easily the fuel is ignited under compression.

Cetane rating: A measure of the combustion quality of diesel fuel.

Chips: Small fragments of wood chopped or broken by mechanical equipment — total tree chips include wood, bark, and foliage, while pulp chips or clean chips are free of bark and foliage.

Chlorofluorocarbon: A family of chemicals composed primarily of carbon, hydrogen, chlorine, and fluorine; they are used principally as refrigerants and industrial cleansers and have the tendency to destroy the protective ozone layer of the Earth.

Clarifier: A tank used to remove solids by gravity, colloidal solids by coagulation, and floating oil and scum through skimming.

Class I area: Any area allotted for the most stringent protection from air quality degradation.

Class II area: Any area where the air is cleaner than required by federal air quality standards and allotted for a moderate degree of protection from air quality degradation; moderate increases in new pollutants may be permitted in Class II areas.

Clean Air Act (CAA): US national law establishing ambient air quality emission standards, to be implemented by participating states; originally enacted in 1963, CAA has been amended several times, the most recent being that in 1990, and contains vehicle emission standards regulating the emission of criteria pollutants (lead, ozone, carbon monoxide, sulfur dioxide, nitrogen oxides, and particulate matter); the 1990 amendments included reformulated gasoline (RFG) requirements and oxygenated gasoline provisions.

Clean fuels: Fuels such as E10 (unleaded) that burn cleaner and produce fewer harmful emissions than ordinary gasoline.

Closed-loop biomass: Crops that are grown in a sustainable manner to optimize their value for bioenergy and bioproduct uses. They include annual crops such as maize and wheat, and perennial crops such as trees, shrubs, and grasses such as switch grass.

Cloud point: The temperature at which paraffin wax and other solid substances begin to crystallize or separate from the solution, imparting a cloudy appearance to the oil when the oil is chilled under prescribed conditions.

Coarse materials: Wood residues that are suitable for chipping, for example, slabs, edgings, and trimmings.

Cogeneration: The sequential production of electricity and useful thermal energy from a common fuel source.

Coking: A thermal method that is used in refineries to convert of bitumen and residua to volatile products and coke. (See delayed coking and fluid coking.)

Colloid: A stable system of small particles dispersed in another phase; a multi-phase system in which one dimension of a dispersed phase is of colloidal size; colloids are the liquid and solid forms of aerosols, foams, emulsions, and suspensions within the colloidal size class.

Colloidal size: 0.001–1 micron in any dimension; dispersions where the particle size is in this range are referred to as colloidal aerosols, colloidal emulsions, colloidal foams, or colloidal suspensions.

Colza: a Eurasian plant cultivated for its seed and used as a forage crop.

Combined heat and power: The use of a power station to generate both heat and electricity simultaneously. The steam or hot water generated in this process is utilized in either industrial processes or community heating.

Combustion (burning): The transformation of biomass fuel into heat, chemicals, and gases through the chemical combination of the hydrogen and carbon present in the fuel with the oxygen in the air.

Combustion gases: The gases released from a combustion process.

Commercial waste: The waste arising from premises that is used wholly or mainly for trade, business, sport, recreation, or entertainment, excluding municipal waste and industrial waste.

Composting: A resource recovery process where biodegradable waste (such as garden and kitchen waste) is converted in the presence of oxygen from the air into a stable granular material, which, when applied to land, improves the soil structure and enriches the nutrient content.

Compound: A chemical term denoting a combination of two or more distinct elements.

Compressed natural gas (CNG): Natural gas that has been compressed under high pressure (typically 2,000–3,600 psi).

Compression ignition engine: An engine in which the fuel is ignited by the high temperature caused by extreme pressure in the cylinder rather than by a spark from a spark plug; see diesel engine.

Concentrated acid hydrolysis: A method for converting biomass into cellulosic ethanol.

Conditional use permit: A permit, with conditions, allowing an approved use on a site located outside the appropriate zoning class.

Conservation: Efficient energy use, production, transmission, or distribution, it results in a decrease of energy consumption while providing the same level of service.

Construction and demolition waste: Waste arising from the construction, repair, maintenance, and demolition of buildings and structures, including roads. It consists mostly of brick, concrete, hardcore, subsoil, and topsoil, but can also contain quantities of timber, metal, plastics, and (occasionally) hazardous waste materials.

Continuous fermentation: A steady-state fermentation system that operates without interruption; each stage of fermentation occurs in a separate section of the fermenter, and flow rates are set to correspond with required residence times.

Continuous flow process: A general term for a number of biodiesel production processes that involve the continuous addition of ingredients to produce biodiesel on a continual, round-the-clock basis, as opposed to the batch process.

Controlled wastes: The UK term for wastes controlled under the Waste Framework Directive; they include household waste, commercial waste, industrial waste, and agricultural waste.

Conventional biofuels: Biofuels such as bioethanol and biodiesel, which are typically made from corn, sugarcane and sugar beet, wheat, or oilseed crops such as soy and rapeseed oil.

Conventional crude oil (conventional petroleum): Crude oil that is pumped from the ground and recovered using the energy inherent in the reservoir; it is also recoverable by the application of secondary recovery techniques.

Conversion efficiency: A comparison of the useful energy output to the potential energy contained in the fuel; the efficiency calculation is related to the type of energy produced and a direct comparison of the efficiencies of different conversion processes can be made only when the processes produce the same type of energy output.

Cooker: A tank or vessel designed to cook a liquid or extract or to digest solids in suspension; a cooker usually contains a source of heat; and is fitted with an agitator.

Cord: A stack of wood of 128 cubic feet (3.62 m^3); standard dimensions are 4 × 4 × 8 feet, including the air space and bark. One cord contains approx. 1.2 US tons (oven-dry) = 2,400 pounds = 1,089 kg.

Corn stover: Residue materials from the harvesting of corn; they consist of the cob, leaves, and stalk; similar to straw, the residue left after any cereal grain or grass has been harvested at maturity for its seed. It can be either directly grazed by cattle or dried for use as fodder. Stover has drawn some attention as a potential fuel source, as biomass for fermentation, or as a feedstock for ethanol production from cellulose.

Cracking: A secondary refining process that uses heat and/or a catalyst to break down high-molecular-weight chemical components into lower molecular weight products that can be used as blending components for fuels.

Cropland: Total cropland includes five components: cropland harvested, crop failure, cultivated summer fallow, cropland used only for pasture, and idle cropland.

Cropland pasture: Land used for long-term crop rotation. However, some cropland pasture is for marginal crop use and may remain in pasture indefinitely. This category includes land that was used for pasture before crops reached maturity and some land used for pasture that could have been cropped without additional improvement.

Crop residue: The "backbone" of sugar and starch crops — the stalks and leaves — is composed mainly of cellulose. The individual six-carbon sugar units in cellulose are linked together in extremely long chains by a stronger chemical bond than exists in starch. As is the case with starch, cellulose must be broken down into sugar units before it can be used by yeast to create ethanol. However, the breaking of the cellulose bonds is much more complex and costly than the breaking of the starch bonds. Breaking the cellulose into individual sugar units is complicated by the presence of lignin (a complex compound surrounding cellulose), which is even more resistant to enzymatic or acidic pretreatment than cellulose. Because of the high cost of converting liquefied cellulose into fermentable sugars, agricultural residues (as well as other crops with a high percentage of cellulose) are not yet a practical feedstock source for small ethanol plants.

Cull tree: A live tree, 5.0 inches in diameter at breast height (d.b.h.) or larger, that is non-merchantable for saw logs at present or prospectively because of rot, roughness, or species.

Cultivated summer fallow: Cropland that is cultivated for one or more seasons in order to control weeds and accumulate moisture before small grains are planted.

DDGS (dried distillers grain with soluble constituents): A by-product of dry mill ethanol production that is fed to livestock.

Delayed coking: A coking process in which thermal reactions are allowed to proceed to completion to produce gaseous, liquid, and solid (coke) products.

Density: The mass (or weight) of one unit volume of any substance at a specified temperature; see also specific gravity.

Desulfurization: The removal of sulfur or sulfur compounds from a feedstock.

Diesel #1 and Diesel #2: Diesel #1, also called kerosene, is generally not used as a fuel oil in diesel vehicles as it has a lower viscosity (it is thinner) than Diesel #2, the typical diesel-vehicle fuel. Biodiesel replaces Diesel #2 or a percentage of it.

Diesel engine: Named after the German engineer Rudolph Diesel, this internal combustion, compression-ignition engine works by heating the fuels and thus causing them to ignite; can run on either petroleum or bio-derived fuel.

Diesel fuel: A distillate of fuel oil that has been historically derived from petroleum for use in internal combustion engines; also derived from plant and animal sources.

Diesel, Rudolph: A German inventor famed for fashioning the diesel engine, which made its debut at the 1900 World Fair and was initially an engine to run on vegetable-derived fuels.

Digester: An airtight vessel or enclosure in which bacteria decompose biomass present in water to produce biogas.

Direct-injection engine: A diesel engine in which the fuel is injected directly into the cylinder.

Distillate: A petroleum product produced by boiling the crude oil and collecting the vapors produced as a condensate in a separate vessel; examples include gasoline (light distillate), gas oil (middle distillate), or fuel oil (heavy distillate).

Distillation: The primary distillation process, which uses high temperature to separate crude oil into vapor and fluids, which can then be fed into a distillation or fractionating tower.

Distiller grains: A by-product of ethanol production; can be used to feed livestock; also called distillers dried grains with soluble constituents (DDGS).

Dispersion: A stable or unstable system of fine particles, larger than colloidal size, evenly distributed in a medium.

Distillate oil: A distilled product of crude oil; a volatile petroleum product used for home heating and other machinery.

Distillation: The process of separating the components of a liquid mixture by boiling the liquid and then condensing the resulting vapor.

Downdraft gasifier: A gasifier in which the product gases pass through a combustion zone at the bottom of the gasifier.

Dutch oven furnace: One of the earliest types of furnaces; has a large, rectangular box lined with firebrick (refractory) on the sides and the top; commonly used for burning wood.

Dry mill: An ethanol production process in which the entire corn kernel is first ground into flour before processing, in addition to ethanol, dry mills also produce dried distiller grains with soluble constituents (DDGS), which is fed to livestock, and carbon dioxide, which is used in food processing and bottling; most recent ethanol plants are dry mill facilities.

Dry ton: 2,000 pounds of material dried to a constant weight.

E10: An alcohol fuel mixture containing 10% ethanol and 90% gasoline by volume.

E85: An alcohol fuel mixture containing 85% ethanol and 15% gasoline by volume and the current alternative fuel of choice of the US government.

Ecology: The study of the interrelationship between organisms and their environment.

E Diesel: A blend of ethanol and diesel fuel plus other additives designed to reduce air pollution from heavy equipment, city buses, and other vehicles that operate on diesel engines.

Effluent: The liquid or gas discharged from a process or chemical reactor; usually containing residues from that process.

Elemental analysis: The determination of the carbon, hydrogen, nitrogen, oxygen, sulfur, chlorine, and ash in a sample as a weight percentage of the whole.

Emission offset: A reduction in the air pollution emissions from existing sources to compensate for emissions from new sources.

Emissions: Substances discharged into the air during combustion; waste substances released into the air or water.

Emulsification: To emulsify: to form an emulsion.

Emulsion: A suspension of small drops of one liquid in a second liquid with which the first liquid will not mix; emulsions can be formed by either mechanical agitation or chemical processes; unstable emulsions will separate with time or temperature, but stable emulsions will not separate.

Energy balance: The difference of the energy produced by a fuel and the energy required to obtain it using agricultural processes, drilling, refining, and transportation.

Energy crops: Crops grown specifically for their value as fuel; include food crops such as corn and sugarcane and non-food crops such as poplar trees and switch grass.

Energy/efficiency ratio: A number indicating the energy stored in a fuel as compared to the energy required to produce, process, transport, and distribute that fuel.

Enhanced recovery: Methods that usually involve the application of thermal energy (e.g. steam flooding) for oil recovery from the reservoir.

Environment: The external conditions that affect organisms and influence their development and survival.

Environmental assessment (EA): A public document analyzing a proposed federal action for the possibility of significant environmental impacts — if the environmental impacts will be significant, then the federal agency must prepare an environmental impact statement.

Environmental impact statement (EIS): A statement of the environmental effects of a proposed action and of alternative actions. Section 102 of the National Environmental Policy Act requires an EIS for all major federal actions.

Enzymatic hydrolysis: A process by which enzymes (biological catalysts) are used to break down starch or cellulose into sugar.

Enzyme: A protein or protein-based molecule that speeds up the chemical reactions occurring in living things; enzymes act as catalysts for a single reaction, converting a specific set of reactants into specific products.

Esters: Any of a large group of organic compounds formed when an acid and alcohol are mixed; methyl acetate (CH_3COOCH_3) is the simplest ester; biodiesel contains methyl stearate.

ETBE: See Ethyl tertiary butyl ether.

Ethanol (ethyl alcohol, alcohol, or grain-spirit): A clear, colorless, flammable oxygenated hydrocarbon; used as a vehicle fuel by itself (E100 is 100% ethanol by volume) or when blended with gasoline (E85 is 85% ethanol by volume), or used as a gasoline octane enhancer and oxygenate (10% by volume).

Ethers: Liquid fuel made by blending an alcohol with isobutylene.

Ethyl tertiary butyl ether (ethyl t-butyl ether): Ether created from ethanol that can increase octane and reduce the volatility of gasoline, decreasing evaporation and smog formation.

Evaporation: The conversion of a liquid to the vapor state by the addition of latent heat or vaporization.

Extractives: Any of different type of compounds present in biomass that are not an integral part of the cellular structure — these compounds can be extracted from wood by means of polar and non-polar solvents including hot or cold water, ether, benzene, methanol, or other solvents that do not degrade the biomass structure. The types of extractives found in biomass samples are entirely dependent on the sample itself.

FAAE: A term for the biodiesel made from any alcohol during its production process.

FAME (fatty acid methyl ester): The ester created by a catalyzed reaction between fatty acids and methanol; the primary constituent of biodiesel is FAMEs; usually obtained from vegetable oils by transesterification.

Fast pyrolysis: Thermal conversion of biomass by rapidly heating it from 450°C to 600°C (842°F to 1,112°F) in the absence of oxygen.

Fatty acid: A carboxylic acid (an acid with a –COOH group) with long hydrocarbon side chains; feedstocks are converted first to fatty acids and then to biodiesel by transesterification.

Fatty acid alkyl ester: See FAAE.

Fatty acid methyl ester: See FAME.

Feedstock: The biomass used in the manufacture of a particular biofuel (e.g. corn or sugarcane for making ethanol, and soybeans or rapeseed for making biodiesel).

Fermentation: Conversion of carbon-containing compounds by micro-organisms for the production of fuels and chemicals such as alcohols, acids, or energy-rich gases.

Fiber products: Products derived from fibers of herbaceous and woody plant materials; examples include pulp, board products, and wood chips for export.

Fine materials: Wood residues not suitable for chipping, such as planar shavings and sawdust.

Fischer–Tropsch Process: Process for producing liquid fuels, usually diesel fuel, from natural gas; or synthetic gas from gasified coal or biomass.

Fixed carbon: The carbonaceous residue remaining after heating in a prescribed manner to decompose thermally unstable components and to distill volatiles; a part of the proximate analysis group.

Flash point: The lowest temperature at which a liquid will produce enough vapor to ignite, if the vapor is flammable.

Flexible-fuel vehicles (flex-fuel vehicles): Vehicles that can run alternatively on two or more sources of fuel; include cars capable of running on gasoline and gasoline/ethanol mixtures as well as cars that can run on both gasoline and natural gas.

Fluid coking: a continuous fluidized solids process that cracks feed and thermally overheated coke particles in a reactor vessel to gas, liquid products, and coke.

Fluidized-bed boiler: A large, refractory-lined vessel with an air distribution member or plate in the bottom, a hot gas outlet in or near the top, and some provisions for introducing fuel; the fluidized bed is formed by blowing air upward through a layer of inert particles (such as sand or limestone) at a rate that causes the particles to go into suspension and continuous motion.

Fly ash: Small ash particles carried in suspension in combustion products.

Foam: A dispersion of a gas in a liquid or solid.

Fodder beets: A promising sugar crop that is presently being developed in New Zealand is the fodder beet. The fodder beet is a high-yielding forage crop obtained by crossing two other beet species: sugar beets and mangolds. It is similar, in most agronomic aspects, to sugar beets. The attraction of this crop lies in its higher yield of fermentable sugars per acre compared to sugar beets and its comparatively high resistance to loss of fermentable sugars during storage. Culture of fodder beets is also less demanding than culture of sugar beets.

Forage crops (e.g. forage sorghum and Sudan grass): Crops are used for ethanol production because, in their early stage of growth, there is very little lignin present and the conversion of the cellulose to sugars is more efficient. In addition, the proportion of carbohydrates in the form of cellulose is less than that in the mature plant. Since forage crops achieve maximum growth in a relatively short period, they can be harvested as many as four times in a single growing season. For this reason, forage crops cut as green chop may have the highest yield of dry material of any storage crop. In addition to cellulose, forage crops contain significant quantities of starch and fermentable sugars, which can also be converted to ethanol. The residues from fermentation containing nonfermentable sugars, protein, and other components may be used to feed livestock.

Forest land: The land stocked at least 10% with forest trees of any size, including land that formerly had such tree cover and that will be naturally or artificially regenerated; includes transition zones, such as areas between heavily forested and non-forested lands that are at least 10% stocked with forest trees and forest areas adjacent to urban and built-up lands; also included are pinyon-juniper and chaparral areas; the minimum area for classification as forest land is 1 acre.

Forest residues: Material that is not harvested or removed from logging sites in commercial hardwood and softwood stands as well as the material resulting from forest management operations, such as precommercial thinning and removal of dead and dying trees.

Forest health: A condition of ecosystem sustainability and attainment of management objectives for a given forest area; usually considered to include green trees, snags, resilient stands growing at a moderate rate, and endemic levels of insects and disease.

Fossil fuel: Solid, liquid, or gaseous fuels formed in the ground after millions of years by chemical and physical changes in plant and animal residues under high temperature and pressure. Oil, natural gas, and coal are fossil fuels.

Fruit crops (e.g. grapes, apricots, peaches, and pears): Another type of feedstock in the sugar crop category. Typically, fruit crops such as grapes are used as the feedstock in wine production. These crops are not likely to be used as feedstocks for the production of fuel-grade ethanol because of their high market value for direct human consumption. However, the coproducts from the processing of fruit crops are likely to be used as feedstocks, because fermentation is an economical method for reducing the potential environmental impacts of untreated wastes containing fermentable sugars

Fuel cell: A device that converts the energy of a fuel directly to electricity and heat without combustion.

Fuel cycle: The series of steps required to produce electricity. The fuel cycle includes mining or otherwise acquiring the raw fuel source, processing and cleaning the fuel, transport, electricity generation, waste management, and plant decommissioning.

Fuel oil: A heavy residue, black in color, used to generate power or heat by burning in furnaces.

Fuel treatment evaluator (FTE): A strategic assessment tool capable of aiding the identification, evaluation, and prioritization of fuel treatment opportunities.

Fuelwood: The wood used for conversion to some form of energy, primarily for residential use.

Furnace: An enclosed chamber or container used to burn biomass in a controlled manner in order to produce heat for use in space or process heating.

Galactan: A polymer of galactose with a repeating unit of $C_6H_{10}O_5$; found in hemicellulose, it can be hydrolyzed to galactose.

Galactose: A six-carbon sugar with the formula $C_6H_{12}O_6$; a product of the hydrolysis of galactan found in the hemicellulose fraction of biomass.

Gas engine: A piston engine that uses natural gas rather than gasoline — fuel and air are mixed before they enter cylinders; ignition occurs with a spark.

Gaseous emissions: Substances discharged into the air during combustion, typically including carbon dioxide, carbon monoxide, water vapor, and hydrocarbons.

Gasification: A chemical or heat process used to convert carbonaceous materials (such as coal, petroleum, and biomass) into gaseous components such as carbon monoxide and hydrogen.

Gasifier: A device for converting solid fuel into gaseous fuel; in biomass systems, this process is referred to as pyrolytic distillation.

Gasohol: A mixture of 10% anhydrous ethanol and 90% gasoline by volume; 7.5% anhydrous ethanol and 92.5% gasoline by volume; or 5.5% anhydrous ethanol and 94.5% gasoline by volume.

Gasoline: A volatile, flammable liquid obtained from petroleum, it has a boiling range of approximately 30–220°C (86–428°F) and is used as fuel for spark-ignition internal combustion engines.

Gas shift process: A process in which carbon monoxide and hydrogen react in the presence of a catalyst to form methane and water.

Gas to liquids (GTL): The process of refining natural gas and other hydrocarbons into longer-chain hydrocarbons; it can be used to convert gaseous waste products into fuels.

Gas turbine (combustion turbine): A turbine that converts the energy of hot compressed gases (produced by the burning of fuel in compressed air) into mechanical power — often fired by natural gas or fuel oil.

Gel point: The point at which a liquid fuel cools to the consistency of petroleum jelly.

Genetically modified organism (GMO): An organism whose genetic material has been modified through recombinant DNA technology, altering the phenotype of the organism to meet desired specifications.

Glycerin ($CH_2OH{\cdot}CHOH{\cdot}CH_2OH$): A by-product of biodiesel production; each of the hydroxyl (OH) functions is one of the three places where an ester is broken off from the triglyceride molecule (e.g. vegetable oil).

Glycerine (glycerin; glycerol): A liquid by-product of biodiesel production; used in the manufacture of dynamite, cosmetics, liquid soaps, inks, and lubricants.

Grain alcohol: See ethanol.

Grassland pasture and range: All open land used primarily for pasture and grazing, including shrub and brush land types of pasture; grazing land with sagebrush and scattered mesquite; and all tame and native grasses, legumes, and other forage used for pasture or grazing; because of the diversity in vegetative composition, grassland pasture and range are not always clearly distinguishable from other types of pastures and ranges; at one extreme, permanent grassland may merge with cropland pasture, or grassland may often be found in transitional areas with forested grazing land.

Grease car: A diesel-powered automobile rigged post production to run on used vegetable oil.

Greenhouse effect: The effect of certain gases in the atmosphere of the Earth in trapping heat from the sun.

Greenhouse gases: Those gases that trap the heat of the sun in the atmosphere of the Earth, thus producing the greenhouse effect. The two major greenhouse gases are water vapor and carbon dioxide. Other greenhouse gases include methane, ozone, chlorofluorocarbons, and nitrous oxide.

Grid: The system used by an electric utility company for distributing power.

Gross heating value (GHV): The maximum potential energy in the fuel as received, considering moisture content (MC).

Growing stock: A classification of timber inventory that includes live trees of commercial species meeting specified standards of quality or vigor; cull trees are excluded.

GTL (gas to liquid): A refinery process that converts natural gas into longer chain hydrocarbons; gas can be converted to liquid fuels via direct conversion or using a process such as the Fischer–Tropsch process.

Habitat: The area where a plant or animal lives and grows under natural conditions. Habitat includes living and non-living attributes and provides all the requirements for food and shelter.

Hardwoods: Usually broad-leaved and deciduous trees; one of the botanical groups of dicotyledonous trees that have broad leaves in contrast to the conifers or softwoods — the term has no reference to the actual hardness of the wood — the botanical name for hardwoods is angiosperms; short-rotation, fast-growing hardwood trees are being developed for use as future energy crops; they are uniquely developed for harvest 5–8 years after planting and examples include hybrid poplars (*Populus* sp.), hybrid willows (*Salix* sp.), silver maple (*Acer saccharinum*), and black locust (*Robinia pseudoacacia*).

Hazardous waste: A broad term for a wide range of waste materials that pose different levels of risk. Some pose a serious and immediate threat to the population and the environment; for example, those that are toxic and could cause cancer or infectious disease. Others, such as fluorescent tubes or cathode ray tubes in television sets, pose little immediate threat but may cause long-term damage over a period of time.

Heating value: The maximum amount of energy that is available after burning a substance.

Heavy oil (heavy crude oil): The oil that is more viscous than conventional crude oil; it has a lower mobility in the reservoir but can be recovered through a well from the reservoir by the application of secondary or enhanced recovery methods.

Hectare: A common metric unit of area, equal to 2.47 acres. 100 hectares = 1 square kilometer.

Hemicellulose: Consists of short, highly branched chains of sugars; in contrast to cellulose, which is a polymer of only glucose, hemicellulose is a polymer of five different sugars; contains five-carbon sugars (usually D-xylose and L-arabinose) and six-carbon sugars (D-galactose, D-glucose, and D-mannose) and uronic acid, which are highly substituted with acetic acid; the branched nature of hemicellulose renders it amorphous and relatively more easy to hydrolyze into its constituent sugars compared to cellulose; when hydrolyzed, the hemicellulose from hardwoods releases products rich in xylose (a five-carbon sugar); by contrast, the hemicellulose from softwoods, yields more six-carbon sugars.

Herbaceous: Non-woody type vegetation, usually lacking permanent, strong stems, for examples, grasses, cereals, and canola (rape).

Herbaceous energy crops: Perennial non-woody crops that are harvested annually, although they may take 2–3 years to reach full productivity; examples include switch grass (*Panicum virgatum*), reed canary grass (*Phalaris arundinacea*), miscanthus (*Miscanthus x giganteus*), and giant reed (*Arundo donax*).

Herbaceous plants: Non-woody species of vegetation, usually of low lignin content, such as grasses.

Heteroatom compounds: Chemical compounds that contain nitrogen and/or oxygen and/or sulfur and/or metals bound within their molecular structure(s).

Hexose: Any of various simple sugars that have six carbon atoms per molecule (e.g. glucose, mannose, and galactose).

Higher heating value (HHV): The potential combustion energy when water vapor from combustion is condensed to recover the latent heat of vaporization. Lower heating value (LHV) is the potential combustion energy when water vapor from combustion is not condensed.

Household waste: It includes domestic waste from household collection rounds, waste from services such as street sweeping, bulky waste collection, litter collection, hazardous household waste collection, and garden waste collection, waste from civic amenity sites and wastes separately collected for recycling or composting through bring recycling schemes and curbside recycling schemes. Household waste is a sub-group of municipal solid waste.

Hydrocarbon: A chemical compound that contains a carbon backbone with hydrogen atoms attached to that backbone.

Hydrocarbonaceous material: A material such as bitumen that is composed of carbon and hydrogen with other elements (heteroelements) such as nitrogen, oxygen, sulfur, and metals chemically combined within the structures of the constituents; even though carbon and hydrogen may be the predominant elements, there may be very few true hydrocarbons (q.v.).

Hydrocarbon compounds: Chemical compounds containing only carbon and hydrogen.

Hydrodesulfurization: The removal of sulfur by hydrotreating (q.v.).

Hydroprocesses: Refinery processes designed to add hydrogen to various products of refining.

Hydrotreating: The removal of heteroatomic (nitrogen, oxygen, and sulfur) species by treatment of a feedstock or product at relatively low temperatures in the presence of hydrogen.

Idle cropland: Land in which no crops were planted; this includes acreage diverted from crops to soil-conserving uses (if not eligible for being used as cropland pasture) under federal farm programs.

Incinerator: Any device used to burn solid or liquid residues or wastes, used as a method for disposal.

Inclined grate: A type of furnace in which the fuel enters the top section of a grate as a continuous ribbon, passes through the upper drying section, where moisture is removed, and then descends to the lower burning section. Ash is removed at the lower part of the grate.

Indirect-injection engine: An older model of the diesel engine in which the fuel is injected into a pre-chamber, partly combusted, and then sent to the fuel-injection chamber.

Indirect liquefaction: Conversion of biomass to a liquid fuel through a synthesis gas as an intermediate step.

Industrial wood: All commercial roundwood products except fuel wood.

Iodine value: A measure of the number of unsaturated carbon–carbon double bonds in a vegetable oil molecule — double bonds can allow polymerization, leading to the formation of lacquers and possibly blockage and damage to engine or fuel train components; in liquid biofuel applications the iodine value gives a lower cold filter plugging point (CFPP) or cloud point.

Jatropha: A non-edible evergreen shrub found in Asia, Africa, and the West Indies; the seeds contain large proportion of oil, which can be used to make biodiesel.

Jerusalem artichoke: A member of the sunflower family, this crop is native to North America and is well adapted to northern climates. Like the sugar beet, the Jerusalem artichoke produces sugar in the top growth and stores it in the roots and tuber. It can grow in a variety of soils, and it is not demanding of soil fertility. It is a perennial; small tubers left in the field can produce the next seasonal crop, so no plowing or seeding is necessary.

Joule: The metric unit of energy, equivalent to the work done by a force of 1 N applied over a distance of 1 m (= 1 kg m^2/s^2). One joule (J) = 0.239 calories (1 calorie = 4.187 J).

Kerosene: A light middle-distillate that in various forms is used as an aviation turbine fuel or for burning in heating boilers or as a solvent, such as white spirit.

Kilowatt (kW): A measure of electrical power equal to 1,000 W. 1 kW = 3412 Btu/h = 1.341 horsepower.

Kilowatt hour (kWh): A measure of energy equivalent to the expenditure of one kilowatt for one hour. For example, 1 kWh will light a 100-W light bulb for 10 hours. 1 kWh = 3412 Btu.

Klason lignin: Lignin obtained from wood after the non-lignin components of the wood have been removed with a prescribed sulfuric acid treatment; a specific type of acid-insoluble lignin analysis.

Knock: The engine sound that results from the ignition of a compressed fuel–air mixture prior to the optimal moment.

KOH: See Potassium hydroxide.

Landfill: Licensed facilities where waste is permanently deposited for subsequent disposal into land. According to the waste hierarchy, the final disposal of waste through a landfill is the least preferred way of managing waste.

Landfill gas: A type of biogas that is generated by the decomposition of organic material at landfill disposal sites. Landfill gas is approximately 50% methane. See also Biogas.

Life cycle assessment (LCA): The systematic identification and evaluation of all the environmental benefits and disadvantages that result, both directly and indirectly, from a product or function throughout its entire life from the extraction of a raw material to its eventual disposal and assimilation into the environment. LCA helps to place the assessment of the environmental costs and benefits of the various options, and the development of appropriate, practical waste management policies, on a sound, and objective basis.

Lignin: Structural constituent of wood and (to a lesser extent) other plant tissues, which encrusts the walls and cements the cells together; the energy-rich material in biomass that can be used as a boiler fuel.

Lignocellulose: Plant material made up of primarily lignin, cellulose, and hemicellulose.

Lipid: Any of a group of organic compounds, including fats, oils, waxes, sterols, and triglycerides, that are insoluble in water but soluble in nonpolar organic solvents, are oily to the touch, and together with carbohydrates and proteins, constitute the principal structural material of living cells.

Live cull: A classification that includes live cull trees; when associated with volume, it is the net volume in those live cull trees that are 5.0 inches in diameter or larger.

Logging residues: The unused portions of growing-stock and non-growing-stock trees cut or killed for logging and left in the woods.

Lower heating value (LLV; net heat of combustion): The heat produced by combustion of one unit of a substance at atmospheric pressure under conditions such that all the water in the products remains in the form of vapor; the net heat of combustion is calculated from the gross heat of combustion at 20°C (68°F) by subtracting 572 cal/g (1030 Btu/lb) of water derived from one unit mass of sample, including both the water originally present as moisture and that formed by combustion — this subtracted amount is not equal to the latent heat of vaporization of water, because the calculation also reduces the data from the gross value at constant volume to the net value at constant pressure and the appropriate factor for this reduction is 572 cal/g.

Lye: see sodium hydroxide.

M85: An alcohol fuel mixture containing 85% methanol and 15% gasoline by volume. Methanol is typically made from natural gas, but can also be derived from the fermentation of biomass.

Megawatt (MW): A measure of electrical power equal to one million watts (1,000 kW).

Methanol: A fuel that is typically derived from natural gas but can be produced by fermentation of the sugars in biomass.

Methoxide (sodium methoxide, sodium methylate, or $CH3O^-$ Na^+): An organic salt; in pure form it is a white powder; in biodiesel production, methoxide is a product obtained by mixing methanol with sodium hydroxide, yielding a solution of sodium methoxide in methanol and a significant amount of heat; creating sodium methoxide is the most dangerous step when making biodiesel.

Methyl alcohol: see Methanol.

Methyl Esters: see Biodiesel.

Million: 1×10^6.

Mill residue: Wood and bark residues produced during processing of logs into lumber, plywood, and paper.

Modified/unmodified diesel engine: Traditional diesel engines must be *modified* to heat the oil before it reaches the fuel injectors in order to handle straight vegetable oil. Modified, any diesel engine can run on veggie oil; without modification, the oil must be first converted to biodiesel.

Moisture content (MC): The weight of the water contained in wood, usually expressed as a percentage of the weight of either oven-dry or as received wood.

Moisture content, dry basis: Moisture content expressed as a percentage of the weight of oven dry wood, i.e. [(Weight of wet sample — Weight of dry sample)/Weight of dry sample] × 100.

Moisture content, wet basis: Moisture content expressed as a percentage of the weight of as-received wood, i.e. [(Weight of wet sample — Weight of dry sample)/Weight of wet sample] × 100.

Moisture-free basis: Biomass composition and chemical analysis data are typically reported on a moisture-free or dry weight basis — moisture (and some volatile matter) is removed prior to analytical testing by heating the sample at 105°C (221°F) to a constant weight; by definition, samples dried in this manner are considered moisture-free.

Monosaccharide: A simple sugar such as a five-carbon sugar (xylose, arabinose) or a six-carbon sugar (glucose, fructose); sucrose, on the other hand, is a disaccharide composed of a combination of two simple sugar units: glucose and fructose.

MTBE: Methyl tertiary butyl ether is a highly refined high-octane light distillate used in the blending of petrol.

Municipal solid waste (MSW): Household waste and other wastes collected by a waste collection authority or its contractors, for example, municipal park and garden waste, beach cleansing waste, and any commercial waste and industrial waste for which the collection authority takes responsibility.

Municipal wastes: Residential, commercial, and institutional post-consumer wastes that contain a significant proportion of plant-derived organic material, a renewable energy resource; waste paper, cardboard, construction and demolition wood waste, and yard wastes are examples of biomass resources in municipal wastes.

Nitrogen fixation: The transformation of atmospheric nitrogen into nitrogen compounds that can then be used by growing plants.

Nitrogen oxides (NO_x): Products of combustion that contribute to the formation of smog and ozone.

Non-forest land: Land that has never supported any forests and lands formerly forested where use of timber management is precluded by development for other uses; if intermingled in forest areas any unimproved roads and non-forest strips must be more than 120 feet wide, and clearings, etc. must be more than 1 acre in area to qualify as non-forest land.

Non-attainment area: Any area that does not meet the national primary or secondary ambient air quality standard established (by the Environmental Protection Agency) for designated pollutants, such as carbon monoxide and ozone.

Non-industrial private: Ownership-class private lands where the owner does not operate wood processing plants.

Particulate emissions: Particles of a solid or liquid suspended in a gas, or the fine particles of carbonaceous soot and other organic molecules discharged into the air during combustion.

Octane number: A measure of the resistance of a fuel to self-ignition; the octane number of a fuel is indicated on the pump — the higher the number, the slower the fuel burns; bioethanol typically adds 2–3 octane numbers when blended with ordinary petroleum, making it a cost-effective octane enhancer; see Knock.

Oil from tar sand: Synthetic crude oil (q.v.).

Oil mining: The application of a mining method to the recovery of bitumen.

OOIP (Oil originally in place or Original oil in place): The quantity of petroleum existing in a reservoir before oil recovery operations begin.

Open-loop biomass: Biomass that can be used to produce energy and bioproducts even though it was not grown specifically for this purpose; includes agricultural livestock waste, and residues from forest harvesting operations and crop harvesting.

Oxygenate: A substance that, when added to gasoline, increases the amount of oxygen in that gasoline blend; includes fuel ethanol, methanol, and methyl tertiary butyl ether (MTBE).

Oxygenated fuels: Ethanol is an oxygenate, meaning that it adds oxygen to the fuel mixture — more oxygen helps the fuel burn more completely, thereby reducing the amount of harmful emissions from the tail pipe; a fuel such as ethanol-blended gasoline that contains high oxygen content is called *oxygenated.*

Palm oil: A form of vegetable oil obtained from the fruit of the oil palm tree; a widely used feedstock for traditional biodiesel production; palm oil and palm kernel oil are composed of fatty acids, esterified with glycerol just like any ordinary fat.

Particulate: A small, discrete mass of solid or liquid matter that remains individually dispersed in gas or liquid emissions.

Particulate emissions: Particles of a solid or liquid suspended in a gas or the fine particles of carbonaceous soot and other organic molecules discharged into the air during combustion.

Pay zone thickness: The depth of a tar sand deposit from which bitumen (or a product) can be recovered.

Perennial: Plant that does not have to be planted every year like traditional row crops.

Petrodiesel: Petroleum-based diesel fuel, usually referred to simply as diesel.

Petroleum: A hydrocarbon-based substance composed of a complex blend of hydrocarbons derived from crude oil through the process of separation, conversion, upgrading, and finishing, including motor fuel, jet oil, lubricants, petroleum solvents, and used oil.

pH: A measure of the acidity and alkalinity of a solution on a scale with seven representing neutrality; lower numbers indicate increasing acidity, and higher numbers indicate increasing alkalinity; each unit of change represents a 10-fold change in acidity or alkalinity.

Photosynthesis: A process by which chlorophyll-containing cells in green plants convert incident light to chemical energy, capturing carbon dioxide in the form of carbohydrates.

Potassium hydroxide (KOH): Used as a catalyst in the transesterification reaction to produce biodiesel.

Pour point: The lowest temperature at which oil will pour or flow when it is chilled without disturbance under definite conditions.

Primary oil recovery: Oil recovery utilizing only naturally occurring forces; recovery of crude oil from a reservoir using the inherent reservoir energy.

Primary wood-using mill: A mill that converts roundwood products into other wood products; common examples are sawmills that convert saw logs into lumber and pulp mills that convert pulpwood of roundwood into wood pulp.

Process heat: Heat used in an industrial process rather than for space heating or other housekeeping purposes.

Producer gas: Gas fuel rich in carbon monoxide (CO) and hydrogen (H_2), produced by burning a solid fuel with insufficient air or by passing a mixture of air and steam through a burning bed of solid fuel.

Protein: A protein molecule is a chain of up to several hundreds of amino acids and is folded into a more or less compact structure; in the biologically active state, proteins function as catalysts in metabolism and to some extent as the structural elements of cells and tissues; the protein content in biomass (in mass percentage) can be estimated by multiplying the mass percentage of nitrogen of the sample by 6.25.

Proximate analysis: The determination, by prescribed methods, of moisture, volatile matter, fixed carbon (by difference), and ash; the term *proximate analysis* does not include determinations of chemical elements or determinations other than those named and the group of analyses is defined in ASTM D3172.

Pulpwood: Roundwood whole-tree chips or wood residues, which are used for the production of wood pulp.

Pyrolysis: The thermal decomposition of biomass at high temperatures [greater than 400°F (or 200°C)] in the absence of air; the end product of pyrolysis is a mixture of solids (char), liquids (oxygenated oils), and gases (methane, carbon monoxide, and carbon dioxide) with proportions determined by operating temperature, pressure, oxygen content, and other conditions. In contrast with incineration, pyrolysis is the thermal degradation of a substance in the absence of oxygen. This process requires an external heat source in order to maintain the temperature required. The products produced by pyrolysis of materials are the following: a solid residue and synthetic gas (syngas). The solid residue (sometimes described as a char) is a combination of non-combustible materials and carbon. Syngas is a mixture of gases (combustible constituents include carbon monoxide, hydrogen, methane, and a wide range of other volatile organic compounds). Some of the products can be condensed to produce oils, waxes, and tars. If required, the condensable fraction can be collected by cooling the syngas, and potentially used as a liquid fuel.

Pyrolysis oil: A bio-oil produced by fast pyrolysis of biomass; typically, a dark brown, mobile liquid containing much of the energy content of the original biomass, with a heating value about half that of conventional fuel oil; conversion of raw biomass to pyrolysis oil represents a considerable increase in energy density and it may thus represent a more efficient form in which to transport it.

Quad: One quadrillion Btu (10^{15} Btu) = 1.055 exajoules (EJ), or approximately 172 million barrels of oil equivalent.

Recovery boiler: A pulp mill boiler in which lignin and spent cooking liquor (black liquor) are burned to generate steam.

Refractory lining: A lining, usually of ceramic, capable of resisting and maintaining high temperatures.

Rapeseed (*Brassica* napus; also called rape, oilseed, rape or canola): A bright yellow flowering member of the family *Brassicaceae* (mustard or cabbage family); a traditional feedstock used for biodiesel production; canola is a name taken from *Canada oil* due to the fact that much of the development of the oil was performed in Canada; see Colza.

Rapeseed oil: The food-grade oil produced from rapeseed is called canola oil; see Colza.

Refuse-derived fuel (RDF): Fuel prepared from municipal solid waste; non-combustible materials such as rocks, glass, and metals are removed, and the remaining combustible portion of the solid waste is chopped or shredded.

Renewable fuels standard (RFS): Legislation enacted by the United States Congress as part of the Energy Policy Act of 2005, requiring that an increasing level of biofuels be used every year, rising to 7.5 billion gallons by 2012.

Residual waste: The waste remaining to be disposed of after re-use, recycling, composting and recovery of materials and energy have been performed.

Residues: Bark and woody materials that are generated in primary wood-using mills when roundwood products are converted to other products.

Residuum (plural: residua; also known as resid or resids): The non-volatile portion of petroleum that remains as residue after refinery distillation; hence, also called atmospheric residuum, or vacuum residuum.

Rotation: Period of years between the establishment of a stand of timber and the time it is considered to be ready for the final harvest and regeneration.

Round wood products: Logs and other round timber generated from harvesting trees for industrial or consumer use.

RTFO (Renewable Transport Fuels Obligation): A United Kingdom policy that places an obligation on fuel suppliers to ensure that a certain percentage of their aggregate sales is composed of biofuels.

Sandstone: A sedimentary rock formed by compaction and cementation of sand grains; can be classified according to the mineral composition of the sand and cement.

Saponification: The reaction of an ester with a metallic base and water (i.e. the creation of soap); can occur when too much lye is used in biodiesel production.

Saturated steam: Steam at boiling temperature for a given pressure.

Secondary oil recovery: The application of energy (e.g. water flooding) to recovery of crude oil from a reservoir after the yield of crude oil from primary recovery has diminished.

Second-generation biofuels: The biofuels produced from biomass or non-edible feedstocks.

Secondary wood processing mills: A mill that uses primary wood products for the manufacture of finished wood products, such as cabinets, moldings, and furniture.

Sodium hydroxide (lye, caustic soda, or NaOH): Strongly alkaline and extremely corrosive; mixing with fluids usually causes heat, and can create enough heat to ignite flammables (such as methanol); one of the main reactants for biodiesel production.

Softwood: Generally, one of the botanical groups of trees that in most cases have needle-like or scale-like leaves; the conifers; also the wood produced by such trees; the term has no reference to the actual softness of the wood; the botanical name for softwoods is gymnosperms.

Soy (soy oil): A vegetable oil pressed from soybeans.

Soybean: A bushy, leguminous plant, *Glycine max*, native to South-East Asia; it is grown for the beans, which are used widely in the food industry, for protein in cattle feed, and for oil production.

Soy diesel: A general term for biodiesel; it emphasizes the renewable nature of biodiesel; a popular term in soy-producing regions.

Specific gravity: The mass (or weight) of a unit volume of any substance at a specified temperature compared to the mass of an equal volume of pure water at a standard temperature.

Stabilized waste: Waste that has been treated so that it is chemically stable.

Stand (of trees): A group of trees that show sufficient uniformity in composition, constitution, age, spatial arrangement, or condition to be distinguishable from adjacent groups.

Starch: A molecule composed of long chains of α-glucose molecules linked together (repeating unit $C_{12}H_{16}O_5$); these linkages occur in chains of α-1,4 linkages with branches formed as a result of α-1,6 linkages; widely distributed in the vegetable kingdom and stored in all grains and tubers (swollen, underground plant stems): this polymer is highly amorphous, so it is more readily attacked by human and animal enzyme systems and then broken down into glucose; gross heat of combustion: Qv(gross) = 7560 Btu/lb.

Starch crops: Crops in which most of the six-carbon sugar units are linked together in long, branched chains (starch). Yeast cannot use these chains to produce ethanol. The starch chains must be broken down into individual six-carbon units or groups of two units. The starch conversion process is relatively simple because the bonds in the starch chain can be broken in an inexpensive manner by the use of heat and enzymes or by a mild acid solution.

Steam turbine: A device for converting the energy of high-pressure steam (produced in a boiler) into mechanical power, which can then be used to generate electricity.

Stover: See Corn stover.

Straight vegetable oil (SVO): Any vegetable oil that has not been optimized through the process of transesterification; to use this type of vegetable oil in a diesel engine requires an engine modification in order to heat the oil before it reaches the fuel injectors.

Sugar beet: A plant whose roots contain a high concentration of sucrose and that is grown commercially for sugar production; grown in many areas of the United States; must be rotated with non-root crops (1 beet crop per 4 year period is the general rule). While beet by-products cannot provide fuel for the distilleries, beet pulp and tops are an excellent feed in either wet or dry form. Or the tops may be left on the field for fertilizer and erosion control.

Sugar cane (also called *sugarcane*, or simply *cane*): Several species of tall perennial true grasses of the genus *Saccharum*, tribe Andropogoneae; used for sugar production. The plant is 6–20 feet tall with stout, jointed, fibrous stalks that are rich in sucrose; belongs to the grass family Poaceae (an economically important seed plant family that includes maize, wheat, rice, and sorghum, and many forage crops). Sugarcane is native to the warm temperate-to-tropical regions of South Asia, Southeast Asia, New Guinea, and South America. Only a minority of US states cultivate sugarcane, but there are hybrids (such as *Saccharum spontaneum*) that can be grown further north. High yields per acre of both sugar and crop residues are strong points of sugarcane production. The crop residue, called *bagasse*, is used in Brazil to provide heat to the distilleries.

Suspension: A dispersion of a solid in a gas, a liquid, or a solid.

Superheated steam: The steam that is hotter than boiling temperature for a given pressure.

Sustainable: An ecosystem condition in which biodiversity, renewability, and resource productivity are maintained over time.

Straight vegetable oil (SVO): Burns well in many diesel engines, but does not start the engine and will coke in the injectors as the hot engine cools; a separate tank of petrodiesel or biodiesel is often used during the starting and stopping of the engine, and an electric valve allows transfer to the straight vegetable oil tank.

Sweet sorghum: The name given to varieties of a species of sorghum: *Sorghum bicolor*. This crop has been cultivated on a small scale in the past for the production of table syrup, but other varieties can be grown for the production of sugar. The most common types of sorghum species are those used for the production of grain.

Switch grass: Prairie grass native to the United States and known for its hardiness and rapid growth; often cited as a potentially abundant feedstock for ethanol.

Syngas: A mixture of carbon monoxide and hydrogen; see Synthesis gas.

Synthesis gas: A mixture of carbon monoxide (CO) and hydrogen (H_2); it is a product of high-temperature gasification of organic materials such as biomass; after being subjected to clean-up to remove impurities such as tar, syngas can be used to synthesize organic molecules such as synthetic natural gas (SNG; methane (CH_4)) or liquid biofuels such as gasoline and diesel fuel via the Fischer–Tropsch process.

Synthetic crude oil (syncrude): A hydrocarbon product yielded by the conversion of coal, oil shale, or tar sand bitumen; resembles conventional crude oil; can be refined in a petroleum refinery.

Synthetic ethanol: The ethanol produced from ethylene (a petroleum by-product).

Tallow: Produced by meat rendering, which evaporates the moisture and enables fat (known as the tallow) to be separated from high-protein solids (the greaves), which are pressed, centrifuged, or subjected to a process of solvent extraction to remove more tallow, before being ground into meat and bone meal (MBM); pure tallow is a creamy white substance.

Tar sand (bituminous sand): A formation in which the bituminous material (bitumen) is found as a filling in the veins and fissures in fractured rocks, or impregnating relatively shallow sand, sandstone, and limestone strata; a sandstone reservoir that is impregnated with a heavy, extremely viscous, black hydrocarbonaceous, petroleum-like material that cannot be retrieved through a well by conventional or enhanced oil recovery techniques; United States law FE 76-4: the various rock types that contain an extremely viscous hydrocarbon that is not recoverable in its natural state by conventional oil-well production methods including currently used enhanced recovery techniques.

Thermal conversion: A process that uses heat and pressure to break apart the molecular structure of organic solids.

Thermal treatment: Waste treatment in which elevated temperatures is used as the primary means to change the chemical, physical, or biological character or composition of the waste. Examples include gasification, incineration, and pyrolysis.

Thermochemical conversion: The use of heat to change substances chemically from one state to another, e.g. to produce useful energy products.

Thermochemical platform: Typically uses a combination of pyrolysis, gasification, and catalysis to transform wood first into syngas — the gaseous constituent of wood — and then into fuels or chemicals.

Timberland: Forest land that is producing or is capable of producing crops of industrial wood and that is not withdrawn from timber utilization by statute or administrative regulations.

Tipping fee: A fee for the disposal of waste.

Titration: Applied to biodiesel, titration is the act of determining the acidity of a sample of waste vegetable oil by drop-wise addition of a known base to the sample while testing with pH paper for the desired neutral reading (pH = 7); the amount of base needed to neutralize an amount of waste vegetable oil helps determine how much base has to be added to the entire batch.

Ton (short ton): 2,000 pounds.

Tonne (Imperial ton, long ton, or shipping ton): 2,240 pounds; equivalent to 1,000 kilograms or, in crude oil terms, about 7.5 barrels of oil.

Topping cycle: A cogeneration system in which electric power is produced at first. The reject heat from power production is then used to produce useful process heat.

Topping and back-pressure turbines: Turbines that operate at exhaust pressures considerably higher than atmospheric (non-condensing turbines); often multistage with relatively high efficiency.

Transesterification: The chemical process in which an alcohol reacts with the triglycerides in vegetable oil or animal fats, separating the glycerin and producing biodiesel.

Traveling grate: A type of furnace in which assembled links of grates are joined together in a perpetual belt arrangement. The fuel is fed in at one end, and ash is discharged at the other.

Trillion: 1×10^{12}.

Turbine: A machine for converting the heat energy in steam or high-temperature gas into mechanical energy. In a turbine, a high-velocity flow of steam or gas passes through successive rows of radial blades fastened to a central shaft.

Turn down ratio: The lowest load at which a boiler will operate efficiently as compared to the maximum design load of the boiler.

Ultimate analysis: The determination of the elemental composition of the organic portion of carbonaceous materials, as well as the total ash and moisture; a description of the elemental composition of a fuel as a percentage of the dry fuel weight as determined by prescribed methods; see the American Society for Testing and Materials.

Ultra-low sulfur diesel (ULSD): Ultra-low sulfur diesel describes a new EPA standard for the sulfur content in diesel fuels sold in the United States starting from in 2006 — the allowable sulfur content (15 ppm) is much lower than the previous US standard (500 ppm), which not only reduces emissions of sulfur compounds (blamed for acid rain), but also allows advanced emission control systems to be fitted that would otherwise be poisoned by these compounds.

Uronic acid: A simple sugar whose terminal $-CH_2OH$ group has been oxidized to an acid, COOH group; uronic acids occur as branching groups bonded to hemicelluloses such as xylan.

Vacuum distillation: A secondary distillation process that uses a partial vacuum to lower the boiling point of residues from primary distillation and to extract further blending components.

Viscosity: A measure of the ability of a liquid to flow or a measure of its resistance to flow; the force required to move a plane surface of area 1 square meter over another parallel plane surface 1 meter away at a rate of 1 meter per second when both surfaces are immersed in the fluid; the higher the viscosity, the more slowly the liquid flows.

VOCs: See Volatile organic compounds.

Volatile organic compounds (VOCs): The name given to the light organic hydrocarbons that escape as vapor from fuel tanks or other source and during the filling of tanks. VOCs contribute to smog.

Volatility: The propensity of a fuel to evaporate.

Waste streams: Unused solid or liquid by-products of a process.

Waste vegetable oil (WVO): Grease from the nearest fryer; it is filtered and used in modified diesel engines or converted to biodiesel through the process of transesterification and used in any diesel-fueled vehicle.

Water-cooled vibrating grate: A boiler grate made up of a tuyere grate surface mounted on a grid of water tubes interconnected with the boiler circulation system for positive cooling; the structure is supported by flexing plates allowing the grid and grate to move in a vibrating action; ash is automatically discharged.

Watershed: The drainage basin contributing water, organic matter, dissolved nutrients, and sediments to a stream or lake.

Watt: The common base unit of power in the metric system; one watt equals one joule per second, or the power developed in a circuit by a current of one ampere flowing through a potential difference of one volt. One Watt = 3.412 Btu/hr.

Wet mill: An ethanol production facility in which corn is first soaked in water before processing; in addition to ethanol, wet mills have the ability to produce co-products such as industrial starch, food starch, high-fructose corn syrup, gluten feed, and corn oils.

Wheeling: The process of transferring electrical energy between the buyer and the seller by way of an intermediate utility or utilities.

Whole-tree chips: Wood chips produced by chipping of whole trees, usually in the forest (where trees contain both bark and wood); frequently produced from low-quality trees or from tops, limbs, and other logging residues.

Whole-tree harvesting: A harvesting method in which the whole tree (above the stump) is removed.

Wood: A solid lignocellulosic material naturally produced in trees and some shrubs; it is made of up to 40–50% cellulose, 20–30% hemicellulose, and 20–30% lignin.

Wood alcohol: see Methanol.

Wort: An oatmeal-like substance consisting of water and mash barley in which soluble starch has been turned into fermentable sugar during the mashing process — the liquid remaining from a brewing mash preparation following the filtration of fermentable beer.

Xylan: A polymer of xylose with a repeating unit of $-C_5H_8O_4$; found in the hemicellulose fraction of biomass; can be hydrolyzed to xylose.

Xylose: A five-carbon sugar, $C_5H_{10}O_5$; a product of hydrolysis of the xylan found in the hemicellulose fraction of biomass.

Yarding: The initial movement of logs from the point of felling to a central loading area, or landing.

Yeast: Any of the various single-cell fungi capable of fermenting carbohydrates; bioethanol is produced by fermenting sugars with yeast.

Yellow grease: A term from the rendering industry, it usually means the used frying oils from deep fryers and restaurants' grease traps; it can also refer to lower-quality grades of tallow from the rendering of plants.

Index

A
ablative pyrolysis, 320
absorption, 430, 442, 449, 465
acetogenesis, 293, 295, 356, 626
acicular coke, 113, 588
acid gases, 197
acid gas removal, 448
acid hydrolysis, 324
acid hydrolysis processes, 272
acidogenesis, 293–294, 626
acidogenic digestate, 627
acid treatment, 326
activated carbon, 449
additives, 94
adsorption, 430, 444, 449
adulteration, 94
advanced biofuels, 121
advanced fuels, 4, 230, 533
aerobic degradation, 355
agricultural biomass energy, 247
agricultural crops, 28, 247
agricultural residues, 124, 148, 317
agricultural waste, 147
agrofuels, 120
agro-industrial residues, 317, 622
air-blown direct gasifiers, 164
alcohols, 143
alcoholysis, 152, 631
alkali metal salt removal, 475
alkaloid derivatives, 123
alkanolamine scrubbers, 452
alkazid process, 459
alkylation, 86
alternate fuels, 4–5, 119, 183, 230, 287, 636
alternate fuel standards, 171
alternative fuel, 4, 26, 230, 533
alternative fuel sources, 17
amine washing, 431
ammonia, 406
ammonia removal, 473
anaerobic digester systems, 628
anaerobic digestion, 167, 291, 294, 298, 625
animal fats, 150
animal wastes, 28
anthracite, 218
anthracite coal, 217
anti-icing additives, 95
anti-knock compounds, 95
antioxidants, 94
aquatic plants, 28
Arge reactors, 509
ash, 368

ash deformation temperature, 574
ash flow temperature, 574
ash fusibility temperature, 574
asphalt, 111
asphalt manufacture, 110
associated gas, 186, 190
Athabasca tar sand bitumen, 214
automotive gasoline, 88
autothermal reforming, 388, 523
aviation gasoline, 87

B

bagasse briquettes, 579
Bakken crude oil, 14
Bakken formation, 12
BASF process, 505
Beavon–Stretford process, 456
bioalcohols, 121
biochemical conversion, 625
biochemical platform, 242
biochemical process, 240, 242
bioconversion platform, 287, 290, 600–601
biocrude, 158, 491
biodiesel, 101, 141–142, 149–150, 170, 330, 557, 561, 639
biodiesel blend specifications, 172
biodiesel production, 151, 168
bio-ethanol, 29
bioethers, 161
biofiltration, 476
biofuel, 119, 126, 234, 570, 598, 637
biofuel production, 6
biogas, 134, 139, 167, 238, 298, 341, 549, 623, 627, 638
biological pre-treatments, 326
biomass, 27–28, 122–123, 125, 145, 230, 417
biomass-based fuels, 551
biomass components, 231
biomass gasification, 141, 513
biomass gasifier, 549, 569
biomass liquefaction, 237
biomass resource base, 288
biomass-to-liquid, 121
biomass types, 289
biomaterials, 119
bio-oil, 158–161, 168, 330, 491–492, 570, 640
biooxidation, 480
biophotolysis, 298
bioprocessing, 120
bioreactor landfill, 336–338
bioreactors, 299
biorefinery, 598–600
bioresidues, 551
bio-scrubbing, 479
biosyngas, 162
biosynthesis gas, 162
bitumen, 19, 211
bituminous coal, 217–218
bituminous sand, 213–214
black liquor, 124, 268
blending, 69
blue water gas, 547–548
boiling point, 542
bottom ash, 305
briquette manufacture, 578
briquettes, 592
briquetting, 327, 578, 584
brown coal, 218
Btu value, 591
bubbling fluidized-bed gasifier, 549
bulk density, 169
butanol, 149, 556

C

calorific value, 591
calorific value of wood, 255
carbohydrate content, 238
carbohydrates, 605
carbonaceous feedstocks, 489
carbonate washing, 432, 458

carbon dioxide emissions, 169
carbon molecular sieves, 460
carbureted water gas, 547–548
casing head gas, 198, 539
catalysts, 516
catalytic cracking, 321
catalytic oxidation, 433
catalytic partial oxidation, 388
catalytic transesterification, 152, 632
cellulose, 124, 130, 232, 235, 239, 256, 262, 325, 354, 573, 291, 605–607
cellulosic biomass, 27, 146
cellulosic feedstocks, 124
cetane number, 590
char, 582
charcoal, 273, 581
chemical absorption, 451
chemical desulfurization, 480
chemicals, 406, 409
chemicals from synthesis gas, 492
chemical solvent processes, 431
chemical waste, 275
chemisorption, 451
circulating fluidized-bed gasifier, 549
classification of biofuels, 126
Claus process, 481
cleats, 193
cloud point, 591
coal, 17, 21, 216
coal-bed methane, 11, 192–193
coal classification systems, 218
coal gasification, 376, 379
coalification process, 193
coal ranks, 218
coal type, 217, 219
cobalt-based catalysts, 510
cocurrent fixed-bed gasifier, 381
co-current gasifier, 549
cofiring, 302
co-firing biomass, 614
co-flow gasifier, 313, 620
coke, 112, 586–587
coke oven gas, 549
cold flow, 591
combined cycle power generation, 388
combustion, 229, 299
combustion plants, 306
combustion process, 300
combustion turbine, 382
comminution, 326
complete mix digester, 629
composition of landfill gas, 346
compound types in shale oil, 225
concentrated acid hydrolysis, 146
constituents of wood, 272
conventional crude oil, 17, 205, 212
conventional fuels, 63
conventional gas, 192
conventional recovery methods, 20
conversion of biomass, 241
cordgrass, 243
corn stover, 148
corrosion inhibitors, 94
corrosion in the refinery, 15
Cosorb process, 524
countercurrent fixed-bed gasifier, 381
counter-current gasifier, 549
countercurrent hydrolysis, 147
covered lagoon digester, 628
cracked gasoline, 87, 89
cracking, 86
crop biorefinery, 600
crop residues, 239, 247, 317, 622
crops, 236
crude oil, 13, 205
crude oil coke, 385
crude oil economics, 44
crude oil family, 13, 204
crude oil from tight formations, 13
crude oil products, 65, 67
crude oil residua, 385
crude oil solvent, 78

crude oil wax, 105
cryogenic expander process, 464
cryogenics, 524
cryogenic upgrading, 446
cyclone collectors, 433

D

dairy farm wastes, 317
dedicated energy crops, 148
definition of tar sand, 18
demolition debris landfill, 338
demulsifiers, 94
densification, 584
density, 126
deodorized solvents, 79
desorption, 430
destructive distillation, 587
diesel, 567
diesel fuel oil, 101
diffusion, 362
digestate, 625
digester, 296
digester gas, 628
digestion process, 295
dilute acid hydrolysis, 327
dimethyl ether, 410, 520
direct combustion, 291, 612
direct-fired gas turbine, 301
direct hydrothermal liquefaction, 625
dissolved natural gas, 190
diterpenes, 266
diterpenoids, 266
domestic waste, 274, 278–279, 299, 609–610
downdraft fixed-bed gasifiers, 164
downdraft gasifier, 313–314, 620
drag reducers, 95
dry milling, 616
dry natural gas, 198, 538
dump, 335
dump site, 340
dust, 368
dust control, 433
dyes, 95

E

Eagle Ford crude oil, 14
Edeleanu process, 80
electronic waste, 276
electrostatic precipitators, 434
electrostatic scrubbers, 434
energy consumption, 10
energy content, 200
energy crops, 124, 148, 230, 237, 240
energy density, 169
energy efficiency, 6
energy flow, 34
energy independence, 6, 49
energy precipice, 8
energy resources, 4
energy risk-premium, 54
energy security, 53
energy supply, 34
enrichment, 467
entrained-bed gasifier, 381
entrained flow, 381
entrained-flow gasifier, 382
enzymatic hydrolysis, 147, 328–329
ETBE, 121
ethanol, 29, 142, 144, 146, 169, 555, 638
ethanolamine process, 452
ethers, 161
evaporation, 202, 542
explosive properties, 542
extra heavy crude oil, 17, 19–20, 37, 41, 212, 214
extra heavy oil, 213
Exxon process, 505

F

fabric filters, 434
factors affecting transesterification, 635

FAME, 150
fast pyrolysis, 320, 625
fatty acid content, 122
fatty acid methyl esters, 150
fermentation, 289, 291, 310, 355
fermentation process, 149, 329, 614, 616
Ferrox process, 456
fire point, 543
firewood, 120, 574
first-generation biofuels, 121, 127
Fischer–Tropsch catalysts, 517–518
Fischer–Tropsch chemistry, 375
Fischer–Tropsch liquids, 409, 490
Fischer–Tropsch process, 374, 376, 489, 494, 501, 569
Fischer–Tropsch process conditions, 490
Fischer–Tropsch product mixtures, 525
Fischer–Tropsch reaction, 505, 512
Fischer–Tropsch reactor designs, 509
Fischer–Tropsch reactors, 515
Fischer–Tropsch synthesis, 140, 507, 514
fish oil, 168
fixed-bed countercurrent gasifier, 313
fixed-bed downdraft, 549
fixed-bed reactor, 514
fixed-bed tubular reactors, 515
fixed-bed updraft, 549
fixed-bed updraft gasifiers, 164
fixed carbon, 220
flammability, 75, 542
flammable range, 544
flash point, 202, 543, 591
fluidized-bed combustor, 304, 613
fluidized-bed Fischer–Tropsch reactors, 510
fluidized-bed gasifier, 165, 381
Fluor process, 458
fly ash, 305
foam scrubber, 434
foamy oil, 15, 209
forest and mill residues, 28
forestry waste, 124, 147
formation of landfill gas, 342, 353
fossil fuels, 3, 229
fouling, 125
fractionation processes, 466
fuel briquettes, 578
fuel oil, 99
fuel oil production, 100
fuel quality, 588
fuel sources, 25
fuel specifications, 170
fuel wood, 573, 574

G

garbage, 274, 610
gas-agitated multiphase reactors, 515
gas analysis, 75
gas cap, 190
gas cleaning, 425, 426, 521
gas collection, 365
gas condensate, 190
gaseous fuels, 72, 406, 535
gaseous fuels from different sources, 537
gas hydrate, 12–13, 192, 194
gasification, 291, 311, 374, 377, 383, 617
gasification chemistry, 617
gasifiers, 380, 619
gas liquids, 191
gas migration, 361
gasoline, 85–86, 564
gasoline blending, 69
gasoline-type jet fuel, 97
gas processing, 521
gas refining, 426
gas streams, 429
gas well gas, 191
geopolitical factors, 44

geothermal energy, 29
Giammarco–Vetrocoke process, 459
Girbotol process, 452
glucan, 167
grass biomass, 234
grasses, 238
green algae, 168
guard beds, 460

H

Haber–Bosch process, 407
Hardwood, 269
hazardous waste, 276
hazardous waste landfill, 339
heat content, 591
heating value, 591
heat of combustion, 200, 541
heat value, 75
heavy crude oil, 16–17, 37, 41, 211–212
heavy fuel oil, 100
heavy metals, 368
heavy oil, 211
heavy residue gasification, 388
hemicellulose, 125, 235, 257, 262, 291, 325, 354, 573, 605
herbaceous crops, 247
herbaceous energy crops, 242
high acid crude oil, 14–15, 209
high-Btu gas, 521
higher heating value, 591
high heat-content gas, 138, 521
horizontal drilling, 8
Hubbert peak theory, 35
hybrid gasification process, 389
hybrid poplar, 148
hybrid willow, 148
hydraulic fracturing, 8, 11, 196
hydrocarbon fuels, 563
hydrocarbon gasification, 390
hydrocarbons, 157
hydrocolloids, 258
hydrocracking, 321
hydroformylation process, 505
hydroformylation synthesis, 504
hydrogen, 30, 141, 385, 408
hydrogen energy, 30
hydrolysis, 293, 355, 615, 626
hydrothermal liquefaction process, 321
hydrothermal upgrading process, 160
hypro process, 390

I

IFPEXOL-1 process, 447–448
IFPEXOL-2 process, 448
incineration, 299
industrial waste, 124, 275, 278, 299, 609–610
inert landfill, 338
in-line blending, 69
inorganic waste, 357
Integrated Gasification Combined-Cycle (IGCC), 382
iodine value, 591
iron-based catalysts, 510
iron oxide process, 454–455
isomerization, 86

J

Jerusalem artichoke, 244–245
jet, 218
jet fuels, 97
junk, 274, 610

K

kerogen, 23, 220–221, 223
kerosene, 96, 98, 567
kerosene-type jet fuel, 97
Koppers–Totzek entrained flow reactor, 499
Kuhlmann-oxo process, 505

L

landfill, 292, 335
landfill classification, 336
landfilled waste, 139
landfill gas, 139, 335, 341, 345, 348, 368, 535
landfill gas migration, 363
landfill gas monitoring, 359
landfill gas production, 343, 359
landfilling, 293
landfill leachate, 357
landfill methane, 346
landfill mining, 360
landfill site, 335
latent heat of vaporization, 542
latex, 157
leachate treatment, 358
lean gas, 190, 538
lean oil extraction, 461
lignin, 125, 129, 235, 260, 262, 273, 325, 573, 577, 606, 609
lignite, 217–218
lignocellulose, 551
lignocellulose biorefinery, 600
lignocellulosic biomass, 598
lignocellulosic feedstock, 235, 601
lignocellulosic fibers, 608
lipids, 233
liquefied natural gas, 192
liquefied petroleum gas, 191
liquid absorption processes, 431
liquid biomass fuels, 141
liquid fuels, 5, 77, 409, 553
liquids from biomass, 13
liquids from coal, 13
liquids from oil shale, 13
liquids removal processes, 466
LNG, 192
logs, 575
low-Btu gas, 136, 521, 621
lower explosion limit, 352
lower explosive limit, 203
lower heating value, 591
low heat-content gas, 136, 521
low-temperature gasification, 301
LPG, 72
lump charcoal, 582
Lurgi fixed-bed reactor, 499

M

maceral, 219
Madison Process, 272
Mahogany Zone, 223
major constituents of landfill gas, 351
manual landfill, 340
manure storage lagoon, 628
markers, 95
marsh gas, 293, 535
mass burn systems, 307
medium-Btu gas, 137, 521
medium heat-content gas, 137, 521
membrane separation, 463
membrane separation processes, 462
Merox process, 458
methanation, 524
methane clathrates, 12
methane hydrate, 12, 194
methane hydrate resources, 13
methane ice, 194
methane-wash cryogenic process, 524
methanogenesis, 293, 356, 626
methanogenic digestate, 627
methanol, 143, 410, 553, 639
methanol synthesis, 495
methanol-to-diesel (MtD) process, 411
methanol-to-gasoline (MtG) process, 411
methanol-to-olefin, 519
methanol-to-olefins (MtO) synthesis, 411
microcrystalline wax, 105
midden, 335

migration of landfill gas, 344
mineral matter content, 125
miscanthus, 245
mixed alcohol synthesis, 412
mixed gas, 73
modular combustion, 308
molasses, 158
molecular sieve process, 459
molecular sieves, 444, 449
monoterpenes, 266
monoterpenoids, 266
moving grate combustor, 302
MSW, 276
MTBE, 121, 376
multi-tubular fixed-bed reactors, 509
municipal and industrial wastes, 28
municipal solid waste, 124, 276, 304, 349, 610
municipal solid waste landfill, 339

N

naphtha, 77, 79, 82, 564
natural gas, 72, 185, 196, 384, 429, 535
natural gas liquids, 12, 189
natural gasoline, 190
natural gas reserves, 11
natural gas reservoirs, 188
needle coke, 113, 588
nitrogen removal, 471
non-associated gas, 191
non-conventional fuels, 4, 230, 533
non-conventional recovery methods, 20
non-edible oils, 122
non-fossil fuel energy sources, 24
non-fossil fuels, 26
non-fossil fuel sources, 287
nuclear energy, 31
nuclear fuels, 230
nuclear power, 31

O

octane number, 92, 590
odorless solvent, 79
oilgae, 132
oil sand, 22, 213–214
oil shale, 17, 22–24, 220, 222–223
oil shale resources, 37
open core gasifier, 314, 620
opportunity crude oil, 14, 208
organic waste, 230
Orinoco tar belt, 37
oxidation, 356
oxidation inhibitors, 94
oxo synthesis, 504
oxygenates, 95

P

paraffin wax, 105
parallel plate reactors, 515
partial oxidation, 523
partial oxidation process, 380, 498
particle size, 126
particulate matter control, 433
particulate matter removal, 473
PCUK-oxo process, 505
peak energy production, 5
peak energy theory, 8
peak oil, 38
peak oil production, 36
peak oil theory, 34, 36, 44
pectin, 262
petrochemical industry, 402–403
petrochemicals, 68, 404
petrol, 85
petroleum, 13
Petroleum Age, 6
petroleum products, 66
petroleum solvent, 78
petro-politics, 6
phase separation, 592
phospholipids, 233
photosynthesis, 28

physical factors, 45
physical solvent processes, 431
pile burner, 303, 613
pitch, 111
plant fibers, 608
plug-flow digesters, 629
polymerization, 86
polysaccharides, 607
pour point, 591
precipice depletion theory, 39
prediction of peak oil, 20
pressure swing adsorption, 524
pressure swing adsorption (PSA) technology, 460
pre-treatment, 326
primary biomass, 231
primary biomass feedstocks, 603
producer gas, 521, 546–547
product distribution, 511
products from synthesis gas, 163, 623
product specifications, 67
propanol, 149, 556
properties of different fuels, 534
properties of methanol, 554
property variation, 166
proximate analysis, 594
PSA process, 471
pyrolysis, 291, 318–319, 323, 624
pyrolysis liquids, 592
pyrolysis of biomass, 141
pyrolysis oil, 491, 571
pyrolysis processes, 391

R

rapid thermal processing process, 160
Rectisol process, 447, 450
Rectisol unit, 502
reed plants, 246
refinery gas, 384, 429
refinery of the future, 597
refuse-derived fuel (RDF) systems, 307
refuse gas, 552
reserve depletion, 35
reserve estimation, 9
reserve growth, 10
reserves estimates, 37
resid, 111
residual herbaceous biomass, 246
residual waste, 276, 610
residue gas, 198, 539
resource depletion, 39
retorting, 224
retort oil, 23, 224
rich gas, 190
rotary kiln combustor, 303
rubbish, 274, 610

S

Sabatier reaction, 497
sanitary landfill, 340
sawdust briquettes, 580
SCOT process, 482–483
scrap tires, 309
secondary biomass feedstocks, 603
second-generation biofuels, 121, 131
Selexol process, 450
separate hydrolysis and fermentation, 310
sesquiterpenes, 266
sesquiterpenoids, 266
shale, 183
shale formations, 184, 210
shale gas, 195
shale oil, 23, 220, 223–224, 415
Shell Claus Off-gas Treating, 482
Shell gasification (partial oxidation) process, 392
shift reaction, 387
short-rotation coppice, 248
short-rotation woody crops, 248
shot coke, 113, 587
siloxane removal, 474
silvichemicals, 273

simple combustor, 302
simultaneous saccharification and fermentation, 147, 329
single-stage gasifiers, 312
slag, 368
slagging, 125
slagging gasifiers, 381
slurry-bed Fischer–Tropsch reactors, 510
slurry-bed reactor, 514
softwood, 270
solar energy, 32
solid-adsorbent dehydrators, 445
solid fuel, 165, 573, 640
sorghum, 248–249
sources of waste, 275
sour gas, 539
soybeans, 151
specifications, 594
sponge coke, 113, 588
starch, 239, 605, 607
steam explosion, 325
steam methane reforming, 141, 408, 523
steam–naphtha reforming, 396
steam reforming, 386–387, 406, 500
steam turbine, 382
steroids, 267
sterol derivatives, 123
stove oil, 101
straw, 246–247, 622
sub-bituminous coal, 217–218
substitute natural gas (SNG), 138, 408
sugar beet, 158
sugar crops, 238
sugars, 607
sugars and starch, 127
sulfinol process, 450
supercritical methanol transesterification, 154
surface impoundment, 341
swamp gas, 535
switchgrass, 148, 243–244
syncline–anticline, 188
synfuels, 183
syngas, 373
synthesis gas, 121, 140, 162, 165, 314, 373–374, 376, 491, 493, 495–496, 623
synthesis gas generation, 396
synthetic crude oil, 223, 491
synthetic fuel, 26, 183, 376
synthetic natural gas, 138, 373, 408, 521

T

tail gas cleaning, 480
tail gas treating, 480
Tall oil, 168
tar removal, 469
tar sand, 22, 213–215
tar sand bitumen, 13, 17, 37, 41, 211–213, 215
TCLP, 358
technological factors, 47
terpene derivatives, 123
terpenoid derivatives, 123
tertiary biomass, 603
tertiary biomass feedstocks, 603
tetraethyl lead, 91
Texaco gasification process, 397
therm, 201
thermal cracking, 323
thermal processes, 242
thermochemical platform, 242, 287, 290
thermochemical process, 241
thermogenic coalbed methane, 192
THIOPAQ® process, 479
third-generation biofuels, 121, 132
tidal energy, 32–33
tight formations, 183, 210
tight gas, 11
tight oil, 210, 212

tight shale formations, 12
tire-derived fuels, 309
town gas, 521
toxicity characteristic leaching procedure, 358
toxic waste, 275
transesterification, 152–154, 156, 330, 332, 630–631
transesterification using supercritical methanol, 634
trans-p-coumaryl alcohol, 261
trans-sinapyl alcohol, 261
trash, 274, 610
traveling grate combustor, 303, 613
traveling grate system, 613
triglyceride derivatives, 123
triterpenes, 266
triterpenoids, 266
type of wood, 267
types of biofuels, 133
types of combustors, 302
types of digesters, 296
types of gasifiers, 164
types of pyrolysis, 237

U

Udex process, 81
ultimate analysis of wood, 253
unconventional gas, 192
unconventional sources, 37
upper explosive limit, 203, 352
urban waste briquettes, 580
uses of landfill gas, 366

V

vapor density, 540
vapor pressure, 203, 543
vegetable fats, 607
vegetable oil, 142, 150, 607
vegetable oil ester, 141
venturi scrubbers, 434
viscosity, 590
viscous crude oil, 211
volatile alkali constituents, 301
volatile matter content, 125
volatility, 542

W

waste, 274
waste-derived fuel, 309
waste-derived solid fuels, 592
water gas, 521
water–gas shift reaction, 408
water removal, 441
water-wall combustion, 308
water washing, 432, 458
wax, 105
wax derivatives, 123
wax recrystallization, 107
wax sweating, 106
wet gas, 538
wet milling, 616
wet natural gas, 198, 538
wet scrubbers, 434
willow biomass crops, 316
Winkler fluidized-bed reactor, 499
wood, 28, 230, 239, 249, 253–254
wood alcohol, 555
wood chips, 575, 577
wood fuel, 573
wood gas, 551
wood pellets, 576
wood wastes, 28
woody biomass, 234
woody feedstocks, 254
world resources, 41

X

xylan, 167

Z

zinc oxide process, 456
Zuata extra heavy crude oil, 214

www.ingramcontent.com/pod-product-compliance
Lightning Source LLC
Chambersburg PA
CBHW060937160926
51501CB00060B/158
9789811203640